T0204008

Cellular Automata
and Complexity

STEPHEN WOLFRAM

Cellular Automata and Complexity

COLLECTED PAPERS

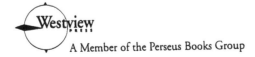
Westview
PRESS
A Member of the Perseus Books Group

Author's mailing address
Stephen Wolfram
Wolfram Research, Inc.
100 Trade Center Drive
Champaign, IL 61820, USA

Library of Congress Cataloging-in-Publication Data

Wolfram, Stephen.
   Cellular automata and complexity : collected papers /
Stephen Wolfram.
       p. cm.
   Includes bibliographical references and index.
   Most papers previously published in various journals.
   ISBN 0-201-62716-7. — ISBN 0-201-62664-0 (pbk.)
   1. Cellular automata. I. Title.
QA267.5.C45W65 1994
511.3—dc20

93-40786
CIP

Copyright © 1994 by Stephen Wolfram

Published by Westview Press, a Member of the Perseus Books Group

All rights reserved. No part of this publication may be reproduced, stored in a retrieval system, or transmitted, in any form or by any means, electronic, mechanical, photocopying, recording, or otherwise, without the prior written permission of the publisher. Printed in the United States of America.

The author and the publisher are grateful to the following for their permission to reproduce material included in this volume: Academic Press, Inc. (*Advances in Applied Mathematics*); the American Physical Society (*Physical Review Letters, Reviews of Modern Physics*); Elsevier Science Publishers B. V. (*Physica D*); Los Alamos National Laboratory (*Los Alamos Science*); Mathematical Association of America (*American Mathematical Monthly*); Plenum Publishing Corp. (*Journal of Statistical Physics*); The Royal Swedish Academy of Sciences (*Physica Scripta*); Scientific American, Inc. (*Scientific American*); Springer-Verlag (*Communications in Mathematical Physics, CRYPTO '85 Proceedings*); University of Illinois Press (*High-Speed Computing: Scientific Applications and Algorithm Design*).

Jacket and text design by André Kuzniarek
Set in Times Roman with Mathematica Symbol font

www.westviewpress.com

EBA 02 03 04 05 10 9 8 7 6 5 4 3

First printing: January 1994

Contents

PART ONE: *Primary Papers*

Statistical Mechanics of Cellular Automata	3
Algebraic Properties of Cellular Automata	71
Universality and Complexity in Cellular Automata	115
Computation Theory of Cellular Automata	159
Undecidability and Intractability in Theoretical Physics	203
Two-Dimensional Cellular Automata	211
Origins of Randomness in Physical Systems	251
Thermodynamics and Hydrodynamics of Cellular Automata	259
Random Sequence Generation by Cellular Automata	267
Approaches to Complexity Engineering	309
Minimal Cellular Automaton Approximations to Continuum Systems	329
Cellular Automaton Fluids: Basic Theory	359

PART TWO: *Additional and Survey Papers*

Cellular Automata	411
Computers in Science and Mathematics	439
Geometry of Binomial Coefficients	451
Twenty Problems in the Theory of Cellular Automata	457
Cryptography with Cellular Automata	487
Complex Systems Theory	491
Cellular Automaton Supercomputing	499

PART THREE: *Appendices*

Tables of Cellular Automaton Properties	513
Scientific Bibliography of Stephen Wolfram	585

Index	591

PART ONE

Primary Papers

Statistical Mechanics
of Cellular Automata

1983

Cellular automata are used as simple mathematical models to investigate self-organization in statistical mechanics. A detailed analysis is given of "elementary" cellular automata consisting of a sequence of sites with values 0 or 1 on a line, with each site evolving deterministically in discrete time steps according to definite rules involving the values of its nearest neighbors. With simple initial configurations, the cellular automata either tend to homogeneous states or generate self-similar patterns with fractal dimensions $\simeq 1.59$ or $\simeq 1.69$. With "random" initial configurations, the irreversible character of the cellular automaton evolution leads to several self-organization phenomena. Statistical properties of the structures generated are found to lie in two universality classes, independent of the details of the initial state or the cellular automaton rules. More complicated cellular automata are briefly considered, and connections with dynamical systems theory and the formal theory of computation are discussed.

1. Introduction

The second law of thermodynamics implies that isolated microscopically reversible physical systems tend with time to states of maximal entropy and maximal "disorder." However, "dissipative" systems involving microscopic irreversibility, or those open to interactions with their environment, may evolve from "disordered" to more "ordered" states. The states attained often exhibit a complicated structure. Examples are outlines of snowflakes, patterns of flow in turbulent fluids, and biological systems. The purpose of this paper is to begin the investigation of cellular automata (introduced in Sec. 2) as a class of mathematical models for such behavior. Cellular automata are sufficiently simple to allow detailed mathematical analysis, yet sufficiently complex to exhibit a wide variety of complicated phenomena. Cellular automata are also of

Originally published in *Reviews of Modern Physics*, volume 55, pages 601–644 (July 1983).

sufficient generality to provide simple models for a very wide variety of physical, chemical, biological, and other systems. The ultimate goal is to abstract from a study of cellular automata general features of "self-organizing" behavior and perhaps to devise universal laws analogous to the laws of thermodynamics. This paper concentrates on the mathematical features of the simplest cellular automata, leaving for future study more complicated cellular automata and details of applications to specific systems. The paper is largely intended as an original contribution, rather than a review. It is presented in this journal in the hope that it may thereby reach a wider audience than would otherwise be possible. An outline of some of its results is given in Wolfram (1982a).

Investigations of simple "self-organization" phenomena in physical and chemical systems (Turing, 1952; Haken, 1975, 1978, 1979, 1981; Nicolis and Prigogine, 1977; Landauer, 1979; Prigogine, 1980; Nicolis et al., 1981) have often been based on the Boltzmann transport differential equations (e.g., Lifshitz and Pitaevskii, 1981) (or its analogs) for the time development of macroscopic quantities. The equations are obtained by averaging over an ensemble of microscopic states and assuming that successive collisions between molecules are statistically uncorrelated. For closed systems (with reversible or at least unitary microscopic interactions) the equations lead to Boltzmann's H theorem, which implies monotonic evolution towards the macroscopic state of maximum entropy. The equations also imply that weakly dissipative systems (such as fluids with small temperature gradients imposed) should tend to the unique condition of minimum entropy production. However, in strongly dissipative systems, several final states may be possible, corresponding to the various solutions of the polynomial equations obtained from the large time limit of the Boltzmann equations. Details or "fluctuations" in the initial state determine which of several possible final states are attained, just as in a system with multiple coexisting phases. Continuous changes in parameters such as external concentrations or temperature gradients may lead to discontinuous changes in the final states when the number of real roots in the polynomial equations changes, as described by catastrophe theory (Thom, 1975). In this way, "structures" with discrete boundaries may be formed from continuous models. However, such approaches become impractical for systems with very many degrees of freedom, and therefore cannot address the formation of genuinely complex structures.

More general investigations of self-organization and "chaos" in dynamical systems have typically used simple mathematical models. One approach (e.g., Ott, 1981) considers dissipative nonlinear differential equations (typically derived as idealizations of Navier-Stokes hydrodynamic equations). The time evolution given particular initial conditions is represented by a trajectory in the space of variables described by the differential equations. In the simplest cases (such as those typical for chemical concentrations described by the Boltzmann transport equations), all trajectories tend at large times to a small number of isolated limit points, or approach simple periodic limit cycle orbits. In other cases, the trajectories may instead concentrate on complicated and apparently chaotic surfaces ("strange attractors"). Nearly linear systems typically exhibit simple limit points or cycles. When nonlinearity is increased by variation of external parameters, the number of limit points or cycles may increase without bound, eventually building up a strange attractor (typically exhibiting a statistically self-similar structure in phase space). A simpler approach (e.g., Ott, 1981) involves discrete time steps, and considers the evolution of numbers on an interval of the real line under iterated mappings. As the nonlinearity is increased, greater numbers of limit points and cycles appear, followed by essentially chaotic behavior. Quantitative features of this approach to chaos are found to be universal to wide classes of mappings. Notice that for both differential equations and iterated mappings, initial conditions are specified by real numbers with a potentially infinite number of significant digits. Complicated or seemingly chaotic behavior is a reflection of sensitive dependence on high-order digits in the decimal expansions of the numbers.

Models based on cellular automata provide an alternative approach, involving discrete coordinates and variables as well as discrete time steps. They exhibit complicated behavior analogous to that found with differential equations or iterated mappings, but by virtue of their simpler construction are potentially amenable to a more detailed and complete analysis.

Section 2 of this paper defines and introduces cellular automata and describes the qualitative behavior of elementary cellular automata. Several phenomena characteristic of self-organization are found. Section 3 gives a quantitative statistical analysis of the states generated in the time evolution of cellular automata, revealing several quantitative universal features. Section 4 describes the global analysis of cellular automata and discusses the results in the context of dynamical systems theory and the formal theory of computation. Section 5 considers briefly extensions to more complicated cellular automata. Finally, Sec. 6 gives some tentative conclusions.

2. Introduction to Cellular Automata

Cellular automata are mathematical idealizations of physical systems in which space and time are discrete, and physical quantities take on a finite set of discrete values. A cellular automaton consists of a regular uniform lattice (or "array"), usually infinite in extent, with a discrete variable at each site ("cell"). The state of a cellular automaton is completely specified by the values of the variables at each site. A cellular automaton evolves in discrete time steps, with the value of the variable at one site being affected by the values of variables at sites in its "neighborhood" on the previous time step. The neighborhood of a site is typically taken to be the site itself and all immediately adjacent sites. The variables at each site are updated simultaneously ("synchronously"), based on the values of the variables in their neighborhood at the preceding time step, and according to a definite set of "local rules."

Cellular automata were originally introduced by von Neumann and Ulam (under the name of "cellular spaces") as a possible idealization of biological systems (von Neumann, 1963, 1966), with the particular purpose of modelling biological self-reproduction. They have been applied and reintroduced for a wide variety of purposes, and referred to by a variety of names, including "tessellation automata," "homogeneous structures," "cellular structures," "tessellation structures," and "iterative arrays."

Physical systems containing many discrete elements with local interactions are often conveniently modelled as cellular automata. Any physical system satisfying differential equations may be approximated as a cellular automaton by introducing finite differences and discrete variables.[1] Nontrivial cellular automata are obtained whenever the dependence on the values at each site is nonlinear, as when the system exhibits some form of "growth inhibition." A very wide variety of examples may be considered; only a few are sketched here. In the most direct cases, the cellular automaton lattice is in position space. At a microscopic level, the sites may represent points in a crystal lattice, with values given by some quantized observable (such as spin component) or corresponding to the types of atoms or units. The dynamical Ising model (with kinetic energy terms included) and other lattice spin systems are simple cellular automata, made nondeterministic by "noise" in the local rules at finite temperature. At a more macroscopic level, each site in a cellular automaton may represent a region containing many molecules (with a scale size perhaps given by an appropriate correlation length), and its value may label one of several discrete possible phases or compositions. In this way, cellular automata may be used as discrete models for nonlinear chemical systems involving a network of reactions coupled with spatial diffusion (Greenberg et al., 1978). They have also been used in a (controversial) model for the evolution of spiral galaxies (Gerola and Seiden, 1978; Schewe, 1981). Similarly, they may provide models for kinetic aspects of phase transitions (e.g., Harvey et al., 1982). For example, it is possible that growth of dendritic crystals (Langer, 1980) may be described by aggregation of discrete "packets" with a local growth inhibition effect associated with local releases of latent heat, and thereby treated as a cellular automaton [Witten and Sander (1981) discuss a probabilistic model of this kind, but there are indications that the probabilistic elements are inessential]. The spatial structure of turbulent fluids may perhaps be modelled using cellular automata by approximating the velocity field as a lattice of cells, each containing one or no eddies, with interactions between neighboring cells. Physical systems may also potentially be described by cellular automata in wave-vector or momentum space, with site values representing excitations in the corresponding modes.

Many biological systems have been modelled by cellular automata (Lindenmayer, 1968; Herman, 1969; Ulam, 1974; Kitagawa, 1974; Baer and Martinez, 1974;

---

[1] The discussion here concentrates on systems first order in time; a more general case is mentioned briefly in Sec. 4.

Rosen, 1981) (cf. Barricelli, 1972). The development of structure and patterns in the growth of organisms often appears to be governed by very simple local rules (Thompson, 1961; Stevens, 1974) and is therefore potentially well described by a cellular automaton model. The discrete values at each site typically label types of living cells, approximated as growing on a regular spatial lattice. Short-range or contact interactions may lead to expression of different genetic characteristics, and determine the cell type. Simple nonlinear rules may lead to the formation of complex patterns, as evident in many plants and animals. Examples include leaf and branch arrangements (e.g., Stevens, 1974) and forms of radiolarian skeletons (e.g., Thompson, 1961). Simple behavior and functioning of organisms may be modelled by cellular automata with site values representing states of living cells or groups of cells [Burks (1973) and Flanigan (1965) discuss an example in heart fibrillation]. The precise mathematical formulation of such models allows the behavior possible in organisms or systems with particular construction or complexity to be investigated and characterized (e.g., von Neumann, 1966). Cellular automata may also describe populations of nonmobile organisms (such as plants), with site values corresponding to the presence or absence of individuals (perhaps of various types) at each lattice point, with local ecological interactions.

Cellular automata have also been used to study problems in number theory and their applications to tapestry design (Miller, 1970, 1980; ApSimon, 1970a, 1970b; Sutton, 1981). In a typical case, successive differences in a sequence of numbers (such as primes) reduced with a small modulus are taken, and the geometry of zero regions is investigated.

As will be discussed in Sec. 4, cellular automata may be considered as parallel processing computers (cf. Manning, 1977; Preston et al., 1979). As such, they have been used, for example, as highly parallel multipliers (Atrubin, 1965; Cole, 1969), sorters (Nishio, 1981), and prime number sieves (Fischer, 1965). Particularly in two dimensions, cellular automata have been used extensively for image processing and visual pattern recognition (Deutsch, 1972; Sternberg, 1980; Rosenfeld, 1979). The computational capabilities of cellular automata have been studied extensively (Codd, 1968; Burks, 1970; Banks, 1971; Aladyev, 1974, 1976; Kosaraju, 1974; Toffoli, 1977b), and it has been shown that some cellular automata could be used as general purpose computers, and may therefore be used as general paradigms for parallel computation. Their locality and simplicity might ultimately permit their implementation at a molecular level.

The notorious solitaire computer game "Life" (Conway, 1970; Gardner, 1971, 1972; Wainwright, 1971–1973; Wainwright, 1974; Buckingham, 1978; Berlekamp et al., 1982; R. W. Gosper, private communications) (qualitatively similar in some respects to the game of "Go") is an example of a two-dimensional cellular automaton, to be discussed briefly in Sec. 5.

Until Sec. 5, we shall consider exclusively one-dimensional cellular automata with two possible values of the variables at each site ("base 2") and in which the neighborhood of a given site is simply the site itself and the sites immediately adjacent to it on the left and right. We shall call such cellular automata elementary. Figure 1 specifies one particular set of local rules for an elementary cellular automaton. On the top row, all $2^3 = 8$ possible values of the three variables in the neighborhood are given, and below each one is given the value achieved by the central site on the next time step according to a particular local rule. Figure 2 shows the evolution of a particular state of the cellular automaton through one time step according to the local rule given in Fig. 1.

The local rules for a one-dimensional neighborhood-three cellular automaton are described by an eight-digit binary number, as in the example of Fig. 1. (In specifying cellular automata, we use this binary number interchangeably with its decimal equivalent.) Since any eight-digit binary number specifies a cellular automaton, there are $2^8 = 256$ possible distinct cellular automaton rules in one dimension with a three-site neighborhood. Two inessential restrictions will usually be imposed on these rules. First, a cellular automaton rule will be considered "illegal" unless a "null" or "quiescent" initial state consisting solely of 0 remains unchanged. This forbids rules whose binary specification ends with a 1 (and removes symmetry in the treatment of 0 and 1 sites). Second, the rules must be reflection symmetric, so that 100 and 001 (and 110 and 011) yield identical values. These restrictions[2] leave 32 possible "legal" cellular automaton rules of the form $\alpha_1\alpha_2\alpha_3\alpha_4\alpha_2\alpha_5\alpha_4 0$.

The local rules for a cellular automaton may be considered as a Boolean function of the sites within the neighborhood. Let $s_n(m)$ be the value of site m at time step

1 1 1	1 1 0	1 0 1	1 0 0	0 1 1	0 1 0	0 0 1	0 0 0
0	1	0	1	1	0	1	0

Figure 1. Example of a set of local rules for the time evolution of a one-dimensional elementary cellular automaton. The variables at each site may take values 0 or 1. The eight possible states of three adjacent sites are given on the upper line. The lower line then specifies a rule for the time evolution of the cellular automaton by giving the value to be taken by the central site of the three on the next time step. The time evolution of the complete cellular automaton is obtained by simultaneous application of these rules at each site for each time step. The rule given is the modulo-two rule: the value of a site at a particular time step is simply the sum modulo two of the values of its two neighbors at the previous time step. Any possible sequence of eight binary digits specifies a cellular automaton.

```
0 1 0 1 1 0 1 1 0 1 0 1 0 1 1 1 0 0 0 1 0
  0 0 1 1 0 1 1 0 0 0 0 0 1 0 1 1 0 1 0
```

Figure 2. Evolution of a configuration in one-dimensional cellular automaton for one time step according to the modulo-two rule given in Fig. 1. The values of the two end sites after the time step depend on the values of sites not shown here.

---

[2] The quiescence condition is required in many applications to forbid "instantaneous propagation" of value-one sites. The reflection symmetry condition guarantees isotropy as well as homogeneity in cellular automaton evolution.

n. As a first example consider the "modulo-two" rule 90 (also used as the example for Fig. 1). According to this rule, the value of a particular site is simply the sum modulo two of the values of its two neighboring sites on the previous time step. The Boolean equivalent of this rule is therefore

$$s_{n+1}(m) = s_n(m-1) \oplus s_n(m+1) \tag{2.1}$$

or schematically $s_+ = s^- \oplus s^+$, where $\oplus$ denotes addition modulo two ("exclusive disjunction" or "inequality"). Similarly, rule 18 is equivalent to $s_+ = s \vee (s^- \oplus s^+)$ [where s denotes $s_n(m)$], rule 22 to $s_+ = s \vee (s^- \wedge s^+)$, rule 54 to $s_+ = s \oplus (s^- \vee s^+)$, rule 150 to $s_+ = s^- \oplus s \oplus s^+$, and so on. Designations s^- and s^+ always enter symmetrically in legal cellular automaton rules by virtue of reflection symmetry. The Boolean function representation of cellular automaton rules is convenient for practical implementation on standard serial processing digital computers.[3]

Some cellular automaton rules exhibit the important simplifying feature of "additive superposition" or "additivity." Evolution according to such rules satisfies the superposition principle

$$s_0 = t_0 \oplus u_0 \leftrightarrow s_n = t_n \oplus u_n, \tag{2.2}$$

which implies that the configurations obtained by evolution from any initial configuration are given by appropriate combinations of those found in Fig. 3 for evolution from a single nonzero site. Notice that such additivity does not imply linearity in the real number sense of Sec. 1, since the addition is over a finite field. Cellular automata satisfy additive superposition only if their rule is of the form $\alpha_1\alpha_20\alpha_3\alpha_2\alpha_1\alpha_30$ with $\alpha_3 = \alpha_1 \oplus \alpha_2$. Only rules 0, 90, 150, and 204 are of this form. Rules 0 and 204 are trivial; 0 erases any initial configuration, and 204 maintains any initial configuration unchanged (performing the identity transformation at each time step). Rule 90 is the modulo-two rule discussed above, and takes a particular site to be the sum modulo two of the values of its two neighbors at the previous time step, as in Eq. (2.1). Rule 150 is similar. It takes a particular site to be the sum modulo two of the values of its two neighbors and its own value at the previous time step ($s_+ = s^- \oplus s \oplus s^+$).

The additive superposition principle of Eq. (2.2) combines values at different sites by addition modulo two (exclusive disjunction). Combining values instead by conjunction (Boolean multiplication) yields a superposition principle for rules 0, 4, 50, and 254. Combining values by (inclusive) disjunction (Boolean addition) yields a corresponding principle for rules 0, 204, 250, and 254. It is found that no other

---

[3] The values of a sequence of (typically 32) sites are represented by bits in a single computer word. Copies of this word shifted one bit to the left and one bit to the right are obtained. Then the cellular automaton rule may be applied in parallel to all bits in the words using single machine instructions for each word-wise Boolean operation. An analogous procedure is convenient in simulation of two-dimensional cellular automata on computer systems with memory-mapped displays, for which application of identical Boolean operations to each display pixel is usually implemented in hardware or firmware.

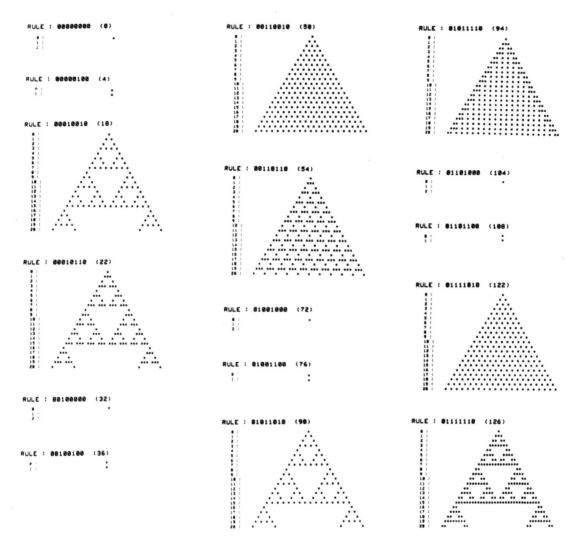

Figure 3. Evolution of one-dimensional elementary cellular automata according to the 32 possible legal sets of rules, starting from a state containing a single site with value 1. Sites with value 1 are represented by stars, and those with value 0 by blanks. The configurations of the cellular automata at successive time steps are shown on successive lines. The time evolution is shown up to the point where the system is

legal cellular automaton rules satisfy superposition principles with any combining function.

The Boolean representation of cellular automaton rules reveals that some rules are "peripheral" in the sense that the value of a particular site depends on the values of its two neighbors at the previous time step, but not on its own previous value. Rules 0, 90, 160, and 250 are of the form $\alpha_1\alpha_2\alpha_1\alpha_2\alpha_2 0\alpha_2 0$ and exhibit this property.

10

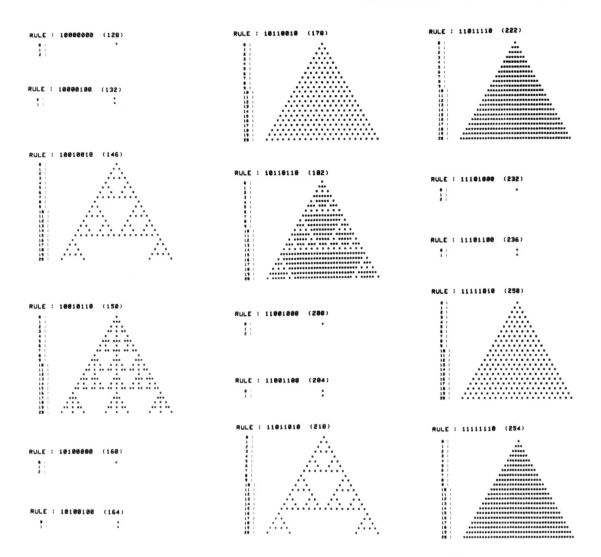

detected to cycle (visiting a particular configuration for the second time), or for at most 20 time steps. The process is analogous to the growth of a crystal from a microscope seed. A considerable variety of behavior is evident. The cellular automata which do not tend to a uniform state yield asymptotically self-similar fractal configurations.

Having discussed features of possible local rules we now outline their consequences for the evolution of elementary cellular automata. Sections 3 and 4 present more detailed quantitative analysis.

Figure 3 shows the evolution of all 32 possible legal cellular automata from an initial configuration containing a single site with value 1 (analogous to the growth of a "crystal" from a microscopic "seed"). The evolution is shown until a particular configuration appears for the second time (a "cycle" is detected), or for at most 20 time steps. Several classes of behavior are evident. In one class, the initial 1 is immediately erased (as in rules 0 and 160), or is maintained unchanged forever (as in rules 4 and 36). Rules of this class are distinguished by the presence of the local rules $100 \rightarrow 0$ and $001 \rightarrow 0$, which prevent any propagation of the initial 1. A second class of rules (exemplified by 50 or 122) copies the 1 to generate a uniform structure which expands by one site in each direction on each time step. These two classes of rules will be termed "simple." A third class of rules, termed "complex," and exemplified by rules 18, 22, and 90, yields nontrivial patterns.

As a consequence of their locality, cellular automaton rules define no intrinsic length scale other than the size of a single site (or of a neighborhood of three sites) and no intrinsic time scale other than the duration of a single time step. The initial state consisting of a single site with value 1 used in Fig. 3 also exhibits no intrinsic scale. The cellular automaton configurations obtained in Fig. 3 should therefore also exhibit no intrinsic scale, at least in the infinite time limit. Simple rules yield a uniform final state, which is manifestly scale invariant. The scale invariance of the configurations generated by complex rules is nontrivial. In the infinite time limit, the configurations are "self-similar" in that views of the configuration with different "magnifications" (but with the same "resolution") are indistinguishable. The configurations thus exhibit the same structure on all scales.

Consider as an example the modulo-two rule 90 (also used as the example for Fig. 1 and in the discussion above). This rule takes each site to be the sum modulo two of its two nearest neighbors on the previous time step. Starting from an initial state containing a single site with value 1, the configuration it yields on successive time steps is thus simply the lines of Pascal's triangle modulo two, as illustrated in Fig. 4 (cf. Wolfram, 1982b). The values of the sites are hence the values of binomial coefficients [or equivalently, coefficients of x^i in the expansion of $(1 + x)^n$] modulo two. In the large time limit, the pattern of sites with value 1 may be obtained by the recursive geometrical construction (cf. Sierpinski, 1916; Abelson and diSessa, 1981, Sec. 2.4) shown in Fig. 5. This geometrical construction manifests the self-similarity

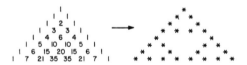

Figure 4. An algebraic construction for the configurations of a cellular automaton starting from a state containing a single site with value 1 and evolving according to the modulo-two rule 90. The rule is illustrated in Fig. 1, and takes the value of a particular site to be the sum modulo two of the values of its two neighboring sites at the previous time step. The value of a site at a given time step is then just the value modulo two of the corresponding binomial coefficient in Pascal's triangle.

Figure 5. Sequence of steps in a geometrical construction for the large time behavior of a cellular automaton evolving according to the modulo-two rule 90. The final pattern is the limit of the sequence shown here. It is a self-similar figure with fractal dimension $\log_2 3$.

(Mandelbrot, 1977, 1982; Geffen et al., 1981) or "scale invariance" of the resulting curve. Figure 3 shows that evolution of other complex cellular automata from a single nonzero site yields essentially identical self-similar patterns. An exception is rule 150, for which the value of each site is determined by the sum modulo two of its own value and the values of its two neighbors on the previous time step. The sequence of binary digits obtained by evolution from a single-site initial state for n time steps with this rule is thus simply the coefficients of x^i in the expansion of $(x^2 + x + 1)^n$ modulo two. A geometrical construction for the pattern obtained is given in Fig. 6.

Figure 7 shows examples of time evolution for some cellular automata with illegal local rules (defined above) which were omitted from Fig. 3. When the quiescence condition is violated, successive time steps involve alternation of 0 and 1 at infinity. When reflection symmetry is violated, the configurations tend to undergo uniform shifting. The self-similar patterns seen in Fig. 3 are also found in cases such as rule 225, but are sheared by the overall shifting. It appears that consideration of illegal as well as "legal" cellular automaton rules introduces no qualitatively new features.

Figure 3 shows the growth of patterns by cellular automaton evolution from a very simple initial state containing a single nonzero site (seed). Figure 8 now illustrates time evolution from a disordered or "random" initial state according to each of the 32 legal cellular automaton rules. A specific "typical" initial configuration was taken, with the value of each site chosen independently, with equal probabilities for values 0 and 1.[4] Just as in Fig. 3, several classes of behavior are evident.

Figure 6. Sequence of steps in a geometrical construction for the large time behavior of a cellular automaton evolving according to the modulo-two rule 150. The final pattern is the limit of the sequence shown here. It is a self-similar figure with fractal dimension $\log_2 2\varphi \simeq 1.69$ [where $\varphi = (1 + \sqrt{5})/2$ is the golden ratio]. An analogous construction for rule 90 was given in Fig. 5.

---

[4] Here and elsewhere a standard linear congruential pseudorandom number generator with recurrence relation $x_{n+1} = (1103515245x_n + 12345) \bmod 2^{31}$ was used. Results were also obtained using other pseudorandom number generation procedures and using random numbers derived from real-time properties of a time-shared computer system.

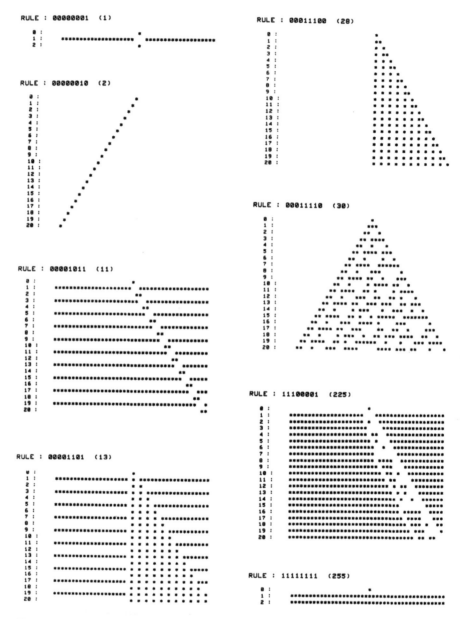

Figure 7. Evolution of a selection of one-dimensional elementary cellular automata obeying illegal rules. Rules are considered illegal if they violate reflection symmetry, which requires identical rules for 100 and 001 and for 110 and 011, or if they violate the quiescence condition which requires that an initial state containing only 0 sites should remain unchanged. For example, rule 2 violates reflection symmetry, and thus yields a uniformly shifting pattern, while rule 1 violates the quiescence condition and yields a pattern which "flashes" from all 0 to all 1 in successive time steps.

The simple rules exhibit trivial behavior, either yielding a uniform final state or essentially preserving the form of the initial state. Complex rules once again yield nontrivial behavior. Figure 8 illustrates the remarkable fact that time evolution according to these rules destroys the independence of the initial sites, and generates correlations between values at separated sites. This phenomenon is the essence of self-organization in cellular automata. An initially random state evolves to a state containing long-range correlations and structure. The bases of the "triangles" visible in Fig. 8 are fluctuations in which a sequence of many adjacent cells have the same value. The length of these correlated sequences is reduced by one site per time step, yielding the distinctive triangular structure. Figure 8 suggests that triangles of all sizes are generated. Section 3 confirms this impression through a quantitative analysis and discusses universal features of the structures obtained.

The behavior of the cellular automata shown in Fig. 8 may be characterized in analogy with the behavior of dynamical systems (e.g., Ott, 1981): simple rules exhibit simple limit points or limit cycles, while complex rules exhibit phenomena analogous to strange attractors.

The cellular automata shown in Fig. 8 were all assumed to satisfy periodic boundary conditions. Instead of treating a genuinely infinite line of sites, the first and last sites are identified, as if they lay on a circle of finite radius. Cellular automata can also be rendered finite by imposing null boundary conditions, under which sites beyond each end are modified to maintain value zero, rather than evolving according to the local rules. Figure 9 compares results obtained with these two boundary conditions in a simple case; no important qualitative differences are apparent.

Finite one-dimensional cellular automata are similar to a class of feedback shift registers (e.g., Golomb, 1967; Berlekamp, 1968).[5] A feedback shift register consists of a sequence of sites ("tubes") carrying values $a(i)$. At each time step, the site values evolve by a shift $a(i) = a(i-1)$ and feedback $a(0) = \mathbf{F}[a(j_1), a(j_2), ...]$ where j_i give the positions of "taps" on the shift register. An elementary cellular automaton of length N corresponds to a feedback shift register of length N with site values 0 and 1 and taps at positions $N-2$, $N-1$, and N. The Boolean function $\mathbf{F}$ defines the cellular automaton rule. [The additive rules 90 and 150 correspond to linear feedback shift registers in which $\mathbf{F}$ is addition modulo two (exclusive disjunction).] At each shift register time step, the value of one site is updated according to the cellular automaton rule. After N time steps, all N sites have been updated, and one cellular automaton time step is complete. All interior sites are treated exactly as in a cellular automaton, but the two end sites evolve differently (their values depend on the two preceding time steps).

---

[5] This similarity may be used as the basis for a simple hardware implementation of one-dimensional cellular automata (Pearson et al., 1981; Hoogland et al., 1982; Toffoli, 1983).

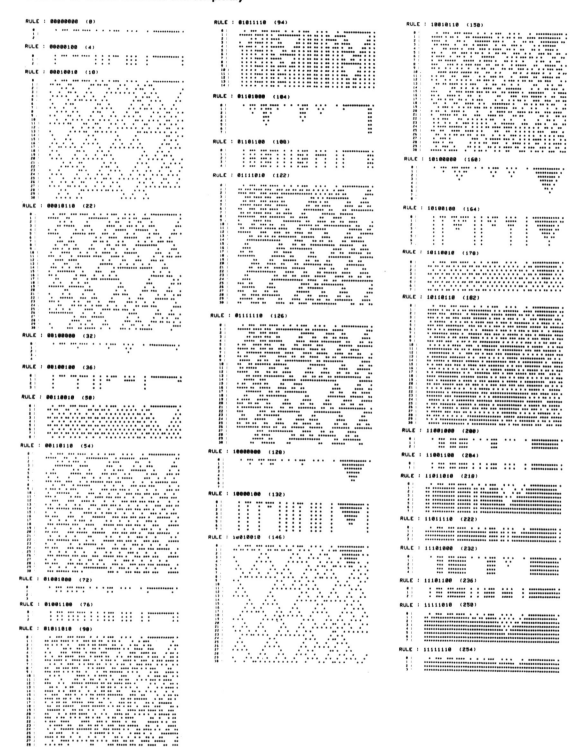

◄ **Figure 8.** Evolution of a disordered (random) initial state in each of the 32 possible legal one-dimensional elementary cellular automata. The value of each site is initially uncorrelated, and is taken to be 0 or 1 with probability $\frac{1}{2}$. Evolution is shown until a particular configuration appears for the second time, or for at most 30 time steps. Just as in Fig. 3, several classes of behavior are evident. In one class, time evolution generates long-range correlations and fluctuations, yielding distinctive "triangular" structures, and exhibiting a simple form of self-organization. All the cellular automata shown are taken to satisfy periodic boundary conditions, so that their sites are effectively arranged on a circle.

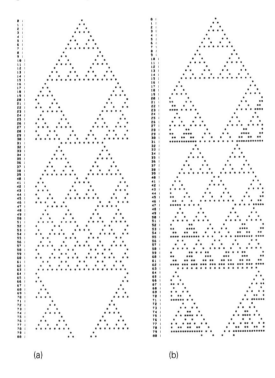

(a)   (b)

Figure 9. Time evolution of a simple initial state according to the modulo-two rule 90, on a line of sites satisfying (a) periodic boundary conditions (so that first and last sites are identified, and the sites are effectively arranged on a circle), and (b) null boundary conditions (so that sites not shown are assumed always to have value 0). Changes in boundary conditions apparently have no significant qualitative effect.

3. Local Properties of Elementary Cellular Automata

We shall examine now the statistical analysis of configurations generated by time evolution of "elementary" cellular automata, as illustrated in Figs. 3 and 8. This section considers statistical properties of individual such configurations; Sec. 4 discusses the ensemble of all possible configurations. The primary purpose is to obtain a quantitative characterization of the "self-organization" pictorially evident in Fig. 8.

A configuration may be considered disordered (or essentially random) if values at different sites are statistically uncorrelated (and thus behave as "independent random variables"). Such configurations represent a discrete form of "white noise." Deviations of statistical measures for cellular automaton configurations from their values for corresponding disordered configurations indicate order, and signal the presence of correlations between values at different sites. An (infinite) disordered

(a) (b) (c)

Figure 10. Examples of sets of disordered configurations in which each site is chosen to have value 1 with independent probability (a) 0.25, (b) 0.5, and (c) 0.75. Successive lines are independent. The configurations are to be compared with those generated by cellular automaton evolution as shown in Fig. 8.

configuration is specified by a single parameter, the independent probability p for each site to have value 1. The description of an ordered configuration requires more parameters.

Figure 10 shows a set of examples of disordered configurations with probabilities $p = 0.25$, 0.5, and 0.75. Such disordered configurations were used as the initial configurations for the cellular automaton evolution shown in Fig. 8. Qualitative comparison of the configurations obtained by this evolution with the disordered configurations of Fig. 10 strongly suggests that cellular automata indeed generate more ordered configurations, and exhibit a simple form of self-organization.

The simplest statistical quantity with which to characterize a cellular automaton configuration is the average fraction (density) of sites with value 1, denoted by ρ. For a disordered configuration, ρ is given simply by the independent probability p for each site to have value 1.

We consider first the density ρ_1 obtained from a disordered configuration by cellular automaton evolution for one time step. When $p = \rho = \frac{1}{2}$ (as in Fig. 8), a disordered configuration contains all eight possible three-site neighborhoods (illustrated in Fig. 1) with equal probability. Applying a cellular automaton rule (specified, say, by a binary sequence $\mathbf{R}$, as in Fig. 1) to this initial state for one time step ($\tau = 1$) yields a configuration in which the fraction of sites with value 1 is given simply by the fraction of the eight possible neighborhoods which yield 1 according to the cellular automaton rule. This fraction is given by

$$\rho_1 = \#_1(\mathbf{R})/(\#_0(\mathbf{R}) + \#_1(\mathbf{R})) = \#_1(\mathbf{R})/8, \tag{3.1}$$

where $\#_d(S)$ denotes the number of occurrences of the digit d in the binary representation of S. Hence, for example, $\#_1(10110110) = \#_1(182) = 5$ and $\#_0(10110110) = \#_0(182) = 3$. With cellular automaton rule 182, therefore, the density ρ after the first time step shown in Fig. 8 is $\frac{5}{8}$ if an infinite number of sites is included. The result (3.1) may be generalized to initial states with $p \neq \frac{1}{2}$ by using the probabilities $p(\sigma) = p^{\#_1(\sigma)}(1-p)^{\#_0(\sigma)}$, for each of the eight possible three-site neighborhoods σ (such as 110) shown in Fig. 1, and adding the probabilities for those σ which yield 1 on application of the cellular automaton rule.

The function $\#_1(n)$ will appear several times in the analysis given below. A graph of it for small n is given in Fig. 11, and is seen to be highly irregular. For any n, $\#_1(n) + \#_0(n)$ is the total number of digits ($\lceil \log_2 n \rceil$) in the binary representation of n, so that $\#_1(n) \leq \log_2 n$. Furthermore, $\#_1(2^k n) = \#_1(n)$ and for $n < 2^k$, $\#_1(n + 2^k) = \#_1(n) + 1$. Finally, one finds that

$$\#_1(n) = n - \sum_{i=1}^{\infty} \lfloor n/2^i \rfloor.$$

References to further results are given in McIlroy (1974) and Stolarsky (1977).

We now consider the behavior of the density ρ_τ obtained after τ time steps in the limit of large τ. When $\tau > 1$, correlations induced by cellular automaton evolution invalidate the approach used in Eq. (3.1), although a similar approach may nevertheless be used in deriving statistical approximations, as discussed below.

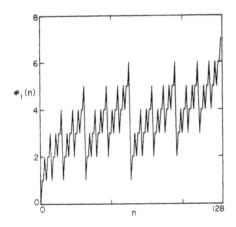

Figure 11. The number of occurrences $\#_1(n)$ of the binary digit 1 in the binary representation of the integer n [$\#_1(1) = 1$, $\#_1(2) = 1$, $\#_1(3) = 2$, $\#_1(4) = 1$, and so on]. The function is defined only for integer n: values obtained for successive integer n have nevertheless been joined by straight lines.

Figure 8 suggests that with some simple rules (such as 0, 32, or 72), any initial configuration evolves ultimately to the null state $\rho = 0$, although the length of transient varies. For rule 0, it is clear that $\rho = 0$ for all $\tau > 0$. Similarly, for rule 72, $\rho = 0$ for $\tau > 1$. For rule 32, infinite transients may occur, but the probability that a nonzero value survives at a particular site for τ time steps assuming an initial disordered state with $\rho = \frac{1}{2}$ is $2^{-3(2\tau+1)}$. Rule 254 yields $\rho_\infty = 1$, with a probability $(1 - \rho_0)^{2\tau+1}$ for a transient of length $\geq \tau$. Rule 204 is the "identity rule," which propagates any initial configuration unchanged and yields $\rho_\infty = \rho_0$. The "disjunctive superposition" principle for rule 250 discussed in Sec. 2 implies $\rho_\infty = 1$. For rule 50, the "conjunctive superposition" principle yields $\rho_\infty = \frac{1}{2}$.

Other simple rules serve as "filters" for specific initial sequences, yielding final densities proportional to the initial density of the sequences to be selected. For rule 4, the final density is equal to the initial density of 101 sequences, so that $\rho_\infty = \rho_0^2(1 - \rho_0)$. For rule 36, ρ_∞ is determined by the density of initial 00100 and ...1010101... sequences and is approximately $\frac{1}{16}$ for $\rho_0 = \frac{1}{2}$.

Exact results for the behavior of ρ_τ with the modulo-two rule 90 may be derived using the additive superposition property discussed in Sec. 2.

Consider first the number of sites $N_\tau^{(1)}$ with value 1 obtained by evolution according to rule 90 from an initial state containing a single site with value 1, as illustrated in Fig. 3. Geometrical considerations based on Fig. 5 yield the result[6]

$$N_\tau^{(1)} = 2^{\#_1(\tau)}, \tag{3.2}$$

where the function $\#_1(\tau)$ gives the number of occurrences of the digit 1 in the binary representation of the integer τ, as defined above, and is illustrated in Fig. 11. Equation (3.2) may be derived as follows. Consider the figure generated by $\lceil \log_2 \tau \rceil$ (the number of digits in the binary representation of τ) steps in the construction of Fig. 5. The configuration obtained after τ time steps of cellular automaton evolution corresponds to a slice through this figure, with a 1 at each point crossed by a line of the figure, and 0 elsewhere. By construction, the slice must lie in the lower half of the figure. Successive digits in the binary representation of τ determine whether the slice crosses the upper (0) or lower (1) halves of successively smaller triangles. The number of lines of the figure crossed is multiplied by a factor each time the lower half is chosen. The total number of sites with value 1 encountered is then given by a product of the factors of two associated with each 1 digit in the binary representation of τ. Inspection of Fig. 5 also yields a formula for the positions of all sites with value 1. With the original site at position 0, the positions of sites with value 1 after τ time steps are given by $\pm(2^{j_1} \pm (2^{j_2} \pm \cdots))$, where all possible combinations of signs are to be taken, and the j_i correspond to the positions at which the digit 1 appears in the binary representation of τ, defined so that $\tau = 2^{j_1} + 2^{j_2} + \cdots$ and $j_1 > j_2 > \cdots$.

Equation (3.2) shows that the density averaged over the region of nonzero sites ("light cone") in the rule 90 evolution of Fig. 3 is given by $\rho_\tau = N_\tau^{(1)}/(2\tau + 1)$ and does not tend to a definite limit for large τ. Nevertheless, the time-average density

$$\bar{\rho}_T = (1/T) \sum_{\tau=0}^{\tau=T} \rho_\tau$$

tends to zero (as expected from the geometrical construction of Fig. 5) like $T^{\log_2 3 - 2} \sim T^{-0.42}$.[7] Results for initial states containing a finite number of sites with value 1 may be obtained by additive superposition. If the initial configuration is one which would

---

[6] This result has also been derived by somewhat lengthy algebraic means in Glaisher (1899), Fine (1947), Roberts (1957), Kimball et al. (1958), and Honsberger (1976).

[7] This form is strictly correct only for $T = 2^k$. For $T = 2^k(1 + \delta)$, there is a correction factor $\simeq (1 + \delta^{\log_2 3})/(1 + \delta)^{\log_2 3}$, which lies between 0.86 and 1, with a broad minimum around $\delta = 0.3$.

be reached by evolution from a single site after, say, τ_0 time steps, then the resulting density is given by Eq. (3.2) with the replacement $\tau \to \tau - \tau_0$. Only a very small fraction of initial configurations may be treated in this way, since evolution from a single site generates only one of the 2^k possible configurations in which the maximum separation between nonzero sites is k. For small or highly regular initial configurations, results analogous to (3.2) may nevertheless be derived. Statistical results for evolution from disordered initial states may also be derived. Equation (3.2) implies that after exactly $\tau = 2^j$ time steps, an initial state containing a single nonzero site evolves to a configuration with only two nonzero sites. At this point, the value of a particular site at position n is simply the sum modulo two of the initial values of sites at positions $n - \tau$ and $n + \tau$. If we start from a disordered initial configuration, the density at such time steps is thus given by $\rho_{\tau=2^j} = 2\rho_0(1 - \rho_0)$. In general, the value of a site at time step τ is a sum modulo two of the initial values of $N_\tau^{(1)} = 2^{\#_1(\tau)}$ sites, which each have value 1 with probability ρ_0. If each of a set of k sites has value 1 with probability p, then the probability that the sum of the values at the sites will be odd (equal to 1 modulo two) is

$$\sum_{i \text{ odd}} \binom{k}{i} p^i (1 - p)^{k-i} = \tfrac{1}{2}[1 - (1 - 2p)^k].$$

Thus the density of sites with value 1 obtained by evolution for τ time steps from an initial state with density ρ_0 according to cellular automaton rule 90 is given by

$$\rho_\tau = \tfrac{1}{2}[1 - (1 - 2\rho_0)^{2^{\#_1(\tau)}}]. \tag{3.3}$$

This result is shown as a function of τ for the case $\rho_0 = 0.2$ in Fig. 12. For large τ, $\#_1(\tau) = O(\log_2 \tau)$, except at a set of points of measure zero, and Eq. (3.3) implies that $\rho_\tau \to \tfrac{1}{2}$ as $\tau \to \infty$ for almost all τ (so long as $\rho_0 \neq 0$).

Cellular automaton rule 150 shares with rule 90 the property of additive superposition. Inspection of the results for rule 150 given in Fig. 3 indicates that the value of a particular site depends on the values of at least three initial sites (this minimum again being achieved when $\tau = 2^k$), so that $|\rho_\tau - \tfrac{1}{2}| \leq |1 - 2\rho_0|^3$. Between the exceptional time steps $\tau = 2^k$, the ρ_τ for rule 150 tends to be much flatter than that for rule 90 (illustrated in Fig. 12). An exact result may be obtained, but is more complicated than in the case of rule 90. The geometrical construction of Fig. 6 shows that for rule 150, $N_\tau^{(1)}$ is a product of factors $\chi(j)$ associated with each sequence of j ones (delimited by zeroes) in the binary representation of τ. The function $\chi(j)$ is given by the recurrence relation $\chi(j) = (2j \pm 1)\chi(j - 1)$ where the upper (lower) sign is taken for j odd (even), and $\chi(1) = 3$ [so that $\chi(2) = 5$, $\chi(3) = 11$ and so on]. [$N_\tau^{(1)}$ thus measures "sequence correlations" in τ.] The density is then given in analogy with Eq. (3.3) by $\rho_\tau = \tfrac{1}{2}[1 - (1 - 2\rho_0)^{N_\tau^{(1)}}]$.

21

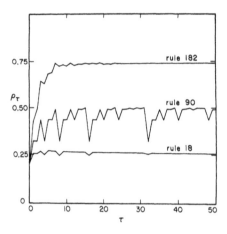

Figure 12. Average density ρ_τ of sites with value 1 obtained by time evolution according to various cellular automaton rules starting from a disordered initial state with $\rho_0 = 0.2$. The additivity of the modulo-two rule 90 may be used to derive the exact result (3.2) for ρ_τ. The irregularities appear for time steps at which the value of each site depends on the values of only a few initial sites. For the nonadditive complex rules exemplified by 18 and 182, the values of sites at time step τ depend on the values of $O(\tau)$ initial sites, and ρ_τ tends smoothly to a definite limit. This limit is independent of the density of the initial disordered state.

Some aspects of the large-time behavior of nonadditive complex cellular automata may be found using a correspondence between nonadditive and additive rules (Grassberger, 1982). Special classes of configurations in nonadditive cellular automata effectively evolve according to additive rules. For example, with the nonadditive complex rule 18, a configuration in which, say, all even-numbered sites have value zero evolves after one time step to a configuration with all odd-numbered sites zero, and with the values of even-numbered sites given by the sums modulo two of their odd-numbered neighbors on the previous time step, just as for the additive rule 90. An arbitrary initial configuration may always be decomposed into a sequence of (perhaps small) "domains," in each of which either all even-numbered sites or all odd-numbered sites have value zero. These domains are then separated by "domain walls" or "kinks." The kinks move in the cellular automaton evolution and may annihilate in pairs. The motion of the kinks is determined by the initial configuration; with a disordered initial configuration, the kinks initially follow approximately a random walk, so that their mean displacement increases with time according to $\langle x^2 \rangle = t$ (Grassberger, 1982), and the paths of the kinks are fractal curves. This implies that the average kink density decreases through annihilation as if by diffusion processes according to the formula $\langle \rho_{kink} \rangle \sim (4\pi t)^{-1/2}$ (Grassberger, 1982). Thus after a sufficiently long time all kinks (at least from any finite initial configuration) must annihilate, leaving a configuration whose alternate sites evolve according to the additive cellular automaton rule 90. Each point on the "front" formed by the kink paths yields a pattern analogous to Fig. 5. The superposition of such patterns, each diluted by the insertion of alternate zero sites, yields configurations with an average density $\frac{1}{4}$ (Grassberger, 1982). The large number of sites on the "front" suppresses the fluctuations found for complete evolution according to additive rule 90. Starting with a disordered configuration of any nonzero density, evolution according to cellular automaton rule 18 therefore yields an asymptotic density $\frac{1}{4}$. The existence of a universal ρ_∞, independent of initial density ρ_0, is characteristic of complex cellular automaton rules.

Straightforward transformations on the case of rule 18 above then yield asymptotic densities $\rho_\infty = \frac{1}{4}$ for the complex nonadditive rules 146, 122, and 126, and an asymptotic density $\frac{3}{4}$ for rule 182, again all independent of the initial density ρ_0 (Grassberger, 1982). No simple domain structure appears with rule 22, and the approach fails. Simulations yield a numerical estimate $\rho_\infty = 0.35 \pm 0.02$ for evolution from disordered configurations with any nonzero ρ_0.

Figure 12 shows the behavior of ρ_τ for the complex nonadditive cellular automata 18 and 182 with $\rho_0 = 0.2$, and suggests that the final constant values $\rho_\infty = 0.25$ and $\rho_\infty = 0.75$ are approached roughly exponentially with time.

One may compare exact results for limiting densities of cellular automata with approximations obtained from a statistical approach (akin to "mean-field theory"). As discussed above, cellular automaton evolution generates correlations between values at different sites. Nevertheless, as a simple approximation, one may ignore these correlations, and parametrize all configurations by their average density ρ, or, equivalently, by the probabilities p and $q = 1 - p$, assumed independent, for each site to have value 1 and 0, respectively. With this approximation, the time evolution of the density is given by a master equation

$$\frac{\delta\rho}{\delta\tau} = \Gamma(0 \to 1) - \Gamma(1 \to 0),$$

$$\Gamma(0 \to 1) = \mathbf{P} \cdot (00110011 \wedge \mathbf{R}),$$

$$\Gamma(1 \to 0) = \mathbf{P} \cdot (11001100 \wedge \sim\mathbf{R}),$$

$$\mathbf{P} = \{p^3, p^2q, p^2q, pq^2, p^2q, pq^2, pq^2, q^3\}.$$

(3.4)

The term $\Gamma(0 \to 1)$ represents the average fraction of sites whose values change from 0 to 1 in each time step, and $\Gamma(1 \to 0)$ the fraction changing from 1 to 0. $\mathbf{R}$ is the binary specification of a cellular automaton rule, and the binary number with which it is "masked" (digitwise conjunction) selects local rules for three-site neighborhoods with appropriate values at the center site. $\mathbf{P}$ is the vector of probabilities for the possible three-site neighborhoods, assuming each site independently to have value 1 with probability $p = \rho$, and to have value 0 with probability $q = 1 - p = 1 - \rho$. The dot indicates that each element of this vector is to be multiplied by the corresponding digit of the binary sequence, and the results are to be added together. The equilibrium density ρ_∞ is achieved when

$$\frac{\delta\rho}{\delta\tau} = 0.$$

This condition yields a polynomial equation for p and thus ρ_∞ for each of the legal cellular automaton rules. For rule 90, the equation is $pq^2 - p^3 = p - 2p^2 = p(1-2p) = 0$, which has solutions $p = 0$ (null state for all time) and $p = \frac{1}{2}$. Rule 18 yields the equation $pq^2 - 2p^2q - p^3 = p(1 - 4p + 2p^2) = 0$, which has the solutions $p = 0$ and $p = 1 - 1/\sqrt{2} \simeq 0.293$, together with the irrelevant solution $p = 1 + 1/\sqrt{2} > 1$. Rule 182 yields $2pq^2 - p^2q = p(2 - 3p)(1 - p) = 0$, giving $p = 0, 1, \frac{2}{3}$. For rules 90 and 18, these approximate results are close to the exact results 0.5 and 0.25. For rule 182, there is a significant discrepancy from the exact value 0.75. Nevertheless, for all complex cellular automaton rules, it appears that the master equation (3.4) yields equilibrium densities within 10–20% of the exact values. The discrepancies are a reflection of the violation of the Markovian approximation required to derive Eq. (3.4) and thus of the presence of correlations induced by cellular automaton evolution.

In the discussion above, a definite value for the density ρ_τ at each time step was found by averaging over all sites of an infinite cellular automaton. If instead the density is estimated by averaging over blocks containing a finite number of sites b, a distribution of density values is obtained. In a disordered state, the central limit theorem ensures that for large b, these density estimates follow a Gaussian distribution with standard deviation $\simeq 1/\sqrt{b}$. Evolution according to any of the complex cellular automaton rules appears accurately to maintain this Gaussian distribution, while shifting its mean as illustrated in Fig. 12. Density in cellular automaton configurations thus obeys the "law of large numbers." Instead of taking many blocks of sites at a single time step, one might estimate the density at "equilibrium" by averaging results for a single block over many time steps. For nonadditive complex cellular automaton rules, it appears that these two procedures yield the same limiting results. However, the large fluctuations in average density visible in Fig. 12 at particular time steps for additive rules (90 and 150) would be lost in a time average.

Cellular automaton evolution is supposed to generate correlations between values at different sites. The very simplest measure of these correlations is the two-point correlation function $C^{(2)}(r) = \langle S(m)S(m+r)\rangle - \langle S(m)\rangle\langle S(m+r)\rangle$, where the average is taken over all possible positions m in the cellular automaton at a fixed time, and $S(k)$ takes on values -1 and $+1$ when the site at position k has values 0 and 1, respectively. A disordered configuration involves no correlations between values at different sites and thus gives $C^{(2)}(r) = 0$ for $r > 0$ $[C^{(2)}(0) = 1 - (2\rho - 1)^2]$. With the single-site initial state of Fig. 3, evolution of complex cellular automata yields configurations with definite periodicities. These periodicities give rise to peaks in $C^{(2)}(r)$. At time step τ, the largest peaks occur when $r = 2^k$ and the digit corresponding to 2^k appears in the binary decomposition of τ; smaller peaks occur when $r = 2^{k_1} \pm 2^{k_2}$, and so on. For the additive cellular automaton rules 90 and 150, a convolution of this result with the correlation function for any initial state gives the form of $C^{(2)}(r)$ after evolution for τ time steps. With these rules, the correlation function obtained by evolution from a disordered initial configuration thus always remains zero. For nonadditive rules, nonzero short-range correlations may nevertheless be generated from disordered initial configurations. The form of $C^{(2)}(r)$ for rule 18 at large times is shown in Fig. 13, and is seen to fall roughly exponentially with a correlation length ~ 2. The existence of a nonzero correlation length in this case is our first indication of the generation of order by cellular automaton evolution.

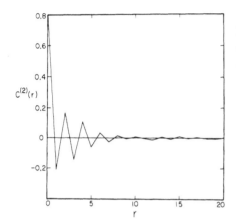

Figure 13. Two-point correlation function $C^{(2)}(r)$ for configurations generated at large times by evolution according to cellular automaton rule 18 from any disordered initial configuration. $C^{(2)}(r)$ is defined as $\langle S(m)S(m+r)\rangle - \langle S(m)\rangle\langle S(m+r)\rangle$, where the average is taken over all sites m of the cellular automaton, and $S(k) = \pm 1$ when site k has values 1 and 0, respectively. No correlations are present in a disordered configuration, so that $C^{(2)}(r) = 0$ for $r > 0$. Evolution according to certain complex cellular automaton rules, such as 18, yield nonzero but exponentially damped correlations.

Figures 3 and 8 show that the evolution of complex cellular automata generates complicated patterns with a distinctive structure. The average density and the two-point correlation function are too coarse as statistical measures to be sensitive to this structure. Individual configurations appear to contain long sequences of correlated sites, punctuated by disordered regions. The two-dimensional picture formed by the succession of configurations in time is characteristically peppered with triangle structures. These triangles are formed when a long sequence of sites which suddenly all attain the same value, as if by a fluctuation, is progressively reduced in length by "ambient noise." Let $T_{(i)}(n)$ denote the density of triangles (in position and time) with base length n and filled with sites of value i. It is convenient to begin by considering the behavior of this density and then to discuss its consequences for the properties of individual configurations, whose long sequences typically correspond to sections through the triangles.

Consider first evolution from a simple initial state containing a single site with value 1. Figure 3 shows that in this case, all complex cellular automata (except rule 150) generate a qualitatively similar pattern, containing many congruent triangles whose bases have lengths 2^k. A geometrical construction for the limiting pattern obtained at large times was given in Fig. 5. At each successive stage in the construction, the linear dimensions (base lengths) of the triangles added are halved, and their number is multiplied by a factor 3. In the limit, therefore, $T(n/2) \sim 3T(n)$, (with $n = 2^k$), and hence

$$T(n) \sim n^{-\log_2 3} \sim n^{-1.59} \tag{3.5}$$

[requiring exactly one triangle of size $\tau/2$ at time step τ fixes the normalization as $T(n) = (2n/\tau)^{-\log_2 3}$]. The result (3.5) demonstrates that the patterns obtained from complex cellular automata in Fig. 3 not only contain structure on all scales (in the form of triangles of all sizes), but also exhibit a scale invariance or self similarity which implies the same structure on all scales (cf. Mandelbrot, 1982; Willson, 1982). The power law form of the triangle density (3.5) is independent of the absolute scale of n.

Self-similar figures on, for example, a plane may in general be characterized as follows. Find the minimum number $N(a)$ of squares with side a necessary to cover all parts of the figure (all sites with nonzero values in the cellular automaton case). The figure is self-similar or scale invariant if rescaling a changes $N(a)$ by a constant factor independent of the absolute size of a. In this case, $N(a) \sim a^{-D}$, where D is defined to be the Hausdorff-Besicovitch or fractal dimension (Mandelbrot, 1977, 1982) of the figure. A figure filling the plane would give $D = 2$, while a line would give $D = 1$. Intermediate values of D indicate clustering or intermittency. According to this definition, the cellular automaton pattern of Fig. 5 has fractal dimension $D = \log_2 3 \simeq 1.59$.

Figure 6 gives the construction analogous to Fig. 5 for the pattern generated by rule 150 in Fig. 3. In this case, the triangle density satisfies the two-term recurrence relation $T(n = 2^k) = 2T(2^{k+1}) + 4T(2^{k+2})$ with, say, $T(1) = 0$ and $T(2) = 2$. For large k, this yields (in analogy with the Fibonacci series)[8]

$$T(n) \sim n^{-\log_2(2\varphi)} = n^{-\log_2(1+\sqrt{5})} \sim n^{1.69}, \tag{3.6}$$

where $\varphi = (1 + \sqrt{5})/2 \simeq 1.618$ is the "golden ratio" which solves the equation $x^2 = x + 1$. The limiting fractal dimension of the pattern in Fig. 6 generated by cellular automaton rule 150 is thus $\log_2(2\varphi) = 1 + \log_2(\varphi) \simeq 1.69$.

The self similarity of the patterns generated by time evolution with complex cellular automaton rules in Fig. 3 is shared by almost all the configurations appearing at particular time steps and corresponding to lines through the patterns. If the fractal dimension of the two-dimensional patterns is D, then the fractal dimension of almost all the individual configurations is $D - 1$. The configurations obtained at, for example, time steps τ of the form 2^k are members of an exceptional set of measure zero, for which no fractal dimension is defined. Almost all configurations generated from a single initial site by complex cellular automaton rules are thus

---

[8] For small k, the triangle density in this case does not behave as a pure power of 2^k. Whereas the solution to any one-term recurrence relation, of the type found for cellular automaton rule 90, is a pure power, the solution to a p-term recurrence relation is in general a sum of p powers, with each exponent given by a root of the characteristic polynomial equation. In the high-order limit, the solutions are dominated by the term with the highest exponent (corresponding to the largest root of the equation). Complex roots yield oscillatory behavior [as in $f(k) = -f(k-1) + f(k-2)$; $f(0) = 0$, $f(1) = 1$].

RULE : 01011010 (90)   RULE : 01111110 (126)   RULE : 11011010 (218)

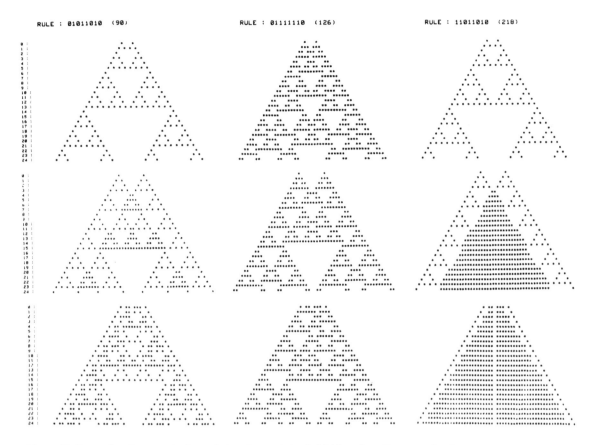

Figure 14. Twenty-five time steps in the evolution of several simple initial configurations according to cellular automaton rules 90, 126, and 218. Configurations generated by rule 90 obey additive superposition (under addition modulo two). The first initial state taken is exceptional for rules 90 and 218, since it occurs in evolution from a single initial site, as shown in Fig. 3, so that the final pattern is a shifted form of that found in Fig. 3. For other initial states, the patterns obtained deviate substantially from those of Fig. 3. However, features with sizes much larger than the extent of the initial state remain unchanged. For complex cellular automaton rules such as 90 and 126, such features share the self-similarity found in Fig. 3.

self-similar, and (except for rule 150) are characterized by a fractal dimension $D = \log_2 3 - 1 = \log_2(\frac{3}{2}) \simeq 0.59$. The second form may be deduced directly from the geometrical construction of Fig. 5. For rule 150, the configurations have fractal dimension $D = \log_2 \varphi$.

Figure 14 shows patterns generated by evolution with a selection of complex cellular automaton rules from initial states containing a few sites with value 1, extending over a region of size n_0. Comparison with Fig. 3 demonstrates that in most cases the patterns obtained even after many time steps differ from those generated with a single initial site. A few exceptional initial configurations (such as the one used for the first rule 90 example in Fig. 14) coincide with configurations reached by evolution from a single initial site and therefore yield a similar pattern, appropriately shifted in time. In the general case, Fig. 14 suggests that the form of the initial state determines the number of triangles with size $n \lesssim n_0$, but does not affect the density of triangles with $n \gg n_0$. As a simple example consider the modulo-two rule 90, whose additive superposition property implies that the final pattern obtained from an arbitrary initial state is simply a superposition of the patterns which would be generated from each of the nonzero initial sites in isolation. These latter patterns were shown in Fig. 5, and involve the generation of a triangle of size 2^k at time step 2^k. The superposition of such patterns yields at time step 2^k a triangle of size at least $2^k - 2n_0$. This conclusion apparently holds also for nonadditive complex cellular automata, so that, in general, for $n \gg n_0$, the density of triangles follows the form (3.5), as for a single site initial state. The patterns thus exhibit self-similarity for features large compared to the intrinsic scale defined by the "size" of the initial state. One therefore concludes that patterns which "grow" from any simple initial state according to any of the "complex" cellular automaton rules (except 150) share the universal feature of self similarity, characterized by a fractal dimension $\log_2 3$. On this basis, one may then conjecture that given suitable geometry (perhaps in more than one dimension, and possibly with more than three sites in a neighborhood), many of the wide variety of systems found to exhibit self-similar structure (Mandelbrot, 1977, 1982) attain this structure through local processes which follow cellular automaton rules.

Having considered the case of simple initial configurations, we now turn to the case of evolution from disordered initial configurations, illustrated in Fig. 8. Figure 15 shows the first 300 time steps in the evolution of cellular automaton 126, starting from a disordered initial state with density $\rho = 0.5$. Triangles of all sizes appear to be generated (the largest appearing in the figure has $n = 27$). Figure 16 shows the density of triangles $T(n)$ obtained at large times by evolution according to rule 126 and all of the other complex cellular automaton rules. The figure reveals the remarkable fact that for large n, all nonadditive rules yield the same $T(n)$, distinct from that for the additive rules (90 and 150). All the results are well fit by the form

$$T(n) \sim \lambda^{-n}. \tag{3.7}$$

For nonadditive rules $\lambda \sim \frac{4}{3}$, while for the additive rules $\lambda \sim 2$. The same results are obtained at large times regardless of the density of the initial state. Thus the spectrum of triangles generated by complex cellular automaton evolution is universal, independent both of the details of the initial state, and of the precise cellular automaton rule used.

The behavior (3.8) of the triangle density with disordered initial states is to be contrasted with that of (3.5) for simple initial states. The precise form of an initial state of finite extent n_0 affects the pattern generated only at length scales $\lesssim n_0$: at larger length scales the pattern takes on a universal self-similar character. A disordered initial state of infinite extent affects the pattern generated at all length scales and for

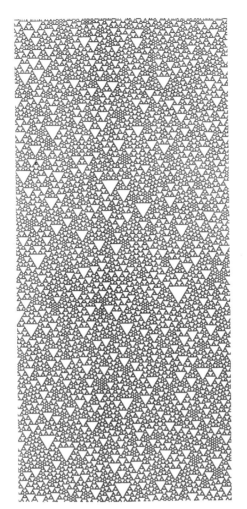

Figure 15. Configurations obtained by evolution for 300 time steps from an initial disordered configuration with $\rho = 0.5$ according to cellular automaton rule 126. The fluctuations visible in the form of triangles and apparent at small scales in Fig. 8 are seen here to occur on all scales. The largest triangle in this sample has a base length of 27 sites.

all times. Triangles of all sizes are nevertheless obtained, so that structure is generated on all scales, as suggested by Fig. 15. However, the pattern is not self-similar, but depends on the absolute scale defined by the spacing between sites.

Disordered configurations are defined to involve no statistical correlations between values at different sites. They thus correspond to a discrete form of white noise and yield a flat spatial frequency spectrum. One may also consider "pseudodisordered" configurations in which the value of each individual site is chosen randomly, but according to a distribution which yields statistical correlations between different sites, and a nontrivial spatial Fourier spectrum. For example, a Brownian configuration (with spatial frequency spectrum $1/k^2$) is obtained by assigning a value to each site in succession, with a certain probability for the value to differ from one site to the next (as in a random walk). The patterns generated by cellular automaton from such initial configurations may differ from those obtained with disordered (white noise) initial configurations. Complex nonadditive cellular automata evolving from a Brownian initial state yield patterns whose triangle density $T(n)$ decreases less rapidly at large n than for disordered initial configurations: the "long-range order" of the initial state leads to the generation of longer-range fluctuations. In the extreme limit of a homogeneous initial state (such as ...11111... or ...10101...), cellular automaton evolution preserves the homogeneity, and no finite structures are generated.

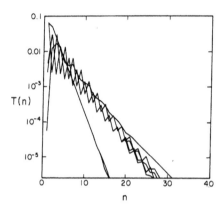

Figure 16. Density $T(n)$ of triangle structures generated in the evolution of all the possible complex cellular automata from disordered initial configurations with density $\rho_0 = 0.5$. Triangles are evident in Figs. 8 and 16. They are formed when a sequence of sites suddenly attain the same value, but the length of the sequence is progressively reduced on subsequent time steps, until the apex of the triangle is reached. The appearance of triangles is a simple indication of self-organization. The triangle density $T(n)$ is defined only at integer values of n, but these points have been joined in the figure. For large n, the triangle densities for all complex cellular automata are seen to tend towards one of two limiting forms. The group tending to the upper curve are the nonadditve complex cellular automata 18, 22, 122, 126, 146, and 182. The additive rules 90 and 150 follow the lower curve. In both cases, $T(n)$ falls off exponentially with n, in contrast to the power law form found for the self-similar patterns of Figs. 3, 5, and 14.

The appearance of triangles over a series of time steps in the evolution of complex cellular automata from disordered initial states reflects the generation of long sequences of correlated sites in individual cellular automaton configurations. This effect is measured by the "sequence density" $Q_{(i)}(n)$, defined as the density of sequences of exactly n adjacent sites with the same value i (bordered by sites with a different value). Thus, for example, $Q_{(0)}(4)$ gives the density of 100001 sequences. $Q_{(0)}(n)$ clearly satisfies the sum rule

$$\sum_{n=1}^{\infty} n Q_{(0)}(n) = 1 - \rho.$$

In a disordered configuration with density $p = 1 - q$, $Q_{(0)}(n) \sim p^2 q^n$ for large n.

Any sequence longer than two sites in a complex cellular automaton must yield a triangle, leading to the sum rule

$$Q(n) \simeq \sum_{i=n}^{\infty} [2T(i)/i].$$

Thus the $Q(n)$ obtained at large times by evolution from a disordered initial state should follow the same exponential form (3.8) as $T(n)$.

Figure 17 shows the sequence density $Q_{(0)}(n)$ obtained at various time steps in the evolution of rule 126 from a disordered initial state, as illustrated in Fig. 15. At each time step, the $Q_{(0)}(n)$ for a disordered configuration (illustrated in Fig. 10) with the same average density has been subtracted. The resulting difference vanishes by definition at $\tau = 0$, but Fig. 17 shows that for $\tau \geq 1$, the cellular automaton evolution yields a nonzero difference. After a few time steps, the cellular automaton tends to an equilibrium state containing an excess of long sequences of sites with value 0, and a deficit of short ones. This final equilibrium $Q_{(0)}(n)$ does not depend on the density of the initial disordered configuration. Starting from any disordered initial state (random noise), repeated application of the local cellular automaton rules is thus seen to generate ordered configurations whose statistical properties, as measured by

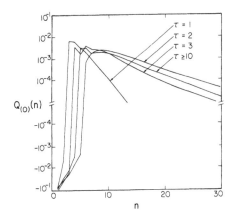

Figure 17. Density $Q_{(0)}(n)$ of sequences of exactly n successive sites with value 0 (delimited by sites with value 1) in configurations generated by τ steps in time evolution according to cellular automaton rule 126, starting from an initial disordered state with density $\rho = 0.5$. [The function $Q_{(0)}(n)$ is defined only for integer n: points are joined for ease of identification.] At each time step, the density of sequences in a disordered configuration with the same average total density has been subtracted. This difference vanishes for $\tau = 0$ by definition. The nonzero value shown in the figure for $\tau \geq 1$ is a manifestation of self-organization in the cellular automaton, suggested qualitatively by comparison of Figs. 8 and 10. For large τ, an equilibrium state is reached, which exhibits an excess of long sequences and a deficit of short ones.

(a)

(c)

(b)

(d)

Figure 18. Configurations generated from a disordered initial state (with $\rho_0 = 0.5$) by the evolution of the complex nonadditive cellular automaton 126, in the presence of noise which causes values obtained at each site to be reversed with probability κ at every time step. (a) is for $\kappa = 0$ (no "noise"), (b) for $\kappa = 0.1$, (c) for $\kappa = 0.2$, and (d) for $\kappa = 0.5$. As κ increases, the structure generated is progressively destroyed. No discontinuity in behavior as a function of κ is found.

sequence densities, differ from those of corresponding disordered configurations. The impression of self-organization in individual configurations given by Fig. 8 is thus quantitatively confirmed.

As suggested by the sum rule, the $Q_{(0)}(n)$ for complex cellular automata with disordered initial states follow the exponential behavior (3.7) found for the $T(n)$. Again, the parameter λ has a universal value $\sim \frac{4}{3}$ for all nonadditive cellular automaton rules and ~ 2 for additive ones. If all configurations of the cellular automata were disordered, then the sequence density would behave at large n as $(1-\rho)^n$ and depend on total average density ρ for the configurations. The form (3.5) yields sequence correlations with the same exponential behavior, but with a fixed λ, universal to all the nonadditive complex cellular automaton rules, and irrespective of the final densities to which they lead. (The universal form may be viewed as corresponding to an "effective density" $\simeq 0.25$.)

Cellular automata are usually defined to evolve according to definite deterministic local rules. In modelling physical or biological systems it is, however, sometimes convenient to consider cellular automata whose local rules involve probabilistic elements or noise (cf. Griffeath, 1970; Schulman and Seiden, 1978; Gach et al., 1978). The simplest procedure is to prescribe that at each time step the value obtained by application of the deterministic rule at each site is to be reversed with a probability κ (and with each site treated independently). (If an energy is associated with the reversal of a site, κ gives the Boltzmann factor corresponding to a finite temperature heat bath.) Figure 18 shows the effects of introducing such noise in the evolution of cellular automaton rule 126. The structures generated are progressively destroyed as κ increases. Investigation of densities and correlation functions indicates that the transition to disorder is a continuous one, and no phenomenon analogous to a "phase transition" is found.

4. Global Properties of Elementary Cellular Automata

Section 3 analyzed the behavior of cellular automata by considering the statistical properties of the set of values of sites in individual cellular automaton configurations. The alternative approach taken in this section considers the statistical properties of the set (ensemble) comprising all possible complete configurations of a cellular automaton (in analogy with the Γ-space approach to classical statistical mechanics). Such an approach provides connections with dynamical systems theory (Ott, 1981) and the formal theory of computation (Minsky, 1967; Arbib, 1969; Manna, 1974; Hopcroft and Ullman, 1979; Beckman, 1980), and yields a view of self-organization phenomena complementary to that developed in Sec. 3. Cellular automaton rules may be considered as a form of "symbolic dynamics" (e.g., Alekseev and Yakobson, 1981), in which the degrees of freedom in the system are genuinely discrete, rather than being continuous but assigned to discrete "bins."

As in Sec. 3, we examine here only elementary cellular automata. Some results on global properties of more complicated cellular automata will be mentioned in Sec. 5.

For most of this section, it will be convenient to consider "finite" cellular automata, containing only a finite number of sites N. There are a total of 2^N possible configurations for such a cellular automaton. Each configuration is uniquely specified by a length N binary integer whose digits give the values of the corresponding sites.[9] (A configuration of an infinite cellular automaton would correspond to a binary real number.) The evolution of a finite cellular automaton depends on the boundary conditions applied. We shall usually assume periodic boundary conditions, in which the first and last sites are identified, as if the sites lay on a circle of circumference N. One could alternatively take an infinite sequence of sites, but assume that all those outside the region of length N have value 0. Results obtained with these two choices were compared in Fig. 9, and no important qualitative differences were found. Most of the re-

---

[9] An alternative specification would take each configuration to correspond to one of the 2^N vertices of an N-dimensional hypercube, labeled by coordinates corresponding to the values of the N sites. Points corresponding to configurations differing by values at a single site are then separated by a unit distance in N-dimensional space.

sults derived in this section are also insensitive to the form of boundary conditions assumed. However, several of the later ones depend sensitively on the value of N taken.

Cellular automaton rules define a transformation from one sequence of binary digits to another. The rules thus provide a mapping from the set of binary numbers of length N onto itself. For the trivial case of rule 0, all binary numbers are mapped to zero. Figure 19 shows the mappings corresponding to evolution for one and five time steps according to cellular automaton rule 90 with $N = 9$. The mapping corresponding to one time step is seen to maintain some nearby sets of configurations. After five time steps, however, the evolution is seen to map configurations roughly

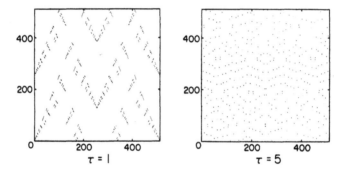

Figure 19. Mapping in the set of 512 possible configurations of a length nine finite cellular automaton corresponding to evolution for τ time steps according to the modulo-two rule 90. Each possible configuration is represented by the decimal equivalent of the binary number whose digits give the values at each of its sites. The horizontal axis gives the number specifying the initial configuration; the vertical axes that for the final configuration. Each initial configuration is mapped to a unique final configuration.

uniformly, so that the final configurations obtained from nearby initial configurations are essentially uncorrelated.

A convenient measure of distance in the space of cellular automaton configurations is the "Hamming distance" $H(s_1, s_2)$ [familiar from the theory of error-correcting codes (Peterson and Weldon, 1972)], defined as the number of digits (bits) which differ between the binary sequences s_1 and s_2. [Thus in Boolean form, $H(s_1, s_2) = \#_1(s_1 \oplus s_2)$.] Particular configurations correspond to points in the space of all possible configurations. Under cellular automaton evolution, each initial configuration traces out a trajectory in time. If cellular automaton evolution is "stochastic," then the trajectories of nearby points (configurations) must diverge (exponentially) with time. Consider first the case of two initial configurations (say, S_1 and S_2) which differ by a change in the value at one site (and are thus separated by unit Hamming distance). After τ time steps of cellular automaton evolution, this initial difference may affect the values of at most 2τ sites (so that $H \le 2\tau$). However, for simple cellular automaton rules, the difference remains localized to a few sites, and the total Hamming distance tends rapidly to a small constant value. The behavior for complex cellular automaton rules differs radically between additive rules (such as 90 and 150) and nonadditive ones. For additive rules, the difference obtained after τ time steps is given simply by the evolution of the initial difference (in this case a single nonzero site) for τ time steps. The Hamming distance at time step τ is thus given by the number of nonzero sites in the configuration obtained by evolution from a single site, and for rule 90 has the form $H_\tau = 2^{\#_1(\tau)}$, as illustrated in Fig. 20(a). The average Hamming distance, smoothed over many time steps, behaves as $H_\tau = \tau^{\log_2 3 - 1} \simeq \tau^{0.59}$. For nonadditive rules, the difference between configurations obtained through cellular automaton evolution no longer depends only on the difference between the initial configurations. Figure 20(c) shows the difference between configurations obtained by evolution according to the nonadditive cellular automaton rule 126. The lack of symmetry in the pattern is a reflection of the dependence on the values of multiple initial sites.

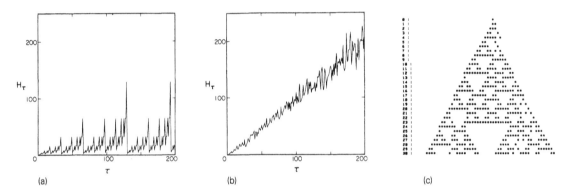

(a)   (b)   (c)

Figure 20. Divergence in behavior of disord red configurations intially differing by a change in the value of a single site under cellular automaton evolution. The Hamming distance H between two configurations is defined as the number of bits (site values) which differ between the configurations. (a) shows the evolution of the Hamming distance between two configurations of the additive cellular automaton 90 (modulo-two rule); (b) shows the corresponding Hamming distance for the nonadditive cellular automaton 126; and (c) gives the actual difference (modulo two) between the configurations of cellular automaton 126 for the first few time steps. For nonadditive rules [case (b)], $H_\tau \sim \tau$, while for additive rules [case (a)], after time averaging, $H_\tau \sim \tau^{0.59}$.

Figure 20(b) shows the Hamming distance corresponding to this difference. Apart from small fluctuations, it is seen to increase linearly with τ, tending at large τ to the form $H_\tau \simeq \tau$. This Hamming distance is the same as would be obtained by comparing sequences of 2τ sites in two disordered configurations with density 0.5. Thus a change in the value of a small number of initial sites is amplified by the evolution of a nonadditive cellular automaton, and leads to configurations with a linearly increasing number of essentially uncorrelated sites. (Changes in single sites may sometimes be eradicated after a single time step; this exceptional behavior occurs for cellular automaton rule 18, but is always absent if more than one adjacent site is reversed.) A bundle of initial trajectories therefore diverges with time into an exponentially increasing volume.

One may specify a statistical ensemble of states for a finite cellular automaton by giving the probability for each of the 2^N possible configurations. In a collection of many disordered states with density $\rho = \frac{1}{2}$, each possible cellular automaton configuration is asymptotically populated with equal probability. Such a collection of states will be termed an "equiprobable ensemble," and may be considered "completely disorganized." Cellular automaton evolution modifies the probabilities for states in an ensemble, thereby generating "organization." Figure 21 shows the probabilities for the 1024 possible configurations of a finite cellular automaton with $N = 10$ obtained after evolution for ten time steps according to rule 126 from an initial equiprobable

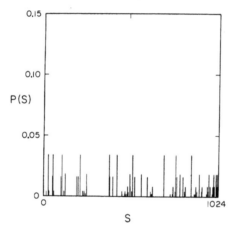

Figure 21. Probabilities for each of the 1024 possible configurations in a finite (circular) cellular automaton with length $N = 10$ obtained by evolution according to rule 126 for ten time steps from an initial ensemble containing each possible configuration with equal probability. On the horizontal axis, each configuration S is labeled by a ten-digit binary integer (marked in decimal form) whose digits give the values of the corresponding sites. The null configuration (with value zero at all sites) is labeled by the integer 0, and occurs with the largest probability ≈ 0.13. The inequality of the probabilities for initially equiprobable configurations is a reflection for self-organization.

ensemble. Figure 22 shows the evolution of these probabilities over ten time steps for several complex cellular automata. At each time step, dots are placed in positions corresponding to configurations occurring with nonzero probabilities. At $\tau = 0$, all configurations are taken to be equally probable. Cellular automaton evolution modifies the probabilities for different configurations, reducing the probabilities for some to zero, and leading to "gaps" in Fig. 22. In the initial ensemble, all configurations were assigned equal *a priori* probabilities. After evolution (or "processing") for a few time steps, an equilibrium ensemble is attained in which different configurations carry

RULE : 00010010 (18)

RULE : 01011010 (90)

RULE : 01111110 (126)

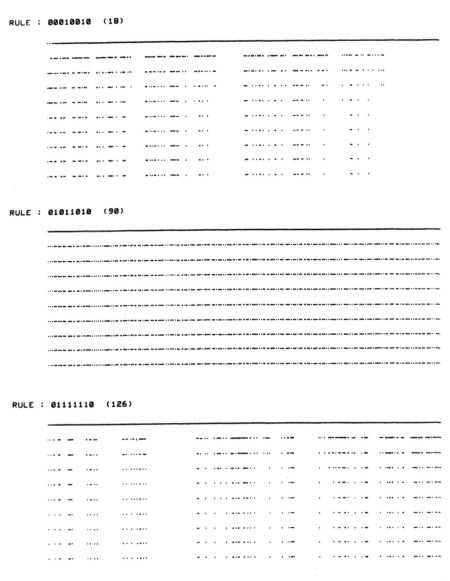

Figure 22. Time evolution of the probabilities for each of the 1024 possible configurations of several length 10 cellular automata starting from an initial ensemble containing all 1024 configurations with equal probabilities. The configurations are specified by binary integers whose digits form the sequence of values at the sites of the cellular automaton. The history of a particular configuration is given on successive lines in a vertical column: a dot appears at a particular time step if the configuration occurs with nonzero probability at that time step. In the initial ensemble, all configurations occur with equal nonzero probabilities, and dots appear in all positions. Cellular automaton evolution modifies the probabilities for the configurations, making some occur with zero probability, yielding gaps in which no dots appear. The probabilities obtained by evolution for ten time steps according to cellular automaton rule 126 were given in Fig. 21: dots appear in the tenth line of the rule 126 part of this figure at the positions corresponding to configurations with nonzero probabilities.

different probabilities, according to a definite distribution. Properties of the more probable configurations dominate statistical averages over the ensemble, giving rise to the distinctive average local features of equilibrium configurations described in Sec. 3.

In the limit $N \to \infty$, a cellular automaton configuration may be specified by real number in the interval 0 to 1 whose binary decomposition consists of a sequence of digits corresponding to the values of the cellular automaton sites. Then the equilibrium ensemble of cellular automaton configurations analogous to those of Fig. 22 corresponds to a set of points on the real line. The unequal probabilities for appearance of 0 and 1 digits, together with higher-order correlations, implies that the points form a Cantor set (Farmer, 1982a, 1982b). The fractal dimensionality of the Cantor set is given by the negative of the entropy discussed below, associated with the ensemble of cellular automaton configurations (and hence real-number binary digit sequences) (Farmer, 1982a, 1982b). For rule 126 the fractal dimension of the Cantor set is then 0.5.

An important feature of the elementary cellular automata considered here and in Sec. 3 is their "local irreversibility." Cellular automaton rules may transform several different initial configurations into the same final configuration. A particular configuration thus has unique descendents, but does not necessarily have unique ancestors (predecessors). Hence the trajectories traced out by the time evolution of several cellular automaton configurations may coalesce, but may never split. A trivial example is provided by cellular automaton rule 0, under which all possible initial configurations evolve after one time step to the unique null configuration. In a reversible system, each state has a unique descendent and a unique ancestor, so that trajectories representing time evolution of different states may never intersect or meet. Thus in a reversible system, the total number of possible configurations must remain constant with time (Liouville's theorem). However, in an irreversible system, the number of possible configurations may decrease with time. This effect is responsible for the "thinning" phenomenon visible in Fig. 22. The trajectories corresponding to the evolution of cellular automaton configurations are found to become concentrated in limited regions, and do not asymptotically fill the available volume densely and uniformly. This behavior makes self-organization possible, by allowing some configurations to occur with larger probabilities than others even in the large-time equilibrium limit.

One consequence of local irreversibility evident from Fig. 22 is that some cellular automaton configurations may appear as initial conditions but may never be reached as descendents of other configurations through cellular automaton time evolution.[10] Such configurations carry zero weight in the ensemble obtained by cellular automaton evolution. In the trivial case of cellular automaton rule 0, only the null state

---

[10] The existence of unreachable or "garden-of-Eden" configurations in cellular automata is discussed in Moore (1962) and Aggarwal (1973), where criteria (equivalent to irreversibility) for their occurrence are given.

with all sites zero may be reached by time evolution; all other configurations are unreachable. Rule 4 generates only those configurations in which no two adjacent sites have the same value. The fraction of the 2^N possible configurations which satisfy this criterion tends to zero as N tends to infinity, so that in this limit, a vanishingly small fraction of the configurations are reached. Cellular automaton rule 204 is an identity transformation, and is unique among cellular automaton rules in allowing all configurations to be reached. (The rule is trivially reversible.) Assuming periodic boundary conditions, one finds that with N odd, the complex additive rule 90 generates only configurations in which an even number of sites have value one, and thus allows exactly half of the 2^N possible configurations to be reached. For even N, $\frac{1}{4}$ of the possible configurations may be reached. A finite fraction of all the configurations are thus reached in the limit $N \rightarrow \infty$. For the complex nonadditive rule 126, inspection of Fig. 8 shows that only configurations in which nonzero sites appear in pairs may be reached. Figure 23 shows the fraction of unreachable configuration for this cellular automaton rule as a function of N. The fraction tends steadily to one as $N \rightarrow \infty$. A complete characterization of the unreachable configurations for this case is given in Martin et al. (1983); these configurations are enumerated there, and their fraction is shown to behave as $1 - \lambda^N$ for large N, where $\lambda \simeq 0.88$ is determined as the root of a cubic equation. Similar behavior is found for other nonadditive rules.

Irreversible behavior in cellular automata may be analyzed by considering the behavior of their "entropy" S or "information content" $-S$. Entropy is defined as usual as the logarithm (here taken to base two) of the average number of possible states of a system, or

$$S = \sum_i p_i \log_2 p_i \qquad (4.1)$$

where p_i is the probability for state i. The entropy may equivalently be considered

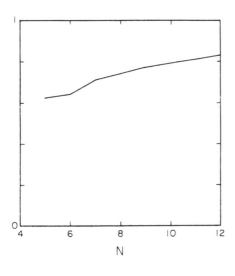

Figure 23. Fraction of the 2^N possible configurations of a length N cellular automaton (with periodic boundary conditions) not reached by evolution from an arbitrary initial configuration according to cellular automaton rule 126. The existence of unreachable configurations is a consequence of the irreversibility of cellular automaton evolution. The fraction of such configurations is seen to increase steadily towards one as N increases.

N

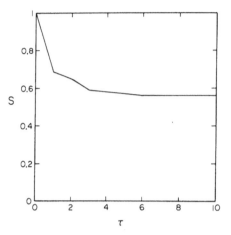

Figure 24. Time evolution of average entropy per site for an ensemble of finite cellular automata with $N = 10$ evolving according to rule 126 from an initial equiprobable ensemble. The entropy gives the logarithm of the average number of possible configurations. Its decrease with time is a reflection of the local irreversibility of the cellular automaton.

as the average number of binary bits necessary to specify one state in an ensemble of possible states. The total entropy of a system is the sum of the entropies of statistically independent subsystems. Entropy is typically maximized when a system is completely disorganized, and the maximum number of subsystems act independently. The entropy of a cellular automaton takes on its maximal value of one bit per site for an equiprobable ensemble. For reversible systems, time evolution almost always leads to an increase in entropy. However, for irreversible systems, such as cellular automata, the entropy may decrease with time. Figure 24 shows the time dependence of the entropy for a finite cellular automaton with $N = 10$, evolving according to rule 126, starting from an initial equiprobable ensemble. The entropy is seen to decrease with time, eventually reaching a constant equilibrium value. The decrease is a direct signal of irreversibility.

The entropy for a finite cellular automaton given in Fig. 24 is obtained directly from Eq. (4.1) by evaluating the probabilities for each of the finite set of 2^N possible configurations. For infinite cellular automata, enumeration of all configurations is no longer possible. However, so long as values of sufficiently separated sites are statistically independent, the average entropy per site may nevertheless be estimated by a limiting procedure. Define a "block entropy" [or "Renyi entropy" (Renyi, 1970; Farmer, 1982a, 1982b)]

$$S_b = (1/b) \sum_i p_i^{(b)} \log p_i^{(b)},$$

where $p_i^{(b)}$ denotes the probability for a sequence i of b values in an infinite cellular automaton configuration. The limit $S_{b \to \infty}$ gives the average total entropy per site. This limit is approached rapidly for almost all cellular automaton configurations, reflecting the exponential decrease of correlations with distance discussed in Sec. 3. [Similar results are obtained in estimating the entropy of printed English from single letter, digram, trigram and so on frequencies (Shannon 1951). Typical results (for example, for the text of this paper) are $S_1 \simeq 4.70$, $S_2 \simeq 4.15$, $S_3 \simeq 3.57$, and $S_\infty \sim 2.3$.]

Irreversibility is not a necessary feature of cellular automata. In the case of the elementary cellular automata considered here, the irreversibility results from the assumption that a configuration S_n at a particular time step n depends only on its immediate predecessor so that its evolution may be represented schematically by $S_n = F[S_{n-1}]$. Except in the trivial case of the identity transformation (rule 204), F is not invertible. The cellular automata are discrete analogs of systems governed by partial differential equations of first order in time (such as the diffusion equation), and exhibit the same local irreversibility. One may construct reversible one-dimensional cellular automata (Fredkin, 1982; Margolus, 1982)[11] by allowing a particular configuration to depend on the previous two configurations, in analogy with reversible second-order differential equations such as the wave equation. The evolution of these cellular automata may be represented schematically by $S_n = F[S_{n-1}] \oplus S_{n-2}$. The invertibility of modulo-two addition allows S_{n-2} to be obtained uniquely from S_n and S_{n-1}, so that all pairs of successive configurations have unique descendants and unique ancestors. For infinite reversible cellular automata, the entropy (4.1) (evaluated for the appropriate successive pairs of configurations) almost always increases with time. Finite reversible cellular automata may exhibit globally irreversible behavior when dissipative boundary conditions are imposed. Such boundary conditions are obtained if sites beyond the boundary take on random values at each time step. If all sites beyond the boundary have a fixed or predictable value as a function of time, the system remains effectively reversible. With simple initial configurations, reversible cellular automata generate self-similar patterns analogous to those found for irreversible ones.[12] A striking difference is that reversible rules yield diamond-shaped structures symmetrical in time, rather than the asymmetrical triangle structures found with irreversible rules.

Since a finite cellular automaton has a total of only 2^N possible configurations, the sequence of configurations reached by evolution from any initial configuration must become periodic after at most 2^N time steps (the "Poincare recurrence time"). After an initial transient, the cellular automaton must enter a cycle in which a set of configurations is generated repeatedly, as illustrated in Fig. 25. Figure 8 suggests that simple cellular automata yield short cycles containing only a few configurations, while complex cellular automata may yield much longer cycles. Simple rules such as 0 or 72 evolve after a fixed small number of time steps from any configuration to the stationary null configuration, corresponding to a trivial length-one cycle. Other simple cellular automaton rules, such as 36, 76, or 104 evolve after $\leqslant N$ time steps to

---

[11] Reversible cellular automata may be constructed in two (or more) dimensions by allowing arbitrary evolution along a line, but generating a sequence of copies ("history") in the orthogonal direction of the configurations on the line at each time step (Toffoli, 1977a, 1980).

[12] For example, evolution from a pair of successive configurations containing zero and one nonzero sites according to the reversible analog of rule 150 yields a self-similar pattern with fractal dimension $\log_2[4/(\sqrt{17} - 3)] \simeq 1.84$.

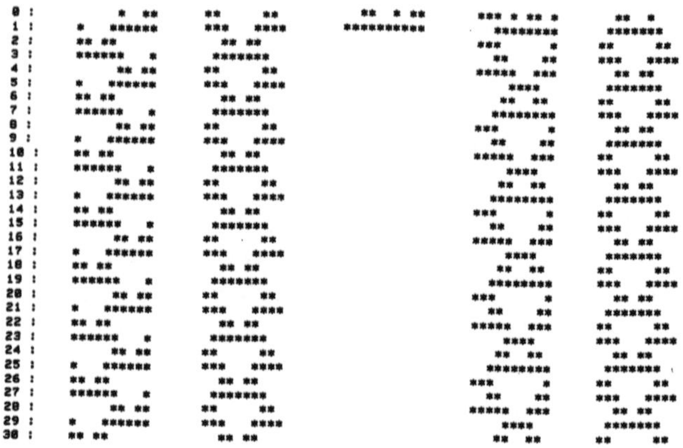

Figure 25. Evolution of typical initial configurations in a finite cellular automaton with $N = 8$ (and periodic boundary conditions) according to rule 126. Evolution from a particular initial state could generate up to $2^8 = 256$ distinct configurations before entering a cycle and returning to a configuration already visited. Much shorter cycles, however, are seen to occur in practice.

nontrivial stationary configurations (with cycle length one). Rules such as 94 or 108 yield (after a transient of $\leq N$ steps) a state consisting of a set of small independent regions, each of which independently follows a short cycle (usually of length one or two and at most of length 2^b, where b is the number of sites in the region). In general, simple cellular automata evolve to cycles whose length remains constant as N increases. On the other hand, complex cellular automata may yield cycles whose length increases without bound as N increases. Figure 26 shows the distribution in the number of time steps before evolution from each possible initial configuration according to the complex rule 126 leads to repetition of a configuration. Only a small fraction of the 2^N possible configurations is seen to be reached in evolution from a particular initial configuration. For example, in the case $N = 8$, a maximum of eight distinct configurations (out of 256) are generated by evolution from any specific initial state. After a transient of at most two time steps, the cellular automaton enters a cycle, which repeats after at most six further time steps. Apart from the trivial one-cycle corresponding to the null configuration, six distinct cycles (containing nonintersecting sets of configurations) occur. Four have length six, and two have length two. A total of 29 distinct "final" configurations appear in these cycles. The number of configurations reached by evolution from a particular initial state increases with N as shown in Fig. 26. For $N = 10$, the maximum is 38 states, while for $N = 32$, it is at least 1547. Similar behavior is found for most other complex nonadditive rules.

Analytical results for transient and cycle lengths may be given for finite cellular automata (with periodic boundary conditions) evolving according to the additive rules 90 and 150 (Martin et al. 1983). A complete and general derivation may be obtained using algebraic methods and is given in Martin et al. (1983). The additive superposition principle implies that the evolution of any initial configuration is a superposition of evolution from single nonzero sites (in each of the N cyclically

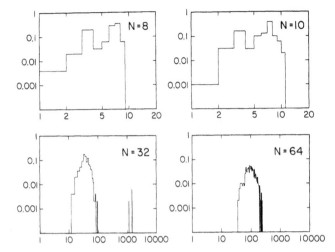

Figure 26. Distribution in the number of time steps required for finite cellular automata of length N (with periodic boundary conditions) evolving according to rule 126 to reach a particular configuration for the second time, signaling the presence of a cycle. The cycle times found are much smaller than the value 2^N obtained if evolution from a particular initial configuration eventually visited all 2^N possible configurations. The results for $N = B$ and $N = 10$ include all 256 and 1024 possible initial configurations; those for $N = 32$ and $N = 64$ are obtained by uniform Monte Carlo sampling from the space of possible initial configurations. In all cases, the number of configurations visited in transients before entering a cycle is very much smaller than the number of configurations in the cycle.

equivalent possible positions). The period of any cycle must therefore divide the period Π_N obtained by evolution from a single nonzero site. Similarly, the length of any transient must divide the length Υ_N obtained with a single nonzero initial site. It is found that Π_N is identical for rules 90 and 150, but Υ_N in general differs. The first few values of Π_N for rules 90 and 150 (for $N = 3$ through $N = 30$) are 1, 1, 3, 2, 7, 1, 7, 6, 31, 4, 63, 14, 15, 1, 15, 14, 511, 12, 63, 62, 2047, 8, 1023, 126, 511, 28, 16383, and 30. Consider rule 90; derivations for rule 150 are similar. Whenever N is of the form 2^α, the cellular automaton ultimately evolves from any initial configuration to the null configuration, so that $\Pi_N = 1$ in this case. When N is odd, it is found that the first configuration in the cycle always consists of two nonzero sites, separated by a single zero site. The nonzero sites may be taken at positions ± 1 modulo N. Equation (3.2) implies that configurations obtained by evolution for 2^j time steps again contain exactly two nonzero sites, at positions $\pm 2^j$ modulo N. A cycle occurs when $2^j \equiv \pm 1 \bmod N$. Π_N then divides Π_N^* given by $2^{\mathrm{sord}_N(2)} - 1$ where $\mathrm{sord}_N(k)$ is defined as the minimum j for which $2^j \equiv \pm 1 \bmod N$, and $\mathrm{sord}_N(k) = \mathrm{ord}_N(k)/2$ or $\mathrm{sord}_N(k) = \mathrm{ord}_N(k)$. The multiplicative order function $\mathrm{ord}_N(k)$ (e.g., MacWilliams and Sloane, 1977) is defined as the minimum j for which $2^j = 1 \bmod N$. It is found in fact that $\Pi_N = \Pi_N^*$ for most N; the first exception occurs for $N = 37$, in which case $\Pi_{37} = \Pi_{37}^*/3$. For $N = k^\alpha - 1$,

$\text{ord}_N(k) = \alpha$, so that when $N = 2^{\alpha} - 1$, $\Pi_N^* = N$. Similarly, when $N = k^{\alpha} + 1$, $k^{\alpha} \equiv -1 \bmod N$ so that $k^{2\alpha} \equiv +1 \bmod N$ and $\text{ord}_N(k) = 2\alpha$, yielding $\Pi_N^* = N - 2$ for $N = 2^{\alpha} + 1$. In general, if $N = p_1^{\alpha_1} p_2^{\alpha_2} \cdots$, where the p_i are primes not equal to k, $\text{ord}_N(k) = \text{lcm}[\text{ord}_{p_1^{\alpha_1}}(k), \text{ord}_{p_2^{\alpha_2}}(k), \ldots]$. $\text{ord}_N(k)$ divides the Euler totient function $\varphi(N)$, defined as the number of integers less than N which are relatively prime to N (e.g., Apostol, 1976; Hardy and Wright, 1979, Sec. 5.5). [$\varphi(N)$ is even for all $N > 1$.] $\varphi(N)$ satisfies the Euler-Fermat relation $k^{\varphi(N)} \equiv 1 \bmod N$. It is clear that $\pi(n) \le \varphi(n) \le n - 1$, where $\pi(n)$ denotes the number of primes less than n, and the upper bound is saturated when n is prime. If $\text{ord}_N(k)$ is even, then $\text{ord}_N(k) \le \varphi(N)$, while for $\text{ord}_N(k)$ odd, $\text{ord}_N(k) \le \varphi(N)/2$. Thus $\Pi_N \le 2^{(N-1)/2} - 1$, where the bound is saturated for some prime N. Such a Π_N is the maximum possible cycle length for configurations with reflection symmetry, but is approximately the square root of the maximum possible length $2^N - 1$ for an arbitrary system with N binary sites.[13] When N is even, $\Pi_N = 2\Pi_{N/2}$. Notice that Π_N is an irregular function of N: its value depends not only on the magnitude of N, but also on its number theoretical properties.

When Π_N is prime, all possible cycles must have a period of one or exactly Π_N. When Π_N is composite, any of its divisors may occur as a cycle period. Thus, for example, with $N = 10$, $\Pi_N = 6$, and in evolution from the $2^{10} - 1$ possible non-null initial configurations, forty distinct cycles of length 6 appear, and five of length 3. In general it appears that for large N, an overwhelming fraction of cycles have the maximal length Π_N.

As mentioned above, for the additive rules 90 and 150, the length of the transients before a cycle is entered in evolution from an arbitrary initial configuration must divide Υ_N, the length of transient with a single nonzero initial site. For rule 90, $\Upsilon_N = 1$ for N odd, and $\Upsilon_N = D_2(N)/2$ otherwise, where $D_2(n)$ is the largest 2^j which divides n. For rule 150, $\Upsilon_N = 0$ if N is not a multiple of three, $\Upsilon_N = 1$ if N is odd, and $\Upsilon_N = D_2(N)$ otherwise. Since, as discussed above, evolution from all 2^N possible initial configurations according to rule 90 visits 2^{N-1} configurations for odd N, the result $\Upsilon_N = 1$ implies that in this case, exactly half of the 2^N possible configurations appear on cycles.

Configurations in cellular automata may be divided into essentially three classes according to the circumstances under which they may be generated. One class discussed above consists of configurations which can appear only as initial states, but can never be generated in the course of cellular automaton evolution. A second class contains configurations which cannot arise except within the first, say τ, time steps. For $\tau = 2$, such configurations have "parents" but no "grandparents." The third class of configurations is those which appear in cycles, and may be visited repeatedly.

---

[13] The result is therefore to be contrasted with the behavior of linear feedback shift registers, analogous to cellular automata except for end effects, in which cycles (de Bruijn sequences) of period $2^N - 1$ may occur (e.g., Golomb, 1967; Berlekamp, 1968).

Such configurations may be generated at any time step (for example, by choosing an initial configuration at the appropriate point in the cycle, and then allowing the necessary number of cycle steps to occur). The second class of configurations appears as transients leading to cycles. The cycles may be considered as attractors eventually attained in evolution from any initial configuration. The 2^N possible configurations of a finite cellular automaton may be represented as nodes in a graph, joined by arcs representing transitions corresponding to cellular automaton evolution. Cycles in the graph correspond to cycles in cellular automaton evolution. As shown in Martin et al. (1983), the transient configurations for the additive rules 90 and 150 appear on balanced quaternary trees, rooted on the cycles. The leaves of the trees correspond to unreachable configurations. The height of the trees is given by Υ_N. The balanced structure of the trees implies that the number of configurations which may appear after τ time steps decreases as $4^{-\tau}$; $4^{-\Upsilon_N}$ configurations appear on cycles and may therefore be generated at arbitrarily large times.

The algebraic techniques of Martin et al. (1983) apply only to additive rules. For nonadditive cellular automaton rules, the periods of arbitrary cycles do not necessarily divide the periods Π_N of cycles generated by evolution from configurations with one nonzero site. Empirical investigations nevertheless reveal many regularities.

Cyclic behavior is inevitable for finite cellular automata which allow only a finite number of possible states. Infinite cellular automata exhibit finite cycles only under exceptional circumstances. For a wide class of initial states, simple cellular automaton rules can yield nontrivial cyclic behavior. Cycles occur in complex cellular automata only with exceptional initial conditions. Any initial configuration with a finite number of nonzero sites either evolves ultimately to the null state, or yields a pattern whose size increases progressively with time. Most infinite initial configurations do not lead to cyclic behavior. However, if the values of the initial sites form an infinite periodic sequence (cf. Miller, 1970, 1980), with period k, then the evolution of the infinite cellular automaton will be identical to that of a finite cellular automaton with $k = N$, and cycles with length $\ll 2^k$ will be found.

The transformation of a finite cellular automaton configuration according to cellular automaton rules defines a mapping in the set of 2^N binary integers representing the cellular automaton configurations. An example of such a mapping was given in Fig. 19. Repeated applications of the mapping yield successive time steps in the evolution of the cellular automaton. One may compare the results with those obtained for a system which evolves by iteration of a random mapping among the 2^N integers (cf. Kauffman, 1969). Random mappings of K elements are obtained by choosing one of the K possible images independently for each integer and with equal probabilities. The mapping is permitted to take an element to itself. In this way, all K^K possible mappings are generated with equal probability. The probability of a particular element's having no preimage (predecessor) under a random mapping between K elements is $(K - 1)^K / K^K = (1 - 1/K)^K$. In the limit $K \to \infty$ this implies that a fraction $1/e \simeq 0.37$ of the possible states are not reached in evolution by iteration of a random mapping. For complex nonadditive cellular automata, it appears that as $N \to \infty$, almost all configurations become unreachable, indicating that cellular automaton evolution is "more irreversible" than iteration of a random mapping would imply. A system evolving according to a random mapping exhibits cycles analogous to those found in actual cellular automata. The probability of a length r cycle's occurring by iteration of a mapping between K elements is found to be

$$\sum_{i=r}^{K} \frac{(K-1)!}{(K-i)!K^i}$$

(Harris, 1960; Knuth, 1981, Sec. 3.1, Ex. 6, 11-16; Levy, 1982). Cycles of the maximum length K occur with finite probability. In the large K limit, the average cycle length becomes $\simeq\sqrt{\pi K/8} \simeq 0.63\sqrt{K}$, while the standard deviation of the cycle length distribution is $\simeq\sqrt{(2/3 - \pi/8)K} \simeq 0.52\sqrt{K}$. The length of transients follows exactly the same distribution. The number of distinct cycles $\sim\sqrt{\pi/2}\log K$. If we take $K = 256$ for comparison with an $N = 8$ cellular automaton, this implies an average cycle length $\simeq 10$, an average transient length $\simeq 10$, $\simeq 94$ unreachable configurations, and $\simeq 7$ distinct cycles. Cellular automaton rule 126 yields in this case an average cycle length $\simeq 3.2$, an average transient length $\simeq 2.5$, 190 unreachable configurations, and 7 distinct cycles. Any agreement with results for random mappings appears to be largely fortuitous: even for large N cellular automata do not behave like random mappings.

This section has thus far considered cellular automata which evolve according to definite deterministic local rules. However, as discussed in Sec. 3, one may introduce probabilistic elements or noise into cellular automata rules—for example, by reversing the value of a site at each time step with probability κ. Section 3 showed that the local properties of cellular automata change continuously as κ is increased from zero. Global properties may, however, change discontinuously when a nonzero κ is introduced. An example of such behavior is shown in Fig. 27, which gives the fraction of configurations visited as a function of time for a cellular automaton evolving

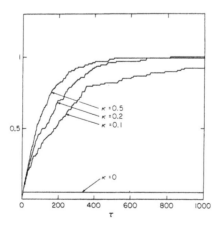

Figure 27. Fraction of configurations visited after τ time steps in a finite cellular automaton (with $N = 7$) evolving from a single typical initial state according to rule 126 in the presence of noise which randomly reverses the values of sites at each time step with probability κ. When $\kappa = 0$, the cellular automaton enters a cycle after visiting only six distinct configurations. When $\kappa \neq 0$, the cellular automaton eventually visits all 128 possible configurations.

according to rule 126 with various values of κ, starting from a single typical initial configuration. When $\kappa = 0$, only six distinct configurations are generated before the cellular automaton enters a cycle. When $\kappa \neq 0$, the cellular automaton ultimately visits every possible configuration (cf. Gach et al., 1978). For $\kappa \simeq 0.5$, one may approximate each configuration as being chosen from the 2^N possible configurations with equal probabilities: in this case, the average number of configurations visited after τ time steps is found to be $1 - ([1 - 2^{-N}]^{2^N})^{\tau/2^N} \simeq 1 - e^{-\tau/2^N}$.

Cellular automata may be viewed as simple idealizations of physical systems. They may also be interpreted as "computers" (von Neumann, 1966; Baer and Martinez, 1974; Burks, 1970; Aladyev, 1974, 1976; Toffoli, 1977b) and analyzed using methods from the formal theory of computation (Minsky, 1967; Arbib, 1969; Manna, 1974; Hopcroft and Ullman, 1979; Beckman, 1980). With this interpretation, the initial configuration of a cellular automaton represents a "program" and "initial data," processed by cellular automaton time evolution to give a configuration corresponding to the "output" or "result" of the "computation." The cellular automaton rules represent the basic mechanism of the computer; different programs may be "run" (or different "functions evaluated") by giving different initial or "input" configurations. This process is analogous to the "evolution" of the sequence of symbols on the tape of a Turing machine (Turing, 1936). However, instead of considering a single "head" which modifies one square of the tape at each time step, the cellular automaton evolution simultaneously affects all sites at each time step. As discussed in Sec. 5, there exist "universal" cellular automata analogous to universal Turing machines, for which changes in the initial configuration alone allow any computable (or "recursive") function to be evaluated. A universal Turing machine may simulate any other Turing machine using an "interpreter program" which describes the machine to be simulated. Each "instruction" of the simulated machine is simulated by running the appropriate part of the interpreter program on the universal machine. Universal cellular automata may similarly simulate any other cellular automata. The interpreter consists of an encoding of the configurations for the cellular automaton to be simulated on the universal automaton. A crucial point is that so long as the encoding defined by the interpreter is sufficiently simple, the statistical characteristics of the evolution of configurations in the universal cellular automaton will be shared by the cellular automaton being simulated. This fact potentially forms the basis for universality in the statistical properties of complicated cellular automata.

The simplest encodings which allow one cellular automaton to represent or simulate others are pure substitution or "linear" ones, under which the value of a single site is represented by a definite sequence of site values. (Such encodings are analogous to the correspondences between complex cellular automaton rules mentioned in Sec. 3.) For example, a cellular automaton A evolving according to rule 22 may be used to simulate another cellular automaton B evolving according to rule 146. For every 0 in the initial configuration of B, a sequence 00 is taken in the initial configuration of

A, and for every 1 in *B*, 01 is taken in *A*. Then after 2τ time steps, the configuration of *A* under this encoding is identical to that obtained by evolution of *B* for τ time steps. If cellular automaton *B* instead evolved according to rule 182, 01 (or 10) in A would correspond to 0 in *B*, and 00 to 1. The simplicity of the interpreter necessary to represent rules 146 and 182 under rule 22 is presumably responsible for the similarities in their statistical behavior found in Sec. 3. Figure 28 gives a network which describes

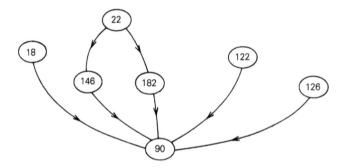

Figure 28. Network describing simulation capabilities of complex elementary cellular automata with length 2 pure substitution or linear encodings. Cellular automata evolving according to the destination rule are simulated by giving an encoded intitial configuration in a cellular automaton evolving according to the source rule. Representability of one cellular automaton by another under a simple encoding implies similar statistical properties for the two cellular automata, and forms potentially the basis for universality in statistical properties of cellular automata.

the simulation capabilities of the complex elementary cellular automaton rules using length two linear encodings and with the simulated rule running at half the speed of the simulator. Many of these complex rules may also simulate simple rules under such an encoding. Simulations possible with longer linear encodings appear to be described by indirection through the network. Not all complex cellular automaton rules are thus related by linear encodings of any length.

As discussed in Sec. 5, the elementary cellular automata considered here and in Secs. 2 and 3 are not of sufficient complexity to be capable of universal computation. However, some of the more complicated cellular automata described in Sec. 5 are "universal," and may therefore in principle represent any other cellular automata. The necessary encoding must be of finite length, but may be very long. The shorter or simpler the encoding, the closer will be the statistical properties of the simulating and simulated cellular automata.

5. Extensions

The results of Secs. 2–4 have for the most part been restricted to elementary cellular automata consisting of a sequence of sites in one dimension with each site taking on two possible values, and evolving at each time step according to the values of its two nearest neighbors. This section gives a brief discussion of the behavior of more complicated cellular automata. Fuller development will be given in future publications.

We consider first cellular automata in which the number of possible values k at each site is increased from two, but whose sites are still taken to lie on a line in one dimension. The evolution of each site at each time step is for now assumed to depend on its own value and on the values of its two nearest neighbors. In this case, the total number of possible sets of local rules is $k^{(k^3)}$. Imposition of the reflection symmetry and quiescence "legality conditions" discussed in Sec. 2 introduces $\frac{1}{2}k^2(k-1)+1$ constraints, yielding $k^{[k^2(1+k)-1]/2}$ "legal" sets of rules. For $k=2$, this implies $2^5 = 32$ legal rules, as considered in Sec. 2. The number of possible legal rules increases rapidly with k. For $k=3$, there are $3^{17} = 129,140,163 \simeq 1.3 \times 10^8$ rules, for $k=4$, $\simeq 3 \times 10^{24}$, and for $k=10$, 10^{549}.

As a very simple example of cellular automata with $k>2$, consider the family of "modulo-k" rules in which at each time step, the value of a site is taken to be the sum modulo k of the values of its two neighbors on the previous time step. This is a generalization of the modulo-two rule (90) discussed on several occasions in Secs. 2–4. Figure 29 shows the evolution of initial states containing a single site with value one according to several modulo-k rules. In all cases, the pattern of nonzero sites is seen to tend to a self-similar fractal figure in the large time limit. The pattern in general depends on the value of the nonzero initial site, but in all cases yields an asymptotically self-similar figure. When k is prime, independent of the value of the initial nonzero site, a very regular pattern is generated, in which the density $T(n)$ of "triangle structures" is found to satisfy a one-term recurrence relation yielding a fractal dimension

$$D_k = \log_k \sum_{i=1}^{k} i = 1 + \log_k\left(\frac{k+1}{2}\right),$$

so that $D_3 = 1 + \log_3 2 \simeq 1.631$, $D_5 \simeq 1.683$, and so on. When k is a composite number, the pattern generated depends on the value s of the initial nonzero site. If the greatest common divisor (s,k) of k and s is greater than one (so that s and k share nontrivial prime factors), then the pattern is identical to that obtained by evolution from an initial site with value one according to a modulo-$k/(s,k)$ rule. In general, the density of triangles satisfies a multiple-term recurrence relation. In all cases, the fractal dimension for large k behaves as $D \sim 2 - 1/\log_2 k$ [assuming $(s,k) \ll k$]. When $k \to \infty$, the values of sites become ordinary integers, all with nonzero values by virtue of the nonvanishing values of binomial coefficients, yielding a figure of dimension two.

All modulo-k rules obey the additive superposition principle discussed for the modulo two in Secs. 2 and 3. The number of sites with value r after evolution for τ steps from a initial state containing a single site with value one is found [on analogy

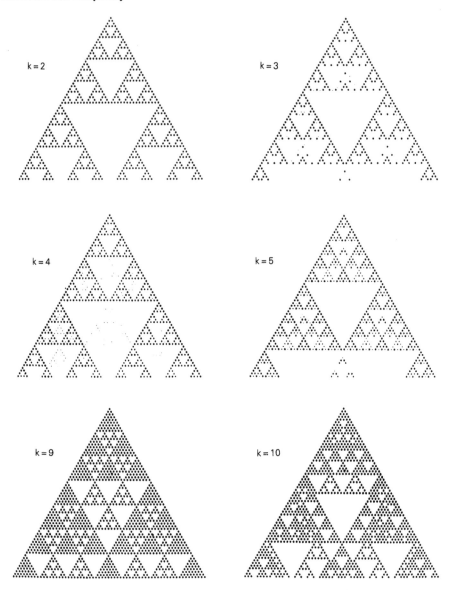

Figure 29. Patterns generated by evolution of one-dimensional cellular automata with k states per site according to a modulo-k rule, starting from an initial configuration containing a single nonzero site with value one. At each time step, the value of a site is the sum of the values of its two nearest neighbors at the previous time step. Configurations obtained at successive time steps are shown on successive lines. Sites with value zero are indicated as blanks; *, +, and − represent, respectively, values one, two, and three, and in the lower two patterns, ρ represents any nonzero value. In the large time limit, all the patterns tend to a self-similar form, with definite fractal dimensions.

to Eq. (3.2)] to be $N_\tau^{(r)} = 2^{\#_r^{[k]}(\tau)}$, where the function $\#_r^{[k]}(\tau)$ gives the number of occurrences of the digit r in the base-k decomposition of the integer τ and generalizes the function $\#_r(\tau)$ introduced in Sec. 3.

Figure 30 shows typical examples of the behavior of some cellular automata with $k = 3$. Considerable diversity is evident. However, with simple initial states, self-similar patterns are obtained at asymptotically large times, just as in the $k = 2$ case of Sec. 3. (Notice that the length and time scales before self-similarity becomes evident are typically longer than those found for $k = 2$: in the limit $k \to \infty$ where each site takes on an arbitrary integer value, self-similarity may not be apparent at any finite time.) Evolution of disordered initial states also again appears to generate nontrivial structure, though several novel phenomena are present. First, alternation of value-one and value-two sites on successive time steps can lead to "half-speed propagation" as in rule

00000000000000001002001010020.

Second, rules such as

00000000000000001011002010010

lead to a set of finite regions containing only sites with values zero and one, separated by "impermeable membranes" of value-two sites. The evolution within each region is independent, with the membranes enforcing boundary conditions, and leading to cycles after a finite number of time steps. Third, even for legal rules such as

00000012102200221002102010 0

and

2110001221210122001120211200,

illustrated in Fig. 30, there exist patterns which display a uniform shifting motion. For example, with rule

2110001221210122001120211200

an isolated 12 shifts to the right by one site every time step, while an isolated 21 shifts to the left; when 21 and 12 meet, they cross without interference. Uniform shifting motion is impossible with legal rules when $k = 2$, since sequences of zero and one sites cannot define suitable directions (evolution of 1101 and 1011 always yield a pattern spreading in both directions).

An important feature of some cellular automata with more than two states per site is the possibility for the formation of a membrane which "protects" sites within

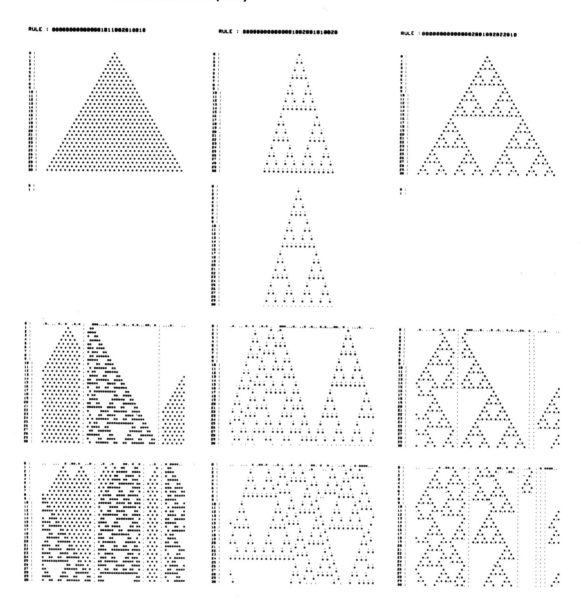

Figure 30. Examples of the evolution of several typical cellular automata with three states per site. Sites with value zero are shown as blanks, while values one and two are indicated by * and · , respectively. The value of a site at each time step is determined in analogy with Fig. 1 by the digit in the ternary specification of the rule corresponding to the values of the site and its two nearest neighbors at the previous time step. The evolution is shown until a configuration is reached for the second time (signaling a cycle) or for at most thirty time steps. The initial configurations in the lower two rows are typical of disordered configurations in which each site is statistically independent and takes on its three possible values with equal probabilities.

52

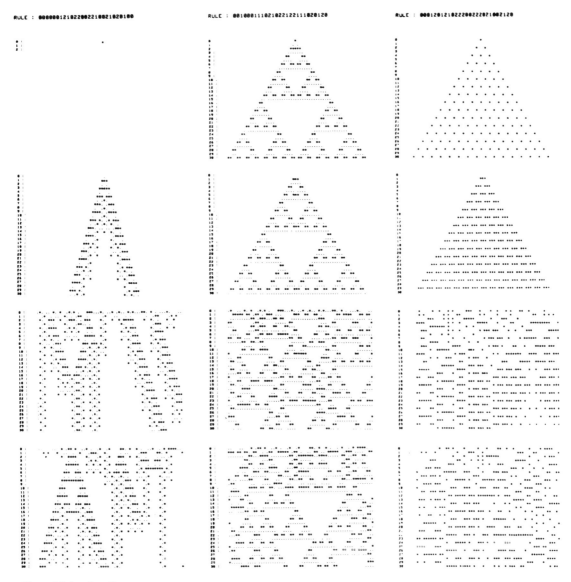

Figure 30 (continued).

it from the effects of noise outside. In this way, there may exist seeds from which
very regular patterns may grow, shielded by membranes from external noise typical
in a disordered configuration. Examples of such behavior are to be found in Fig. 30.
Only when two protective membranes meet is the structure they enclose potentially
destroyed. The size of the region affected by a particular seed may grow linearly with
time. Even if seeds occur with very low probability, any sufficiently long disordered

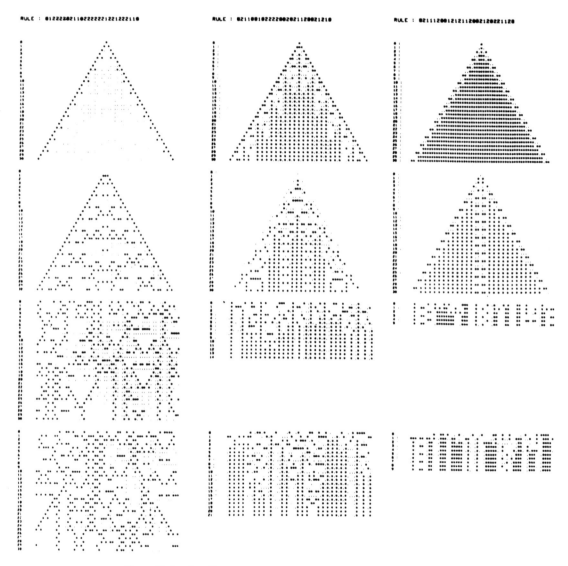

Figure 30 (continued).

configuration will contain at least one, and the large time behavior of the cellular automaton will be radically affected by its presence.

In addition to increasing the number of states per site, the cellular automata discussed above may be generalized by increasing the number of sites whose values affect the evolution of a particular site at each time step. For example, one may take the neighborhood of each site to contain the site itself, its nearest neighbors, and its next-nearest neighbors. With two states per site, the number of possible sets of legal local rules for such cellular automata is $2^{26} \simeq 7 \times 10^7$ (for $k = 3$, this number increases to $3^{174} \simeq 10^{83}$). Figure 31 shows patterns generated by these cellular

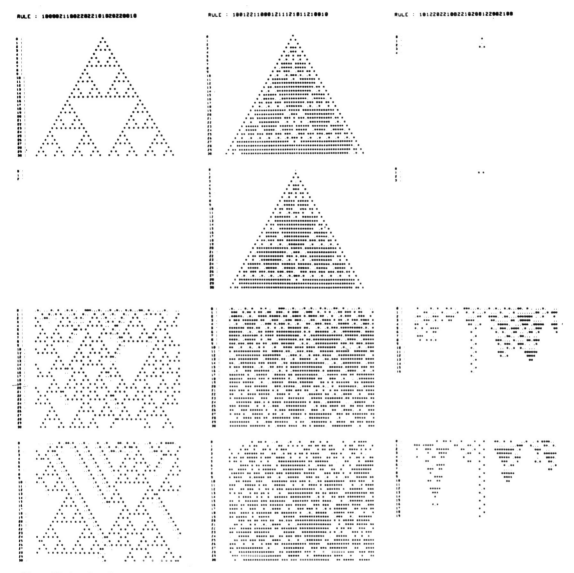

Figure 30 (continued).

automata for two typical sets of local rules. With simple initial states, self-similar patterns are obtained at large times. With disordered initial states, less structure is apparent than in the three-site neighborhood cellular automata discussed above. The patterns obtained with such cellular automata are again qualitatively similar to those shown in Sec. 2.

The cellular automata discussed so far have all involved a line of sites in one dimension. One may also consider cellular automata in which the sites lie on a regular square or (hyper)cubic lattice in two or more space dimensions. As usual, the value

55

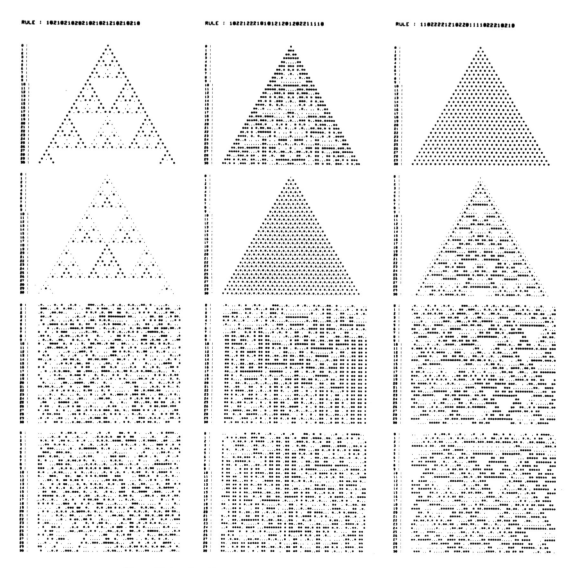

Figure 30 (continued).

of each site is determined by the values of a neighborhood of sites at the previous time step. In the simplest case, the neighborhood includes a site and its nearest neighbors. However, in $d > 1$ dimensions two possible identifications of nearest neighbors can be made. First, sites may be considered neighbors if one of their coordinates differ by one unit, and all others are equal, so that the sites are "orthogonally" adjacent. In this case, a "type-I" cellular automaton neighborhood containing $2d + 1$ sites is obtained. Second, sites may be considered neighbors if none of their coordinates differ by more than one unit, so that the sites are "orthogonally" or "diagonally"

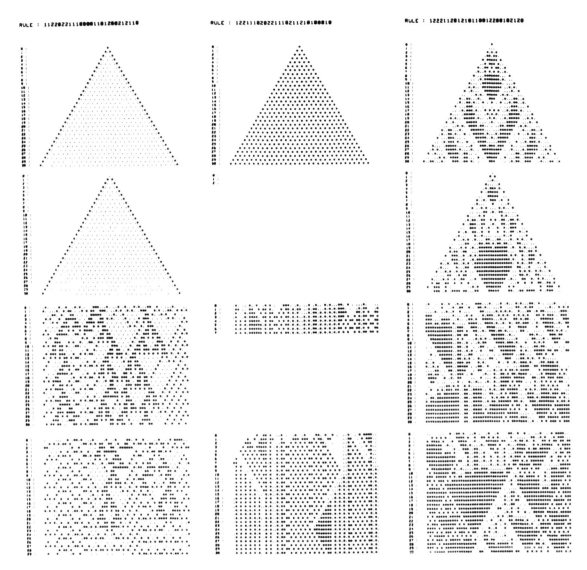

Figure 30 (continued).

adjacent. This case yields a "type-II" cellular automaton neighborhood containing 3^d sites. When $d = 1$, type-I and -II neighborhoods are identical and each contains three sites. For $d = 2$, the type-I neighborhood contains five sites, while the type-II neighborhood contains nine sites.[14] Cellular automaton rules may be considered legal if they satisfy the quiescence condition and are invariant under the rotation and reflection symmetries of the lattice. For $d = 2$, the number of possible legal type-I

---

[14] In the case $d = 2$, neighborhoods of types I and II are known as von Neumann and Moore neighborhoods, respectively.

RULE : 2012220010202120222101021120   RULE : 2110001221210122001120212200

Figure 30 (continued).

rules with k states per site is found to be $k^{(k^5+k^3+2k^2-4)/4}$, yielding 2^{11} = 2048 rules for k = 2 and $3^{71} \simeq 8 \times 10^{33}$ for k = 3. The number of type-II rules with k = 2 in two dimensions is found to be $2^{59} \simeq 6 \times 10^{17}$ (or 2^{71} if reflection symmetries are not imposed).

Figure 32 shows the evolution of an initial configuration containing a single nonzero site according to two-dimensional (type-I) modulo-two rules. In case (a) the value of a site is taken to be the sum modulo two of the values of its four neighbors on the previous time step, in analogy with one-dimensional cellular automaton rule 90. In case (b), the previous value of the site itself is included in the sum (and the comple-

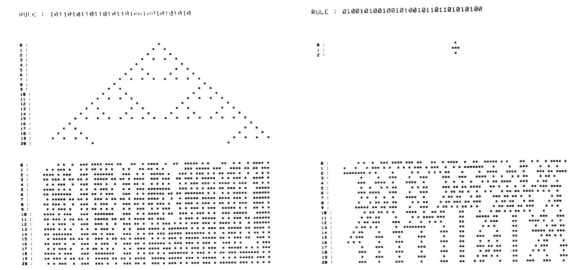

Figure 31. Evolution of two typical one-dimensional cellular automata with two states per site in which the value of a site at a particular time step is determined by the preceding values of a neighborhood of five sites containing the site, its nearest neighbors, and its next-nearest neighbors. The initial configurations in the lower row are typical of disordered configurations, in which each site has value one with probability $\frac{1}{2}$.

ment is taken), in analogy with rule 150. The sequence of patterns obtained at successive time steps may be "stacked" to form pyramidal structures in three-dimensional space. These structures become self-similar at large times: in case (a) they exhibit a fractal dimension $\log_2 5 \simeq 2.32$, and in case (b) a dimension $1 + \log_2(1 + \sqrt{3}) \simeq 2.45$. The patterns found on vertical slices containing the original nonzero site through the pyramids (along one of the two lattice directions) are the same as those generated by the one-dimensional modulo-two rules discussed in Secs. 2 and 3. The patterns obtained at each time step in Fig. 31 are almost always self-similar in the large time limit. For case (a), the number of sites with value one generated after τ time steps in Fig. 31 is found to be $4^{\#_1(\tau)}$, where $\#_1(\tau)$ gives the number of occurrences of the digit one in the binary decomposition of the integer τ, as discussed in Sec. 3 (cf. Butler and Ntafos, 1977). The type-I modulo-two rules may be generalized to d-dimensional cellular automata. In case (a) the patterns obtained by evolution from a single nonzero initial site have fractal dimension $\log_2(2d + 1)$ and give $(2d)^{\#_1(\tau)}$ nonzero sites at time step τ. In case (b), the asymptotic fractal dimension is found to be $\log_2[d(\sqrt{1 + 4/d} + 1)]$. Once again, simple initial states always yield self-similar structures in the large time limit.

A particular type-II two-dimensional cellular automaton whose evolution has been studied extensively is the game of "Life" (Conway, 1970; Gardner, 1971, 1972; Wainwright, 1971–1973; Wainwright, 1974; Buckingham, 1978; Berlekamp et al., 1982, Chap. 25; R. W. Gosper, private communications). The local rules take a site

Figure 32. Evolution of an initial state containing a single nonzero site in a two-dimensional cellular automaton satisfying type-I modulo-two rules. In case (a) the value of each site is taken to be the sum modulo two of the values of its four (orthogonally adjacent) neighbors at the previous time step, while in case (b) the previous value of the site itself is included in the sum, and the complement is taken. Case (a) is the two-dimensional analog of a one-dimensional cellular automaton evolving according to local rule 90, and case (b) of one evolving according to rule 150. The pyramidal structure obtained in each case by stacking the patterns generated at successive time steps is self-similar in the large time limit.

to "die" (attain value zero) unless two or three of its neighbors are "alive" (have value one). If two neighbors are alive, the value of the site is left unchanged; if three are alive, the site always takes on the value one. Many configurations exhibiting particular properties have been found. The simplest isolated configurations invariant under time evolution are the "square" (or "block") consisting of four adjacent live sites, and the "hexagon" (or "beehive") containing six live sites. "Oscillator" configurations which cycle through a sequence of states are also known. The simplest is the "blinker" consisting of a line of three live sites, which cycles with a period of two time steps. Oscillators with periods 3, 5, and 7 are also known; other periods may be obtained by composition. So long as they are separated by four or more unfilled sites, many of these structures may exist without interference in the configurations of a cellular automaton, and their effects are localized. There also exist configurations which "move" uniformly across the lattice, executing a cycle of a few internal states. The simplest example is the "glider" which contains five live sites and undergoes a cycle of length two. The number of filled sites in all the configurations mentioned so far is bounded as a function of time. However, "glider gun" configurations have been found which generate infinite streams of gliders, yielding a continually increasing number of live sites. The simplest known glider gun configuration evolves from a configuration containing 26 live cells. Monte Carlo simulation suggests that a disordered state of N^2 cells usually evolves to a steady state within about N^2 time steps (and typically an order of magnitude quicker); very few of the 2^{N^2} possible configurations are visited. Complicated structures such as glider guns are very rarely produced. Rough empirical investigation suggests that the density of structures containing L live sites generated from a disordered initial state (cf. Buckingham, 1981) decreases like e^{-L_-}/L, where L_- is the size of the minimal distinct configuration which evolves to the required structure in one time step. Just as for the one-dimensional cellular automata discussed in Sec. 4, the irreversibility of "Life" leads to configurations which cannot be reached by evolution from any other configurations, and can appear only as initial states. However, the simplest known "unreachable" configuration contains around 300 sites (Wainwright, 1971–1973; Hardouin-Duparc, 1974; Berlekamp et al., 1982, Chap. 25).

The game of "Life" is an example of a special class of "totalistic" cellular automata, in which the value of a site depends only on the sum of the values of its neighbors at the previous time step, and not on their individual values. Such cellular automata may arise as models of systems involving additive local quantities, such as chemical concentrations. In one dimension with $k = 2$ (and three sites in each neighborhood) all cellular automaton rules are totalistic. In general, the number of totalistic (legal) sets of rules for cellular automata with v neighbors for each site is $k^{(k-1)(vk+1)}$. In one dimension with $k = 3$, $\approx 5 \times 10^6$ of the $\approx 10^8$ possible rules are therefore totalistic. Only 243 of the totalistic rules are also peripheral in the sense defined in Sec. 2. With $k = 2$ in two dimensions, 2^9 of the 2^{11} possible rules in a type-I neighborhood are totalistic (and 32 are also peripheral), and 2^{17} of the 2^{59} in a type-II neighborhood.

A potentially important feature of cellular automata is the capability for "self-reproduction" through which the evolution of a configuration yields several separated identical copies of the configuration. Figure 33 illustrates a very simple form of self-reproduction with the elementary one-dimensional modulo-two rule (see Waksman, 1969; Amoroso and Cooper, 1971; Fredkin, 1981). With a single nonzero site in the initial state, a configuration containing exactly two nonzero sites is obtained

Figure 33. Evolution of a simple pattern according to the modulo-two cellular automaton rule (number 90), exhibiting a simple self-reproduction phenomenon. The additive superposition property of the cellular automaton leads to the generation of two exact copies of the initial 1011 pattern at time steps 8, 16, 32, ... "Geometrical overcrowding" prevents exponential increase in the number of copies produced.

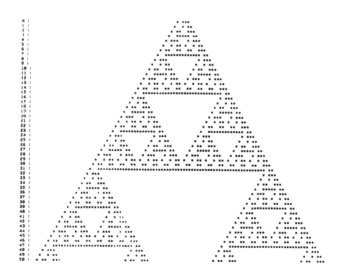

after 2^j time steps[15] as indicated by Eq. (3.2). The additive superposition property of the modulo-two rule implies that results for more complicated initial states are obtained by superposition of those for single-site initial states. Thus after $\tau = 2^j$ time steps, for sufficiently large j, the cellular automaton generates two exact copies of any initial sequence of site values. After a further 2^{j-1} time steps, four copies are obtained. However, after another 2^{j-1} time steps, the innermost pair of these copies meet again, and annihilate, leaving only two copies when $\tau = 2^{j+1}$. Purely geometrical "overcrowding" thus prevents exponential multiplication of copies by self-reproduction in this case. An exactly analogous phenomenon occurs with the two-dimensional modulo-two rule illustrated in Fig. 32, and its higher-dimensional analogs. In general, the number of sites in a d-dimensional cellular automaton configuration grows with time at most as fast as $(2\tau)^d$, which is asymptotically slower than the number $> (2d)^{\alpha\tau}$ required for an exponentially increasing number of copies to be generated. Exponential self-reproduction can thus occur only if the copies generated are not precisely identical, but exhibit variability, and for example execute a random walk motion in response to external noise or contain a "counter" which causes later generations to "live" longer before reproducing.

Section 4 mentioned the view of cellular automata as computers. An important class of computers is those with the property of "computational universality," for which changes in input alone allow any "computable function" to be evaluated,

---

[15] An analogous result holds for all modulo-k rules with k prime by virtue of the relation $\binom{k^j}{i} \bmod k = 0, 0 < i < k^j$ valid for all primes k. The relation is a special case of the general result (Knuth, 1973, Sec. 1.2.6, Ex. 10)

$$\binom{j}{i} = \binom{\lfloor j/k \rfloor}{\lfloor i/k \rfloor}\binom{j \bmod k}{i \bmod k} \bmod k.$$

without any change in internal construction. Universal computers can simulate the operation of any other computer if their input is suitably encoded. Many Turing machines have been shown to be computationally universal. The simplest has seven internal states, and allows four possible "symbols" in each square of its tape. One method for demonstrating computational universality of cellular automata shows correspondence with a universal Turing machine. The head of the Turing machine is typically represented by a phononlike structure which propagates along the cellular automaton. It may be shown (Smith, 1971) that an eighteen-state one-dimensional cellular automaton with a three-site neighborhood can simulate the seven-state four-symbol Turing machine in this way, and is therefore computationally universal. Simpler computationally universal cellular automata must be found by other methods. The most straightforward method is to show correspondence with a standard digital computer or electronic circuit by identifying cellular automaton structures which act like "wires," carrying signals without dissipation and crossing without interference, and structures representing NAND gates at intersections between wires. "Memories" which maintain the same state for all time are also required. In the Life-game cellular automaton discussed above, streams of gliders generated by glider guns may be used as wires, with bits in the signal represented by the presence or absence of gliders. At the points where "glider streams" meet, other structures determine whether the corresponding wires cross or interact through a "NAND gate." The Life-game cellular automaton is thus computationally universal. "Circuits" such as binary adders (Buckingham, 1978) may be constructed from Life configurations. It appears that such circuits run at a speed slower than the digital computers to which they correspond only by a constant multiplicative factor. The "Life game" is a type-II two-dimensional cellular automaton with two states per site. A computationally universal type-I two-dimensional cellular automaton has been constructed with three states per site (Banks, 1971); only two states are required if the initial configuration is permitted to contain an infinite "background" of nonzero sites (Toffoli, 1977a). In one dimension, with a neighborhood of three sites, there are some preliminary indications that a universal cellular automaton may be constructed with five states per site. The details and implications of this cellular automaton will be described in a future publication.

6. Discussion

This paper represents a first step in the investigation of cellular automata as mathematical models for self-organizing statistical systems. The bulk of the paper consisted in a detailed analysis of elementary cellular automata involving a sequence of sites on a line, with a binary variable at each site evolving in discrete time steps according to the values of its nearest neighbors. Despite the simplicity of their construction, these systems were found to exhibit very complicated behavior.

The 32 possible (legal) elementary cellular automata were found to fall into two broad classes. The first class consisted of simple cellular automata whose time evolution led eventually to simple, usually homogeneous, final states. The second class contained complex cellular automata capable of generating quite complicated structures even from simple initial states. Figure 3 showed the patterns of growth obtained with the very simplest initial state in which only one site had a nonzero value. The complex rules were found to yield self-similar fractal patterns. For all but one of the rules, the patterns exhibited the same fractal dimension $\log_2 3 \simeq 1.59$ (the remaining rule gave a fractal dimension $\log_2 2\varphi \simeq 1.69$). With more complicated initial states, the patterns obtained after evolution for many time steps remained self-similar—at least on scales larger than the region of nonzero initial sites. The generation of self-similar patterns was thus found to be a generic feature of complex cellular automata evolving from simple initial states. This result may provide some explanation for the widespread occurrence of self-similarity in natural systems.

Section 3 discussed the evolution of cellular automata from general initial states, in which a finite fraction of the infinite number of initial sites carried value one. Regardless of the initial density of nonzero sites, definite densities were found in the large time limit. Markovian master equation approximations to the density development were found inadequate because of the importance of "feedback" in the cellular automaton evolution. Even with disordered or random initial states, in which the values of different sites are statistically uncorrelated, the evolution of complex cellular automata was found to lead to the formation of definite structures, as suggested by Figs. 8 and 15. One characteristic of this self-organization was the generation of long sequences of correlated sites. The spectrum of these sequences was found to reach an equilibrium form after only a few time steps, extending to arbitrarily large scales, but with an exponential damping. The exponents were again found to be universal for all initial states and almost all complex cellular automata (with the exception of two special additive cellular automata).

Any initial cellular automaton state was found to lead at large times to configurations with the same statistical structures. However, in complex cellular automata, the trajectories of almost all specific nearby initial configurations (differing by changes in the values at a few sites) were found to diverge exponentially with time in the phase space of possible configurations. After a few time steps, the mapping from initial to final configurations becomes apparently random (although there are quantitative deviations from a uniform random mapping). Cellular automaton rules may map several initial configurations into the same final configuration, and thus lead to microscopically irreversible time evolution in which trajectories of different states may merge. In the limit of an infinite number of sites, a negligible fraction of all the possible cellular automaton configurations are reached by evolution from any of the possible initial states after a few time steps. Starting even from an ensemble in which each possible configuration appears with equal probability, the cellular automaton evolution concentrates the probabilities for particular configurations, thereby reducing entropy. This phenomenon allows for the possibility of self-organization by enhancing the probabilities of organized configurations and suppressing disorganized configurations.

Many of the qualitative features found for elementary cellular automata appear to survive in more complicated cellular automata (considered briefly in Sec. 5), although several novel phenomena may appear. For example, in one-dimensional cellular automata with three or more possible values at each site, protective membranes may be generated which shield finite regions from the effects of external noise, and allow very regular patterns to grow from small seeds.

Cellular automata may be viewed as computers, with initial configurations considered as input programs and data processed by cellular automaton time evolution. Sufficiently complicated cellular automata are known to be universal computers, capable of computing any computable function given appropriate input. Such cellular automata may be considered as capable of the most complicated behavior conceivable and are presumably capable of simulating any physical system given a suitable input encoding and a sufficiently long running time. In addition, they may be used to simulate the evolution of any other cellular automaton. If the necessary encoding is sufficiently simple, the statistical properties of the simulated cellular automaton should follow those of the universal cellular automaton. Although not capable of universal simulation, simpler cellular automata may often simulate each other. This capability may well form a basis for the universality found in the statistical properties of various cellular automata.

Cellular automata have been developed in this paper as general mathematical models. One may anticipate their application as simple models for a wide variety of natural processes. Their nontrivial features are typically evident only when some form of growth inhibition is present. Examples are found in aggregation processes in which aggregation at a particular point prevents further aggregation at the same point on the next time step.

Acknowledgments

I am grateful for suggestions and assistance from J. Ambjorn, N. Margolus, O. Martin, A. Odlyzko, and T. Shaw, and for discussions with J. Avron, C. Bennett, G. Chaitin, J. D. Farmer, R. Feynman, E. Fredkin, M. Gell-Mann, R. W. Gosper, A. Hoogland, T. Toffoli, and W. Zurek. I thank S. Kauffman, R. Landauer, P. Leyland, B. Mandelbrot, and A. Norman for suggesting references. The symbolic manipulation computer language SMP (Wolfram et al., 1981) was used in some of the calculations. Some of this work was done before I resigned from Caltech; computer calculations performed at Caltech were supported in part by the U.S. Department of Energy under Contract Number DE-AC-03-81-ER40050.

References

Abelson, H., and A. A. diSessa, 1981, *Turtle Geometry: The Computer as a Medium for Exploring Mathematics* (MIT Press, Cambridge).

Aggarwal, S., 1973, "Local and global Garden of Eden theorems," University of Michigan technical report No. 147.

Aladyev, V., 1974, "Survey of research in the theory of homogeneous structures and their applications," Math. Biosci. **22**, 121.

Aladyev, V., 1976, "The behavioural properties of homogeneous structures," Math. Biosci. **29**, 99.

Alekseev, V. M. and M. V. Yakobson, 1981, "Symbolic dynamics and hyperbolic dynamic systems," Phys. Rep. **75**, 287.

Amoroso, S., and G. Cooper, 1971, "Tessellation structures of reproduction of arbitrary patterns," J. Comput Syst. Sci. **5**, 455.

Apostol, T. M., 1976, *Introduction to Analytic Number Theory* (Springer, Berlin).

ApSimon, H. G., 1970a, "Periodic forests whose largest clearings are of size 3," Philos. Trans. R. Soc. London, Ser. A **266**, 113.

ApSimon, H. G., 1970b, "Periodic forests whose largest clearings are of size $n \geq 4$," Proc. R. Soc. London, Ser. A **319**, 399.

Arbib, M. A., 1969, *Theories of Abstract Automata* (Prentice-Hall, Englewood Cliffs).

Atrubin, A. J., 1965, "A one-dimensional real-time iterative multiplier," IEEE Trans. Comput. **EC-14**, 394.

Baer, R. M., and H. M. Martinez, 1974, "Automata and biology," Ann. Rev. Biophys. **3**, 255.

Banks, E. R., 1971, "Information processing and transmission in cellular automata," MIT Project MAC report No. TR-81.

Barricelli, N. A., 1972, "Numerical testing of evolution theories," J. Statist. Comput. Simul. **1**, 97.

Berlekamp, E. R., 1968, *Algebraic Coding Theory* (McGraw-Hill, New York).

Berlekamp, E. R., J. H. Conway, and R. K. Guy, 1982, *Winning Ways for Your Mathematical Plays* (Academic, New York), Vol. 2, Chap. 25.

Buckingham, D. J., 1978, "Some facts of life," Byte **3**, 54.

Burks, A. W., 1970, *Essays on Cellular Automata* (University of Illinois, Urbana).

Burks, A. W., 1973, "Cellular Automata and Natural Systems," Proceedings of the 5th Congress of the Deutsche Gessellschaft für Kybernetik, Nuremberg.

Butler, J. T., and S. C. Ntafos, 1977, "The vector string descriptor as a tool in the analysis of cellular automata systems," Math. Biosci. **35**, 55.

Codd, E. F., 1968, *Cellular Automata* (Academic, New York).

Cole, S. N., 1969, "Real-time computation by n-dimensional iterative arrays of finite-state machines," IEEE Trans. Comput. **C-18**, 349.

Conway, J. H., 1970, unpublished.

Deutsch, E. S., 1972, "Thinning algorithms on rectangular, hexagonal and triangular arrays," Commun. ACM **15**, 827.

Farmer, J. D., 1982a, "Dimension, fractal measures, and chaotic dynamics," in *Evolution of Order and Chaos in Physics, Chemistry and Biology*, edited by H. Haken (Springer, Berlin).

Farmer, J. D., 1982b, "Information dimension and the probabilistic structure of chaos," Z. Naturforsch. **37a**, 1304.

Fine, N. J., 1947, "Binomial coefficients modulo a prime," Am. Math. Mon. **54**, 589.

Fischer, P. C., 1965, "Generation of primes by a one-dimensional real-time iterative array," J. ACM **12**, 388.

Flanigan, L. K., 1965, "An experimental study of electrical conduction in the mammalian atrioventricular node," Ph.D. thesis (University of Michigan).

Fredkin, E., 1981, unpublished, and PERQ computer demonstration (Three Rivers Computer Corp.).

Gach, P., G. L. Kurdyumov, and L. A. Levin, 1978, "One-dimensional uniform arrays that wash out finite islands," Probl. Peredachi. Info., **14**, 92.

Gardner, M., 1971, "Mathematical games," Sci. Amer. **224**, February, 112; March, 106; April, 114.

Gardner, M., 1972, "Mathematical games," Sci. Amer. **226**, January, 104.

Geffen, Y., A. Aharony, B. B. Mandelbrot, and S. Kirkpatrick, 1981, "Solvable fractal family, and its possible relation to the backbone at percolation," Phys. Rev. Lett. **47**, 1771.

Gerola, H., and P. Seiden, 1978, "Stochastic star formation and spiral structure of galaxies," Astrophys. J. **223**, 129.

Glaisher, J. W. L., 1899, "On the residue of a binomial-theorem coefficient with respect to a Prime Modulus," Q. J. Math. **30**, 150.

Golomb, S. W., 1967, *Shift Register Sequences* (Holden-Day, San Francisco).

Grassberger, P., 1982, "A new mechanism for deterministic diffusion," Wuppertal preprint WU B 82-18.

Greenberg, J. M., B. D. Hassard, and S. P. Hastings, 1978, "Pattern formation and periodic structures in systems modelled by reaction-diffusion equations," Bull. Am. Math. Soc. **84**, 1296.

Griffeath, D., 1970, *Additive and Cancellative Interacting Particle Systems* (Springer, Berlin).

Haken, H., 1975, "Cooperative phenomena in systems far from thermal equilibrium and in nonphysical systems," Rev. Mod. Phys. **47**, 67.

Haken, H., 1978, *Synergetics*, 2nd ed. (Springer, Berlin).

Haken, H., 1979, *Pattern Formation by Dynamic Systems and Pattern Recognition* (Springer, Berlin).

Haken, H., 1981, *Chaos and Order in Nature* (Springer, Berlin).

Hardouin-Duparc, J., 1974, "Paradis terrestre dans l'automate cellulaire de conway," R.A.I.R.O. 8 R-3, 63.

Hardy, G. H., and E. M. Wright, 1979, *An Introduction to the Theory of Numbers*, 5th ed. (Oxford University Press, Oxford).

Harris, B., 1960, "Probability distributions related to random mappings," Ann. Math. Stat. **31**, 1045.

Harvey, J. A., E. W. Kolb, and S. Wolfram, 1982, unpublished.

Herman, G. T., 1969, "Computing ability of a developmental model for filamentous organisms," J. Theor. Biol. **25**, 421.

Honsberger, R., 1976, "Three surprises from combinatorics and number theory," in *Mathematical Gems II*, Dolciani Math. Expositions (Mathematical Association of America, Oberlin), p. 1.

Hoogland, A., et al., 1982, "A special-purpose processor for the Monte Carlo simulation of Ising spin systems," Delft preprint.

Hopcroft, J. E., and J. D. Ullman, 1979, *Introduction to Automata Theory, Languages and Computation* (Addison-Wesley, Reading).

Kauffman, S. A., 1969, "Metabolic stability and epigenesis in randomly constructed genetic nets," J. Theor. Biol. **22**, 437.

Kimball, S. H., et al., 1958, "Odd binomial coefficients," Am. Math. Mon. **65**, 368.

Kitagawa, T., 1974, "Cell space approaches in biomathematics," Math. Biosci. **19**, 27.

Knuth, D. E., 1973, *Fundamental Algorithms* (Addison-Wesley, Reading).

Knuth, D. E., 1981, *Seminumerical Algorithms*, 2nd ed. (Addison-Wesley, Reading).

Kosaraju, S. R., 1974, "On some open problems in the theory of cellular automata," IEEE Trans. Comput. **C-23**, 561.

Landauer, R., 1979, "The role of fluctuations in multistable systems and in the transition to multistability," Ann. N.Y. Acad. Sci. **316**, 433.

Langer, J. S., 1980, "Instabilities and pattern formation in crystal growth," Rev. Mod. Phys. **52**, 1.

Levy, Y. E., 1982, "Some remarks about computer studies of dynamical systems," Phys. Lett. A **88**, 1.

Lifshitz, E. M., and L. P. Pitaevskii, 1981, *Physical Kinetics* (Pergamon, New York).

Lindenmayer, A., 1968, "Mathematical models for cellular interactions in development," J. Theoret. Biol. **18**, 280.

MacWilliams, F. J., and N. J. A. Sloane, *Theory of Error-Correcting Codes* (North-Holland, Amsterdam).

Mandelbrot, B., 1977, *Fractals: Form, Chance and Dimension* (Freeman, San Francisco).

Mandelbrot, B., 1982, *The Fractal Geometry of Nature* (Freeman, San Francisco).

Manna, Z., 1974, *Mathematical Theory of Computation* (McGraw-Hill, New York).

Manning, F. B., 1977, "An approach to highly integrated, computer-maintained cellular arrays," IEEE Trans. Comput. **C-26**, 536.

Martin, O., A. Odlyzko, and S. Wolfram, 1983, "Algebraic properties of cellular automata," Bell Laboratories report (January 1983).

McIlroy, M. D., 1974, "The numbers of 1's in binary integers: bounds and extremal properties," SIAM J. Comput. **3**, 255.

Miller, J. C. P., 1970, "Periodic forests of stunted trees," Philos. Trans. R. Soc. London, Ser. A **266**, 63.

Miller, J. C. P., 1980, "Periodic forests of stunted trees," Philos. Trans. R. Soc. London Ser. A **293**, 48.

Minsky, M. L., 1967, *Computation: Finite and Infinite Machines* (Prentice-Hall, Englewood Cliffs).

Moore, E. F., 1962, "Machine Models of Self-Reproduction," Proceedings of a Symposium on Applied Mathematics **14**, 17, reprinted in *Essays on Cellular Automata*, edited by A. W. Burks (University of Illinois, Urbana, 1970), p. 187.

Nicolis, G., and Prigogine, I., 1977, *Self-Organization in Nonequilibrium Systems*, (Wiley, New York).

Nicolis, G., G. Dewel, and J. W. Turner, editors, 1981, *Order and Fluctuations in Equilibrium and Nonequilibrium Statistical Mechanics*, Proceedings of the XVIIth International Solvay Conference on Physics (Wiley, New York).

Nishio, H., 1981, "Real time sorting of binary numbers by 1-dimensional cellular automata," Kyoto University report.

Ott, E., 1981, "Strange attractors and chaotic motions of dynamical systems," Rev. Mod. Phys. **53**, 655.

Pearson, R., J. Richardson, and D. Toussaint, 1981, "A special purpose machine for Monte-Carlo simulation," Santa Barbara preprint NSF-ITP-81-139.

Peterson, W. W., and E. J. Weldon, *Error-Correcting Codes*, 2nd ed. (MIT Press, Cambridge).

Preston, K., M. J. B. Duff, S. Levialdi, Ph. E. Norgren, and J.-I. Toriwaki, 1979, "Basics of Cellular Logic with Some Applications in Medical Image Processing," Proc. IEEE **67**, 826.

Prigogine, I., 1980, *From Being to Becoming* (Freeman, San Francisco).

Renyi, A., 1970, *Probability Theory* (North-Holland, Amsterdam).

Roberts, J. B., 1957, "On binomial coefficient residues," Can. J. Math. **9**, 363.

Rosen, R., 1981, "Pattern generation in networks," Prog. Theor. Biol. **6**, 161.

Rosenfeld, A., 1979, *Picture Languages* (Academic, New York).

Schewe, P. F., editor, 1981, "Galaxies, the Game of Life, and percolation," in *Physics News*, Amer. Inst. Phys. Pub **R-302**, 61.

Schulman, L. S., and P. E. Seiden, 1978, "Statistical mechanics of a dynamical system based on Conway's game of life," J. Stat. Phys. **19**, 293.

Shannon, C. E., 1951, "Prediction and entropy of printed English," Bell Syst. Tech. J., **30**, 50.

Sierpinski, W., 1916, "Sur une courbe dont tout point est un point de ramification," Pr. Mat.-Fiz. **27**, 77; *Oeuvres Choisis*, (Państowe Wydawnictwo Naukowe, Warsaw) Vol. II, p. 99.

Smith, A. R., 1971, "Simple computation-universal cellular spaces," J. ACM **18**, 339.

Sternberg, S. R., 1980, "Language and architecture for parallel image processing," in *Pattern Recognition in Practice*, edited by E. S. Gelesma and L. N. Kanal (North-Holland, Amsterdam), p. 35.

Stevens, P. S., 1974, *Patterns in Nature* (Little, Brown, Boston).

Stolarsky, K. B., 1977, "Power and exponential sums of digital sums related to binomial coefficient parity," SIAM J. Appl. Math. **32**, 717.

Sutton, C., 1981, "Forests and numbers and thinking backwards," New Sci. **90**, 209.

Thom, R., 1975, *Structural Stability and Morphogenesis* (Benjamin, New York).

Thompson, D'A. W., 1961, *On Growth and Form*, abridged ed. edited by J. T. Bonner (Cambridge University, Cambridge, England).

Toffoli, T., 1977a, "Computation and construction universality of reversible cellular automata," J. Comput. Sys. Sci. **15**, 213.

Toffoli, T., 1977b, "Cellular automata mechanics," Ph.D. thesis, Logic of Computers Group, University of Michigan.

Toffoli, T., 1980, "Reversible computing," MIT report MIT/LCS/TM-151.

Toffoli, T., 1983, "Squareland: a hardware cellular automaton simulator," MIT LCS preprint, in preparation.

Turing, A. M., 1936, "On computable numbers, with an application to the Entscheidungsproblem," Proc. London Math. Soc. Ser. 2 **42**, 230; **43**, 544E, reprinted in *The Undecidable*, edited by M. Davis (1965; Hewlett, New York), p. 115.

Turing, A. M., 1952, "The chemical basis of morphogenesis," Philos. Trans. R. Soc. London, Ser. B **237**, 37.

Ulam, S., 1974, "Some ideas and prospects in biomathematics," Ann. Rev. Bio., **255**.

von Neumann, J., 1963, "The general and logical theory of automata," in J. von Neumann, *Collected Works*, edited by A. H. Taub, **5**, 288.

von Neumann, J., 1966, *Theory of Self-Reproducing Automata*, edited by A. W. Burks (University of Illinois, Urbana).

Wainwright, R. T., 1971–73, Lifeline, 1–11.

Wainwright, R. T., 1974, "Life is Universal!," Proceedings of the Winter Simulation Conference, Washington, D.C., ACM, p. 448.

Waksman, A., 1969, "A model of replication," J. ACM **16**, 178.

Willson, S., 1982, "Cellular automata can generate fractals," Iowa State University, Department of Mathematics, preprint.

Witten, T. A., and L. M. Sander, 1981, "Diffusion-limited aggregation, a kinetic critical phenomenon," Phys. Rev. Lett. **47**, 1400.

Wolfram, S., et al., 1981, "SMP Handbook," Caltech.

Wolfram, S., 1982a, "Cellular automata as simple self-organizing systems," Caltech preprint CALT-68-938 (submitted to Nature).

Wolfram, S., 1982b, "Geometry of binomial coefficients," to be published in Am. Math. Monthly.

Algebraic Properties
of Cellular Automata

1984

Cellular automata are discrete dynamical systems, of simple construction but complex and varied behaviour. Algebraic techniques are used to give an extensive analysis of the global properties of a class of finite cellular automata. The complete structure of state transition diagrams is derived in terms of algebraic and number theoretical quantities. The systems are usually irreversible, and are found to evolve through transients to attractors consisting of cycles sometimes containing a large number of configurations.

1. Introduction

In the simplest case, a cellular automaton consists of a line of sites with each site carrying a value 0 or 1. The site values evolve synchronously in discrete time steps according to the values of their nearest neighbours. For example, the rule for evolution could take the value of a site at a particular time step to be the sum modulo two of the values of its two nearest neighbours on the previous time step. Figure 1 shows the pattern of nonzero sites generated by evolution with this rule from an initial state containing a single nonzero site. The pattern is found to be self-similar, and is characterized by a fractal dimension $\log_2 3$. Even with an initial state consisting of a random sequence of 0 and 1 sites (say each with probability $\frac{1}{2}$), the evolution of such a cellular automaton leads to correlations between separated sites and the appearance of structure. This behaviour contradicts the second law of thermodynamics for systems with reversible dynamics, and is made possible by the irreversible nature

Coauthored with Olivier Martin and Andrew M. Odlyzko. Originally published in *Communications in Mathematical Physics*, volume 93, pages 219–258 (March 1984).

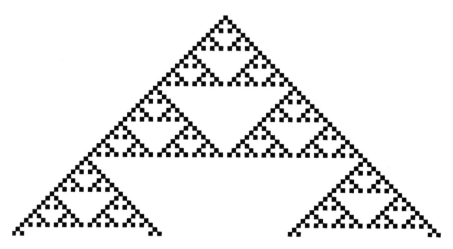

Figure 1. Example of evolution of a one-dimensional cellular automaton with two possible values at each site. Configurations at successive time steps are shown as successive lines. Sites with value one are black; those with value zero are left white. The cellular automaton rule illustrated here takes the value of a site at a particular time step to be the sum modulo two of the values of its two nearest neighbours on the previous time step. This rule is represented by the polynomial $T(x) = x + x^{-1}$, and is discussed in detail in Sect. 3.

of the cellular automaton evolution. Starting from a maximum entropy ensemble in which all possible configurations appear with equal probability, the evolution increases the probabilities of some configurations at the expense of others. The configurations into which this concentration occurs then dominate ensemble averages and the system is "organized" into having the properties of these configurations. A finite cellular automaton with N sites (arranged for example around a circle so as to give periodic boundary conditions) has 2^N possible distinct configurations. The global evolution of such a cellular automaton may be described by a state transition graph. Figure 2 gives the state transition graph corresponding to the cellular automaton described above, for the cases $N = 11$ and $N = 12$. Configurations corresponding to nodes on the periphery of the graph are seen to be depopulated by transitions; all initial configurations ultimately evolve to configurations on one of the cycles in the graph. Any finite cellular automaton ultimately enters a cycle in which a sequence of configurations are visited repeatedly. This behaviour is illustrated in Fig. 3.

Cellular automata may be used as simple models for a wide variety of physical, biological and computational systems. Analysis of general features of their behaviour may therefore yield general results on the behaviour of many complex systems, and may perhaps ultimately suggest generalizations of the laws of thermodynamics appropriate for systems with irreversible dynamics. Several aspects of cellular automata were recently discussed in [1], where extensive references were given. This paper details and extends the discussion of global proper-

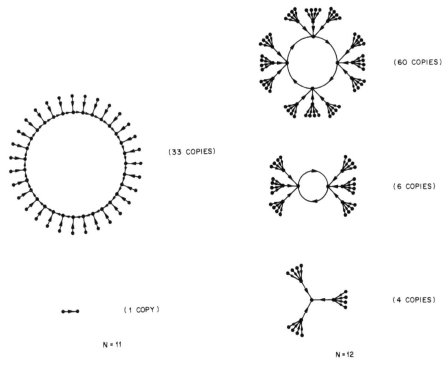

(60 COPIES)

(33 COPIES)

(6 COPIES)

(1 COPY)

(4 COPIES)

N = 11

N = 12

Figure 2. Global state transition diagrams for finite cellular automata with size N and periodic boundary conditions evolving according to the rule $\mathbb{T}(x) = x + x^{-1}$, as used in Fig. 1, and discussed extensively in Sect. 3. Each node in the graphs represents one of the 2^N possible configurations of the N sites. The directed edges of the graphs indicate transitions between these configurations associated with single time steps of cellular automaton evolution. Each cycle in the graph represents an "attractor" for the configurations corresponding to the nodes in trees rooted on it.

ties of cellular automata given in [1]. These global properties may be described in terms of properties of the state transition graphs corresponding to the cellular automata.

This paper concentrates on a class of cellular automata which exhibit the simplifying feature of "additivity". The configurations of such cellular automata satisfy an "additive superposition" principle, which allows a natural representation of the configurations by characteristic polynomials. The time evolution of the configurations is represented by iterated multiplication of their characteristic polynomials by fixed polynomials. Global properties of cellular automata are then determined by algebraic properties of these polynomials, by methods analogous to those used in the analysis of linear feedback shift registers [2, 3]. Despite their amenability to algebraic analysis, additive cellular automata exhibit many of the complex features of general cellular automata.

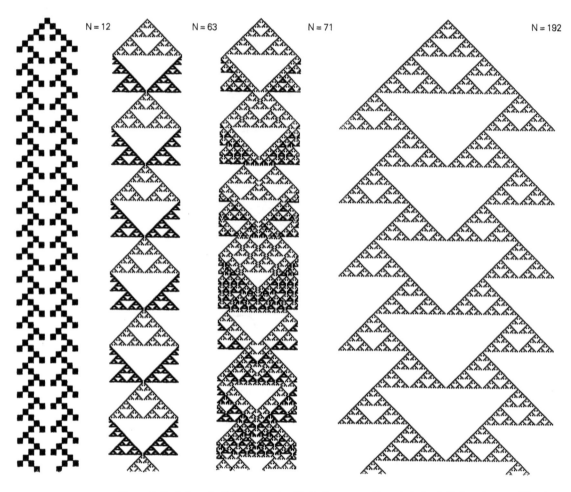

Figure 3. Evolution of cellular automata with N sites arranged in a circle (periodic boundary conditions) according to the rule $\mathbb{T}(x) = x + x^{-1}$ (as used in Fig. 1 and discussed in Sect. 3). Finite cellular automata such as these ultimately enter cycles in which a sequence of configurations are visited repeatedly. This behaviour is evident here for $N = 12$, 63, and 192. For $N = 71$, the cycle has length $2^{35} - 1$.

Having introduced notation in Sect. 2, Sect. 3 develops algebraic techniques for the analysis of cellular automata in the context of the simple cellular automaton illustrated in Fig. 1. Some necessary mathematical results are reviewed in the appendices. Section 4 then derives general results for all additive cellular automata. The results allow more than two possible values per site, but are most complete when the number of possible values is prime. They also allow influence on the evolution of a site from sites more distant than its nearest neighbours. The results are extended in Sect. 4D to allow cellular automata in which the sites are arranged in a square or cubic lattice in two, three or more dimensions, rather than just on a line. Section 4E then discusses generalizations in which the cellular automaton time evolution rule involves several preceding time steps. Section 4F considers alternative boundary conditions. In all cases, a characterization of the global structure of the state transition diagram is found in terms of algebraic properties of the polynomials representing the cellular automaton time evolution rule.

Section 5 discusses non-additive cellular automata, for which the algebraic techniques of Sects. 3 and 4 are inapplicable. Combinatorial methods are nevertheless used to derive some results for a particular example.

Section 6 gives a discussion of the results obtained, comparing them with those for other systems.

2. Formalism

We consider first the formulation for one-dimensional cellular automata in which the evolution of a particular site depends on its own value and those of its nearest neighbours. Section 4 generalizes the formalism to several dimensions and more neighbours.

We take the cellular automaton to consist of N sites arranged around a circle (so as to give periodic boundary conditions). The values of the sites at time step t are denoted $a_0^{(t)}, \ldots, a_{N-1}^{(t)}$. The possible site values are taken to be elements of a finite commutative ring $\mathbb{R}_k$ with k elements. Much of the discussion below concerns the case $\mathbb{R}_k = \mathbb{Z}_k$, in which site values are conveniently represented as integers modulo k. In the example considered in Sect. 3, $\mathbb{R}_k = \mathbb{Z}_2$, and each site takes on a value 0 or 1.

The complete configuration of a cellular automaton is specified by the values of its N sites, and may be represented by a characteristic polynomial (generating function) (cf. [2, 3])

$$A^{(t)}(x) = \sum_{i=0}^{N-1} a_i^{(t)} x^i, \tag{2.1}$$

where the value of site i is the coefficient of x^i, and all coefficients are elements of the ring $\mathbb{R}_k$. We shall often refer to configurations by their corresponding characteristic polynomials.

It is often convenient to consider generalized polynomials containing both positive and negative powers of x: such objects will be termed "dipolynomials". In general, $H(x)$ is a dipolynomial if there exists some integer m such that $x^m H(x)$ is an ordinary polynomial in x. As discussed in Appendix A, dipolynomials possess divisibility and congruence properties analogous to those of ordinary polynomials.

Multiplication of a characteristic polynomial $A(x)$ by $x^{\pm j}$ yields a dipolynomial which represents a configuration in which the value of each site has been transferred (shifted) to a site j places to its right (left). Periodic boundary conditions in the cellular automaton are implemented by reducing the characteristic dipolynomial modulo the fixed polynomial $x^N - 1$ at all stages, according to

$$\sum_i a_i x^i \quad \mathrm{mod}\,(x^N - 1) = \sum_{i=0}^{N-1} \left(\sum_j a_{i+jN} \right) x^i. \tag{2.2}$$

Note that any dipolynomial is congruent modulo $(x^N - 1)$ to a unique ordinary polynomial of degree less than N.

In general, the value $a_i^{(t)}$ of a site in a cellular automaton is taken to be an arbitrary function of the values $a_{i-1}^{(t-1)}$, $a_i^{(t-1)}$, and $a_{i+1}^{(t-1)}$ at the previous time step. Until Sect. 5, we shall consider a special class of "additive" cellular automata which evolve with time according to simple linear combination rules of the form (taking the site index i modulo N)

$$a_i^{(t)} = \alpha_{-1} a_{i-1}^{(t-1)} + \alpha_0 a_i^{(t-1)} + \alpha_{+1} a_{i+1}^{(t-1)}, \tag{2.3}$$

where the α_j are fixed elements of $\mathbb{R}_k$, and all arithmetic is performed in $\mathbb{R}_k$. This time evolution may be represented by multiplication of the characteristic polynomial by a fixed dipolynomial in x,

$$\mathbb{T}(x) = \alpha_{-1} x + \alpha_0 + \alpha_{+1} x^{-1}, \tag{2.4}$$

according to

$$A^{(t)}(x) \equiv \mathbb{T}(x) A^{(t-1)}(x) \quad \mathrm{mod}\,(x^N - 1), \tag{2.5}$$

where arithmetic is again performed in $\mathbb{R}_k$. Additive cellular automata obey an additive superposition principle which implies that the configuration obtained by evolution for t time steps from an initial configuration $A^{(0)}(x) + B^{(0)}(x)$ is identical to $A^{(t)}(x) + B^{(t)}(x)$, where $A^{(t)}(x)$ and $B^{(t)}(x)$ are the results of separate evolution of $A^{(0)}(x)$ and $B^{(0)}(x)$, and all addition is performed in $\mathbb{R}_k$. Since any initial configuration can be represented as a sum of "basis" configurations $\Delta(x) = x^j$ containing single nonzero sites with unit values, the additive superposition principle determines the evolution of all configurations in terms of the evolution of $\Delta(x)$. By virtue of the cyclic symmetry between the sites it suffices to consider the case $j = 0$.

3. A Simple Example

A. Introduction

This section introduces algebraic techniques for the analysis of additive cellular automata in the context of a specific simple example. Section 4 applies the techniques to more general cases. The mathematical background is outlined in the appendices.

The cellular automaton considered in this section consists of N sites arranged around a circle, where each site has value 0 or 1. The sites evolve so that at each time step the value of a site is the sum modulo two of the values of its two nearest neighbours at the previous time step:

$$a_i^{(t)} = a_{i-1}^{(t-1)} + a_{i+1}^{(t-1)} \qquad \mathrm{mod}\, 2. \tag{3.1}$$

This rule yields in many respects the simplest non-trivial cellular automaton. It corresponds to rule 90 of [1], and has been considered in several contexts elsewhere (e.g. [4]).

The time evolution (3.1) is represented by multiplication of the characteristic polynomial for a configuration by the dipolynomial

$$\mathbb{T}(x) = x + x^{-1} \tag{3.2}$$

according to Eq. (2.5). At each time step, characteristic polynomials are reduced modulo $x^N - 1$ (which is equal to $x^N + 1$ since all coefficients are here, and throughout this section, taken modulo two). This procedure implements periodic boundary conditions as in Eq. (2.2) and removes any inverse powers of x.

Equation (3.2) implies that an initial configuration containing a single nonzero site evolves after t time steps to a configuration with characteristic dipolynomial

$$\mathbb{T}(x)^t 1 = (x + x^{-1})^t = \sum_{i=0}^{t} \binom{t}{i} x^{2i-t}. \tag{3.3}$$

For $t < N/2$ (before "wraparound" occurs), the region of nonzero sites grows linearly with time, and the values of sites are given simply by binomial coefficients modulo two, as discussed in [1] and illustrated in Fig. 1. (The positions of nonzero sites are equivalently given by $\pm 2^{j_1} \pm 2^{j_2} \pm \ldots$, where the j_i give the positions of nonzero digits in the binary decomposition of the integer t.) The additive superposition property implies that patterns generated from initial configurations containing more than one nonzero site may be obtained by addition modulo two (exclusive disjunction) of the patterns (3.3) generated from single nonzero sites.

B. Irreversibility

Every configuration in a cellular automaton has a unique successor in time. A configuration may however have several distinct predecessors, as illustrated in the state transition diagram of Fig. 2. The presence of multiple predecessors implies that the time evolution mapping is not invertible but is instead "contractive". The cellular automaton thus exhibits irreversible behaviour in which information on initial states is lost through time evolution. The existence of configurations with multiple predecessors implies that some configurations have no predecessors[1]. These configurations occur only as initial states, and may never be generated in the time evolution of the cellular automaton. They appear on the periphery of the state transition diagram of Fig. 2. Their presence is an inevitable consequence of irreversibility and of the finite number of states.

Lemma 3.1. Configurations containing an odd number of sites with value 1 can never be generated in the evolution of the cellular automaton defined in Sect. 3A, and can occur only as initial states.

Consider any configuration specified by characteristic polynomial $A^{(0)}(x)$. The successor of this configuration is $A^{(1)}(x) = T(x)A^{(0)}(x) = (x + x^{-1})A^{(0)}(x)$, taken, as always, modulo $x^N - 1$. Thus

$$A^{(1)}(x) = (x^2 + 1)B(x) + R(x)(x^N - 1)$$

for some dipolynomials $R(x)$ and $B(x)$. Since $x^2 + 1 = x^N - 1 = 0$ for $x = 1$, $A^{(1)}(1) = 0$. Hence $A^{(1)}(x)$ contains an even number of terms, and corresponds to a configuration with an even number of nonzero sites. Only such configurations can therefore be reached from some initial configuration $A^{(0)}(x)$.

An extension of this lemma yields the basic theorem on the number of unreachable configurations:

Theorem 3.1. The fraction of the 2^N possible configurations of a size N cellular automaton defined in Sect. 3A which can occur only as initial states, and cannot be reached by evolution, is $1/2$ for N odd and $3/4$ for N even.

A configuration $A^{(1)}(x)$ is reachable after one time step of cellular automaton evolution if and only if for some dipolynomial $A^{(0)}(x)$,

$$A^{(1)}(x) \equiv T(x)A^{(0)}(x) \equiv (x + x^{-1})A^{(0)}(x) \quad \mathrm{mod}\,(x^N - 1), \tag{3.4}$$

so that

$$A^{(1)}(x) = (x^2 + 1)B(x) + R(x)(x^N - 1) \tag{3.5}$$

for some dipolynomials $R(x)$ and $B(x)$. To proceed, we use the factorization of $(x^N - 1)$ given in Eq. (A.7), and consider the cases N even and N odd separately.

(a) N even. Since by Eq. (A.4), $(x^2 + 1) = (x + 1)^2 = (x - 1)^2$ (taken, as always, modulo 2), and by Eq. (A.7),

$$(x - 1)^2 \,|\, (x^{N/2} - 1)^2 = (x^N - 1)$$

for even N, Eq. (3.5) shows that

$$(x - 1)^2 \,|\, A^{(1)}(x)$$

in this case. But since $(x - 1)^2$ contains a constant term, $A^{(1)}(x)/(x - 1)^2$ is thus an

---

[1] Such configurations have been termed "Gardens of Eden" [5].

ordinary polynomial if $A^{(1)}(x)$ is chosen as such. Hence all reachable configurations represented by a polynomial $A^{(1)}(x)$ are of the form

$$A^{(1)}(x) = (x - 1)^2 C(x),$$

for some polynomial $C(x)$. The predecessor of any such configuration is $xC(x)$, so any configuration of this form may in fact be reached. Since deg $A(x) < N$, deg $C(x) < N - 2$. There are thus exactly 2^{N-2} reachable configurations, or $1/4$ of all the 2^N possible configurations.

(b) N odd. Using Lemma 3.1 the proof for this case is reduced to showing that all configurations containing an even number of nonzero sites have predecessors. A configuration $A^{(1)}(x)$ with an even number of nonzero sites can always be written in the form $(x + 1)D(x)$. But

$$
\begin{aligned}
A^{(1)}(x) &= (x + 1)D(x) \\
&\equiv (x + x^{-1})(x^2 + x^4 + \ldots + x^{N-1})D(x) \quad \mathrm{mod}\,(x^N - 1) \\
&\equiv \mathbb{T}(x)(x^2 + x^4 + \ldots + x^{N-1})D(x) \quad \mathrm{mod}\,(x^N - 1),
\end{aligned}
$$

giving an explicit predecessor for $A^{(1)}(x)$.

The additive superposition principle for the cellular automaton considered in this section yields immediately the result:

Lemma 3.2. Two configurations $A^{(0)}(x)$ and $B^{(0)}(x)$ yield the same configuration $C(x) \equiv \mathbb{T}(x)A^{(0)}(x) \equiv \mathbb{T}(x)B^{(0)}$ after one time step in the evolution of the cellular automaton defined in Sect. 3A if and only if $A^{(0)}(x) = B^{(0)}(x) + Q(x)$, where $\mathbb{T}(x)Q(x) \equiv 0$.

Theorem 3.2. Configurations in the cellular automaton defined in Sect. 3A which have at least one predecessor have exactly two predecessors for N odd and exactly four for N even.

This theorem is proved using Lemma 3.2 by enumeration of configurations $Q(x)$ which evolve to the null configuration after one time step. For N odd, only the configurations 0 and $1 + x + \ldots + x^{N-1} = \frac{x^N - 1}{x - 1}$ (corresponding to site values $11111\ldots$) have this property. For N even, $Q(x)$ has the form

$$(1 + x^2 + \ldots + x^{N-2})S_i(x) = \frac{x^N - 1}{x^2 - 1}S_i(x),$$

where the $S_i(x)$ are the four polynomials of degree less than two. Explicitly, the possible forms for $Q(x)$ are 0, $1 + x^2 + \ldots + x^{N-2}$, $x + x^3 + \ldots + x^{N-1}$, and $1 + x + x^2 + \ldots + x^{N-1}$.

C. Topology of the State Transition Diagram

This subsection derives topological properties of the state transition diagrams illustrated in Fig. 2. The results determine the amount and rate of "information loss" or "self organization" associated with the irreversible cellular automaton evolution.

The state transition network for a cellular automaton is a graph, each of whose nodes represents one of the possible cellular automaton configurations. Directed arcs join the nodes to represent the transitions between cellular automaton configurations at each time step. Since each cellular automaton configuration has a unique successor, exactly one arc must leave each node, so that all nodes have out-degree one. As discussed in the previous subsection, cellular automaton configurations may have several or no predecessors, so that the in-degrees of nodes in the state transition graph may differ. Theorems 3.1 and 3.2 show that for N odd, $1/2$ of all nodes have zero in-degree and the rest have in-degree two, while for N even, $3/4$ have zero in-degree and $1/4$ in-degree four.

As mentioned in Sect. 1, after a possible "transient", a cellular automaton evolving from any initial configuration must ultimately enter a loop, in which a sequence of configurations are visited repeatedly. Such a loop is represented by a cycle in the state transition graph. At every node in this cycle a tree is rooted; the transients consist of transitions leading towards the cycle at the root of the tree.

Lemma 3.3. The trees rooted at all nodes on all cycles of the state transition graph for the cellular automaton defined in Sect. 3A are identical.

This result is proved by showing that trees rooted on all cycles are identical to the tree rooted on the null configuration. Let $A(x)$ be a configuration which evolves to the null configuration after exactly t time steps, so that $\mathbb{T}(x)^t A(x) \equiv 0 \bmod (x^N - 1)$. Let $R(x)$ be a configuration on a cycle, and let $R^{(-t)}(x)$ be another configuration on the same cycle, such that $\mathbb{T}(x)^t R^{(-t)}(x) \equiv R(x) \bmod (x^N - 1)$. Then define

$$\Psi_{R(x)}[A(x)] = A(x) + R^{(-t)}(x).$$

We first show that as $A(x)$ ranges over all configurations in the tree rooted on the null configuration, $\Psi_{R(x)}[A(x)]$ ranges over all configurations in the tree rooted at $R(x)$. Since

$$\mathbb{T}(x)^t \Psi_{R(x)}[A(x)] = \mathbb{T}(x)^t A(x) + \mathbb{T}(x)^t R^{(-t)}(x) \equiv R(x) \qquad \bmod (x^N - 1),$$

it is clear that all configurations $\Psi_{R(x)}[A(x)]$ evolve after t time steps [where the value of t depends on $A(x)$] to $R(x)$. To show that these configurations lie in the tree rooted at $R(x)$, one must show that their evolution reaches no other cycle configurations for any $s < t$. Assume this supposition to be false, so that there exists some $m \neq 0$ for which

$$R^{(-m)}(x) \equiv \mathbb{T}(x)^s \Psi_{R(x)}[A(x)] = \mathbb{T}(x)^s A(x) + R^{(s-t)}(x) \qquad \bmod (x^N - 1).$$

Since $\mathbb{T}(x)^t A(x) \equiv 0 \bmod (x^N - 1)$, this would imply $R^{(t-s-m)}(x) = R^{(0)}(x) = R(x)$, or $R^{(-m)}(x) = R^{(s-t)}(x)$. But $R^{(-m)}(x) - R^{(s-t)}(x) \equiv \mathbb{T}(x)^s A(x)$, and by construction $\mathbb{T}(x)^s A(x) \neq 0$ for any $s < t$, yielding a contradiction. Thus $\Psi_{R(x)}$ maps configurations at height t in the tree rooted on the null configuration to configurations at height t in the tree rooted at $R(x)$, and the mapping Ψ is one-to-one. An analogous argument shows that Ψ is onto. Finally one may show that Ψ preserves the time evolution structure of the trees, so that if $\mathbb{T}(x)A^{(0)}(x) = A^{(1)}(x)$, then

$$\mathbb{T}(x)\Psi_{R(x)}[A^{(0)}(x)] = \Psi_{R(x)}[A^{(1)}(x)],$$

which follows immediately from the definition of Ψ. Hence Ψ is an isomorphism, so that trees rooted at all cycle configurations are isomorphic to that rooted at the null configuration.

Notice that this proof makes no reference to the specific form (3.2) chosen for $\mathbb{T}(x)$ in this section; Lemma 3.3 thus holds for any additive cellular automaton.

Theorem 3.3. For N odd, a tree consisting of a single arc is rooted at each node on each cycle in the state transition graph for the cellular automaton defined in Sect. 3A.

By virtue of Lemma 3.3, it suffices to show that the tree rooted on the null configuration consists of a single node corresponding to the configuration $111 \ldots 111$. This configuration has no predecessors by virtue of Lemma 3.1.

Corollary. For N odd, the fraction of the 2^N possible configurations which may occur in the evolution of the cellular automaton defined in Sect. 3A is $1/2$ after one or more time steps.

The "distance" between two nodes in a tree is defined as the number of arcs which are visited in traversing the tree from one node to the other (e.g. [6]). The "height" of a (rooted) tree is defined as the maximum number of arcs traversed in a descent from any leaf or terminal (node with zero in-degree) to the root of the tree (formally node with zero out-degree). A tree is "balanced" if all its leaves are at the same distance from its root. A tree is termed "quaternary" ("binary") if each of its non-terminal nodes has in-degree four (two).

Let $D_2(N)$ be the maximum 2^j which divides N (so that for example $D_2(12) = 4$).

Theorem 3.4. For N even, a balanced tree with height $D_2(N)/2$ is rooted at each node on each cycle in the state transition graph for the cellular automaton defined in Sect. 3A; the trees are quaternary, except that their roots have in-degree three.

Theorem 3.2 shows immediately that the tree is quaternary. In the proof of Theorem 3.1, we showed that a configuration $Q_1(x)$ can be reached from some configuration $Q_0(x)$ if and only if $(1 + x^2) \mid Q_1(x)$; Theorem 3.2 then shows that if $Q_1(x)$ is reachable, it is reachable from exactly four distinct configurations $Q_0(x)$. We now extend this result to show that a configuration $Q_m(x)$ can be reached from some configuration $Q_0(x)$ by evolution for m time steps, with $m \leq D_2(N)/2$, if and only if $(1 + x^2)^m \mid Q_m(x)$. To see this, note that if

$$Q_m(x) \equiv \mathbb{T}(x)^m Q_0(x) \qquad \mod (x^N - 1), \tag{3.6}$$

then

$$(x^N - 1) \mid Q_m(x) + (x^2 + 1)^m x^{N-m} Q_0(x), \tag{3.7}$$

and so, since by Eq. (A.7), $(x^2 + 1)^m \mid (x^N - 1)$ for $m \leq D_2(N)/2$, it follows that

$$(x^2 + 1)^m \mid Q_m(x) \tag{3.8}$$

for $m \leq D_2(N)/2$. On the other hand, if $(x^2 + 1)^m \mid Q_m(x)$, say $Q_m(x) = (x^2 + 1)^m Q_0(x)$, then $Q_m(x) \equiv \mathbb{T}(x)^m x^m Q_0(x)$, which shows that $Q_m(x)$ is reachable in m steps.

The balance of the trees is demonstrated by showing that for $m < D_2(N)/2$, if $(x^2 + 1)^m \mid Q_m(x)$, then $Q_m(x)$ can be reached from exactly 4^m initial configurations $Q_0(x)$. This may be proved by induction on m. If

$$(1 + x^2)^m \mid Q_m(x) \qquad (1 \leq m < D_2(N)/2),$$

then all of the four states $Q_{m-1}(x)$ from which $Q_m(x)$ may be reached in one step satisfy $(x^2 + 1)^{m-1} \mid Q_{m-1}(x)$. Consider now the configurations $Q(x)$ which satisfy

$$(x^2 + 1)^{D_2(N)/2} \mid Q(x). \tag{3.9}$$

If we write $Q(x) = (x + 1)^{D_2(N)} R(x)$, then as in Theorem 3.2, the four predecessors of $Q(x)$ are exactly

$$Q_{-1}(x) = (x + 1)^{D_2(N)-2} R^*(x) + \left(\frac{x^{N/2} - 1}{x - 1} \right)^2 S_i(x), \tag{3.10}$$

where $x R(x) \equiv R^*(x) \bmod (x^N - 1)$. $S_i(x)$ ranges over the four polynomials of degree less than two, as in Theorem 3.2. Exactly one of these polynomials satisfies Eq. (3.9), whereas the other three satisfy only

$$(x + 1)^{D_2(N)-2} \mid Q_{-1}(x).$$

Any state satisfying Eq. (3.9) thus belongs to a cycle, since it can be reached after an arbitrary number of steps. Conversely, since any cycle configuration must be reachable after $D_2(N)/2$ time steps, any and all configurations $Q_{-1}(x)$ satisfying Eq. (3.9) are indeed on cycles. But, as shown above, the three $Q_{-1}(x)$ which do not satisfy Eq. (3.9) are roots of balanced quaternary trees of height $D_2(N)/2 - 1$. The proof of the theorem is thus completed.

Corollary. For N even, a fraction 4^{-t} of the 2^N possible configurations appear after t steps in the evolution of the cellular automaton defined in Sect. 3A for $t \leq D_2(N)/2$. A fraction $2^{-D_2(N)}$ of the configurations occur in cycles, and are therefore generated at arbitrarily large times.

Corollary. All configurations $A(x)$ on cycles in the cellular automaton of Sect. 3A are divisible by $(1 + x)^{D_2(N)}$.

This result follows immediately from the proof of Theorems 3.3 and 3.4.

Entropy may be used to characterize the irreversibility of cellular automaton evolution (cf. [1]). One may define a set (or topological) entropy for an ensemble of configurations i occurring with probabilities p_i according to

$$s = \frac{1}{N} \log_2 \sum_i \theta(p_i),$$ (3.11)

where $\theta(p) = 1$ for $p > 0$, and 0 otherwise. One may also define a measure entropy

$$s_\mu = -\frac{1}{N} \sum_i p_i \log_2 p_i.$$ (3.12)

For a maximal entropy ensemble in which all 2^N possible cellular automaton configurations occur with equal probabilities,

$$s = s_\mu = 1.$$

These entropies decrease in irreversible cellular automaton evolution, as the probabilities for different configurations become unequal. However, the balance property of the state transition trees implies that configurations either do not appear, or occur with equal nonzero probabilities. Thus the set and measure entropies remain equal in the evolution of the cellular automaton of Sect. 3A. Starting from a maximal entropy ensemble, both nevertheless decrease with time t according to

$$s(t) = s_\mu(t) = 1 - 2t/N, \qquad 0 \le t \le D_2(N)/2,$$

$$s(t) = s_\mu(t) = 1 - D_2(N)/N, \qquad t \ge D_2(N)/2.$$

D. Maximal Cycle Lengths

Lemma 3.4. The lengths of all cycles in a cellular automaton of size N as defined in Sect. 3A divide the length Π_N of the cycle obtained with an initial configuration containing a single site with value one.

This follows from additivity, since any configuration can be considered as a superposition of configurations with single nonzero initial sites.

Lemma 3.5. For the cellular automaton defined in Sect. 3A, with N of the form 2^j, $\Pi_N = 1$.

In this case, any initial configuration evolves ultimately to a fixed point consisting of the null configuration, since

$$(x + x^{-1})^{2^j} 1 \equiv (x^{2^j} + x^{-2^j}) \equiv (x^N + x^{-N}) \equiv 0 \qquad \mathrm{mod}\,(x^N - 1).$$

Lemma 3.6. For the cellular automaton defined in Sect. 3A, with N even but not of the form 2^j, $\Pi_N = 2\Pi_{N/2}$.

A configuration $A(x)$ appears in a cycle of length π if and only if

$$T(x)^\pi A(x) \equiv A(x) \qquad \mathrm{mod}\,(x^N - 1),$$

and therefore

$$(x^N - 1) \,|\, [(x^2 + 1)^\pi + x^\pi]A(x).$$

After t time steps, the configuration obtained by evolution from an initial state containing a single nonzero site is $(x + x^{-1})^t$; by Theorems 3.3 and 3.4 and the additive superposition principle, the configuration

$$A(x) \equiv (x + x^{-1})^{D_2(N)/2}$$

is therefore on the maximal length cycle. Thus the maximal period Π_N is given by the minimum π for which

$$(x^N - 1) \,|\, [(x^2 + 1)^\pi + x^\pi](x + 1)^{D_2(N)},$$

and so

$$\left(\frac{x^n - 1}{x + 1}\right)^{D_2(N)} \,|\, [(x^2 + 1)^{\Pi_N} + x^{\Pi_N}], \qquad (3.13)$$

with $N = D_2(N)n$, n odd. Similarly,

$$(x^{N/2} - 1) \,|\, [(x^2 + 1)^{\Pi_{N/2}} + x^{\Pi_{N/2}}](x + 1)^{D_2(N/2)},$$
$$\left(\frac{x^n - 1}{x + 1}\right)^{D_2(N)/2} \,|\, [(x^2 + 1)^{\Pi_{N/2}} + x^{\Pi_{N/2}}]. \qquad (3.14)$$

Squaring this yields

$$\left(\frac{x^n - 1}{x + 1}\right)^{D_2(N)} \,|\, [(x^2 + 1)^{2\Pi_{N/2}} + x^{2\Pi_{N/2}}],$$

from which it follows that

$$\Pi_N \,|\, 2\Pi_{N/2}. \qquad (3.15)$$

Since $x^N - 1$ divides $[(x^2 + 1)^{\Pi_N} + x^{\Pi_N}](x + 1)^{D_2(N)}$, so does its square root, $x^{N/2} - 1$, and therefore

$$\Pi_{N/2} \,|\, \Pi_N. \qquad (3.16)$$

Combining Eqs. (3.15) and (3.16) implies that either $\Pi_N = 2\Pi_{N/2}$ or $\Pi_N = \Pi_{N/2}$. To exclude the latter possibility, we use derivatives. Using Eq. (A.6), and the fact that the derivative of $x^2 + 1$ vanishes over $GF(2)$, one obtains from (3.13),

$$\left(\frac{x^n - 1}{x + 1}\right) \,|\, \Pi_N x^{\Pi_N - 1}.$$

If Π_N were odd, the right member would be non-trivial, and the divisibility condition could not hold. Thus Π_N must be even. But then the right member of (3.13) is a perfect square, so that

$$\left(\frac{x^{N/2} - 1}{(x + 1)^{D_2(N)/2}}\right)^2 \,|\, [(x^2 + 1)^{\Pi_N/2} + x^{\Pi_N/2}]^2.$$

Thus $\Pi_{N/2} \,|\, \Pi_N/2$, and the proof is complete.

84

Theorem 3.5. For the cellular automaton defined in Sect. 3A, with N odd, $\Pi_N \mid \Pi_N^* = 2^{\mathrm{sord}_N(2)} - 1$ where $\mathrm{sord}_N(2)$ is the multiplicative "suborder" function of 2 modulo N, defined as the least integer j such that $2^j = \pm 1 \bmod N$. (Properties of the suborder functions are discussed in Appendix B.)

By Lemma 3.1, an initial configuration containing a single nonzero site cannot be reached in cellular automaton evolution. The configuration $(x + x^{-1}) \bmod (x^N - 1)$ obtained from this after one time step can be reached, and in fact appears again after $2^{\mathrm{sord}_N(2)} - 1$ time steps, since

$$\mathbb{T}(x)^{2^{\mathrm{sord}_N(2)}} 1 \equiv (x + x^{-1})^{2^{\mathrm{sord}_N(2)}} \equiv (x^{2^{\mathrm{sord}_N(2)}} + x^{-2^{\mathrm{sord}_N(2)}})$$
$$\equiv (x^{\pm 1} + x^{\mp 1}) \equiv (x + x^{-1}) \qquad \bmod (x^N - 1).$$

The maximal cycle lengths Π_N for the cellular automaton considered in this section are given in the first column of Table 1. The values are plotted as a function of N in Fig. 4. Table 1 together with Table 4 show that $\Pi_N = \Pi_N^*$ for almost all odd N. The first exception appears for $N = 37$, where $\Pi_N = \Pi_N^*/3$; subsequent exceptions are $\Pi_{95} = \Pi_{95}^*/3$, $\Pi_{101} = \Pi_{101}^*/3$, $\Pi_{141} = \Pi_{141}^*/3$, $\Pi_{197} = \Pi_{197}^*/3$, $\Pi_{199} = \Pi_{199}^*/7$, $\Pi_{203} = \Pi_{203}^*/105$ and so on.

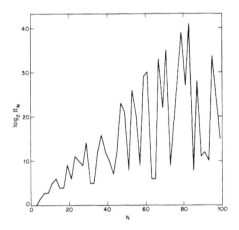

Figure 4. The maximal length Π_N of cycles generated in the evolution of a cellular automaton with size N and $\mathbb{T}(x) = x + x^{-1}$, as a function of N. Only values for integer N are plotted. The irregular behaviour of Π_N as a function of N is a consequence of the dependence of Π_N on number theoretical properties of N.

As discussed in Appendix B, $\mathrm{sord}_N(2) \le (N - 1)/2$. This bound can be attained only when N is prime. It implies that the maximal period is $2^{(N-1)/2} - 1$. Notice that this period is the maximum that could be attained with any reflection symmetric initial configuration (such as the single nonzero site configuration to be considered by virtue of Lemma 3.4).

E. Cycle Length Distribution

Lemma 3.4 established that all cycle lengths must divide Π_N and Theorems 3.3 and 3.4 gave the total number of states in cycles. This section considers the number of distinct cycles and their lengths.

N	k = 2		k = 3			k = 4			
3	1	1	6	1	3	2	2	1	1
4	1	2	2	2	2	1	4	1	4
5	3	3	8	8	4	6	6	3	6
6	2	1	6	6	3	2	2	2	2
7	7	7	26	26	13	14	14	7	14
8	1	4	4	8	8	1	8	1	8
9	7	7	18	1	9	14	14	7	14
10	6	6	8	8	8	6	12	6	12
11	31	31	242	121	121	62	62	31	62
12	4	2	6	6	6	4	4	4	4
13	63	21	26	13	13	126	42	63	42
14	14	14	26	26	13	14	28	14	28
15	15	15	24	24	12	30	30	15	30
16	1	8	16	80	80	1	16	1	16
17	15	15	1,640	6,560	820	30	30	15	30
18	14	14	18	18	9	14	28	14	28
19	511	511	19,682	19,682	9,841	1,022	1,022	511	1,022
20	12	12	16	40	40	12	24	12	24
21	63	63	78	78	39	126	126	63	126
22	62	62	242	242	242	62	124	62	124
23	2,047	2,047	177,146	88,573	88,573	4,094	4,094	2,047	4,094
24	8	4	12	24	24	8	8	8	8
25	1,023	1,023	59,048	59,048	29,524	2,046	2,046	1,023	2,046
26	126	42	26	26	26	126	84	126	84
27	511	511	54	1	27	1,022	1,022	511	1,022
28	28	28	26	26	26	28	56	28	56
29	16,383	16,383	4,782,968	4,782,968	2,391,484	32,766	32,766	16,383	32,766
30	30	30	24	24	24	30	60	30	60
31	31	31	1,103,762	14,348,906	551,881	62	62	31	62
32	1	16	160	6,560	6,560	1	32	1	32
33	31	31	726	363	363	62	62	31	62
34	30	30	1,640	6,560	6,560	30	60	30	60
35	4,095	4,095	265,720	265,720	132,860	8,190	8,190	4,095	8,190
36	28	28	18	18	18	28	56	28	56
37	87,381	29,127	19,682	19,682	9,841	174,762	58,254	87,381	58,254
38	1,022	1,022	19,682	19,682	9,841	1,022	2,044	1,022	2,044
39	4,095	4,095	78	39	39	8,190	8,190	4,095	8,190
40	24	24	80	40	40	24	48	24	48

Table 1. Maximal cycle lengths Π_N for one-dimensional nearest-neighbour additive cellular automata with size N and k possible values at each site. Results for all possible nontrivial symmetrical rules with $k \leq 4$ are given. For $k = 2$, the fixed time evolution polynomials are $T(x) = x + x^{-1}$ and $x + 1 + x^{-1}$ (corresponding to rules 90 and 150 of [1], respectively). For $k = 3$, the polynomials are $x + x^{-1}$, $x + 1 + x^{-1}$, and $x + 2 + x^{-1}$, while for $k = 4$, they are $x + x^{-1}$, $x + 1 + x^{-1}$, $x + 2 + x^{-1}$, and $x + 3 + x^{-1}$.

N		
3	4×1	4
4	1×1	1
5	$1 \times 1; 5 \times 3$	6
6	$4 \times 1; 6 \times 2$	10
7	$1 \times 1; 9 \times 7$	10
8	1×1	1
9	$4 \times 1; 36 \times 7$	40
10	$1 \times 1; 5 \times 3; 40 \times 6$	46
11	$1 \times 1; 33 \times 31$	34
12	$4 \times 1; 6 \times 2; 60 \times 4$	70
13	$1 \times 1; 65 \times 63$	66
14	$1 \times 1; 9 \times 7; 288 \times 14$	298
15	$4 \times 1; 20 \times 3; 1,088 \times 15$	1,112
16	1×1	1
17	$1 \times 1; 51 \times 5; 4,352 \times 15$	4,404
18	$4 \times 1; 6 \times 2; 36 \times 7; 4,662 \times 14$	4,708
19	$1 \times 1; 513 \times 511$	514
20	$1 \times 1; 5 \times 3; 40 \times 6; 5,440 \times 12$	5,486
21	$4 \times 1; 36 \times 7; 16,640 \times 63$	16,680
22	$1 \times 1; 33 \times 31; 16,896 \times 62$	16,930
23	$1 \times 1; 2,049 \times 2,047$	2,050
24	$4 \times 1; 6 \times 2; 60 \times 4; 8,160 \times 8$	8,230
25	$1 \times 1; 5 \times 3; 16,400 \times 1,023$	16,406
26	$1 \times 1; 65 \times 63; 133,120 \times 126$	133,186
27	$4 \times 1; 36 \times 7; 131,328 \times 511$	131,368
28	$1 \times 1; 9 \times 7; 288 \times 14; 599,040 \times 28$	599,338
29	$1 \times 1; 16,385 \times 16,383$	16,386
30	$4 \times 1; 6 \times 2; 20 \times 3; 670 \times 6; 1,088 \times 15; 8,947,168 \times 30$	8,948,956
31	$1 \times 1; 34,636,833 \times 31$	34,636,834
32	1×1	1
33	$4 \times 1; 138,547,332 \times 31$	138,547,336
34	$1 \times 1; 51 \times 5; 6,528 \times 10; 4,352 \times 15; 143,161,216 \times 30$	143,172,148
35	$1 \times 1; 5 \times 3; 9 \times 7; 45 \times 21; 4,195,328 \times 4,095$	4,195,388
36	$4 \times 1; 6 \times 2; 60 \times 4; 36 \times 7; 4,662 \times 14; 153,389,340 \times 28$	153,394,108
37	$1 \times 1; 786,435 \times 87,381$	786,436
38	$1 \times 1; 513 \times 511; 67,239,936 \times 1,022$	672,340,450
39	$4 \times 1; 260 \times 63; 49,164 \times 1,365; 67,108,860 \times 4,095$	67,158,288
40	$1 \times 1; 5 \times 3; 40 \times 6; 5,440 \times 12; 178,954,240 \times 24$	178,959,726

Table 2. Multiplicities and lengths of cycles in the cellular automaton of Sect. 3A with size N. The notation $g_i \times \pi_i$ indicates the occurrence of g_i distinct cycles each of length π_i. The last column of the table gives the total number of distinct cycles or "attractors" in the system.

Lemma 3.7. For the cellular automaton defined in Sect. 3A, with N a multiple of 3, there are four distinct fixed points (cycles of length one); otherwise, only the null configuration is a fixed point.

For $N = 3n$, the only stationary configurations are $000000\ldots$ (null configuration), $0110110\ldots$, $1011011\ldots$, and $1101101\ldots$.

Table 2 gives the lengths and multiplicities of cycles in the cellular automaton defined in Sect. 3A, for various values of N. One result suggested by the table is that the multiplicity of cycles for a particular N increases with the length of the cycle, so that for large N, an overwhelming fraction of all configurations in cycles are on cycles with the maximal length.

When Π_N is prime, the only possible cycle lengths are Π_N and 1. Then, using Lemma 3.7, the number of cycles of length Π_N is $(2^{(N-1)} - 4)/\Pi_N$ for $N = 3n$, and is $(2^{(N-1)} - 1)/\Pi_N$ otherwise.

When Π_N is not prime, cycles may exist with lengths corresponding to various divisors of Π_N. It has not been possible to express the lengths and multiplicities of cycles in this case in terms of simple functions. We nevertheless give a computationally efficient algorithm for determining them.

Theorems 3.3 and 3.4 show that any configuration $A(x)$ on a cycle may be written in the form

$$A(x) = (1 + x)^{D_2(N)} B(x),$$

where $B(x)$ is some polynomial. The cycle on which $A(x)$ occurs then has a length given by the minimum π for which

$$\mathbb{T}(x)^\pi B(x) \equiv (x + x^{-1})^\pi B(x) \equiv B(x) \quad \mathrm{mod} \left(\frac{x^n - 1}{x + 1}\right)^{D_2(N)}, \tag{3.17}$$

where $N = D_2(N)n$ with n odd, and $(x^n - 1)^{D_2(N)} = x^N - 1$. Using the factorization [given in Eq. (A.8)]

$$x^n - 1 = (x - 1) \prod_{\substack{d\,|\,n \\ d \neq 1}} \prod_{i=1}^{\frac{\phi(d)}{\mathrm{ord}_d(2)}} C_{d,i}(x), \tag{3.18}$$

where the $C_{d,i}(x)$ are the irreducible cyclotomic polynomials over $\mathbb{Z}_2$ of degree $\mathrm{ord}_d(2)$, Eq. (3.17) can be rewritten as

$$(x + x^{-1})^\pi B(x) \equiv B(x) \quad \mathrm{mod}\, C_{d,i}(x)^{D_2(N)} \tag{3.19}$$

for all $d\,|\,n$, $d \neq 1$, and for all i such that $1 \leq i \leq \phi(d)/\mathrm{ord}_d(2)$. Let $\pi_{d,i}[B(x)]$ denote the smallest π for which (3.19) holds with given d, i. Then the length of the cycle on which $A(x)$ occurs is exactly the least common multiple of all the $\pi_{d,i}[B(x)]$. If $C_{d,i}(x)^{D_2(N)}\,|\,B(x)$, then clearly Eq. (3.19) holds for $\pi = 1$, and $\pi_{d,i}[B(x)] = 1$. If $C_{d,i}(x)^{r_{d,i}[B(x)]}\|B(x)$ (and $0 \leq r_{d,i}[B(x)] < D_2(N)$), then Eq. (3.19) is equivalent to

$$(x + x^{-1})^\pi \equiv 1 \quad \mathrm{mod}\, C_{d,i}(x)^{D_2(N) - r_{d,i}[B(x)]}. \tag{3.20}$$

The values of $\pi_{d,i}$ for configurations with $r_{d,i}[B(x)] = s$ are therefore equal, and will be denoted $\pi_{d,i,s}$ $(0 \leq s \leq D_2(N))$. Since $C_{d,i}(x) \mid (x^d - 1)/(x + 1)$ $(d \neq 1)$, the value of $\pi_{d,i,1}$ divides the minimum π for which $(x + x^{-1})^\pi \equiv 1 \bmod (x^d - 1)/(x + 1)$. This equation is the same as the one for the maximal cycle length of a size d cellular automaton: the derivation of Theorem 3.5 then shows that

$$\pi_{d,i,1} \mid 2^{\mathrm{sord}_d(2)} - 1. \tag{3.21}$$

It can also be shown that $\pi_{d,i,2s} = \pi_{d,i,s}$ or $\pi_{d,i,2s} = 2\pi_{d,i,s}$.

As an example of the procedure described above, consider the case $N = 30$. Here,

$$x^{30} + 1 = (x^{15} + 1)^2 = C_{1,1}(x)^2 C_{3,1}(x)^2 C_{5,1}(x)^2 C_{15,1}(x)^2 C_{15,2}(x)^2, \tag{3.22}$$

where

$$C_{1,1}(x) = x + 1,$$
$$C_{3,1}(x) = x^2 + x + 1,$$
$$C_{5,1}(x) = x^4 + x^3 + x^2 + x + 1,$$
$$C_{15,1}(x) = x^4 + x + 1,$$
$$C_{15,2}(x) = x^4 + x^3 + 1.$$

Then

$$\pi_{d,i,2} = 1,$$
$$\pi_{3,1,1} = 1, \qquad \pi_{3,1,0} = 2,$$
$$\pi_{5,1,1} = 3, \qquad \pi_{5,1,0} = 6, \tag{3.23}$$
$$\pi_{15,1,1} = \pi_{15,2,1} = 15,$$
$$\pi_{15,1,0} = \pi_{15,2,0} = 30.$$

Thus the cycles which occur in the case $N = 30$ have lengths 1, 2, 3, 6, 15, and 30.

To determine the number of distinct cycles of a given length, one must find the number of polynomials $B(x)$ with each possible set of values $r_{d,i}[B(x)]$. This number is given by

$$\prod_{\substack{d \mid n \\ d \neq 1}} \prod_i V(r_{d,i}, d, D_2(N)),$$

where $V(D_2(N), d, D_2(N)) = 1$ and

$$V(r, d, D_2(N)) = 2^{\mathrm{ord}_d(2)(D_2(N)-r)} - 2^{\mathrm{ord}_d(2)(D_2(N)-r-1)}$$

for $0 \leq r < D_2(N)$. The cycle lengths of these polynomials are determined as above by the least common multiple of the $\pi_{d,i,r_{d,i}}$.

In the example $N = 30$ discussed above, one finds that configurations on cycles of length 3 have $(r_{3,1}, r_{5,1}, r_{15,1}, r_{15,2}) = (1, 1, 2, 2)$ or $(2, 1, 2, 2)$, implying that 60 such configurations exist, in 20 distinct cycles.

4. Generalizations

A. Enumeration of Additive Cellular Automata

We consider first one-dimensional additive cellular automata, whose configurations may be represented by univariate characteristic polynomials. We assume that the time evolution of each site depends only on its own value and the value of its two nearest neighbours, so that the time evolution dipolynomial $T(x)$ is at most of degree two. Cyclic boundary conditions on N sites are implemented by reducing the characteristic polynomial at each time step modulo $x^N - 1$ as in Eq. (2.2). There are taken to be k possible values for each site. With no further constraints imposed, there are k^3 possible $T(x)$, and thus k^3 distinct cellular automaton rules. If the coefficients of x and x^{-1} in $T(x)$ both vanish, then the characteristic polynomial is at most multiplied by an overall factor at each time step, and the behaviour of the cellular automaton is trivial. Requiring nonzero coefficients for x and x^{-1} in $T(x)$ reduces the number of possible rules to $k^3 - 2k^2 + k$. If the cellular automaton evolution is assumed reflection symmetric, then $T(x) = T(x^{-1})$, and only $k^2 - k$ rules are possible. Further characterisation of possible rules depends on the nature of k.

(a) k Prime. In this case, integer values $0, 1, \ldots, k-1$ at each site may be combined by addition and multiplication modulo k to form a field (in which each nonzero element has a unique multiplicative inverse) $\mathbb{Z}_k$. For a symmetrical rule, $T(x)$ may always be written in the form

$$T(x) = x + s + x^{-1} \tag{4.1}$$

up to an overall multiplicative factor. For $k = 2$, the rule $T(x) = x + x^{-1}$ was considered above; the additional rule $T(x) = x + 1 + x^{-1}$ is also possible (and corresponds to rule 150 of [1]).

(b) k Composite.

Lemma 4.1. For $k = p_1^{\alpha_1} p_1^{\alpha_2} \ldots$, with p_i prime, the value $a^{[k]}$ of a site obtained by evolution of an additive cellular automaton from some initial configuration is given uniquely in terms of the values $a^{[p^\alpha]}$ attained by that site in the evolution of the set of cellular automata obtained by reducing $T(x)$ and all site values modulo $p_i^{\alpha_i}$.

This result follows from the Chinese remainder theorem for integers (e.g. [8, Chap. 8]), which states that if k_1 and k_2 are relatively prime, then the values n_1 and n_2 determine a unique value of n modulo $k_1 k_2$ such that $n \equiv n_i \bmod k_i$ for $i = 1, 2$.

Lemma 4.1 shows that results for any composite k may be obtained from those for k a prime or a prime power.

When k is composite, the ring $\mathbb{Z}_k$ of integers modulo k no longer forms a field, so that not all commutative rings $\mathbb{R}_k$ are fields. Nevertheless, for k a prime power, there exists a Galois field $GF(k)$ of order k, unique up to isomorphism (e.g. [9, Chap. 4]). For example, the field $GF(4)$ may be taken to act on elements $0, 1, \kappa, \kappa^2$ with multiplication taken modulo the irreducible polynomial $\kappa^2 + \kappa + 1$. Time evolution for a cellular automaton with site values in this Galois field can be reduced to that given by $x + \sigma + x^{-1}$, where σ is any element of the field. The behaviour of this subset of cellular automata with k composite is directly analogous to those over $\mathbb{Z}_p$ for prime p.

It has been assumed above that the value of a site at a particular time step is determined solely by the values of its nearest neighbours on the previous time step. One generalization allows dependence on sites out to a distance $r > 1$, so that the evolution of the cellular automaton corresponds to multiplication by a fixed dipolynomial $\mathbb{T}(x)$ of degree $2r$. Most of the theorems to be derived below hold for any r.

B. Cellular Automata over $\mathbb{Z}_p$ (p Prime)

Lemma 4.2. The lengths of all cycles in any additive cellular automaton over $\mathbb{Z}_p$ of size N divide the length Π_N of the cycle obtained for an initial configuration containing a single site with value 1.

This lemma is a straightforward generalization of Lemma 3.4, and follows directly from the additivity assumed for the cellular automaton rules.

Lemma 4.3. For N a multiple of p, $\Pi_N \mid p\Pi_{N/p}$ for an additive cellular automaton over $\mathbb{Z}_p$.

Remark. For N a multiple of p, but not a power of p, it can be shown that $\Pi_N = p\Pi_{N/p}$ for an additive cellular automaton over $\mathbb{Z}_p$ with $\mathbb{T}(x) = x + x^{-1}$. In addition, $\Pi_{p^j} = 1$ in this case.

Theorem 4.1. For any N not a multiple of p, $\Pi_N \mid \Pi_N^* = p^{\mathrm{ord}_N(p)} - 1$, and $\Pi_N \mid \Pi_N^* = p^{\mathrm{sord}_N(p)} - 1$ if $\mathbb{T}(x)$ is symmetric, for any additive cellular automaton over $\mathbb{Z}_p$.

The period Π_N divides Π_N^* if

$$[\mathbb{T}(x)]^{\Pi_N^*+1} \equiv \mathbb{T}(x) \quad \mathrm{mod}\,(x^N - 1). \tag{4.2}$$

Taking

$$\mathbb{T}(x) = \sum_i \alpha_i x^{\gamma_i},$$

Eq. (A.3) yields

$$[\mathbb{T}(x)]^{p^{\mathrm{ord}_N(p)}} \equiv \sum_i \alpha_i x^{\gamma_i p^{\mathrm{ord}_N(p)}} \equiv \sum_i \alpha_i x^{\gamma_i} = \mathbb{T}(x) \quad \mathrm{mod}\,(x^N - 1),$$

since $\alpha^{p^\lambda} \equiv \alpha \bmod p$ and $p^{\mathrm{ord}_N(p)} \equiv 1 \bmod N$, and the first part of the theorem follows. Since $x^{p^{\mathrm{sord}_N(p)}} \equiv x^{\pm 1} \bmod p$, Eq. (4.2) holds for

$$\Pi_N^* = p^{\mathrm{sord}_N(p)} - 1$$

if $\mathbb{T}(x)$ is symmetric, so that $\mathbb{T}(x) = \mathbb{T}(x^{-1})$.

91

This result generalizes Theorem 3.5 for the particular $k = 2$ cellular automaton considered in Sect. 3.

Table 1 gives the values of Π_N for all non-trivial additive symmetrical cellular automata over $\mathbb{Z}_2$ and $\mathbb{Z}_3$. Just as in the example of Sect. 3 (given as the first column of Table 1), one finds that for many values of N not divisible by p

$$\Pi_N = p^{\text{sord}_N(p)} - 1. \tag{4.3}$$

When $p = 2$, all exceptions to (4.3) when $\mathbb{T}(x) = x + x^{-1}$ are also exceptions for $\mathbb{T}(x) = x+1+x^{-1}$ [19]. We outline a proof for the simplest case, when N is relatively prime to 6 (as well as 2). Let $\Pi_N(x + x^{-1})$ be the maximal period obtained with $\mathbb{T}(x) = x + x^{-1}$, equal to the minimum integer π for which

$$(x + 1)^{2\pi} \equiv x^\pi \quad \mod \left(\frac{x^N - 1}{x + 1} \right). \tag{4.4}$$

We now show that $\Pi_N(x + x^{-1})$ is a multiple of the maximum period $\Pi_N(x + 1 + x^{-1})$ obtained with $\mathbb{T}(x) = x + 1 + x^{-1}$. Since the mapping $x \to x^3$ is a homomorphism in the field of polynomials with coefficients in $GF(2)$, one has

$$(x^3 + 1)^{2\pi} \equiv x^{3\pi} \quad \mod \left(\frac{x^N - 1}{x + 1} \right)$$

for any π such that $\Pi_N(x + x^{-1}) \mid \pi$. Dividing by Eq. (4.4), and using the fact that N is odd to take square roots, yields

$$\left(\frac{x^3 + 1}{x + 1} \right)^\pi \equiv x^\pi \quad \mod \left(\frac{x^N - 1}{x + 1} \right) \tag{4.5}$$

for any π such that $\Pi_N(x + x^{-1}) \mid \pi$. But since $x + 1 + x^{-1} = x^{-1} \left(\frac{x^3+1}{x+1} \right)$, Eq. (4.5) is the analogue of Eq. (4.4) for $\mathbb{T}(x) = x + 1 + x^{-1}$, and the result follows.

More exceptions to Eq. (4.3) are found with $p = 3$ than with $p = 2$.

Lemma 4.4. A configuration $A(x)$ is reachable in the evolution of a size N additive cellular automaton over $\mathbb{Z}_p$, as described by $\mathbb{T}(x)$ if and only if $A(x)$ is divisible by $\Lambda_1(x) = (x^N - 1, \mathbb{T}(x))$.

Appendix A.A gives conventions for the greatest common divisor $(A(x), B(x))$. If $A^{(1)}(x)$ can be reached, then

$$A^{(1)}(x) = \mathbb{T}(x)A^{(0)}(x) \quad \mod (x^N - 1)$$

for some $A^{(0)}(x)$, so that

$$(x^N - 1) \mid A^{(1)}(x) - \mathbb{T}(x)A^{(0)}(x).$$

But $\Lambda_1(x) \mid x^N - 1$ and $\Lambda_1(x) \mid \mathbb{T}(x)$, and hence if $A^{(1)}(x)$ is reachable,

$$\Lambda_1(x) \mid A^{(1)}(x). \tag{4.6}$$

We now show by an explicit construction that all $A^{(1)}(x)$ satisfying (4.6) in fact have predecessors $A^{(0)}(x)$. Using Eq. (A.10), one may write

$$\Lambda_1(x) = r(x)\mathbb{T}(x) + \xi(x)(x^N \to 1)$$

for some dipolynomials $r(x)$ and $\xi(x)$, so that

$$\Lambda_1(x) \equiv r(x)\mathbb{T}(x) \quad \mathrm{mod}\,(x^N - 1).$$

Then taking $A^{(1)}(x) = \Lambda_1(x)B(x)$, the configuration given by the polynomial obtained by reducing the dipolynomial $r(x)B(x)$ satisfies

$$\mathbb{T}(x)r(x)B(x) \equiv \Lambda_1(x)B(x) \equiv A^{(1)}(x) \quad \mathrm{mod}\,(x^N - 1)$$

and thus provides an explicit predecessor for $A^{(1)}(x)$.

Corollary. $A(x)$ is reachable in j steps if and only if $\Lambda_j(x) = (x^N - 1,\ \mathbb{T}^j(x))$ divides $A(x)$.

This is a straightforward extension of the above lemma.

Theorem 4.2. The fraction of possible configurations which may be reached by evolution of an additive cellular automaton over $\mathbb{Z}_p$ of size N is $p^{-\deg\Lambda_1(x)}$, where $\Lambda_1(x) = (x^N - 1,\ \mathbb{T}(x))$.

By Lemma 4.4, only configurations divisible by $\Lambda_1(x)$ may be reached. The number of such configurations is $p^{N-\deg\Lambda_1(x)}$, while the total number of possible configurations is p^N.

Let $D_p(N)$ be the maximum p^j which divides N and let v_i denote the multiplicity of the ith irreducible factor of $\Lambda_1(x)$ in $\mathbb{T}^*(x)$, where $\mathbb{T}^*(x) = x^r\mathbb{T}(x)$ is a polynomial with a nonzero constant term. We further define $\chi = \min_i v_i$, so that $0 \leq \chi \leq D_p(N)$.

Theorem 4.3. The state transition diagram for an additive cellular automaton of size N over $\mathbb{Z}_p$ consists of a set of cycles at all nodes of which are rooted identical $p^{\deg\Lambda_1(x)}$-ary trees. A fraction $p^{-D_p(N)\deg\Lambda_1(x)}$ of the possible configurations appear on cycles. For $\chi > 0$, the height of the trees is $\lceil D_p(N)/\chi \rceil$. The trees are balanced if and only if (a) $v_i \geq D_p(N)$ for all i, or (b) $v_i = v_j$ for all i and j, and $v_i \mid D_p(N)$.

To determine the in-degrees of nodes in the trees, consider a configuration $A(x)$ with predecessors represented by the polynomials $B_1(x)$ and $B_2(x)$, so that

$$A(x) \equiv \mathbb{T}(x)B_i(x) \quad \mathrm{mod}\,(x^N - 1).$$

Then since

$$\mathbb{T}(x)(B_1(x) - B_2(x)) \equiv 0 \quad \mathrm{mod}\,(x^N - 1),$$

and $\Lambda_1(x) \mid x^N - 1$, it follows that

$$B_1(x) - B_2(x) \equiv 0 \quad \mathrm{mod} \left(\frac{x^N - 1}{\Lambda_1(x)} \right).$$

Since $C(x) = (x^N - 1)/\Lambda_1(x)$ has a non-zero constant term, $(B_1(x) - B_2(x))/C(x)$ is an ordinary polynomial. The number of solutions to this congruence and thus the number of predecessors $B_i(x)$ of $A(x)$ is $p^{\deg\Lambda_1(x)}$.

The proof of Lemma 3.3 demonstrates the identity of the trees. The properties of the trees are established by considering the tree rooted on the null configuration. A configuration $A(x)$ evolves to the null configuration after j steps if $\mathbb{T}(x)^j A(x) \equiv 0 \bmod (x^N - 1)$, so that

$$\left. \frac{x^N - 1}{\Lambda_j(x)} \right| A(x). \tag{4.7}$$

Hence all configurations on the tree are divisible by $(x^N - 1)/\Lambda_\infty(x)$, where $\Lambda_\infty(x) = \lim_{j\to\infty} \Lambda_j(x)$. All configurations in the tree evolve to the null configuration after at most $\lceil D_p(N)/\chi \rceil$ steps, which is thus an upper bound on the height of the trees. But since the configuration $(x^N - 1)/\Lambda_\infty(x)$ evolves to the null configuration after exactly $\lceil D_p(N)/\chi \rceil$ steps, this quantity gives the height of the trees. The tree of configurations which evolve to the null configuration (and hence all other trees in the state transition diagram) is balanced if and only if all unreachable (terminal) configurations evolve to the null configuration after the same number of steps. First suppose that neither condition (a) nor (b) is true. One possibility is that some irreducible factor $\sigma(x)$ of $\Lambda_1(x)$ satisfies $\sigma^v(x) \| \Lambda_1(x)$ with $v < D_p(N)$ but v does not divide $D_p(N)$. The configuration $(x^N - 1)/\sigma^{D_p(N)}(x)$ reaches 0 in $\lceil D_p(N)/v \rceil$ steps whereas $(x^N - 1)/\sigma^{D_p(N)+1-v}(x)$ reaches 0 in one step fewer, yet both are unreachable, so that the tree cannot be balanced. The only other possibility is that there exist two irreducible factors $\sigma_1(x)$ and $\sigma_2(x)$ of multiplicities v_1 and v_2, respectively, with v_1 and v_2 dividing $D_p(N)$ but $v_1 \neq v_2$. Then $(x^N - 1)/\sigma_1^{D_p(N)}(x)$ reaches 0 in $D_p(N)/v_1$ steps, whereas $(x^N - 1)/\sigma_2^{D_p(N)}(x)$ reaches 0 in $D_p(N)/v_2$ steps. Neither of these configurations is reachable, so again the trees cannot be balanced. This establishes that in all cases either condition (a) or (b) must hold. The sufficiency of condition (a) is evident. If the condition (b) is true, then

$$\Lambda_1(x) = \left[\prod \sigma(x) \right]^v, \qquad \Lambda_\infty(x) = \left[\prod \sigma(x) \right]^{D_p(N)},$$

and $\Lambda_j(x) = \Lambda_1^j(x)$. Equation (4.7) shows that any configuration $A(x)$ which evolves to the null configuration after j steps is of the form

$$A(x) = \frac{x^N - 1}{\Lambda_1^j(x)} R(x),$$

where $R(x)$ is some polynomial. The proof is completed by showing that all such configurations $A(x)$ with $j < D_p(N)/v$ are indeed reachable. To construct an explicit predecessor for $A(x)$, define the dipolynomial $S(x)$ by $\mathbb{T}(x) = \Lambda_1(x)S(x)$, so that $(S(x),\ x^N - 1) = 1$. Then there exist dipolynomials $r(x)$ and $\xi(x)$ such that

$$r(x)S(x) + \xi(x)(x^N - 1) = 1.$$

The configuration given by the dipolynomial

$$B(x) = \frac{x^N - 1}{\Lambda_1^{j+1}(x)} r(x)R(x)$$

then provides a predecessor for $A(x)$.

Notice that whenever the balance condition fails, the set and measure entropies of Eqs. (3.11) and (3.12) obtained by evolution from an initial maximal entropy ensemble become unequal.

The results of Theorems 4.2 and 4.3 show that if $\deg\Lambda_1(x) = 0$, then the evolution of an additive cellular automaton is effectively reversible, since every configuration has a unique predecessor.

In general,

$$\deg\Lambda(x) \leq \deg\mathbb{T}^*(x),$$

so that for the one-dimensional additive cellular automata considered so far, the maximum decrease in entropy starting from an initial equiprobable ensemble is $D_p(N)$.

Note that for a cellular automaton over $\mathbb{Z}_p$ $(p > 2)$ of length N with $\mathbb{T}(x) = x + x^{-1}$, $\deg\Lambda(x) = 2$ if $4 \mid N$ and $\deg\Lambda(x) = 0$ otherwise. Such cellular automata are thus effectively reversible for $p > 2$ whenever N is not a multiple of 4.

Remark. A configuration $A(x)$ lies on a cycle in the state transition diagram of an additive cellular automaton if and only if $\Lambda_\infty(x) \mid A(x)$.

This may be shown by the methods used in the proof of Theorem 4.3.

C. Cellular Automata over $\mathbb{Z}_k$ (k Composite)

Theorem 4.4. For an additive cellular automaton over $\mathbb{Z}_k$,

$$\Pi_N(\mathbb{Z}_k;\ \mathbb{T}_k(x)) = \mathrm{lcm}(\Pi_N(\mathbb{Z}_{p_1^{\alpha_1}};\ \mathbb{T}_{p_1^{\alpha_1}}(x)),\ \Pi_N(\mathbb{Z}_{p_2^{\alpha_2}};\ \mathbb{T}_{p_2^{\alpha_2}}(x)),\ \ldots),$$

where $k = p_1^{\alpha_1} p_2^{\alpha_2} \ldots$, and in $\mathbb{T}_j(x)$ all coefficients are reduced modulo j.

This result follows immediately from Lemma 4.1.

Theorem 4.5. $\Pi_N(\mathbb{Z}_{p^{\alpha+1}};\ \mathbb{T}_{p^{\alpha+1}}(x))$ is equal to either (a) $p\Pi_N(\mathbb{Z}_{p^\alpha};\ \mathbb{T}_{p^\alpha}(x))$ or (b) $\Pi_N(\mathbb{Z}_{p^\alpha};\ \mathbb{T}_{p^\alpha}(x))$ for an additive cellular automaton.

First, it is clear that

$$\Pi_N(\mathbb{Z}_{p^\alpha};\ \mathbb{T}_{p^\alpha}(x) \mid \Pi_N(\mathbb{Z}_{p^{\alpha+1}};\ \mathbb{T}_{p^{\alpha+1}}(x)).$$

To complete the proof, one must show that in addition

$$\Pi_N(\mathbb{Z}_{p^{\alpha+1}};\, \mathbb{T}_{p^{\alpha+1}}(x)) \mid p\Pi_N(\mathbb{Z}_{p^\alpha};\, \mathbb{T}_{p^\alpha}(x)).$$

$\Pi_N(\mathbb{Z}_{p^\alpha};\, \mathbb{T}_{p^\alpha}(x))$ is the smallest positive integer π for which a positive integer m and dipolynomials $U(x)$ and $V(x)$ satisfying

$$\mathbb{T}(x)^{m+\pi} = \mathbb{T}(x)^m + (x^N - 1)U(x) + p^\alpha V(x) \tag{4.8}$$

exist, where all dipolynomial coefficients (including those in $\mathbb{T}(x)$) are taken as ordinary integers in $\mathbb{Z}$, and irrelevant powers of x on both sides of the equation have been dropped. Raising both sides of Eq. (4.8) to the power p, one obtains

$$\begin{aligned}
\mathbb{T}(x)^{mp+\pi p} &= (x^N - 1)W(x) + (\mathbb{T}(x)^m + p^\alpha V(x))^p \\
&= (x^N - 1)W(x) + \mathbb{T}(x)^{mp} + p^{\alpha+1}Q(x).
\end{aligned}$$

Reducing modulo $p^{\alpha+1}$ yields the required result.

For $p = 2$ and $\alpha = 1$, it can be shown that case (a) of Theorem 4.5 always obtains if $\mathbb{T}(x) = x + x^{-1}$, but case (b) can occur when $\mathbb{T}(x) = x + 1 + x^{-1}$.

Theorem 4.6. With $k = k_1 k_2 \ldots$ (all k_i relatively prime), the number of configurations which can be reached by evolution of an additive cellular automaton over $\mathbb{Z}_k$ is equal to the product of the numbers reached by evolution of cellular automata with the same $\mathbb{T}(x)$ over each of the $\mathbb{Z}_{k_i}$. The state transition diagram for the cellular automaton over $\mathbb{Z}_k$ consists of a set of identical trees rooted on cycles. The in-degrees of non-terminal nodes in the trees are the product of those for each of the $\mathbb{Z}_{k_i}$ cases. The height of the trees is the maximum of the heights of trees for the $\mathbb{Z}_{k_i}$ cases, and the trees are balanced only if all these heights are equal.

These results again follow directly from Lemma 4.1.

Theorem 4.6 gives a characterisation of the state transition diagram for additive cellular automata over $\mathbb{Z}_k$ when k is a product of distinct primes. No general results are available for the case of prime power k. However, for example, with $\mathbb{T}(x) = x + x^{-1}$, one may obtain the fraction of reachable states by direct combinatorial methods. With $k = 2^\alpha$ one finds in this case that the fraction is $1/2$ for N odd, $1/4$ for $N \equiv 2$ mod 4, and $2^{-2\alpha}$ for $4 \mid N$. With $k = p^\alpha$ ($p \neq 2$) the systems are reversible (all configurations reachable) unless $4 \mid N$, in which case a fraction $p^{-2\alpha}$ may be reached.

D. Multidimensional Cellular Automata

The cellular automata considered above consist of a sequence of sites on a line. One generalization takes the sites instead to be arranged on a square lattice in two dimensions. The evolution of a site may depend either on the values of its four orthogonal neighbours (type I neighbourhood) or on the values of all eight neighbours including those diagonally adjacent (type II neighbourhood) (e.g. [1]). Configurations of two-dimensional cellular automata may be represented by bivariate characteristic polynomials $A(x_1, x_2)$. Time evolution for additive cellular automaton rules is obtained by multiplication of these characteristic polynomials by a fixed bivariate dipolynomial $\mathbb{T}(x_1, x_2)$. For a type I neighbourhood, $\mathbb{T}(x_1, x_2)$ contains no $x_1 x_2$ cross-terms; such terms may be present for a type II neighbourhood. Periodic boundary conditions with periods N_1 and N_2 may be implemented by reduction modulo $x_1^{N_1} - 1$ and modulo $x_2^{N_2} - 1$ at each time step. Cellular automata may be generalized to an arbitrary d-dimensional cubic or hypercubic lattice. A type I neighbourhood in d dimensions contains $2d + 1$ sites, while a type II neighbourhood contains 3^d sites. As before, we consider cellular automata with k possible values for each site.

Theorem 4.7. For an additive cellular automaton over $\mathbb{Z}_k$ on a d-dimensional cubic lattice, with a type I or type II neighbourhood, and with periodicities $N_1, N_2, \ldots, N_d$, $\mathrm{lcm}(\Pi_{N_1}(\mathbb{Z}_k; \mathbb{T}(x_1, 1, \ldots, 1)), \ldots, \Pi_{N_d}(\mathbb{Z}_k; \mathbb{T}(1, \ldots, 1, x_d))) \mid \Pi_{N_1, \ldots, N_d}(\mathbb{Z}_k; \mathbb{T}(x_1, \ldots, x_d))$.

The result may be proved by showing that

$$\Pi_{N_i}(\mathbb{Z}_i; \mathbb{T}(1, \ldots, 1, x_i, 1, \ldots, 1)) \mid \Pi_{N_1, \ldots, N_d}(\mathbb{Z}_k; \mathbb{T}(x_1, \ldots, x_d)) \tag{4.9}$$

for all i (such that $1 \le i \le d$). The right member of Eq. (4.9) is given by the smallest integer π for which there exists a positive integer m such that

$$[\mathbb{T}(x_1, \ldots, x_d)]^{\pi+m} = [\mathbb{T}(x_1, \ldots, x_d)]^m + \sum_{j=1}^{d} (x_j^{N_j} - 1) U_j(x_1, \ldots, x_d) \tag{4.10}$$

for some dipolynomials U_j. Taking $x_j = 1$ with $j \ne i$ in Eq. (4.10), all terms in the sum vanish except for the one associated with x_i, and the resulting value of π corresponds to the left member of Eq. (4.9).

Theorem 4.8. For an additive cellular automaton over $\mathbb{Z}_p$ on a d-dimensional cubic lattice (type I or type II neighbourhood) with periodicities $N_1, N_2, \ldots, N_d$ none of which are multiples of p,

$$\Pi_{N_1, \ldots, N_d}(\mathbb{Z}_p; \mathbb{T}(x_1, \ldots, x_d)) \mid \Pi_{N_1, \ldots, N_d}^* = p^{\mathrm{ord}_{N_1, \ldots, N_d}(p)} - 1.$$

If $\mathbb{T}(x_1, \ldots, x_d)$ is symmetrical, so that

$$\mathbb{T}(x_1, \ldots, x_i, \ldots, x_d) = \mathbb{T}(x_1, \ldots, x_i^{-1}, \ldots, x_d)$$

for all i, then

$$\Pi_{N_1, \ldots, N_d}^* = p^{\mathrm{sord}_{N_1, \ldots, N_d}(p)} - 1.$$

The $\mathrm{ord}_{n_1, \ldots, n_d}(p)$ and $\mathrm{sord}_{n_1, \ldots, n_d}(p)$ are multidimensional generalizations of the multiplicative order and suborder functions, described in Appendix B.

This theorem is proved by straightforward extension of the one-dimensional Theorem 4.1.

Using the result (B.13), one finds for symmetrical rules

$$\Pi^*_{N_1,\ldots,N_d} = p^{\text{lcm}(\text{sord}_{N_1}(p),\ldots,\text{sord}_{N_d}(p))} - 1.$$

The maximal cycle length is thus bounded by

$$\Pi_{N_1,\ldots,N_d} \leq p^{\text{lcm}((N_1-1)/2,\ldots,(N_d-1)/2)} - 1 \leq p^{(N_1-1)\ldots(N_d-1)/2^d} - 1,$$

with the upper limits achieved only if all the N_i are prime. (For example,

$$\Pi_{83,59} = 2^{1189} \simeq 10^{358}$$

saturates the upper bound.)

Algebraic determination of the structure of state transition diagrams is more complicated for multi-dimensional cellular automata than for the one dimensional cellular automata considered above[2]. The generalization of Lemma 4.4 states that a configuration $A(x_1,\ldots,x_d)$ is reachable only if $A(z_1,\ldots,z_d)$ vanishes whenever the z_i are simultaneous roots of $\mathbb{T}(x_1,\ldots,x_d)$, $x^{N_1}-1,\ldots,x^{N_d}-1$. The root sets z_i form an algebraic variety over $\mathbb{Z}_k$ (cf. [9]).

E. Higher Order Cellular Automata

The rules for cellular automaton evolution considered above took configurations to be determined solely from their immediate predecessors. One may in general consider higher order cellular automaton rules, which allow dependence on say s preceding configurations. The time evolution for additive one-dimensional higher-order cellular automata (with N sites and periodic boundary conditions) may be represented by the order s recurrence relation

$$A^{(t)}(x) = \sum_{j=1}^{s} \mathbb{T}_j(x) A^{(t-j)}(x) \qquad \text{mod}\,(x^N - 1). \tag{4.11}$$

This may be solved in analogy with order s difference equations to yield

$$A^{(t)}(x) = \sum_{j=1}^{s} c_j(x)[U_j(x)]^t,$$

where the $U_j(x)$ are solutions to the equation

$$[U(x)]^s = \sum_{j=1}^{s} [U(x)]^{s-j} \mathbb{T}_j(x),$$

and the $c_j(x)$ are analogous to "constants of integration" and are determined by the initial configurations $A^{(0)}(x),\ldots,A^{(s-1)}(x)$. The state of an order s cellular

---

[2] In the specific case $\mathbb{T}(x_1,x_2) = x_1 + x_1^{-1} + x_2 + x_2^{-1}$, one finds that the in-degrees I_{N_1,N_2} of trees in the state transition diagrams for a few $N_1 \times N_2$ cellular automata are: $I_{2,2} = 16$, $I_{2,3} = 4$, $I_{2,4} = 16$, $I_{2,5} = 4$, $I_{2,6} = 16$, $I_{3,3} = 32$, $I_{3,4} = 4$, $I_{3,5} = 2$, $I_{4,4} = 256$.

automaton depends on the values of its N sites over a sequence of s time steps; there are thus a total k^{Ns} possible states. The transition diagram for these states can in principle be derived by algebraic methods starting from Eq. (4.11). In practice, however, the $U_j(x)$ are usually not polynomials, but elements of a more general function field, leading to a somewhat involved analysis not performed here.

For first-order additive cellular automata, any configuration may be obtained by superposition of the configuration 1 (or its translates x^j). For higher-order cellular automata, several "basis" configurations must be included. For example, when $s = 2$, $\{0, 1\}$, $\{1, 0\}$, and $\{x^j, 1\}$ are all basis configurations, where in $\{A_1(x), A_2(x)\}$, $A_1(x)$, and $A_2(x)$ represent configurations at successive time steps.

As discussed in Sect. 4B, some first-order cellular automata over $\mathbb{Z}_p$ ($p > 2$) are effectively reversible for particular values of N, so that all states are on cycles. The class of second-order cellular automata with $\mathbb{T}_2(x) = -1$ is reversible for all N and k, and for any $\mathbb{T}_1(x)$ [10]. In the simple case $\mathbb{T}_1(x) = x + x^{-1}$, one finds $U_1(x) = x$, $U_2(x) = x^{-1}$. It then appears that

$$\Pi_N = kN/2 \qquad (k \text{ even, } N \text{ even})$$
$$= kN \qquad (\text{otherwise}).$$

(The proof is straightforward when $k = 2$.) In the case $\mathbb{T}_1(x) = x + 1 + x^{-1}$, the $U_j(x)$ are no longer polynomials. For the case $k = 2$, the results for Π_N with N between 3 and 30 are: 6, 6, 15, 12, 9, 12, 42, 30, 93, 24, 63, 18, 510, 24, 255, 84, 513, 60, 1170, 186, 6141, 48, 3075, 126, 3066, 36, 9831, 1020.

F. Other Boundary Conditions

The cellular automata discussed above were taken to consist of N indistinguishable sites with periodic boundary conditions, as if arranged around a circle. This section considers briefly cellular automata with other boundary conditions. The discussion is restricted to the case of symmetric time evolution rules $\mathbb{T}(x) = \mathbb{T}(x^{-1})$.

The periodic boundary conditions considered above are not the only possible choice which preserve the translation invariance of cellular automata (or the indistinguishability of their sites)[3]. One-dimensional cellular automata may in general be viewed as $\mathbb{R}_k$ bundles over $\mathbb{Z}_N$. Periodic boundary conditions correspond to trivial bundles. Non-trivial bundles are associated with "twisted" boundary conditions. Explicit realizations of such boundary conditions require a twist to be introduced at a particular site. The evolution of particular configurations then depends on the position of the twist, but the structure of the state transition diagram does not.

A twist of value R at position $i = \sigma$ causes sites with $i \geq \sigma$ to appear multiplied by R in the time evolution of sites with $i < \sigma$, and correspondingly, for sites with $i < \sigma$ to appear multiplied by R^{-1} in the evolution of sites with $i \geq \sigma$. In the

---

[3] We are grateful to L. Yaffe for emphasizing this point.

presence of a twist taken at position $\sigma = 0$, the time evolution formula (2.5) becomes

$$A^{(t)}(x) = \mathbb{T}(x)A^{(t-1)}(x) \qquad \mathrm{mod}\,(x^N - R). \tag{4.12}$$

Multiple twists are irrelevant; only the product of their values R_j is significant for the structure of the state transition diagram. If $\mathbb{R}_k = \mathbb{Z}_p$ with p prime, then $\mathbb{R}_k$ (with the zero element removed) forms a multiplicative group, and twists with any value R not equal to 0 or 1 yield equivalent results. When $\mathbb{R}_k = \mathbb{Z}_k$ with k composite, several equivalence classes of R values may exist.

Using Eq. (4.12) one may obtain general results for twisted boundary conditions analogous to those derived above for the case of periodic boundary conditions (corresponding to $R = 1$). When $\mathbb{R}_k = \mathbb{Z}_p$ (p prime), one finds for example,

$$\Pi_N^{[R\neq 1]} \mid \Pi_{N(p-1)}^{[R=1]}.$$

An alternative class of boundary conditions introduces fixed values at particular cellular automaton sites. One may consider cellular automata consisting of N sites with values $a_1, \ldots, a_N$ arranged as if along a line, bounded by sites with fixed values a_0 and a_{N+1}. Maximal periods obtained with such boundary conditions will be denoted $\Pi_N^{(a_0, a_{N+1})}$. The case $a_0 = a_{N+1} = 0$ is simplest. In this case, configurations

$$A(x) = \sum_{i=1}^{N} a_i x^i$$

of the length N system with fixed boundary conditions may be embedded in configurations

$$\tilde{A}(x) = \sum_{i=1}^{N} a_i x^i + \sum_{i=1}^{N} (k - a_{N+1-i})x^{N+1+i} \tag{4.13}$$

of a length $\tilde{N} = 2N + 2$ system with periodic boundary conditions. The condition $a_0 = a_{N+1} = 0$ is preserved by time evolution, so that one must have

$$\Pi_N^{(0,0)} \mid \Pi_{2N+2}.$$

The periods are equal if the configurations obtained by evolution from a single nonzero initial site have the symmetry of Eq. (4.13). (The simplest cellular automaton defined in Sect. 3A satisfies this condition.)

Fixed boundary conditions $a_0 = r$, $a_{N+1} = 0$, may be treated by constructing configurations $\tilde{A}(x)$ of the form (4.13), with periodic boundary conditions, but now with time evolution

$$\tilde{A}^{(t)}(x) \equiv [\mathbb{T}(x)\tilde{A}^{(t-1)}(x) + r(1 - \alpha_0)] \qquad \mathrm{mod}\,(x^{\tilde{N}} - 1),$$

where $\mathbb{T}(x)$ is taken of the form $x + \alpha_0 + x^{-1}$. Iteration generates a geometric series in $\mathbb{T}(x)$, which may be summed to yield a rational function of x. For $k = 2$, $r = 1$, one may then show that with $\mathbb{T}(x) = x + 1 + x^{-1}$, $\Pi_N^{(0,1)} = \Pi_{2N+2}$, while with $\mathbb{T}(x) = x + x^{-1}$ (the case of Sect. 3A), $\Pi_N^{(0,1)} \mid \Pi_{2(2N+2)}$.

5. Non-Additive Cellular Automata

Equation (2.3) defines the time evolution for a special class of "additive" cellular automata, in which the value of a site is given by a linear combination (in $\mathbb{R}_k$) of the values of its neighbours on the previous time step. In this section we discuss "non-additive" cellular automata, which evolve according to

$$a_i^{(t)} = \mathbb{F}[a_{i-1}^{(t-1)}, a_i^{(t-1)}, a_{i+1}^{(t-1)}], \tag{5.1}$$

where $\mathbb{F}[a_{-1}, a_0, a_{+1}]$ is an arbitrary function over $\mathbb{R}_k$, not reducible to linear form. The absence of additivity in general prevents use of the algebraic techniques developed for additive cellular automata in Sects. 3 and 4. The difficulties in the analysis of non-additive cellular automata are analogous to those encountered in the analysis of non-linear feedback shift registers (cf. [11]). In fact, the possibility of universal computation with sufficiently complex non-additive cellular automata demonstrates that a complete analysis of these systems is fundamentally impossible. Some results are nevertheless available (cf. [12]). This section illustrates some methods which may be applied to the analysis of non-additive cellular automata, and some of the results which may be obtained.

As in [1], most of the discussion in this section will be for the case $k = 2$. In this case, there are 32 possible functions $\mathbb{F}$ satisfying the symmetry condition

$$\mathbb{F}[a_{-1}, a_0, a_{+1}] = \mathbb{F}[a_{+1}, a_0, a_{-1}]$$

and the quiescence condition

$$\mathbb{F}[0, 0, 0] = 0.$$

Reference [1] showed the existence of two classes of these "legal" cellular automata. The "simple" class evolved to fixed points or short cycles after a small number of time steps. The "complex" class (which included the additive rules discussed above) exhibited more complicated behaviour.

We consider as an example the complex non-additive $k = 2$ rule defined by

$$\mathbb{F}[1, 0, 0] = \mathbb{F}[0, 0, 1] = 1,$$
$$\mathbb{F}[a_{-1}, a_0, a_{+1}] = 0 \qquad \text{otherwise,} \tag{5.2}$$

and referred to as rule 18 in [1]. This function yields a time evolution rule equivalent to

$$a_i^{(t)} \equiv (1 + a_i^{(t-1)})(a_{i-1}^{(t-1)} + a_{i+1}^{(t-1)}) \qquad \text{mod } 2. \tag{5.3}$$

The rule does not in general satisfy any superposition principle. However, for the special class of configurations with $a_{2j} = 0$ or $a_{2j+1} = 0$, Eq. (5.3) implies that

101

the evolution of even (odd) sites on even (odd) time steps is given simply by the rule defined in Sect. 3A. Any configuration may be considered as a sequence of "domains" in which all even (or odd) sites have value zero, separated by "domain walls" or "kinks" [13]. In the course of time the kinks annihilate in pairs. If sites are nonzero only in some finite region, then at sufficiently large times in an infinite cellular automaton, all kinks (except perhaps one) will have annihilated, and an effectively additive system will result. However, out of all 2^N possible initial configurations for a cellular automaton with N sites and periodic boundary conditions, only a small fraction are found to evolve to this form before a cycle is reached: in most cases, "kinks" are frozen into cycles, and contribute to global behaviour in an essential fashion.

Typical examples of the state transition diagrams obtained with the rule (5.3) are shown in Fig. 5. They are seen to be much less regular than those for additive rules illustrated in Fig. 2. In particular, not all transient trees are identical, and few of the trees are balanced. Just as for the additive rules discussed in Sects. 3 and 4, only a fraction of the 2^N possible configurations may be reached by evolution according to Eq. (5.3); the rest are unreachable and appear as nodes with zero in-degree on the periphery of the state transition diagram of Fig. 5. An explicit characterization of these unreachable configurations may be found by lengthy but straightforward analysis.

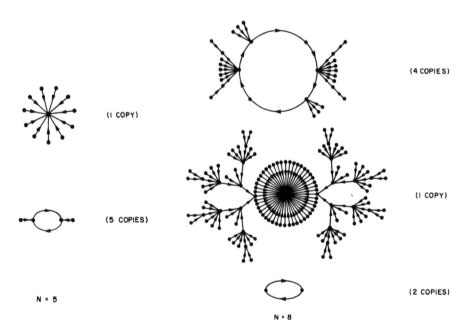

Figure 5. Global state transition diagrams for a typical finite non-additive cellular automaton discussed in Sect. 5.

102

Lemma 5.1. A configuration is unreachable by cellular automaton time evolution according to Eq. (5.3) if and only if one of the following conditions holds:

(a) The sequence of site values 111 appears.

(b) No sequence 11 appears, but the total number of 1 sites is odd.

(c) A sequence $11a_1a_2\ldots a_n11$ appears, with an odd number of the a_i having value 1. The two 11 sequences may be cyclically identified.

The number of reachable configurations may now be found by enumerating the configurations defined by Lemma 5.1. This problem is analogous to the enumeration of legal sentences in a formal language. As a simple example of the techniques required (e.g. [14]), consider the enumeration of strings of N symbols 0 or 1 in which no sequence 111 appears (no periodicity is assumed). Let the number of such strings be α. In addition, let β_N be the number of length N strings containing no 111 sequences in their first $N-1$ positions, but terminating with the sequence 111. Then

$$\beta_0 = \beta_1 = \beta_2 = 0, \quad \beta_3 = 1, \quad \alpha_0 = 1, \quad \alpha_1 = 2, \tag{5.4a}$$

and

$$2\alpha_N = \alpha_{N+1} + \beta_{N+1} \qquad (N \geq 0), \tag{5.4b}$$

$$\alpha_N = \beta_{N+1} + \beta_{N+2} + \beta_{N+3} \qquad (N \geq 0). \tag{5.4c}$$

The recurrence relations (5.4) may be solved by a generating function technique. With

$$A(z) = \sum_{n=0}^{\infty} \alpha_n z^n, \qquad B(z) = \sum_{n=0}^{\infty} \beta_n z^n, \tag{5.5a}$$

Eq. (5.4) may be written as

$$2A(z) = z^{-1}(A(z) - 1) + z^{-1}B(z),$$

$$A(z) = z^{-3}B(z) + z^{-2}B(z) + z^{-1}B(z).$$

Solving these equations yields the result

$$A(z) = \frac{1 + z + z^2}{1 - z - z^2 - z^3}. \tag{5.5b}$$

Results for specific N are obtained as the coefficients of z^N in a series expansion of $A(z)$. Taking

$$A(z) = \frac{A_N(z)}{A_D(z)},$$

Eq. (5.5a) may be inverted to yield

$$\alpha_N = \sum_i \left(\frac{-A_N(z_i)}{z_i A_D'(z_i)} \right) (1/z_i)^N, \tag{5.5c}$$

where the z_i are the roots of $A_D(z)$ (all assumed distinct), and prime denotes differentiation. This yields finally

$$\alpha_N \simeq 1.14(1.84)^N + 0.283(0.737)^N \cos(2.176N + 2.078). \tag{5.6}$$

The behaviour of the coefficients for large N is dominated by the first term, associated with the smallest root of $A_D(N)$. The first ten values of α_N are 1, 2, 4, 7, 13, 24, 44, 81, 149, 274, 504.

A lengthy calculation shows that the number of possible strings of length N which do not satisfy the conditions in Lemma 5.1, and may therefore be reached by evolution of the cellular automaton defined by Eq. (5.3), is given as the coefficient of z^N in the expansion of the generating function

$$P(z) = \frac{z - 3z^2 + 6z^3 - 8z^4 + 4z^5 - z^7}{1 - 4z + 6z^2 - 5z^3 + 2z^4 + z^5 - z^6 + z^7}$$

$$= \frac{3 - 4z + z^2}{1 - 2z + z^2 - z^3} - \frac{2 - z}{2(1 - z + z^2)} + \frac{2 - z}{2(-1 + z + z^2)} - 1. \tag{5.7}$$

Inverting according to Eq. (5.5c), the number of reachable configurations of length N is given by

$$\rho_N = \kappa^N - (\phi^N + (-\phi)^{-N}) - \cos(N\pi/3) + 2\mu^N \cos(N\theta), \tag{5.8}$$

where $\kappa \simeq 1.7548$ is the real root of $z^3 - z^2 + 2z - 1 = 0$, $\phi = (1 + \sqrt{5})/2 = 1.6182$, and $\mu \simeq 0.754$, $\theta \simeq 1.408$. The first ten values of ρ_N are 1, 1, 4, 7, 11, 19, 36, 67, 121, 216. For large N, $\rho_N \sim \kappa^N$. Equation (5.8) shows that corrections decrease rapidly and smoothly with N. This behaviour is to be contrasted with the irregular behaviour as a function of N found for additive cellular automata in Theorems 3.1 and 4.2.

Equation (5.8) shows that the fraction of all 2^N possible configurations which are reachable after one time step in the evolution of the cellular automaton of Eq. (5.2) is approximately $(\kappa/2)^N \simeq 0.92^N$. Thus, starting from an initial maximal entropy ensemble with $s = 1$, evolution for one time step according to Eq. (5.2) yields a set entropy

$$s(t = 1) \simeq \log_2 \kappa \simeq 0.88. \tag{5.9}$$

The irregularity of the transient trees illustrated in Fig. 5 implies a measure entropy $s_\mu < s$.

The result (5.9) becomes exact in the limit $N \to \infty$. A direct derivation in this limit is given in [17, 18], where it is also shown that the set of infinite configurations generated forms a regular formal language. The set continues to contract with time, so that the set entropy decreases below the value given by Eq. (5.9) [18].

Techniques similar to those used in the derivation of Eq. (5.5) may in principle be used to deduce the number of configurations reached after any given number of steps in the evolution of the cellular automaton (5.2). The fraction of configurations

N	ρ_N^∞
4	0.3125
5	0.3438
6	0.1094
7	0.0078
8	0.1133
9	0.1426
10	0.0791
11	0.0435
12	0.0466
13	0.0350
14	0.0163
15	0.00308
16	0.00850
17	0.00857

Table 3. Fraction of configurations appearing in cycles for the non-additive cellular automaton of Eq. (5.2).

which appear in cycles is an irregular function of N; some results for small N are given in Table 3.

6. Discussion

The analysis of additive cellular automata in Sects. 3 and 4 yielded results on the global behaviour of additive cellular automata more complete than those available for most other dynamical systems. The extensive analysis was made possible by the discrete nature of cellular automata, and by the additivity property which led to the algebraic approach developed in Sect. 3. Similar algebraic techniques should be applicable to some other discrete dynamical systems.

The analysis of global properties of cellular automata made in this paper complements the analysis of local properties of ref. [1].

One feature of the results on additive cellular automata found in Sects. 3 and 4, is the dependence of global quantities not only on the magnitude of the size parameter N, but also on its number theoretical properties. This behaviour is shared by many dynamical systems, both discrete and continuous. It leads to the irregular variation of quantities such as cycle lengths with N, illustrated in Table 1 and Fig. 3. In physical realizations of cellular automata with large size N, an average is presumably performed over a range of N values, and irregular dependence on N is effectively smoothed out. A similar irregular dependence is found on the number k of possible values for each site: simple results are found only when k is prime.

Despite such detailed dependence on N, results such as Theorems 4.1–4.3 show that global properties of additive cellular automata exhibit a considerable universality, and independence of detailed aspects of their construction. This property is again shared by many other dynamical systems. It potentially allows for generic results, valid both in the simple cases which may easily be analysed, and in the presumably complicated cases which occur in real physical systems.

105

The discrete nature of cellular automata makes possible an explicit analysis of their global behaviour in terms of transitions in the discrete phase space of their configurations. The results of Sect. 4 provide a rather complete characterization of the structure of the state transition diagrams for additive cellular automata. The state transition diagrams consist of trees corresponding to irreversible "transients", leading to "attractors" in the form of distinct finite cycles. The irreversibility of the cellular automata is explicitly manifest in the convergence of several distinct configurations to single configurations through motion towards the roots of the trees. This irreversibility leads to a decrease in the entropy of an initially equiprobable ensemble of cellular automaton configurations; the results of Sect. 4 show that in most cases the entropy decreases by a fixed amount at each time step, reflecting the balanced nature of the trees. Theorem 4.3 gives an algebraic characterization of the magnitude of the irreversibility, in terms of the in-degrees of nodes in the trees. The length of the transients during which the entropy decreases is given by the height of the trees in Theorem 4.3, and is found always to be less than N. After these transients, any initial configurations evolve to configurations on attractors or cycles. Theorem 4.3 gives the total number of configurations on cycles in terms of N and algebraic properties of the cellular automaton time evolution polynomial. At one extreme, all configurations may be on cycles, while at the other extreme, all initial configurations may evolve to a single limit point consisting simply of the null configuration.

Theorem 4.1 gives a rather general result on the lengths of cycles in additive cellular automata. The maximum possible cycle length is found to be of order the square root of the total number of possible configurations. Rather long cycles are therefore possible. No simple results on the total number of distinct cycles or attractors were found; however, empirical results suggest that most cycles have a length equal to the maximal length for a particular cellular automaton.

The global properties of additive cellular automata may be compared with those of other mathematical systems. One closely related class of systems are linear feedback shift registers. Most results in this case concentrate on analogues of the cellular automaton discussed in Sect. 3, but with the values at a particular time step in general depending on those of a few far-distant sites. The boundary conditions assumed for feedback shift registers are typically more complicated than the periodic ones assumed for cellular automata in Sect. 3 and most of Sect. 4. The lack of symmetry in these boundary conditions allows for maximal length shift register sequences, in which all $2^N - 1$ possible configurations occur on a single cycle [2, 3].

A second mathematical system potentially analogous to cellular automata is a random mapping [15]. While the average cycle length for random mappings is comparable to the maximal cycle length for cellular automata, the probability for a node in the state transition diagram of a random mapping to have in-degree d is $\sim 1/d!$, and is much more sharply peaked at low values than for a cellular automaton, leading to many differences in global properties.

Non-additive cellular automata are not amenable to the algebraic techniques used in Sects. 3 and 4 for the additive case. Section 5 nevertheless discussed some properties of non-additive cellular automata, concentrating on a simple one-dimensional example with two possible values at each site. Figure 5 indicates that the state transition diagrams for such non-additive cellular automata are less regular than those for additive cellular automata. Combinatorial methods were nevertheless used to derive the fraction of configurations with no predecessors in these diagrams, giving the irreversibility and thus entropy decrease associated with one time step in the cellular automaton evolution. Unlike the case of additive cellular automata, the result was found to be a smooth function of N.

Appendix A:
Notations and Elementary Results on Finite Fields

Detailed discussion of the material in this appendix may be found in [8].

A. Basic Notations

$a \bmod b$ denotes a reduced modulo b, or the remainder of a after division by b.

(a, b) or $\gcd(a, b)$ denotes the greatest common divisor of a and b. When a and b are polynomials, the result is taken to be a polynomial with unit leading coefficient (monic).

$a \mid b$ represents the statement that a divides b (with no remainder).

$a^n \| b$ indicates that a^n is the highest power of a which divides b.

Exponentiation is assumed right associative, so that a^{b^c} denotes $a^{(b^c)}$ not $(a^b)^c$.

p usually denotes a prime integer.

$\mathbb{R}_k$ denotes an arbitrary commutative ring of k elements.

$\mathbb{Z}_k$ denotes the ring of integers modulo k.

$\deg P(x)$ denotes the highest power of x which appears in $P(x)$.

B. Finite Fields

There exists a finite field unique up to isomorphism with any size p^α (p prime), denoted GF(p^α). p is termed the characteristic of the field.

The ring $\mathbb{Z}_k$ of integers modulo k forms a field only when k is prime, since only in this case do unique inverses under multiplication modulo k exist for all nonzero elements. (For example, in $\mathbb{Z}_4$, 2 has no inverse.) GF(p) is therefore isomorphic to $\mathbb{Z}_p$.

The field GF(p^α) is conveniently represented by the set of polynomials of degree less than α with coefficients in $\mathbb{Z}_p$, with all polynomial operations performed modulo a fixed irreducible polynomial of degree α over GF(p). For example, GF(4) may be represented by elements 0, 1, κ, $\kappa + 1$ with operations performed modulo 2 and modulo $\kappa^2 + \kappa + 1$. In this case for example $\kappa \times \kappa \equiv \kappa + 1$. Notice that, as mentioned in Sect. A.C below, polynomials over a field form a unique factorization domain.

107

Any field of size q yields a group of size $q - 1$ under multiplication if the zero element is removed. Thus for any element of GF(q),

$$x^q = x, \tag{A.1}$$

and $x^{q-1} = 1$ for $x \neq 0$. Notice that if $x \in$ GF(p^α) and $x^{p^\beta} = x$, then $x \in$ GF(p^β).

C. Polynomials over Finite Fields

Polynomials in any number of variables with coefficients in GF(q) form a unique factorization domain. For such polynomials, therefore, $A(x)B(x) \equiv A(x)C(x) \bmod P(x)$ implies $B(x) \equiv C(x) \bmod P(x)$ if $(A(x), P(x)) = 1$.

For any polynomials $A(x)$ and $B(x)$ with coefficients in GF(q), there exist polynomials $\alpha(x)$ and $\beta(x)$ such that

$$C(x) = (A(x), B(x)) = \alpha(x)A(x) + \beta(x)B(x). \tag{A.2}$$

There are exactly q^n univariate polynomials over GF(q) with degree less than n. With a polynomial $Q(x)$ of degree m, the number of polynomials $P(x)$ with degree not exceeding n for which $Q(x) \mid P(x)$ is q^{n-m} for $m \leq n$.

For any prime p, and for elements a_i of GF(p^β),

$$\left(\sum a_i x^i \right)^{p^\alpha} = \sum (a_i x^i)^{p^\alpha}. \tag{A.3}$$

Thus for example,

$$(x^{2^\alpha} + 1) \equiv (x + 1)^{2^\alpha} \quad \bmod 2, \tag{A.4}$$

a result used extensively in Sect. 3.

If $P(x) \mid Q(x)$, then every root of $P(x)$ must be a root of $Q(x)$. If $\lambda \geq 2$ and

$$[P(x)]^\lambda \mid Q(x), \tag{A.5}$$

then

$$P(x) \mid Q'(x), \tag{A.6}$$

where $Q'(x)$ is the formal derivative of $Q(x)$, obtained by differentiation of each term in the polynomial. [Note that integration is not defined for polynomials over GF(q).]

The number of roots (not necessarily distinct) of a polynomial over GF(q) is equal to the degree of the polynomial. The roots may lie in an extension of GF(q).

Over the field GF(p),

$$x^N - 1 = (x^n - 1)^{D_p(N)}, \tag{A.7}$$

where $N = D_p(N)n$, with $D_p(N)$ defined in Sects. 3 and 4 as the maximum power of p which divides N. The polynomial $x^n - 1$ with n not a multiple of p then factorizes over GF(p) according to

$$x^n - 1 = (x - 1) \prod_{\substack{d \mid n \\ d \neq 1}} \prod_{i=1}^{\frac{\phi(d)}{\text{ord}_d(p)}} C_{d, i}(x), \tag{A.8}$$

where the $C_{d,i}(x)$ are irreducible cyclotomic polynomials of degree $\text{ord}_d(p)$. Note that the multiplicity of any irreducible factor of $x^N - 1$ is exactly $D_p(N)$, and that

$$C_{d,i}(x) \mid x^d - 1. \tag{A.9}$$

D. Dipolynomials over Finite Fields

A dipolynomial $A(x)$ is taken to divide a dipolynomial $B(x)$ if there exists a dipolynomial $C(x)$ such that $B(x) = A(x)C(x)$. Hence if $A(x)$ and $B(x)$ are polynomials, with $A(0) \neq 0$, and if $A(x) \mid B(x)$ are dipolynomials, then $A(x) \mid B(x)$ are polynomials.

Congruence in the ring of dipolynomials is defined as follows: $A(x) \equiv B(x) \bmod C(x)$ for dipolynomials $A(x)$, $B(x)$, and $C(x)$ if $C(x) \mid A(x) - B(x)$.

The greatest common divisor of two nonzero dipolynomials $A_1(x)$ and $A_2(x)$ is defined as the ordinary polynomial $(A_1^*(x), A_2^*(x))$, where $A_i^*(x) = x^{m_i} A_i(x)$ and m_i is chosen to make $A_i^*(x)$ a polynomial with nonzero constant term. Note that by analogy with Eq. (A.2), for any dipolynomials $A_1(x)$ and $A_2(x)$, there exist dipolynomials $\alpha_1(x)$ and $\alpha_2(x)$ such that

$$(A_1(x), A_2(x)) = \alpha_1(x)A_1(x) + \alpha_2(x)A_2(x). \tag{A.10}$$

Appendix B:
Properties and Values of Some Number Theoretical Functions

A. Euler Totient Function $\phi(N)$

$\phi(N)$ is defined as the number of integers less than N which are relatively prime to N [7]. $\phi(N)$ is a multiplicative function, so that

$$\phi(mn) = \phi(m)\phi(n), \qquad (m, n) = 1. \tag{B.1}$$

For p prime,

$$\phi(p^\alpha) = p^{\alpha-1}(p - 1). \tag{B.2}$$

Hence

$$\phi(n) = \prod_{p^\alpha \| n} p^{\alpha-1}(p - 1), \tag{B.3}$$

providing a formula by which $\phi(N)$ may be computed. Some values of $\phi(N)$ are given in Table 4.

$\phi(N)$ is bounded (for $N > 1$) by

$$cN / \log \log N \leq \phi(N) \leq N - 1, \tag{B.4}$$

where c is some positive constant, and the upper bound is achieved if and only if N is prime. For large N, $\phi(N)/N$ tends on average to a constant value.

$\phi(n)$ satisfies the Euler-Fermat theorem

$$k^{\phi(n)} = 1 \qquad \bmod n \qquad (k, n) = 1. \tag{B.5}$$

109

B. Multiplicative Order Function ord$_N(k)$

The multiplicative order function ord$_N(k)$ is defined as the minimum positive integer j for which [8]

$$k^j = 1 \quad \mod N. \tag{B.6}$$

This condition can only be satisfied if $(k, N) = 1$.

By the Euler-Fermat theorem (B.5),

$$\text{ord}_N(k) \,|\, \phi(N) \tag{B.7}$$

In addition, ord$_{mn}(k) = \text{lcm}(\text{ord}_n(k), \text{ord}_m(k))$, $(n, k) = (m, k) = (n, m) = 1$. Some special cases are

$$\text{ord}_{k^\alpha-1}(k) = \alpha,$$
$$\text{ord}_{k^\alpha+1}(k) = 2\alpha.$$

A rigorous bound on ord$_N(k)$ is

$$\log_k(N) \le \text{ord}_N(k) \le N - 1, \tag{B.8}$$

where the upper bound is attained only if N is prime. It can be shown that on average, for large N, $\text{ord}_N(k) \gtrsim \sqrt{N}$; the actual average is presumably closer to N. Nevertheless, for large N, $\text{ord}_N(k)/N$ tends to zero on average.

Some values of the multiplicative order function are given in Table 4.

The multidimensional generalization $\text{ord}_{N_1,\dots,N_d}(k)$ of the multiplicative order function is defined as the minimum positive integer j for which $k^j = 1$ simultaneously modulo $N_1, N_2, \dots,$ and N_d. It is clear that

$$\text{ord}_{N_1,\dots,N_d}(k) = \text{lcm}(\text{ord}_{N_1}(k), \dots, \text{ord}_{N_d}(k)) = \text{ord}_{\text{lcm}(N_1,\dots,N_d)}(k),$$
$$(k, N_1) = \dots = (k, N_d) = 1. \tag{B.9}$$

C. Multiplicative Suborder Function sord$_N(k)$

The multiplicative suborder function is defined as the minimum j for which

$$k^j = \pm 1 \mod N, \tag{B.10}$$

again assuming $(k, N) = 1$. Comparison with (B.6) yields

$$\text{sord}_N(k) = \text{ord}_N(k), \tag{B.11a}$$

or

$$\text{sord}_N(k) = \tfrac{1}{2}\text{ord}_N(k). \tag{B.11b}$$

The second case becomes comparatively rare for large N; the fraction of integers less than X for which it is realised may be shown to be asymptotic to $c/[\log X]^\lambda$ [16], where c and λ are constants determined by k.

N	k = 2		k = 3		k = 4		k = 5		φ(N)
1									1
2			1	1			1	1	1
3	2	1			1	1	2	1	2
4			2	1			1	1	2
5	4	2	4	2	2	1			4
6							2	1	2
7	3	3	6	3	3	3	6	3	6
8			2	2			2	2	4
9	6	3			3	3	6	3	6
10			4	2					4
11	10	5	5	5	5	5	5	5	10
12							2	2	4
13	12	6	3	3	6	3	4	2	12
14			6	3			6	3	6
15	4	4			2	2			8
16			4	4			4	4	8
17	8	4	16	8	4	2	16	8	16
18							6	3	6
19	18	9	18	9	9	9	9	9	18
20			4	4					8
21	6	6			3	3	6	3	12
22			5	5			5	5	10
23	11	11	11	11	11	11	22	11	22
24							2	2	8
25	20	10	20	10	10	5			20
26			3	3			4	2	12
27	18	9			9	9	18	9	18
28			6	3			6	6	12
29	28	14	28	14	14	7	14	7	28
30									8
31	5	5	30	15	5	5	3	3	30
32			8	8			8	8	16
33	10	5			5	5	10	10	20
34			16	8			16	8	16
35	12	12	12	12	6	6			24
36							6	6	12
37	36	18	18	9	18	9	36	18	36
38			18	9			9	9	18
39	12	12			6	6	4	4	24
40			4	4					16

Table 4. Values of the multiplicative order $\text{ord}_N(k)$ and suborder $\text{sord}_N(k)$ functions defined in Eqs. (B.6) and (B.10), respectively, together with values of the Euler totient function $\phi(N)$. Each column gives values of the pair $\text{ord}_N(k)$, $\text{sord}_N(k)$.

In general,

$$\log_k(N) \le \mathrm{sord}_N(k) \le (N-1)/2, \tag{B.12}$$

the upper limit again being achieved only if N is prime. For large N, $\mathrm{sord}_N(k)/N \to 0$ on average.

The multidimensional generalization $\mathrm{sord}_{N_1,\ldots,N_d}(k)$ of the multiplicative suborder function is defined as the minimum positive integer j for which $k^j = \pm 1$ simultaneously modulo $N_1,\ldots,N_d$, with $+1$ and -1 perhaps taken variously for the different N_i. The analogue of Eq. (B.9) for this function is

$$\mathrm{sord}_{N_1,\ldots,N_d}(k) = \mathrm{lcm}(\mathrm{sord}_{N_1}(k),\ldots,\mathrm{sord}_{N_d}(k)), \tag{B.13a}$$

and

$$\mathrm{lcm}(\mathrm{sord}_{N_1}(k),\ldots,\mathrm{sord}_{N_d}(k)) = \mathrm{sord}_{\mathrm{lcm}(N_1,\ldots,N_d)}(k), \tag{B.13b}$$

or

$$\mathrm{lcm}(\mathrm{sord}_{N_1}(k),\ldots,\mathrm{sord}_{N_d}(k)) = \tfrac{1}{2}\mathrm{sord}_{\mathrm{lcm}(N_1,\ldots,N_d)}(k). \tag{B.13c}$$

Acknowledgement. We are grateful to O. E. Lanford for several suggestions.

References

1. Wolfram, S.: Statistical mechanics of cellular automata. Rev. Mod. Phys. **55**, 601 (1983).
2. Golomb, S.W.: Shift register sequences. San Francisco: Holden-Day 1967.
3. Selmer, E.S.: Linear recurrence relations over finite fields. Dept. of Math., Univ. of Bergen, Norway (1966).
4. Miller, J.C.P.: Periodic forests of stunted trees. Phil. Trans. R. Soc. Lond. **A266**, 63 (1970); **A293**, 48 (1980).
   ApSimon, H.G.: Periodic forests whose largest clearings are of size 3. Phil. Trans. R. Soc. Lond. **A266**, 113 (1970).
   ApSimon, H.G.: Periodic forests whose largest clearings are of size $n \ge 4$. Proc. R. Soc. Lond. **A319**, 399 (1970).
   Sutton, C.: Forests and numbers and thinking backwards. New Sci. **90**, 209 (1981).
5. Moore, E.F.: Machine models of self-reproduction. Proc. Symp. Appl. Math. **14**, 17 (1962) reprinted in: Essays on cellular automata, A. W. Burks. Univ. of Illinois Press (1966).
   Aggarwal, S.: Local and global Garden of Eden theorems. Michigan University technical rept. 147 (1973).
6. Knuth, D.: Fundamental algorithms, Reading, MA: Addison-Wesley 1968.
7. Hardy, G.H., Wright, E.M.: An introduction to the theory of numbers. Oxford: Oxford University Press 1968.
8. MacWilliams, F.J., Sloane, N.J.A.: The theory of error-correcting codes. Amsterdam: North-Holland 1977.
9. Griffiths, P., Harris, J.: Principles of algebraic geometry. New York: Wiley 1978.
10. Fredkin, E., Margolus, N.: Private communications.

11. Ronse, C.: Non-linear shift registers: A survey. MBLE Research Lab. report, Brussels (May 1980).

12. Harao, M. and Noguchi, S.: On some dynamical properties of finite cellular automaton. IEEE Trans. Comp. C-**27**, 42 (1978).

13. Grassberger, P.: A new mechanism for deterministic diffusion. Phys. Rev. A (to be published).

14. Guibas, L.J., Odlyzko, A.M.: String overlaps, pattern matching, and nontransitive games. J. Comb. Theory (A) **30**, 83 (1981).

15. Knuth, D.: Seminumerical algorithms. 2nd ed. Reading, MA: Addison-Wesley 1981.
    Gelfand, A.E.: On the cyclic behavior of random transformations on a finite set. Tech. rept. 305, Dept. of Statistics, Stanford Univ. (August 1981).

16. Odlyzko, A.M.: Unpublished.

17. Lind, D.A.: Applications of ergodic theory and sofic systems to cellular automata. Physica D **10** (to be published).

18. Wolfram, S.: Computation theory of cellular automata. Institute for Advanced Study preprint (January 1984).

19. Lenstra, H.W., Jr.: Private communication.

113

Universality and Complexity in Cellular Automata

1984

Cellular automata are discrete dynamical systems with simple construction but complex self-organizing behaviour. Evidence is presented that all one-dimensional cellular automata fall into four distinct universality classes. Characterizations of the structures generated in these classes are discussed. Three classes exhibit behaviour analogous to limit points, limit cycles and chaotic attractors. The fourth class is probably capable of universal computation, so that properties of its infinite time behaviour are undecidable.

1. Introduction

Cellular automata are mathematical models for complex natural systems containing large numbers of simple identical components with local interactions. They consist of a lattice of sites, each with a finite set of possible values. The values of the sites evolve synchronously in discrete time steps according to identical rules. The value of a particular site is determined by the previous values of a neighbourhood of sites around it.

The behaviour of a simple set of cellular automata was discussed in ref. 1, where extensive references were given. It was shown that despite their simple construction, some cellular automata are capable of complex behaviour. This paper discusses the nature of this complex behaviour, its characterization, and classification. Based on investigation of a large sample of cellular automata, it suggests that many (perhaps all) cellular automata fall into four basic behaviour classes. Cellular automata within each class exhibit qualitatively similar behaviour. The small number of classes implies considerable universality in the qualitative behaviour of cellular automata. This universality implies that many details of the construction of a cellular automaton

Originally published in *Physica D*, volume 10, pages 1–35 (January 1984).

are irrelevant in determining its qualitative behaviour. Thus complex physical and biological systems may lie in the same universality classes as the idealized mathematical models provided by cellular automata. Knowledge of cellular automaton behaviour may then yield rather general results on the behaviour of complex natural systems.

Cellular automata may be considered as discrete dynamical systems. In almost all cases, cellular automaton evolution is irreversible. Trajectories in the configuration space for cellular automata therefore merge with time, and after many time steps, trajectories starting from almost all initial states become concentrated onto "attractors". These attractors typically contain only a very small fraction of possible states. Evolution to attractors from arbitrary initial states allows for "self-organizing" behaviour, in which structure may evolve at large times from structureless initial states. The nature of the attractors determines the form and extent of such structures.

The four classes mentioned above characterize the attractors in cellular automaton evolution. The attractors in classes 1, 2 and 3 are roughly analogous respectively to the limit points, limit cycles and chaotic ("strange") attractors found in continuous dynamical systems. Cellular automata of the fourth class behave in a more complicated manner, and are conjectured to be capable of universal computation, so that their evolution may implement any finite algorithm.

The different classes of cellular automaton behaviour allow different levels of prediction of the outcome of cellular automaton evolution from particular initial states. In the first class, the outcome of the evolution is determined (with probability 1), independent of the initial state. In the second class, the value of a particular site at large times is determined by the initial values of sites in a limited region. In the third class, a particular site value depends on the values of an ever-increasing number of initial sites. Random initial values then lead to chaotic behaviour. Nevertheless, given the necessary set of initial values, it is conjectured that the value of a site in a class 3 cellular automaton may be determined by a simple algorithm. On the other hand, in class 4 cellular automata, a particular site value may depend on many initial site values, and may apparently be determined only by an algorithm equivalent in complexity to explicit simulation of the cellular automaton evolution. For these cellular automata, no effective prediction is possible; their behaviour may be determined only by explicit simulation.

This paper describes some preliminary steps towards a general theory of cellular automaton behaviour. Section 2 below introduces notation and formalism for cellular automata. Section 3 discusses general qualitative features of cellular automaton evolution illustrating the four behaviour classes mentioned above. Section 4 introduces entropies and dimensions which characterize global features of cellular automaton evolution. Successive sections consider each of the four classes of cellular automata in turn. The last section discusses some tentative conclusions.

This paper covers a broad area, and includes many conjectures and tentative results. It is not intended as a rigorous mathematical treatment.

2. Notation and Formalism

$a_i^{(t)}$ is taken to denote the value of site i in a one-dimensional cellular automaton at time step t. Each site value is specified as an integer in the range 0 through $k-1$. The site values evolve by iteration of the mapping

$$a_i^{(t)} = \mathbf{F}[a_{i-r}^{(t-1)}, a_{i-r+1}^{(t-1)}, \ldots, a_i^{(t-1)}, \ldots, a_{i+r}^{(t-1)}]. \tag{2.1}$$

$\mathbf{F}$ is an arbitrary function which specifies the cellular automaton rule.

The parameter r in eq. (2.1) determines the "range" of the rule: the value of a given site depends on the last values of a neighbourhood of at most $2r+1$ sites. The region affected by a given site grows by at most r sites in each direction at every time step; propagating features generated in cellular automaton evolution may therefore travel at most r sites per time step. After t time steps, a region of at most $1+2rt$ sites may therefore be affected by a given initial site value.

The "elementary" cellular automata considered in ref. 1 have $k = 2$ and $r = 1$, corresponding to nearest-neighbour interactions.

An alternative form of eq. (2.1) is

$$a_i^{(t)} = \mathbf{f}\left[\sum_{j=-r}^{j=r} \alpha_j a_{i+j}^{(t-1)}\right], \tag{2.2}$$

where the α_j are integer constants, and the function $\mathbf{f}$ takes a single integer argument. Rules specified according to (2.1) may be reproduced directly by taking $\alpha_j = k^{r-j}$.

The special class of additive cellular automaton rules considered in ref. 2 correspond to the case in which $\mathbf{f}$ is a linear function of its argument modulo k. Such rules satisfy a special additive superposition principle. This allows the evolution of any initial configuration to be determined by superposition of results obtained with a few basis configurations, and makes possible the algebraic analysis of ref. 2.

"Totalistic" rules defined in ref. 1, and used in several examples below, are obtained by taking

$$\alpha_j = 1 \tag{2.3}$$

in eq. (2.2). Such rules give equal weight to all sites in a neighbourhood, and imply that the value of a site depends only on the total of all preceding neighbourhood site values. The results of section 3 suggest that totalistic rules exhibit behaviour characteristic of all cellular automata.

Cellular automaton rules may be combined by composition. The set of cellular automaton rules is closed under composition, although composition increases the number of sites in the neighbourhood. Composition of a rule with itself yields patterns corresponding to alternate time steps in time evolution according to the rule. Compositions of distinct rules do not in general commute. However, if a composition $\mathbf{F}_1\mathbf{F}_2$ of rules generates a sequence of configurations with period π, then the rule $\mathbf{F}_2\mathbf{F}_1$ must also allow a sequence of configurations with period π. As discussed

117

below, this implies that the rules F_1F_2 and F_2F_1 must yield behaviour of the same class.

The configuration $a_i = 0$ may be considered as a special "null" configuration ("ground state"). The requirement that this configuration remain invariant under time evolution implies

$$F[0, 0, \ldots, 0] = 0 \qquad\qquad\qquad (2.4a)$$

and

$$f[0] = 0. \qquad\qquad\qquad (2.4b)$$

All rules satisfy this requirement if iterated at most k times, at least up to a relabelling of the k possible values.

It is convenient to consider symmetric rules, for which

$$F[a_{i-r}, \ldots, a_{i+r}] = F[a_{i+r}, \ldots, a_{i-r}]. \qquad\qquad\qquad (2.5)$$

Once a cellular automaton with symmetric rules has evolved to a symmetric state (in which $a_{n+i} = a_{n-i}$ for some n and all i), it may subsequently generate only symmetric states (assuming symmetric boundary conditions), since the operation of space reflection commutes with time evolution in this case.

Rules satisfying the conditions (2.4) and (2.5) will be termed "legal".

The cellular automaton rules (2.1) and (2.2) may be considered as discrete analogues of partial differential equations of order at most $2r + 1$ in space, and first order in time. Cellular automata of higher order in time may be constructed by allowing a particular site value to depend on values of a neighbourhood of sites on a number s of previous time steps. Consideration of "effective" site values $\sum_{n=0}^{s-1} m^n a_i^{(t-n)}$ always allows equivalent first-order rules with $k = m^s - 1$ to be constructed.

The form of the function F in the time evolution rule (2.1) may be specified by a "rule number" [1]

$$R_F = \sum_{\{a_{i-r}, a_{i+r}\}} F[a_{i-r}, \ldots, a_{i+r}] k^{\sum_{j=-r}^{r} k^{r-j} a_{i+j}}. \qquad\qquad\qquad (2.6)$$

The function f in eq. (2.2) may similarly be specified by a numerical "code"

$$C_f = \sum_{n=0}^{(2r+1)(k-1)} k^n f[n]. \qquad\qquad\qquad (2.7)$$

The condition (2.4) implies that both R_F and C_f are multiples of k.

In general, there are a total of $k^{(k^{(2r+1)})}$ possible cellular automaton rules of the form (2.1) or (2.2). Of these, $k^{k^{r+1}(k^r+1)/2-1}$ are legal. The rapid growth of the number of possible rules with r implies that an exponentially small fraction of rules may be obtained by composition of rules with smaller r.

A few cellular automaton rules are "reducible" in the sense that the evolution of sites with particular values, or on a particular grid of positions and times, are independent of other site values. Such cellular automata will usually be excluded from the classification described below.

Very little information on the behaviour of a cellular automaton can be deduced directly from simple properties of its rule. A few simple results are nevertheless clear.

First, necessary (but not sufficient) conditions for a rule to yield unbounded growth are

$$\mathbf{F}[a_{i-r}, a_{i-r+1}, \ldots, a_{i-1}, 0, 0, \ldots, 0] \neq 0,$$
$$\mathbf{F}[0, \ldots, 0, 0, a_{i+1}, \ldots, a_{i+r}] \neq 0 \tag{2.8}$$

for some set of a_i. If these conditions are not fulfilled then regions containing nonzero sites surrounded by zero sites can never grow, and the cellular automaton must exhibit behaviour of class 1 or 2. For totalistic rules, the condition (2.8) becomes

$$\mathbf{f}[n] \neq 0 \tag{2.9}$$

for some $n < r$.

Second, totalistic rules for which

$$\mathbf{f}[n_1] \geq \mathbf{f}[n_2] \tag{2.10}$$

for all $n_1 > n_2$ exhibit no "growth inhibition" and must therefore similarly be of class 1 or 2.

One may consider cellular automata both finite and infinite in extent.

When finite cellular automata are discussed below, they are taken to consist of N sites arranged around a circle (periodic boundary conditions). Such cellular automata have a finite number k^N of possible states. Their evolution may be represented by finite state transition diagrams (cf. [2]), in which nodes representing each possible configuration are joined by directed arcs, with a single arc leading from a particular node to its successor after evolution for one time step. After a sufficiently long time (less than k^N), any finite cellular automaton must enter a cycle, in which a sequence of configurations is visited repeatedly. These cycles represent attractors for the cellular automaton evolution, and correspond to cycles in the state transition graph. At nodes in the cycles may be rooted trees representing transients. The transients are irreversible in the sense that nodes in the tree have a single successor, but may have several predecessors. In the course of time evolution, all states corresponding to nodes in the trees ultimately evolve through the configurations represented by the roots of the trees to the cycles on which the roots lie. Configurations corresponding to nodes on the periphery of the state transition diagram (terminals or leaves of the transient trees) are never reached in the evolution: they may occur only as initial states. The fraction of configurations which may be reached after one time step in cellular automaton evolution, and which are therefore not on the periphery of the state transition diagram, gives a simple measure of irreversibility.

The configurations of infinite cellular automata are specified by (doubly) infinite sequences of site values. Such sequences are naturally identified as elements of a Cantor set (e.g. [3]). (They differ from real numbers through the inequivalence of configurations such as .111111 . . . and 1.0000 . . .). Cellular automaton rules define mappings from this Cantor set to itself. The mappings are invariant under shifts by virtue of the identical treatment of each site in eqs. (2.1) and (2.2). With natural measures of distance in the Cantor set, the mappings are also continuous. The typical irreversibility of cellular automaton evolution is manifest in the fact that the mapping is usually not injective, as discussed in section 4.

Equations (2.1) and (2.2) may be generalized to several dimensions. For $r = 1$, there are at least two possible symmetric forms of neighbourhood, containing $2d + 1$ (type I) and 3^d (type II) sites respectively; for larger r other "unit cells" are possible.

3. Qualitative Characterization of Cellular Automaton Behaviour

This section discusses some qualitative features of cellular automaton evolution, and gives empirical evidence for the existence of four basic classes of behaviour in cellular automata. Section 4 introduces some methods for quantitative analysis of cellular automata. Later sections use these methods to suggest fundamental characterizations of the four cellular automaton classes.

Figure 1 shows the pattern of configurations generated by evolution according to each of the 32 possible legal totalistic rules with $k = 2$ and $r = 2$, starting from a "disordered" initial configuration (in which each site value is independently chosen as 0 or 1 with probability $\frac{1}{2}$). Even with such a structureless initial state, many of the rules are seen to generate patterns with evident structure. While the patterns obtained with different rules all differ in detail, they appear to fall into four qualitative classes:

1. Evolution leads to a homogeneous state (realized for codes 0, 4, 16, 32, 36, 48, 54, 60 and 62).

2. Evolution leads to a set of separated simple stable or periodic structures (codes 8, 24, 40, 56 and 58).

3. Evolution leads to a chaotic pattern (codes 2, 6, 10, 12, 14, 18, 22, 26, 28, 30, 34, 38, 42, 44, 46 and 50).

4. Evolution leads to complex localized structures, sometimes long-lived (codes 20 and 52).

Some patterns (e.g. code 12) assigned to class 3 contain many triangular "clearings" and appear more regular than others (e.g. code 10). The degree of regularity is related to the degree of irreversibility of the rules, as discussed in section 7.

Figure 2 shows patterns generated from several different initial states according to a few of the cellular automaton rules of fig. 1. Patterns obtained with different

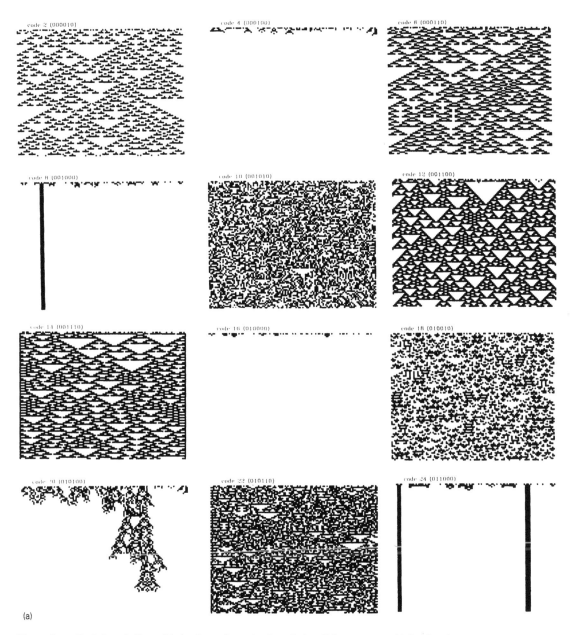

(a)

Figure 1a–c. Evolution of all possible legal one-dimensional totalistic cellular automata with $k = 2$ and $r = 2$. k gives the number of possible values for each site, and r gives the range of the cellular automaton rules. A range $r = 2$ allows the nearest and next-nearest neighbours of a site to affect its value on the next time step. Time evolution for totalistic cellular automata is defined by eqs. (2.2) and (2.7). The initial state is taken disordered, each site having values 0 and 1 with independent equal probabilities. Configurations obtained at successive time steps in the cellular automaton evolution are shown on successive horizontal lines. Black squares represent sites with value 1; white squares sites with value 0. All the cellular automaton rules illustrated are seen to exhibit one of four qualitative classes of behaviour.

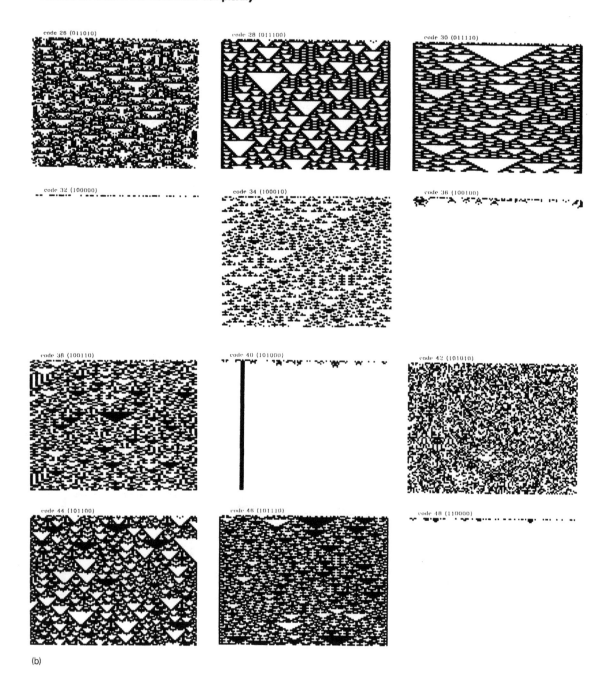

(b)

Figure 1 (continued).

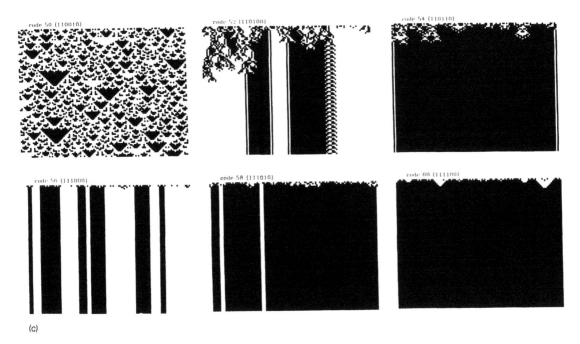

(c)

Figure 1 (continued).

initial states are seen to differ in their details, but to exhibit the same characteristic qualitative features. (Exceptional initial states giving rise to different behaviour may exist with low or zero probability.) Figure 3 shows the differences between patterns generated by various cellular automaton rules from initial states differing in the value of a single site.

Figures 4, 5 and 6 show examples of various sets of totalistic cellular automata. Figure 4 shows some $k = 2, r = 3$ rules, fig. 5 some $k = 3, r = 1$ rules, and fig. 6 some $k = 5, r = 1$ rules. The patterns generated are all seen to be qualitatively similar to those of fig. 1, and to lie in the same four classes.

Patterns generated by all possible $k = 2, r = 1$ cellular automata were given in ref. 1, and are found to lie in classes 1, 2 and 3. Totalistic $k = 2, r = 1$ rules are found to give patterns typical of all $k = 2, r = 1$ rules. In general, totalistic rules appear to exhibit no special simplifications, and give rise to behaviour typical of all cellular automaton rules with given k and r.

An extensive sampling of many other cellular automaton rules supports the general conjecture that the four classes introduced above cover all one-dimensional cellular automata.*

---

* This sampling and many other investigations reported in this paper were performed using the C language computer program [4].

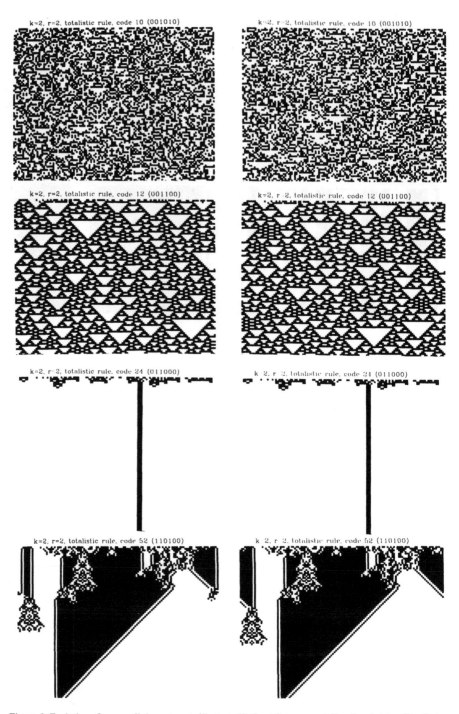

Figure 2. Evolution of some cellular automata illustrated in fig. 1 from several disordered states. The first two initial states shown differ by a change in the values of two sites, the next by a change in the values of ten sites. The last state is completely different.

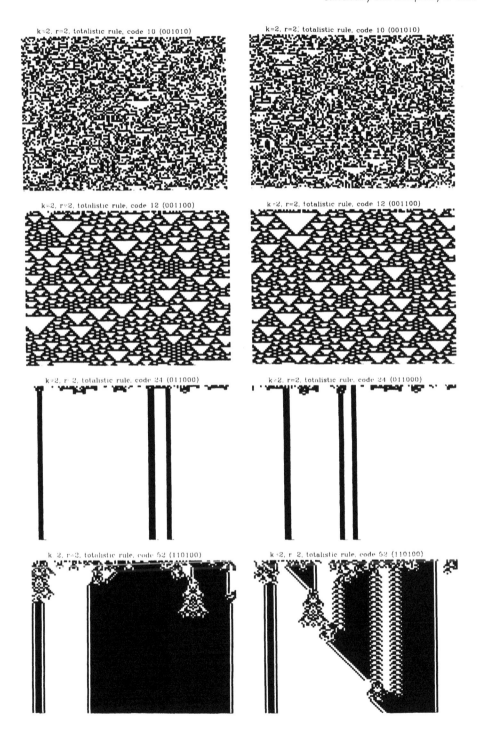

Figure 2 (continued).

Class	$k = 2$ $r = 1$	$k = 2$ $r = 2$	$k = 2$ $r = 3$	$k = 3$ $r = 1$
1	0.50	0.25	0.09	0.12
2	0.25	0.16	0.11	0.19
3	0.25	0.53	0.73	0.60
4	0	0.06	0.06	0.07

Table 1. Approximate fractions of legal totalistic cellular automaton rules in each of the four basic classes.

Table 1 gives the fractions of various sets of cellular automata in each of the four classes. With increasing k and r, class 3 becomes overwhelmingly the most common. Classes 1 and 2 are decreasingly common. Class 4 is comparatively rare, but becomes more common for larger k and r.

"Reducible" cellular automata (mentioned in section 2) may generate patterns which contain features from several classes. In a typical case, fixed or propagating "membranes" consisting of sites with a particular value may separate regions containing patterns from classes 3 or 4 formed from sites with other values.

This paper concerns one-dimensional cellular automata. Two-dimensional cellular automata also appear to exhibit a few distinct classes of behaviour. Superficial investigations [5] suggest that these classes may in fact be identical to the four found in one-dimensional cellular automata.

4. Quantitative Characterizations of Cellular Automaton Behaviour

This section describes quantitative statistical measures of order and chaos in patterns generated by cellular automaton evolution. These measures may be used to distinguish the four classes of behaviour identified qualitatively above.

Consider first the statistical properties of configurations generated at a particular time step in cellular automaton evolution. A disordered initial state, in which each site takes on its k possible values with equal independent probabilities, is statistically random. Irreversible cellular automaton evolution generates deviations from statistical randomness. In a random sequence, all k^X possible subsequences ("blocks") of length X must occur with equal probabilities. Deviations from randomness imply unequal probabilities for different subsequences. With probabilities $p_i^{(x)}$ for the k^X possible sequences of site values in a length X block, one may define a specific "spatial set entropy"

$$s^{(x)}(X) = \frac{1}{X} \log_k \left(\sum_{j=1}^{k^X} \theta(p_j^{(x)}) \right), \tag{4.1}$$

where $\theta(p) = 1$ for $p > 0$ and $\theta(0) = 0$, and a specific "spatial measure entropy"

$$s_\mu^{(x)}(X) = -\frac{1}{X} \sum_{j=1}^{k^X} p_j^{(x)} \log_k p_j^{(x)}. \tag{4.2}$$

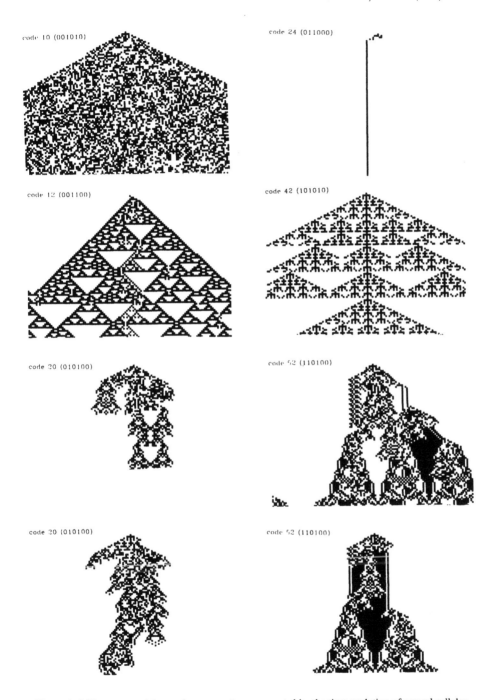

Figure 3. Differences modulo two between patterns generated by the time evolution of several cellular automata illustrated in fig. 1 with disordered states differing by a change in the value of a single site.

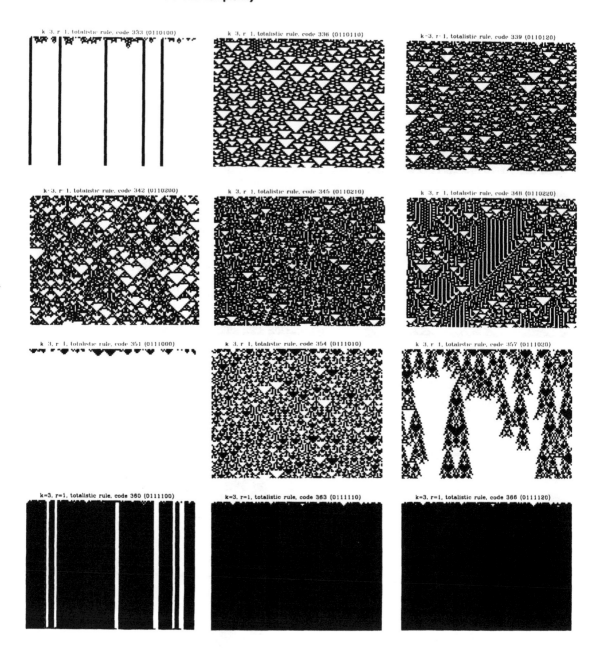

Figure 4. Examples of the evolution of typical cellular automata with $k = 3$ (three possible site values) and $r = 1$ (only nearest neighbours included in time evolution rules). White squares represent value 0, grey squares value 1, and black squares value 2. The initial state is taken disordered, with each site having values 0, 1 and 2 with equal independent probabilities.

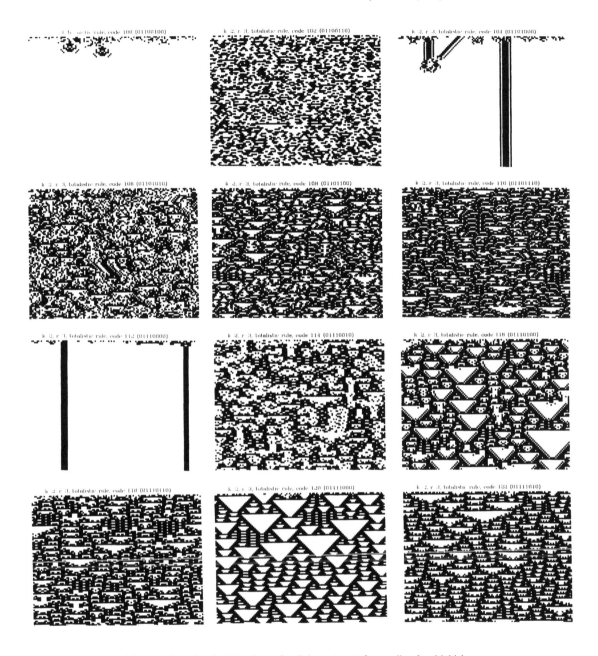

Figure 5. Examples of the evolution of typical $k = 2$, $r = 3$ cellular automata from a disordered initial state.

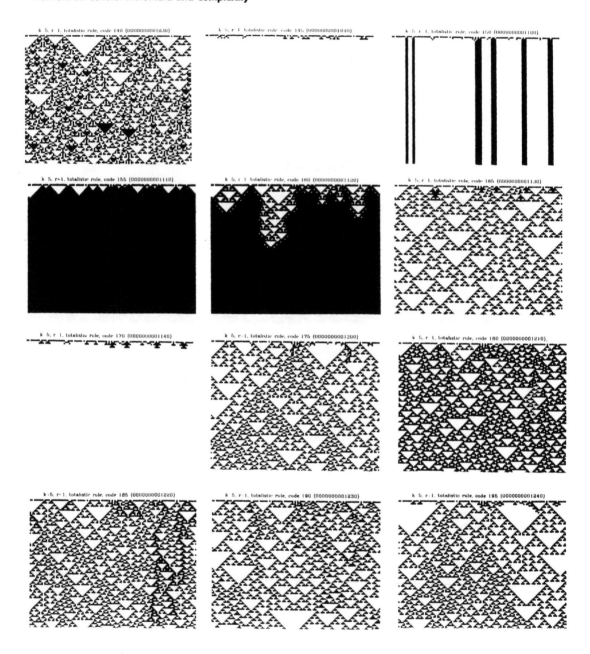

Figure 6. Examples of the evolution of typical $k = 5$, $r = 1$ cellular automata from a disordered initial state. Darker squares represent sites with larger values.

In both cases, the superscript (x) indicates that "spatial" sequences (obtained at a particular time step) are considered. The "set entropy" (4.1) is determined directly by the total number $N^{(x)}(X)$ of length X blocks generated (with any nonzero probability) in cellular automaton evolution, according to

$$s^{(x)}(X) = \frac{1}{X} \log_k N^{(x)}(X). \tag{4.3}$$

In the "measure entropy" (4.2) each block is weighted with its probability, so that the result depends explicitly on the probability measure for different cellular automaton configurations, as indicated by the subscript μ. Set entropy is often called "topological entropy"; measure entropy is sometimes referred to as "metric entropy"* (e.g. [6]). For blocks of length 1, the measure entropy $s_\mu^{(x)}(1)$ is related to the densities ρ_i, of sites with each of the k possible values i. $s_\mu^{(x)}(2)$ is related to the densities of "digrams" (blocks of length 2), and so on. In general, the measure entropy gives the average "information content" per site computed by allowing for correlations in blocks of sites up to length X. Note that the entropies (4.1) and (4.2) may be considered to have units of (k-ary) bits per unit distance.

In the equations below, $s_{(\mu)}^{(x)}$ stands for either set entropy $s^{(x)}$ or for measure entropy $s_\mu^{(x)}$.

The definitions (4.1) and (4.2) yield immediately

$$s_\mu^{(x)}(X) \le s^{(x)}(X) \le 1. \tag{4.4}$$

The first inequality is saturated (equality holds) only for "equidistributed" systems, in which all nonzero block probabilities $p_i^{(x)}$ are equal. The second inequality is saturated if all possible length X blocks of site values occur, but perhaps with unequal probabilities. $s_\mu(X) = 1$ only for "X-random" sequences [7], in which all k^X possible sequences of X site values occur with equal probabilities. In addition to (4.4), the definitions (4.1) and (4.2) imply

$$0 \le s_\mu^{(x)}(X) \le s^{(x)}(X). \tag{4.5}$$

$s_\mu^{(x)}(X) = 0$ if and only if just one length X block occurs with nonzero probability, so that $s^{(x)}(X) = 0$ also. As discussed below, the inequality (4.5) is saturated for class 1 cellular automata.

Both set and measure entropies satisfy the subadditivity condition

$$(X_1 + X_2)s_{(\mu)}^{(x)}(X_1 + X_2) \le X_1 s_{(\mu)}^{(x)}(X_1) + X_2 s_{(\mu)}^{(x)}(X_2). \tag{4.6}$$

The inequality is saturated if successive blocks of sites are statistically uncorrelated. In general, it implies some decrease in $s_{(\mu)}^{(x)}(X)$ with X (for example, $s_{(\mu)}^{(x)}(2X) \le s_{(\mu)}^{(x)}(X)$). For cellular automata with translation invariant initial probability measures, stronger constraints may be obtained (analogous to those for "stationary" processes

---

* The terms "set" and "measure" entropy, together with "set" and "measure" dimension, are introduced here to rationalize nomenclature.

in communication theory [8]). First, note that bounds on $s^{(x)}_{(\mu)}(X)$ valid for any set of probabilities $p^{(x)}_i$ also apply to $s^{(x)}(X)$, since $s^{(x)}(X)$ may formally be reproduced from the definition (4.2) for $s^{(x)}_\mu(X)$ by a suitable (extreme) choice of the $p^{(x)}_i$. The probability $p^{(x)}[a_1, \ldots, a_X]$ for the sequence of site values $a_1, \ldots, a_X$ is given in general by

$$p^{(x)}[a_1, \ldots, a_X] = p^{(x)}[a_1, \ldots, a_{X-1}]p^{(x)}[a_X | a_1, \ldots, a_{X-1}], \tag{4.7}$$

where $p^{(x)}[a_X | a_1, \ldots, a_{X-1}]$ denotes the conditional probability for a site value a_X, preceded by site values $a_1, \ldots, a_{X-1}$. Defining a total entropy

$$S^{(x)}_\mu[a_1, \ldots, a_X] = -\sum p^{(x)}[a_1, \ldots, a_X] \log_k p^{(x)}[a_1, \ldots, a_X], \tag{4.8}$$

and corresponding conditional total entropy

$$S^{(x)}_\mu[a_X | a_1, \ldots, a_{X-1}] = -\sum p^{(x)}[a_1, \ldots, a_X] \log_k p^{(x)}[a_X | a_1, \ldots, a_{X-1}]$$
$$\leq S^{(x)}_\mu[a_1, \ldots, a_X], \tag{4.9}$$

one obtains

$$X s^{(x)}_\mu(X) = S^{(x)}_\mu(X) \leq \frac{X-1}{X} S^{(x)}_\mu(X-1) + \frac{1}{X} S^{(x)}_\mu(X). \tag{4.10}$$

Hence

$$s^{(x)}_{(\mu)}(X) \leq s^{(x)}_{(\mu)}(X-1), \tag{4.11}$$

so that the set and measure entropies for a translationally invariant system decrease monotonically with the block size X. One finds in addition in this case that

$$\Delta^2_X \left(X s^{(x)}_{(\mu)}(X) \right) = (X+1) s^{(x)}_{(\mu)}(X+1) - 2X s^{(x)}_{(\mu)}(X)$$
$$+ (X-1) s^{(x)}_{(\mu)}(X-1) \leq 0, \tag{4.12}$$

so that $X s^{(x)}_{(\mu)}(X)$ is a convex function of X.

With the definition $s^{(x)}(0) = 1$, this implies that there exists a critical block size X_c, such that

$$s^{(x)}(X) = 1, \qquad \text{for } X < X_c,$$
$$s^{(x)}(X) < 1, \qquad \text{for } X \geq X_c. \tag{4.13}$$

The significance and values of the critical block size X_c will be discussed in section 7.

The entropies $s^{(x)}$ and $s^{(x)}_\mu$ may be evaluated either for many blocks in a single cellular automaton configuration, or for blocks in an ensemble of different configurations. For smooth probability measures on the ensemble of possible initial configurations, the results obtained in these two ways are almost always the same. (A probability measure will be considered "smooth" if changes in the values of a few sites in an infinite configuration lead only to infinitesimal changes in the probability for the configuration.) The set entropy $s^{(x)}$ is typically independent of the probability measure on the ensemble, for any smooth measure. The measure entropy $s_\mu^{(x)}$ in general depends on the probability measure for initial configurations, although for class 3 cellular automata, it is typically the same for at least a large class of smooth measures. Notice that with smooth measures, the values of $s^{(x)}(X)$ and $s_\mu^{(x)}(X)$ are the same whether the length X blocks used in their computation are taken disjoint or overlapping.

The entropies (4.1) and (4.2) are defined for infinite cellular automata. A corresponding definition may be given for finite cellular automata, with a maximum block length given by the total number of sites N in the cellular automaton. The entropies $s^{(x)}(N)$ and $s_\mu^{(x)}(N)$ are related to global properties of the state transition diagram for the finite cellular automaton. The value of $s^{(x)}(N)$ at a particular time is determined by the fraction of possible configurations which may be reached at that time by evolution from any initial configuration. The limiting value of $s^{(x)}(N)$ at large times is determined by the fraction of configuration on cycles in the state transition graph. Starting from an initial ensemble in which all k^N configurations occur with equal probabilities, the limiting value of $s_\mu^{(x)}(N)$ is equal to the limiting value of $s^{(x)}(N)$ if all transient trees in the state transition graph for the finite cellular automaton are identical, so that all configurations with nonzero probabilities are generated with the same probability (cf. [2]).

As mentioned in section 2, the configurations of an infinite cellular automaton may be considered as elements of a Cantor set. For an ensemble of disordered configurations (in which each site takes on its k possible values with equal independent probabilities), this Cantor set has fractal dimension 1. Irreversible cellular automaton evolution may lead to an ensemble of configurations corresponding to elements of a Cantor set with dimension less than one. The limiting value of $s^{(x)}(X)$ as $X \to \infty$ gives the fractal or "set" dimension of this set.

Relations between entropy and dimension may be derived in many ways (e.g. [6, 9]). Consider a set of numbers in the interval [0, 1] of the real line. Divide this interval into k^b bins of width k^{-b}, and let the fraction of bins containing numbers in the set be $N(b)$. For large b (small bin width), this number grows as k^{db}. The exponent d is the Kolmogorov dimension (or "capacity" (cf. [8])) of the set. If the set contains all real numbers in the interval [0, 1], then $N(b) = k^b$, and $d = 1$, as expected. If the set contains only a finite number of points, then $N(b)$ must tend to a constant for large b, yielding $d = 0$. The classic Cantor set consists of real numbers in the interval [0, 1], whose ternary decomposition contains only the digits 0 and 2. Dividing the interval into 3^b equal bins, it is clear that 2^b of these bins contain points in the set. The dimension of the set is thus $\log_3 2$. This dimension may also be found by an explicit recursive geometrical construction, using the fact that the set is "self-similar", in the sense that with appropriate magnification, its parts are identical to the whole.

The example above suggests that one may define a "set dimension" d according to

$$d = \lim_{b \to \infty} \frac{1}{b} \log_k N(b),$$
(4.14)

where $N(b)$ is the number of bins which contain elements of the set. The bins are of equal size, and their total number is taken as k^b. Except in particularly pathological examples,* the dimension obtained with this definition is equal to the more usual Hausdorff (or "fractal") dimension (e.g. [11]) obtained by considering the number of patches at arbitrary positions required to cover the set (rather than the number of fixed bins containing elements of the set).

The definition (4.14) may be applied directly to cellular automaton configurations. The k^b "bins" may be taken to consist of cellular automaton configurations in which a block of b sites has a particular sequence of values. The definition (4.3) of set entropy then shows that the set dimension is given by

$$d^{(x)} = \lim_{X \to \infty} s^{(x)}(X).$$
(4.15)

A disordered cellular automaton configuration, in which all possible sequences of site values occur with nonzero probability (or an ensemble of such configurations), gives $d^{(x)} = 1$, as expected. Similarly, a homogeneous configuration, such as the null configuration, gives $d^{(x)} = 0$.

The set of configurations which appear at large times in the evolution of a cellular automaton constitute the attractors for the cellular automaton. The set dimension of these attractors is given in terms of the entropies for configurations appearing at large times by eq. (4.15).

Accurate direct evaluation of the set entropy $s^{(x)}(X)$ from cellular automaton configurations typically requires sampling of many more than k^X length X blocks. Inadequate samples yield systematic underestimates of $s^{(x)}(X)$. Direct estimates are most accurate when all nonzero probabilities for length X blocks are equal. In this case, a sample of k^b blocks yields an entropy underestimated on average by approximately

$$\log_k(1 - \exp(-k^{b-Xs(X)})).$$
(4.16)

Unequal probabilities increase the magnitude of this error, and typically prevent the generation of satisfactory estimates of $d^{(x)}$ from direct simulations of cellular automaton evolution. (If the probabilities follow a log normal distribution, as in many continuous chaotic dynamical systems [12], then the exponential in eq. (4.16) is apparently replaced by a power [13].)

The dimension (4.15) is given as the limiting exponent with which $N^{(x)}(X)$ increases for large X. In the formula (4.15), this exponent is obtained as the limit of

---

* Such as the set formed from the end points of the intervals at each stage in the geometrical construction of the classic Cantor set. This set has zero Hausdorff dimension, but Kolmogorov dimension $\log_3 2$ [9].

$\log_k[N(X)^{1/X}]$ for large X. If $N^{(x)}(X)$ indeed increases roughly exponentially with X, then the alternative formula

$$d^{(x)} = \lim_{X\to\infty} \frac{Xs^{(x)}(X)}{(X-1)s^{(x)}(X-1)} = \lim_{X\to\infty} \log_k\left[\frac{N^{(x)}(X)}{N^{(x)}(X-1)}\right] \qquad (4.17)$$

is typically more accurate if entropy values are available only for small X.

The set dimension (4.15) may be used to characterize the set of configurations occurring on the attractor for a cellular automaton, without regard to their probabilities. One may also define a "measure dimension" $d_\mu^{(x)}$ which characterizes the probability measure for the configurations (cf. [12]):

$$d_\mu^{(x)} = \lim_{X\to\infty} s_\mu^{(x)}(X). \qquad (4.18)$$

It is clear that

$$0 \le d_\mu^{(x)} \le d^{(x)} \le 1. \qquad (4.19)$$

The measure dimension $d_\mu^{(x)}$ is equal to the "average information per symbol" contained in the sequence of site values in a cellular automaton configuration. If the sequence is completely random (or "∞-random" [7]), then the probabilities $p_i^{(x)}$ for all k^X sequences of length X must be equal for all X, so that $d_\mu^{(x)} = 1$. In this case, there is no redundancy or pattern in the sequence of site values, so that determination of each site value represents accquisition of one (k-ary) bit of information. A cellular automaton configuration with any structure or pattern must give $d_\mu^{(x)} < 1$.

In direct simulations of cellular automaton evolution, the probabilities $p_i^{(x)}$ for each possible length X block are estimated from the frequencies with which the blocks occur. These estimated probabilities are thus subject to Gaussian errors. Although the individual estimated probabilities are unbiased, the measure entropy deduced from them according to eq. (4.2), is systematically biased. Its mean typically yields a systematic underestimate of the true measure entropy, and with fixed sample size, the underestimate deteriorates rapidly with increasing X, making an accurate estimate of $d_\mu^{(x)}$ impossible. However, since an unbiased estimate may be given for any polynomial function of the $p_i^{(x)}$, unbiased estimated upper and lower bounds for the measure entropy may be obtained from estimates for polynomials in $p_i^{(x)}$ just larger and just smaller than $-p_i^{(x)} \log_k p_i^{(x)}$ for $0 \le p_i^{(x)} \le 1$ [14]. In this way, it may be possible to obtain more accurate estimates of $s_\mu^{(x)}(X)$ for large X, and thus of $d_\mu^{(x)}$.

The "spatial" entropies (4.1) and (4.2) were defined in terms of the sequence of site values in a cellular automaton configuration at a particular time step. One may also define "temporal" entropies which characterize the sequence of values taken on by a particular site through many time steps of cellular automaton evolution, as illustrated in fig. 7. With probabilities $p_i^{(t)}$ for the k^T possible sequences

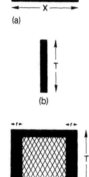

(a)

(b)

Figure 7. Space-time regions sampled in the computation of (a) spatial entropies, (b) temporal entropies and (c) patch or mapping entropies. In case (c), the values of sites in the cross-hatched area are completely determined by values in the black "rind".

(c)

of values for a site at T successive time steps, one may define a specific temporal set entropy in analogy with eq. (4.1) by

$$s^{(t)}(T) = \frac{1}{T} \log_k \left(\sum_{j=1}^{k^T} \theta(p_j^{(t)}) \right),$$ (4.20)

and a specific temporal measure entropy in analogy with eq. (4.2) by

$$s_\mu^{(t)}(T) = -\frac{1}{T} \sum_{j=1}^{k^T} p_j^{(t)} \log_k p_j^{(t)}.$$ (4.21)

These entropies satisfy relations directly analogous to those given in eqs. (4.3) through (4.6) for spatial entropies. They obey relations analogous to (4.11) and (4.12) only for cellular automata in "equilibrium", statistically independent of time. The temporal entropies (4.20) and (4.21) may be considered to have units of (k-ary) bits per unit time.

Sequences of values in particular cellular automaton configurations typically have little similarity with the "time series" of values attained by a particular site under cellular automaton evolution. The spatial and temporal entropies for a cellular automaton are therefore in general quite different. Notice that the spatial entropy of a cellular automaton configuration may be considered as the temporal entropy of a pure shift mapping applied to the cellular automaton configuration.

Just as dimensions may be assigned to the set of spatial configurations generated in cellular automaton evolution, so also one may assign dimensions to the set of temporal sequences generated by the evolution. The temporal set dimension may be defined in analogy with eq. (4.15) by

$$d^{(t)} = \lim_{T \to \infty} s^{(t)}(T),$$ (4.22)

and the temporal measure dimension may be defined by

$$d_\mu^{(t)} = \lim_{T \to \infty} s_\mu^{(t)}(T).$$
(4.23)

If the evolution of a cellular automaton is periodic, so that each site takes on a fixed cycle of values, then

$$d^{(t)} = d_\mu^{(t)} = 0.$$
(4.24)

As discussed in section 6 below, class 2 cellular automata yield periodic structures at large times, so that the correspondingly temporal entropies vanish.

As a generalization of the spatial and temporal entropies introduced above, one may consider entropies associated with space-time "patches" in the patterns generated by cellular automaton evolution, as illustrated in fig. 7. With probabilities $p_i^{(t,x)}$ for the k^{XT} possible patches of spatial width X and temporal extent T, one may define a set entropy

$$s^{(t;x)}(T; X) = \frac{1}{T} \log_k \left(\sum_{j=1}^{k^{XT}} \theta(p_j^{(t,x)}) \right),$$
(4.25)

and a measure entropy

$$s_\mu^{(t;x)}(T; X) = -\frac{1}{T} \sum_{j=1}^{k^{XT}} p_j^{(t,x)} \log_k p_j^{(t,x)}.$$
(4.26)

Clearly

$$s_{(\mu)}^{(t)}(T) = s_{(\mu)}^{(t;x)}(T; 1),$$
(4.27)

$$s_{(\mu)}^{(x)}(X) = \frac{1}{X} s_{(\mu)}^{(t;x)}(1; X).$$

If no relation existed between configurations at successive time steps then the entropies (4.25) and (4.26) would be bounded simply by

$$s_\mu^{(t;x)}(T; X) \le s^{(t;x)}(T; X) \le X.$$
(4.28)

The cellular automaton rules introduce definite relations between successive configurations and tighten this bound. In fact, the values of all sites in a $T \times X$ space-time patch are determined according to the cellular automaton rules by the values in the "rind" of the patch, as indicated in fig. 7. The rind contains only $X + 2r(T - 1)$ sites (where r is the "range" of the cellular automaton rule, defined in section 2), so that

$$s_\mu^{(t;x)}(T; X) \le s^{(t;x)}(T; X) \le [X + 2r(T - 1)]/T.$$
(4.29)

For large T (and fixed X), therefore

$$s_\mu^{(t;x)}(T; X) \le s^{(t;x)}(T; X) \le 2r.$$
(4.30)

If both X and T tend to infinity with T/X fixed, eq. (4.30) implies that the "information per site" $s_\mu^{(t;x)}(T; X)/X$ in a $T \times X$ patch must tend to zero. The evolution of cellular automata can therefore never generate random space-time patterns.

With $T \to \infty$, X fixed, the length X horizontal section of the rind makes a negligible contribution to the entropies. The entropy is maximal if the $2r$ vertical columns in the rind are statistically independent, so that

$$s_{(\mu)}^{(t;x)}(\infty; X) \le 2r s_{(\mu)}^{(t)}(\infty) = 2r d_{(\mu)}^{(t)}. \tag{4.31}$$

In addition,

$$s_{(\mu)}^{(t;x)}(\infty; X) \le s_{(\mu)}^{(t;x)}(\infty; X + 1), \tag{4.32}$$

where the bounds are saturated for large X if the time series associated with different sets of sites are statistically uncorrelated.

The limiting set entropy

$$\mathbf{h} = \lim_{\substack{T \to \infty \\ X \to \infty \\ T/X \to \infty}} s^{(t;x)}(T; X) \tag{4.33}$$

for temporally-extended patches is a fundamental quantity equivalent to the set (or topological) entropy of the cellular automaton mapping in symbolic dynamics. $\mathbf{h}$ may be considered as a dimension for the mapping. It specifies the asymptotic rate at which the number of possible histories for the cellular automaton increases with time. The limiting measure entropy

$$\mathbf{h}_\mu = \lim_{\substack{T \to \infty \\ X \to \infty \\ T/X \to \infty}} s_\mu^{(t;x)}(T; X) \tag{4.34}$$

gives the average amount of "new information" contained in each cellular automaton configuration, and not already determined from previous configurations. Equations (4.31) and (4.32) show that

$$d_{(\mu)}^{(t)} \le \mathbf{h}_{(\mu)} \le 2r d_{(\mu)}^{(t)}. \tag{4.35}$$

In addition,

$$\mathbf{h}_{(\mu)} \le 2r d_{(\mu)}^{(x)}. \tag{4.36}$$

The basic cellular automaton time evolution rule (2.1) implies that the value a_i of a site i at a particular time step depends on sites a maximum distance r away on the previous time step according to the function $\mathbf{F}[a_{i-r}, \dots, a_{i+r}]$. After T time steps, the values of the site could depend on sites at distances up to rT, so that features in patterns generated by cellular automaton evolution could propagate at "speeds" up to r sites per time step. For many rules, however, the value of a site after many time steps depends on fewer initial site values, and features may propagate only at lower speeds. In general, let $\|\mathbf{F}^T\|$ denote the minimum R for which the value of site i depends only on the initial values of sites $i - R, \ldots, i + R$. Then the maximum propagation speed associated with the cellular automaton rule $\mathbf{F}$ may be defined as

$$\lambda_+ = \overline{\lim_{T \to \infty}} \|\mathbf{F}^T\|/T. \tag{4.37}$$

(The rule is assumed symmetric; for nonsymmetric rules, distinct left and right propagation speeds may be defined.) Clearly,

$$\lambda_+ \leq r. \tag{4.38}$$

When $\lambda_+ = 0$, finite regions of the cellular automaton must ultimately become isolated, so that

$$d_{(\mu)}^{(t)} = \mathbf{h}_{(\mu)}^{(t)} = 0. \tag{4.39}$$

The construction of fig. 8 shows that for any T,

$$s_{(\mu)}^{(t)}(T) \leq 2r s_{(\mu)}^{(x)}(2rT). \tag{4.40}$$

In the limit $T \to \infty$, the construction implies

$$d_{(\mu)}^{(t)} \leq 2\lambda_+ d_{(\mu)}^{(x)}. \tag{4.41}$$

The ratio of temporal to spatial entropy is thus bounded by the maximum propagation speed in the cellular automaton. The relation is consistent with the assignment of units to the spatial and temporal entropies mentioned above.

The corresponding inequalities for mapping entropies are:

$$d_{(\mu)}^{(t)} \leq \mathbf{h}_{(\mu)} \leq 2\lambda_+ d_{(\mu)}^{(x)},$$

$$\mathbf{h}_{(\mu)} \leq 2r d_{(\mu)}^{(t)}. \tag{4.42}$$

The quantity λ_+ defined by eq. (4.37) gives the maximum speed with which any feature in a cellular automaton may propagate. With many cellular automaton rules, however, almost all "features" propagate much more slowly. To define an appropriate maximum average propagation speed, consider the effect after many time steps of changes in the initial state. Let $G(|x - x'|; t)$ denote the probability that the value of a site at position x' is changed when the value of a site at position x is changed t time steps before. The form of $G(|x - x'|; t)$ for various cellular automaton rules

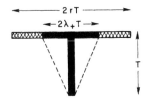

Figure 8. Pattern of dependence of temporal sequences on spatial sequences, used in the proof of inequalities between spatial and temporal entropies.

is suggested by fig. 3. $G(|x - x'|; t)$ may be considered as a Green function for the cellular automaton evolution. For large t, $G(|x - x'|; t)$ typically vanishes outside a "cone" defined by $|x - x'| = \bar{\lambda}_+ t$. $\bar{\lambda}_+$ may then be considered as a maximum average propagation speed. In analogy with eqs. (4.41) and (4.42), one expects

$$d^{(t)}_{(\mu)} \le \mathbf{h}_{(\mu)} \le 2\bar{\lambda}_+ d^{(t)}_{(\mu)}. \tag{4.43}$$

Mapping and temporal entropies thus vanish for cellular automata with zero maximum average propagation speed. Cellular automata in class 2 have this property.

The maximum average propagation speed $\bar{\lambda}_+$ specifies a cone outside which $G(|x - x'|; t)$ almost always vanishes. One may also define a minimum average propagation speed $\bar{\lambda}_-$, such that $G(|x - x'|; t) > 0$ for almost any $|x - x'| < \bar{\lambda}_-$.

The Green function $G(|x - x'|; t)$ gives the probability that a particular site is affected by changes in a previous configuration. The total effect of changes may be measured by the "Hamming distance" $H(t)$ between configurations before and after the changes, defined as the total number of site values which differ between the configurations after t time steps. ($H(t)$ is analogous to Lyapunov exponents for continuous dynamical systems.) Changing the values of initial sites in a small region, $H(t)$ may be given as a space integral of the Green function, and for large t obeys the inequality

$$H(t)/t \le 2\bar{\lambda}_+, \tag{4.44}$$

to be compared with the result (4.43) obtained above.

The definitions and properties of dimension given above suggest that the behaviour of these quantities determines the degree of "chaotic" behaviour associated with cellular automaton evolution. "Spatial chaos" occurs when $d^{(x)}_{(\mu)} > 0$, and "temporal chaos" when $d^{(t)}_{(\mu)} > 0$. Temporal chaos requires a nonzero maximum average propagation speed for features in cellular automaton patterns, and implies that small changes in initial conditions lead to effects ever-increasing with time.

5. Class 1 Cellular Automata

Class 1 cellular automata evolve after a finite number of time steps from almost all initial states to a unique homogeneous state, in which all sites have the same value. Such cellular automata may be considered to evolve to simple "limit points" in phase space; their evolution completely destroys any information on the initial state. The spatial and temporal dimensions for such attractors are zero.

Rules for class 1 cellular automata typically take the function $\mathbf{F}$ of eq. (2.1) to have the same value for almost all of its $k^{(2r+1)}$ possible sets of arguments.

Some exceptional configurations in finite class 1 cellular automata may not evolve to a homogeneous state, but may in fact enter non-trivial cycles. The fraction of such exceptional configurations appears to decrease very rapidly with the size N, suggesting that for infinite class 1 cellular automata the set of exceptional configurations is always of measure zero in the set of all possible configurations. For (legal) class

1 cellular automata whose usual final state has $a_i = n$, $n \neq 0$ (such as code 60 in fig. 1), the null configuration is exceptional for any size N, and yields $a_i = 0$.

6. Class 2 Cellular Automata

Class 2 cellular automata serve as "filters" which generate separated simple structures from particular (typically short) initial site value sequences.* The density of appropriate sequences in a particular initial state therefore determines the statistical properties of the final state into which it evolves. (There is therefore no unique large-time (invariant) probability measure on the set of possible configurations.) Changes of site values in the initial state almost always affect final site values only within a finite range, typically of order r. The maximum average propagation speed $\bar{\lambda}_+$ defined in section 4 thus vanishes for class 2 cellular automata. The temporal and mapping (but not spatial) dimensions for such automata therefore also vanish.

Although $\bar{\lambda} = 0$ for all class 2 cellular automata, λ is often nonzero. Thus exceptional initial states may exist, from which, for example, unbounded growth may occur. Such initial states apparently occur with probability zero for ensembles of (spatially infinite) cellular automata with smooth probability measures.

The simple structures generated by class 2 cellular automata are either stable, or are periodic, typically with small periods. The class 2 rules with codes 8, 24, 40 and 56 illustrated in fig. 1 all apparently exhibit only stable persistent structures. Examples of class 2 cellular automata which yield periodic, rather than stable, persistent structures include the $k = 2$, $r = 1$ cellular automaton with rule number 108 [1], and the $k = 3$, $r = 1$ totalistic cellular automaton with code 198. The periods of persistent structures generated in the evolution of class 2 cellular automata are usually less than $k!$. However, examples have been found with larger periods. One is the $k = 2$, $r = 3$ totalistic cellular automata with code 228, in which a persistent structure with period 3 is generated.

The finiteness of the periods obtained at large times in class 2 cellular automata implies that such systems have $d_{(\mu)}^{(t)} = \mathbf{h}_{(\mu)} = 0$, as deduced above from the vanishing of $\bar{\lambda}_+$. The evolution of class 2 cellular automata to zero (temporal) dimension attractors is analogous to the evolution of some continuous dynamical systems to limit cycles.

The set of persistent structures generated by a given class 2 cellular automaton is typically quite simple. For some rules, there are only a finite number of persistent structures. For example, for the code 8 and code 40 rules of fig. 1, only the sequence 111 (surrounded by 0 sites) appears to be persistent. For code 24, 111 and 1111 are both persistent. Other rules yield an infinite sequence of persistent structures, typically constructed by a simple process. For example, with code 56 in fig. 1, any sequence of two or more consecutive 1 sites is persistent.

---

* They are thus of direct significance for digital image processing.

In general, it appears that the set of persistent structures generated by any class 2 cellular automaton corresponds to the set of words generated by a regular grammar. A regular grammar [15–18] (or "sofic system" [19]) specifies a regular language, whose legal words may be recognized by a finite automaton, represented by a finite state transition graph. A sequence of symbols (site values) specifies a particular traversal of the state transition graph. The traversal begins at a special "start" node; the symbol sequence represents a legal word only if the traversal does not end at an absorbing "stop" node. Each successive symbol in the sequence causes the automaton to make a transition from one state (node) to one of k others, as specified by the state transition graph. At each step, the next state of the automaton depends only on its current state, and the current symbol read, but not on its previous history.

The set of configurations (symbol sequences) generated from all possible initial configurations by one time step of cellular automaton evolution may always be specified by a regular grammar. To determine whether a particular configuration $a^{(1)}$ may be generated after one time step of cellular automaton evolution, one may attempt to construct an explicit predecessor $a^{(0)}$ for it. Assume that a predecessor configuration has been found which reproduces all site values up to position i. Definite values $a_j^{(0)}$ for all $j \leq i - r$ are then determined. Several of the total of k^{2r} sequences of values $a_{i-r+1}^{(0)}, \ldots, a_{i+r+1}^{(0)}$ may be possible. Each sequence may be specified by an integer $q = \sum_{j=0}^{2r} k^j a_{i-r+j+1}^{(0)}$. An integer ψ_i between 0 and $2^{k^{2r}}$ may then be defined, with the qth binary bit in ψ_i equal to one if sequence q is allowed, and 0 otherwise. Each possible value of ψ may be considered to correspond to a state in a finite automaton. $\psi = 0$ corresponds to a "stop" state, which is reached if and only if $a^{(1)}$ has no predecessors. Possible values for $a_{i+r+1}^{(0)}$ are then found from ψ_i and the value of $a_{i+1}^{(0)}$. These possible values then determine the value of ψ_{i+1}. A finite state transition graph, determined by the cellular automaton rules, gives the possible transitions $\psi_i \to \psi_{i+1}$. Configurations reached after one time step of cellular automaton evolution may thus be recognized by a finite automaton with at most $2^{k^{2r}}$ states. The set of such configurations is thus specified by a regular grammar.

In general, if the value of a given site after t steps of cellular automaton evolution depends on m initial site values, then the set of configurations generated by this evolution may be recognized by a finite automaton with at most 2^{k^m} states. The value of m may increase as $2rt$, potentially requiring an infinite number of states in the recognizing automaton, and preventing the specification of the set of possible configurations by a regular grammar. However, as discussed above, the value of m for a class 2 cellular automaton apparently remains finite as $t \to \infty$. Thus the set of configurations which may persist in such a cellular automaton may be recognized by a finite automaton, and are therefore specified by a regular grammar. The complexity of this grammar (measured by the minimum number of states required in the state transition graph for the recognizing automaton) may be used to characterize the complexity of the large time behaviour of the cellular automaton.

Finite class 2 cellular automata usually evolve to short period cycles containing the same persistent structures as are found in the infinite case. The fraction of exceptional initial states yielding other structures decreases rapidly to zero as N increases.

7. Class 3 Cellular Automata

Evolution of infinite class 3 cellular automata from almost all possible initial states leads to aperiodic ("chaotic") patterns. After sufficiently many time steps, the statistical properties of these patterns are typically the same for almost all initial states. In particular, the density of nonzero sites typically tends to a fixed nonzero value (often close to $1/k$). In infinite cellular automata, "equilibrium" values of statistical quantities are approached roughly exponentially with time, and are typically attained to high accuracy after a very few time steps. For a few rules (such as the $k = 2, r = 1$ rule with rule number 18 [20]), however, "defects" consisting of small groups of sites may exist, and may execute approximate random walks, until annihilating, usually in pairs. Such processes lead to transients which decrease with time only as $t^{-1/2}$.

Figure 1 showed examples of the patterns generated by evolution of some typical class 3 cellular automata from disordered initial states. The patterns range from highly irregular (as for code 10), to rather regular (as for code 12). The most obvious regularity is the appearance of large triangular "clearings" in which all sites have the same value. These clearings occur when a "fluctuation" in which a sequence of consequence of consecutive sites have the same value, is progressively destroyed by the effects of other sites. The rate at which "information" from other sites may "flow" into the fluctuation, and thus the slope of the boundaries of the clearing, may range from $1/k$ to r sites per time step. The qualitative regularity of patterns generated by some class 3 rules arises from the high density of long sequences of correlated site values, and thus of triangular clearings. In general, however, it appears that the density of clearings decreases with their size n roughly as σ^{-n}. Different cellular automata appear to yield a continuous range of σ values. Those with larger σ yield more regular patterns, while those with smaller σ yield more irregular patterns. No sharp distinction appears to exist between class 3 cellular automata yielding regular and irregular patterns.

The first column of fig. 9 shows patterns obtained by evolution with typical class 3 cellular automaton rules from initial states containing a single nonzero site. Unbounded growth, leading to an asymptotically infinite number of nonzero sites, is evident in all cases. Some rules are seen to give highly regular patterns, others lead to irregular patterns.

The regular patterns obtained with rules such as code 2 are asymptotically self-similar fractal curves (cf. [11]). Their form is identical when viewed at different magnifications, down to length scales of order r sites. The total number of nonzero sites in such patterns after t time steps approaches t^d, where d gives the fractal dimension of the pattern. Many class 3 $k = 2$ rules generate a similar pattern, illustrated by codes 2 and 34 in fig. 9, with $d = \log_2 3 \approx 1.59$. Some rules yield

143

Figure 9. Evolution of some class 3 totalistic cellular automata with $k = 2$ and $r = 2$ (as illustrated in fig. 1) from initial states containing one or a few nonzero sites. Some cases yield asymptotically self-similar patterns, while others are seen to give irregular patterns.

self-similar patterns with other fractal dimensions (for example, code 38 yields $d \approx 1.75$), but all self-similar patterns have $d < 2$, and lead to an asymptotic density of sites which tends to zero as t^{d-2}.

Rules such as code 10 are seen to generate irregular patterns by evolution even from a single site initial state. The density of nonzero sites in such patterns is found to tend asymptotically to a nonzero value; in some, but not all, cases the value is the same as would be obtained by evolution from a disordered initial state. The patterns appear to exhibit no large-scale structure.

Cellular automata contain no intrinsic scale beyond the size of neighbourhood which appears in their rules. A configuration containing a single nonzero site is also scale invariant, and any pattern obtained by evolution from it with cellular automaton rules must be scale invariant. The regular patterns in fig. 9 achieve this scale invariance by their self-similarity. The irregular patterns presumably exhibit correlations only over a finite range, and are therefore effectively uniform and scale invariant at large distances.

The second and third columns in fig. 11 show the evolution of several typical class 3 cellular automata from initial states with nonzero sites in a small region. In some cases (such as code 12), the regular fractal patterns obtained with single nonzero sites are stable under addition of further nonzero initial sites. In other cases (such as code 2) they are seen to be unstable. The numbers of rules yielding stable and unstable fractal patterns are found to be roughly comparable.

Many but not all rules which evolve to regular fractal patterns from simple initial states generate more regular patterns in evolution from disordered initial states. Similarly, many but not all rules which produce stable fractal patterns yield more regular patterns from disordered initial states. For example, code 42 in figs. 1 and 9 generates stable fractal patterns from a simple initial state, but leads to an irregular pattern under evolution from a disordered state. (Although not necessary for such behaviour, this rule possesses the additivity property mentioned in section 2.)

The methods of section 4 may be used to analyse the general behaviour of class 3 cellular automata evolving from typical initial states, in which all sites have nonzero values with nonzero probability. Class 3 cellular automata apparently always exhibit a nonzero minimum average propagation speed $\overline{\lambda}_-$. Small changes in initial states thus almost always lead to increasingly large changes in later states. This suggests that both spatial and temporal dimensions $d_{(\mu)}^{(x)}$ and $d_{(\mu)}^{(t)}$ should be nonzero for all class 3 cellular automata. These dimensions are determined according to eqs. (4.15), (4.18), (4.22) and (4.23) by the limiting values of spatial and temporal entropies.

A disordered or statistically random initial state, in which each site takes on its k possible values with equal independent probabilities, has maximal spatial entropy $s_{(\mu)}^{(x)}(X) = 1$ for all block lengths X. Figure 10 shows the behaviour of $s_{\mu}^{(x)}(X)$ as a function of time for several block lengths X in the evolution of a typical class 3 cellular automaton from a disordered (maximal entropy) initial state. The entropies

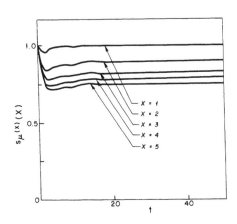

Figure 10. Evolution of spatial measure entropies $s_{\mu}^{(x)}(X)$ as a function of time for evolution of the class 3 cellular automaton with code 12 illustrated in fig. 1 from a disordered initial state. The irreversibility of cellular automaton evolution results in a decrease of the entropies with time. Rapid relaxation to an "equilibrium" state is nevertheless seen.

145

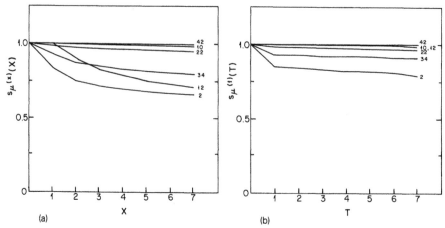

Figure 11. Evolution of (a) spatial and (b) temporal measure entropies $s_\mu^{(x)}(X)$ and $s_\mu^{(t)}(T)$ obtained at equilibrium by evolution of several class 3 cellular automata illustrated in fig. 1, as a function of the spatial and temporal block lengths X and T. The entropies are evaluated for the regions indicated in figs. 7(a) and 7(b). The limit of $s_\mu^{(x)}(X)$ as $X \to \infty$ is the spatial measure dimension of the attractor for the system; the limit of $s_\mu^{(t)}(T)$ as $T \to \infty$ is the temporal measure dimension.

are seen to decrease for a few time steps, and then to reach "equilibrium" values. The "equilibrium" values of $s_\mu^{(x)}(X)$ for class 3 cellular automata are typically independent of the probability measure on the ensemble of possible initial states, at least for "smooth" measures. The decrease in entropy with time manifests the irreversible nature of the cellular automaton evolution. The decrease is found to be much greater for class 3 cellular automata which generate regular patterns (with many triangular clearings) than for those which yield irregular patterns. The more regular patterns require a higher degree of self-organization, with correspondingly greater irreversibility, and larger entropy decrease.

As discussed in section 4, the dependence of $s_{(\mu)}^{(x)}(X)$ on X measures spatial correlations in cellular automaton configurations. $s_{(\mu)}^{(x)}(X)$ therefore tends to a constant if X is larger than the range of any correlations between site values. In the presence of correlations, $s_{(\mu)}^{(x)}(X)$ always decreases with X. Available data from simulations provide reliable accurate estimates for $s_{(\mu)}^{(x)}(X)$ only for $0 \le X \lesssim 8$. Figure 11 shows the behaviour of the equilibrium value of $s_\mu^{(x)}(X)$ as a function of X over this range for several typical class 3 cellular automata. For rules which yield irregular patterns the equilibrium value of $s_\mu^{(x)}(X)$ typically remains $\gtrsim 0.9$ for $X \lesssim 8$. $s_\mu^{(x)}(X)$ at equilibrium typically decreases much more rapidly for class 3 cellular automata which generate more regular patterns. At least for small X, $s_\mu^{(x)}(X)$ for such cellular automata typically decreases roughly as $X^{-\eta}$ with $\eta \approx 0.1$.

The values of the spatial set entropy $s^{(x)}(X)$ provide upper bounds on the spatial measure entropy $s_\mu^{(x)}(X)$. The distribution of nonzero probabilities $p_i^{(x)}$ for possible length X blocks is typically quite broad, yielding an $s_\mu^{(x)}(X)$ significantly smaller than $s^{(x)}(X)$. Nevertheless, the general behaviour of $s_\mu^{(x)}(X)$ with X usually roughly follows $s^{(x)}(X)$, but with a slight X delay.

As discussed in section 4, the set entropy $s^{(x)}(X)$ attains its maximum value of 1 if and only if all k^X sequences of length X appear (with nonzero probability) in evolution from some initial state. Notice that if $s^{(x)}(X) = 1$ after one time step, then $s^{(x)}(X) = 1$ at any time. In general, $s^{(x)}(X)$ takes on value 1 for blocks up to some critical length X_c (perhaps infinite), as defined in eq. (4.13).

Since a block of length X is completely determined by a sequence of length $X + 2r$ in the previous configuration, any predecessors for the block may in principle be found by an exhaustive search of all k^{X+2r} possible length $X + 2r$ sequences. The procedure for progressive construction of predecessors outlined in section 6 provides a more efficient procedure [21]. The critical block length X_c is determined by the minimum number of nodes in the finite automaton state transition graph visited on any path from the "start" to "stop" node. The state transition graph is determined by the set of transition rules $\Psi_i \rightarrow \Psi_{i+1}$. Starting with length 1 blocks, these transition rules may be found by considering construction of all possible progressively longer blocks, but ignoring blocks associated with values Ψ_i for which the transition rules have already been found. If X_c is finite, the "stop" node $\Psi = 0$ is reached in the construction of length X_c blocks. Alternatively, the state transition graph may be found to consist of closed cycles, not including $\Psi = 0$. In this case, X_c is determined to be infinite. Since the state transition graph contains at most $2^{k^{2r}}$ nodes, the value of X_c may be found after at most this many tests. The procedure thus provides a finite algorithm for determining whether all possible arbitrarily long sequences of site values may be generated by evolution with a particular cellular automaton rule.

Table 2 gives the critical block lengths X_c for the cellular automata illustrated in fig. 1. Class 3 cellular automata with smaller X_c tend to generate more regular patterns. Those with larger X_c presumably give systematically larger entropies and their evolution is correspondingly less irreversible.

For additive cellular automata (such as code 42 in fig. 1 and table 2), all possible blocks of any length X may be reached, and have exactly k^{2r} predecessors of length $X + 2r$. In this case, therefore, evolution from a disordered initial state gives $s^{(x)}(X) = 1$ for all X (hence $X_c = \infty$). The equality of the number of predecessors for each block implies in addition in this case that $s_\mu^{(x)}(X) = 1$, at least for evolution from disordered initial states. Hence for additive cellular automata

$$d^{(x)} = d_\mu^{(x)} = 1. \tag{7.1}$$

The configurations generated by additive cellular automata are thus maximally chaotic.

In general cellular automata evolving according to eq. (2.1) yield $s^{(x)}(X) = 1$ for all X, so that $d^{(x)} = 1$, if $\mathbf{F}$ is an injective (one-to-one) function of either its first or last argument (or can be obtained by composition of functions with such a property). This may be proved by induction. Assume that all blocks of length X are reachable,

Code	X_c	Code	X_c
2	5	32	3
4	12	34	5
6	7	36	12
8	12	38	7
10	36	40	12
12	5	42	∞
14	5	44	5
16	5	46	5
18	5	48	5
20	36	50	5
22	12	52	22
24	7	54	12
26	12	56	7
28	5	58	12
30	3	60	5

Table 2. Values of critical block length X_c for legal totalistic $k = 2, r = 2$ cellular automata as illustrated in fig. 1. For $X < X_c$, all k^X possible blocks of X site values appear with nonzero probability in configurations generated after any number of time steps in evolution from disordered initial states, while for $X \geq X_c$, some blocks are absent, so that the spatial set entropy $s^{(x)}(X) < 1$.

with predecessors of lengths $X + 2r$. Then form a block of length $X + 1$ by adding a site at one end. To obtain all possible length $X + 1$ blocks, the value a' of this additional site must range over k possibilities. Any predecessors for length $X + 1$ blocks must be obtained by adding a $(X + 2r + 1)$-th site (with value a) at one end. For all length $X + 1$ blocks to be reachable, all values of a' must be generated when a runs over its k possible values, and the result follows. Notice that not all length $X + 1$ blocks need have the same (nonzero) number of predecessors, so that the measure entropy $s_\mu^{(x)}(X)$ may be less than the set entropy $s^{(x)}(X)$.

While injectivity of the rule function $\mathbf{F}$ for a cellular automaton in its first or last arguments is sufficient to give $d^{(x)} = 1$, it is apparently not necessary. A necessary condition is not known.

In section 6 it was shown that the set of configurations obtained by cellular automaton evolution for a finite number of time steps from any initial state could be specified by a regular grammar. In general the complexity of the grammar may increase rapidly with the number of time steps, potentially leading at infinite time to a set not specifiable by a regular grammar. Such behaviour may generically be expected in class 3 cellular automata, for which the average minimum propagation speed $\bar{\lambda} > 0$.

As discussed in section 4, one may consider the statistics of temporal as well as spatial sequences of site values. The temporal aperiodicity of the patterns generated by evolution of class 3 cellular automata from almost all initial states suggests that these systems should have nonvanishing temporal entropies $s_{(\mu)}^{(t)}(T)$ and nonvanishing temporal dimensions $d_{(\mu)}^{(t)}$. Once again, the temporal entropies for blocks starting at progressively later times quickly relax to equilibrium values. Notice that the dimension $d_{(\mu)}^{(t)}$ obtained from the large T limit of the $s_{(\mu)}^{(t)}(T)$ is always independent

148

of the starting times for the blocks. This is to be contrasted with the spatial dimensions $d^{(x)}_{(\mu)}$, which depend on the time at which they are evaluated. Just as for spatial entropies, it found that the equilibrium temporal entropies are essentially independent of probability measure for initial configurations.

The temporal entropies $s^{(t)}_{(\mu)}(T)$ decrease slowly with T. In fact, it appears that in all cases

$$s^{(t)}_{(\mu)}(Z) \geq s^{(x)}_{(\mu)}(Z). \tag{7.2}$$

The ratio $s^{(t)}_{(\mu)}(Z)/s^{(x)}_{(\mu)}(Z)$ is, however, typically much smaller than its maximum value (4.38) equal to the maximum propagation speed λ_+. Notice that the value of λ_+ determines the slopes of the edges of triangular clearings in the patterns generated by cellular automaton evolution.

At least for the class 3 cellular automata in fig. 1 which generate irregular patterns, the equilibrium set entropy $s^{(t)}(T) = 1$ for all $T \leq 8$ for which data are available. Note that the result $s^{(t)}(T) = 1$ holds for all T for any additive cellular automaton rule. One may speculate that class 3 cellular automata which generate apparently irregular patterns form a special subclass, characterized by temporal dimension $d^{(t)} = 1$.

For class 3 cellular automata which generate more regular patterns, $s_{(t)}(T)$ appears to decrease, albeit slowly, with T. Just as for spatial sequences, one may consider whether the temporal sequences which appear form a set described by a regular grammar. For the particular case of the $k = 2$, $r = 1$ cellular automaton with rule number 18, there is some evidence [21] that all possible temporal sequences which contain no 11 subsequences may appear, so that $N^{(t)}(T) = F_T$ where F_T is the Tth Fibonacci number ($F_T = F_{T-1} + F_{T-2}$, $F_0 = F_1 = 1$). This implies that $N_{(t)}(T) \sim \phi^T$ ($\phi = (\sqrt{5}+1)/2 \simeq 1.618$) for large T, suggesting a temporal set dimension $d^{(t)} = \log_2 \phi \approx 0.694$. In general, however, the set of possible temporal sequences is not expected to be described by a regular grammar.

The nonvanishing value of the average minimum propagation speed $\bar{\lambda}_-$ for class 3 cellular automata suggests that in all cases the value of a particular site depends on an ever-increasing number of initial site values. However, the complexity of this dependence is not known. The value of a site after t time steps can always be specified by a table with an entry for each of $k^{2\lambda_+ t}$ relevant initial sequences. Nevertheless, it is possible that a finite state automaton, specified by a finite state transition graph, could determine the values of sites at any time.

The behaviour of finite class 3 cellular automata with additive rules was analysed in some detail in ref. 2. It was shown there that the maximal cycle length for additive cellular automata grows on average exponentially with the size N of the cellular automaton. Most cycles were found to have maximal length, and the number of distinct cycles was found also to grow on average exponentially with N. The lengths of transients leading to cycles was found to grow at most linearly with N. The fraction of states on cycles was found on average to tend to a finite limit.

For most class 3 cellular automata, the average cycle length grows quite slowly with N, although in some cases, the absolute maximum cycle length appears to grow rapidly. The lengths of transients are typically short for cellular automata which generate more regular patterns, but often become very long as N increases for cellular automata which generate more irregular patterns. The fractions of states on cycles are typically much larger for finite class 3 cellular automata which generate irregular patterns than for those which generate more regular patterns. This is presumably a reflection of the lower irreversibility and larger attractor dimension found for the former case in the infinite size limit.

8. Class 4 Cellular Automata

Figure 12 shows the evolution of the class 4 cellular automaton with $k = 2, r = 2$ and code number 20, from several disordered initial configurations. In most cases, all sites are seen to "die" (attain value zero) after a finite time. However, in a few cases, stable or periodic structures which persist for an infinite time are formed. In addition, in some cases, propagating structures are formed. Figure 13 shows the persistent structures generated by this cellular automaton from all initial configurations whose nonzero sites lie in a region of length 20 (reflected versions of the last three structures are also found). Table 3 gives some characteristics of these structures. An important feature, shared by other class 4 cellular automata, is the presence of propagating structures. By arranging for suitable reflections of these propagating structures, final states with any cycle lengths may be obtained.

The behaviour of the cellular automata illustrated in fig. 13, and the structures shown in fig. 14 are strongly reminiscent of the two-dimensional (essentially totalistic) cellular automaton known as the "Game of Life"* (for references see [1]). The Game of Life has been shown to have the important property of computational universality. Cellular automata may be viewed as computers, in which data represented by initial configurations is processed by time evolution. Computational universality (e.g. [15–18]) implies that suitable initial configurations can specify arbitrary algorithmic procedures. The system can thus serve as a general purpose computer, capable of evaluating any (computable) function. Given a suitable encoding, the system may therefore in principle simulate any other system, and in this sense may be considered capable of arbitrarily complicated behaviour.

The proof of computational universality for the Game of Life [22] uses the existence of cellular automaton structures which emulate components (such as "wires" and "NAND gates") of a standard digital computer. The structures shown in fig. 14 represent a significant fraction of those necessary. A major missing element is a configuration (dubbed the "glider gun" in the Game of Life) which acts like a clock,

---

* Each site in this cellular automaton can take on one of two possible values; the time evolution rule involves nine site (type II) neighbourhoods. If the values of less than 2 or more than 3 of the eight neighbours of a particular site are nonzero then the site takes on value 0 at the next time step; if 2 neighbouring sites are nonzero the site takes the same value as on the previous time steps; if exactly 3 neighbouring sites are nonzero, the site takes on value 1.

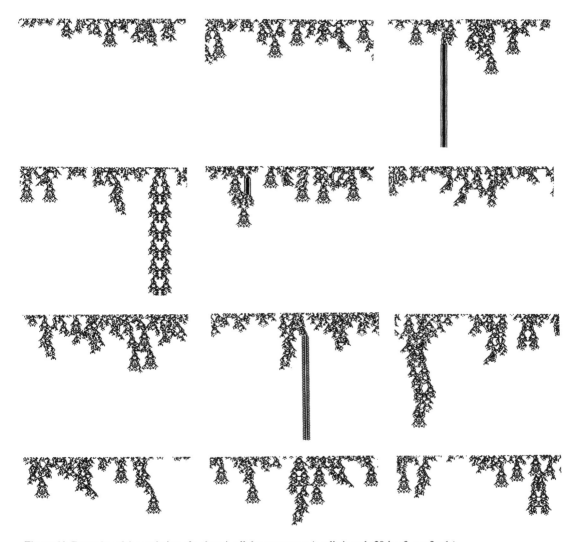

Figure 12. Examples of the evolution of a class 4 cellular automaton (totalistic code 20 $k = 2$, $r = 2$ rule) from several disordered initial states. Persistent structures are seen to be generated in a few cases. The evolution is truncated after 120 time steps.

and generates an infinite sequence of propagating structures. Such a configuration would involve a finite number of initial nonzero sites, but would lead to unbounded growth, and an asymptotically infinite number of nonzero sites. There are however indications that the required initial configuration is quite large, and is very difficult to find.

These analogies lead to the speculation that class 4 cellular automata are characterized by the capability for universal computation. $k = 2$, $r = 1$ cellular automata are too simple to support universal computation; the existence of class 4 cellular

151

Period	Minimal predecessor	$\phi(10)$	$\phi(20)$
2	10010111 (151)	0.027	0.024
9R	10111011 (187)	0.012	0.0061
1	10111101 (189)	0.014	0.0075
22	11000011 (195)	0.018	0.017
9L	11011101 (221)	0.012	0.0061
1R	1001111011 (635)	0.0020	0.00066
1L	1101111001 (889)	0.0020	0.00066
38	11110100100101111 (125231)	0	2.9×10^{-5}
4	10010001011011110111 (595703)	0	7.6×10^{-6}
4	10010101001010110111 (610999)	0	7.6×10^{-6}
4	10011000011111101111 (624623)	0	7.6×10^{-6}

Table 3. Persistent structures arising from initial configurations with length less than 20 sites in the class 4 totalistic cellular automaton with $k = 2$, $r = 2$ and code number 20, illustrated in figs. 12, 13 and 14. $\phi(X)$ gives the fraction of initial configurations with nonzero sites in a region less than X sites in length which generate a particular structure. When an initial configuration yields multiple structures, each is included in this fraction.

automata with $k = 2$, $r = 2$ (cf. figs. 13 and 14) and $k = 3$, $r = 1$ suggests that with suitable time evolution rules even such apparently simple systems may be capable of universal computation.

There are important limitations on predictions which may be made for the behaviour of systems capable of universal computation. The behaviour of such systems may in general be determined in detail essentially only by explicit simulation of their time evolution. It may in general be predicted using other systems only by procedures ultimately equivalent to explicit simulation. No finite algorithm or procedure may be devised capable of predicting detailed behaviour in a computationally universal system. Hence, for example, no general finite algorithm can predict whether a particular initial configuration in a computationally universal cellular automaton will evolve to the null configuration after a finite time, or will generate persistent structures, so that sites with nonzero values will exist at arbitrarily large times. (This is analogous to the insolubility of the halting problem for universal Turing machines (e.g. [15–18]).) Thus if the cellular automaton of figs. 12 and 13 is indeed computationally universal, no finite algorithm could predict whether a particular initial state would ultimately "die", or whether it would ultimately give rise to one of the persistent structures of fig. 13. The result could not be determined by explicit simulation, since an arbitrarily large time might elapse before one of the required states was reached. Another universal computer could also in general determine the result effectively only by simulation, with the same obstruction.

If class 4 cellular automata are indeed capable of universal computation, then their evolution involves an element of unpredictability presumably not present in other classes of cellular automata. Not only does the value of a particular site after many time steps potentially depend on the values of an increasing number of initial site values; in addition, the value cannot in general be determined by any "short-cut" procedure much simpler than explicit simulation of the evolution. The behaviour of a class 4 cellular automaton is thus essentially unpredictable, even given complete initial information: the behaviour of the system may essentially be found only by explicitly running it.

Only infinite cellular automata may be capable of universal computation; finite cellular automata involve only a finite number of internal states, and may therefore evaluate only a subset of all computable functions (the "space-bounded" ones).

The computational universality of a system implies that certain classes of general predictions for its behaviour cannot be made with finite algorithms. Specific predictions may nevertheless often be made, just as specific cases of generally non-computable function may often be evaluated. Hence, for example, the behaviour of all configurations with nonzero sites in a region of length 20 or less evolving according to the cellular automaton rules illustrated in figs. 12 and 13 has been completely determined. Figure 14 shows the fraction of initial configurations which evolve to the null state within T time steps, as a function of T, for various sizes X of the region of nonzero sites. For large X and large T, it appears that the fraction of configurations which generate no persistent structures (essentially the "halting probability") is approximately 0.93. It is noteworthy that the curves in fig. 14 as a function of T appear to approach a fixed form at large X. One may speculate that some aspects of the form of such curves may be universal to all systems capable of universal computation.

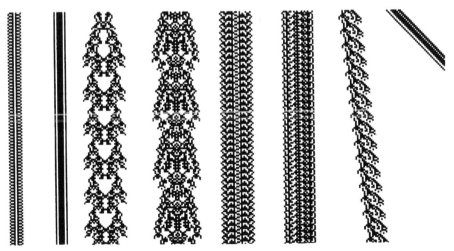

Figure 13. Persistent structures found in the evolution of the class 4 cellular automaton illustrated in fig. 12 from initial states with nonzero sites in a region of 20 or less sites. Reflected versions of the last three structures are also found. Some properties of the structures are given in Table 3. These structures are almost sufficient to provide components necessary to demonstrate a universal computation capability for this cellular automaton.

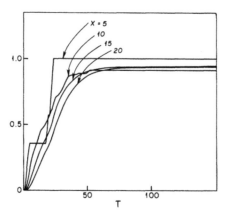

Figure 14. Fraction of configurations in the class 4 cellular automaton of figs. 12 and 13 which evolves to the null configuration after T time steps, from initial states with nonzero sites in a region of length less than X (translates of configurations are not included). The asymptotic "halting probability" is around 0.93; 7% of initial configurations generate the persistent structures of fig. 13 and never evolve to the null configuration.

The sets of persistent structures generated by class 4 cellular automata typically exhibit no simple patterns, and do not appear to be specified, for example, by regular grammars. Specification of persistent structures by a finite procedure is necessarily impossible if class 4 cellular automata are indeed capable of universal computation. Strong support of the conjecture that class 4 cellular automata are capable of universal computation would be provided by a demonstration of the equivalence of systematic enumeration of all persistent structures in particular class 4 cellular automata to the systematic enumeration of solutions to generally insoluble Diophantine equations or word problems.

Although one may determine by explicit construction that specific cellular automata are capable of universal computation, it is impossible to determine in general whether a particular cellular automaton is capable of universal computation. This is a consequence of the fact that the structures necessary to implement universal computation may be arbitrarily complicated. Thus, for example, the smallest propagating structure might involve an arbitrarily long sequence of site values.

For class 1, 2 and 3 cellular automata, fluctuations in statistical quantities are typically found to become progressively smaller as larger numbers of sites are considered. Such systems therefore exhibit definite properties in the "infinite volume" limit. For class 4 cellular automata, it seems likely that fluctuations do not decrease as larger number of sites are considered, and no simple smooth infinite volume limit exists. Important qualitative effects can arise from special sequences appearing with arbitrarily low probabilities in the initial state. Consider for example the class 4 cellular automaton illustrated in figs. 12 and 13. The evolution of the finite sequences in this cellular automaton shown in fig. 12 (and many thousands of other finite sequences tested) suggests that the average density of nonzero sites in configurations of this cellular automaton should tend to a constant at large times. However, in a sufficiently long finite initial sequence, there should exist a subsequence from which a "glider gun" structure evolves. This structure would generate an increasing number of nonzero sites at large times, and its presence would completely change the average large time density. As a more extreme example, it seems likely that a sufficiently long

154

(but finite) initial sequence should evolve to behave as a self-reproducing "organism", capable of eventually taking over its environment, and leading to completely different large time behaviour. Very special, and highly improbable, initial sequences may thus presumably result in large changes in large time properties for class 4 cellular automata. These sequences must appear in a truly infinite (typical) initial configuration. Although their density is perhaps arbitrarily low, the sequences may evolve to structures which come to dominate the statistical properties of the system. The possibility of such phenomena suggest that no smooth infinite volume exists for class 4 cellular automata.

Some statistical results may be obtained from large finite class 4 cellular automata, although the results are expected to be irrelevant in the truly infinite volume limit. The evolution of most class 4 cellular automata appears to be highly irreversible.* This irreversibility is reflected in the small set of persistent structures usually generated as end-products of the evolution. Changes in small regions of the initial state may affect many sites at large times. There are however very large fluctuations in the propagation speed, and no meaningful averages may be obtained. It should be noted that groups of class 4 cellular automata with different rules often yield qualitatively similar behaviour, and similar sets of persistent structures, suggesting further classification.

The frequency with which a particular structure is generated after an infinite time by the evolution of a universal computer from random (disordered) input gives the "algorithmic probability" p_A [24] for that structure. This algorithmic probability has been shown to be invariant (up to constant multiplicative factors) for a wide class of universal computers. In general, one may define an "evolutionary probability" $p_E(t)$ which gives the probability for a structure to evolve after t time steps from a random initial state. Complex structures formed by cellular automata will typically have evolutionary probabilities which are initially small, but later grow. As a simple example, the probability for the sequence which yields a period 9 propagating structure in the cellular automaton of figs. 12 and 13 begins small, but later increases to a sufficiently large value that such structures are almost always generated from disordered states of 2000 or more sites. In a much more complicated example, one may imagine that the probability for a self-reproducing structure begins small, but later increases to a substantial value. Structures whose evolutionary probability becomes significant only after a time $>T$ may be considered to have "logical depth" [25] T.

9. Discussion

Cellular automata are simple in construction, but are capable of very complex behaviour. This paper has suggested that a considerable universality exists in this complex behaviour. Evidence has been presented that all one-dimensional cellular automata fall into only four basic classes. In the first class, evolution from almost

---

* This feature allows practical simulation of such cellular automata to be made more efficient by storing information on the evolution of the specific sequences of sites which occur with larger probabilities (cf. [23]).

all initial states leads ultimately to a unique homogeneous state. The second class evolves to simple separated structures. Evolution of the third class of cellular automata leads to chaotic patterns, with varying degrees of structure. The behaviours of these three classes of cellular automata are analogous to the limit points, limit cycles and chaotic ("strange") attractors found in continuous dynamical systems. The fourth class of cellular automata exhibits still more complicated behaviour, and its members are conjectured to be capable of universal computation.

Even starting from disordered or random initial configurations, cellular automata evolve to generate characteristic patterns. Such self-organizing behaviour occurs by virtue of the irreversibility of cellular automaton evolution. Starting from almost any initial state, the evolution leads to attractors containing a small subset of all possible states. At least for the first three classes of cellular automata, the states in these attractors form a Cantor set, with characteristic fractal and other dimensions. For the first and second classes, the states in the attractor may be specified as sentences with a regular grammar. For the fourth class, the attractors may be arbitrarily complicated, and no simple statistical characterizations appear possible.

The four classes of cellular automata may be distinguished by the level of predictability of their "final" large time behaviour given their initial state. For the first class, all initial states yield the same final state, and complete prediction is trivial. In the second class, each region of the final state depends only on a finite region of the initial state; knowledge of a small region in the initial state thus suffices to predict the form of a region in the final state. In the evolution of the third class of cellular automata, the effects of changes in the initial state almost always propagate forever at a finite speed. A particular region thus depends on a region of the initial state of ever-increasing size. Hence any prediction of the "final" state requires complete knowledge of the initial state. Finally, in the fourth class of cellular automata, regions of the final state again depend on arbitrarily large regions of the initial state. However, if cellular automata in the class are indeed capable of universal computation, then this dependence may be arbitrarily complex, and the behaviour of the system can be found by no procedure significantly simpler than direct simulation. No meaningful prediction is therefore possible for such systems.

Acknowledgements

I am grateful to many people for discussions, including C. Bennett, J. Crutchfield, D. Friedan, P. Gacz, E. Jen, D. Lind, O. Martin, A. Odlyzko, N. Packard, S^2. Shenker, W. Thurston, T. Toffoli and S. Willson. I am particularly grateful to J. Milnor for extensive discussions and suggestions.

References

1. S. Wolfram, "Statistical mechanics of cellular automata", Rev. Mod. Phys. 55 (1983) 601.
2. O. Martin, A. M. Odlyzko and S. Wolfram, "Algebraic properties of cellular automata", Bell Laboratories report (January 1983); Comm. Math. Phys., to be published.

3. D. Lind, "Applications of ergodic theory and sofic systems to cellular automata", University of Washington preprint (April 1983); Physica 10D (1984) 36 (these proceedings).

4. S. Wolfram, "CA: an interactive cellular automaton simulator for the Sun Workstation and VAX", presented and demonstrated at the Interdisciplinary Workshop on Cellular Automata, Los Alamos (March 1983).

5. T. Toffoli, N. Margolus, G. Vishniac, private demonstrations.

6. P. Billingsley, Ergodic Theory and Information (Wiley, New York, 1965).

7. D. Knuth, Seminumerical Algorithms, 2nd. ed. (Addison-Wesley, New York, 1981), section 3.5.

8. R. G. Gallager, Information Theory and Reliable Communications (Wiley, New York, 1968).

9. J. D. Farmer, "Dimension, fractal measures and the probabilistic structure of chaos", in Evolution of Order and Chaos in Physics, Chemistry and Biology, H. Haken, ed. (Springer, Berlin, 1982).

10. J. D. Farmer, private communication.

11. B. Mandelbrot, The Fractal Geometry of Nature (Freeman, San Francisco, 1982).

12. J. D. Farmer, "Information dimension and the probabilistic structure of chaos", Z. Naturforsch. 37a (1982) 1304.

13. P. Grassberger, to be published.

14. P. Diaconis, private communication; C. Stein, unpublished notes.

15. F. S. Beckman, Mathematical Foundations of Programming (Addison-Wesley, New York, 1980).

16. J. E. Hopcroft and J. D. Ullman, Introduction to Automata Theory, Languages, and Computation (Addison-Wesley, New York, 1979).

17. Z. Manna, Mathematical Theory of Computation (McGraw-Hill, New York, 1974).

18. M. Minsky, Computation: Finite and Infinite Machines (Prentice-Hall, London, 1967).

19. B. Weiss, "Subshifts of finite type and sofic systems", Monat. Math. 17 (1973) 462; E. M. Coven and M. E. Paul, "Sofic systems", Israel J. Math. 20 (1975) 165.

20. P. Grassberger, "A new mechanism for deterministic diffusion", Wuppertal preprint WU B 82-18 (1982).

21. J. Milnor, unpublished notes.

22. R. W. Gosper, unpublished; R. Wainwright, "Life is universal!", Proc. Winter Simul. Conf., Washington D.C., ACM (1974). E. R. Berlekamp, J. H. Conway and R. K. Guy, Winning Ways, for Your Mathematical Plays, vol. 2 (Academic Press, New York, 1982), chap. 25.

23. R. W. Gosper, "Exploiting regularities in large cellular spaces", Physica 10D (1984) 75 (these proceedings).

24. G. Chaitin, "Algorithmic information theory", IBM J. Res. & Dev., 21 (1977) 350; "Toward a mathematical theory of life", in The Maximum Entropy Formalism, R. D. Levine and M. Tribus, ed. (MIT press, Cambridge, MA, 1979).

25. C. Bennett, "On the logical "depth" of sequences and their reducibilities to random sequences", IBM report (April 1982) (to be published in Info. & Control).

Computation Theory
of Cellular Automata

1984

Self-organizing behaviour in cellular automata is discussed as a computational process. Formal language theory is used to extend dynamical systems theory descriptions of cellular automata. The sets of configurations generated after a finite number of time steps of cellular automaton evolution are shown to form regular languages. Many examples are given. The sizes of the minimal grammars for these languages provide measures of the complexities of the sets. This complexity is usually found to be non-decreasing with time. The limit sets generated by some classes of cellular automata correspond to regular languages. For other classes of cellular automata they appear to correspond to more complicated languages. Many properties of these sets are then formally non-computable. It is suggested that such undecidability is common in these and other dynamical systems.

1. Introduction

Systems that follow the second law of thermodynamics evolve with time to maximal entropy and complete disorder, destroying any order initially present. Cellular automata are examples of mathematical systems which may instead exhibit "self-organizing" behaviour[1]. Even starting from complete disorder, their irreversible evolution can spontaneously generate ordered structure. One coarse indication of such self-organization is a decrease of entropy with time. This paper discusses an approach to a more complete mathematical characterization of self-organizing processes in cellular automata, and possible quantitative measures of the "complexity" generated by them. The evolution of cellular automata is viewed as a computation

---

Originally published in *Communications in Mathematical Physics*, volume 96, pages 15–57 (November 1984).

[1] An introduction to cellular automata in this context, together with many references is given in [1]. Further results are given in [2, 3], and are surveyed in [4, 5].

which processes information specified as the initial state. The structure of the output from such information processing is then described using the mathematical theory of formal languages (e.g. [6–8]). Detailed results and examples for simpler cases are presented, and some general conjectures are outlined. Computation and formal language theory may in general be expected to play a role in the theory of non-equilibrium and self-organizing systems analogous to the role of information theory in conventional statistical mechanics.

A one dimensional cellular automaton consists of a line of sites, with each site taking on a finite set of possible values, updated in discrete time steps according to a deterministic rule involving a local neighbourhood of sites around it. The value of site i at time step t is denoted $a_i^{(t)}$ and is a symbol chosen from the alphabet

$$S = \{0, 1, \ldots, k-1\}. \tag{1.1}$$

The possible sequences of these symbols form the set Σ of cellular automaton configurations $A^{(t)}$. Most of this paper concerns the evolution of infinite sequences $\Sigma = S^Z$; finite sequences $\Sigma = S^N$ flanked by quiescent sites (with say value 0) may also be considered. At each time step each site value is updated according to the values of a neighbourhood of $2r + 1$ sites around it by a local rule

$$\phi : S^{2r+1} \rightarrow S \tag{1.2}$$

of the form[2]

$$a_i^{(t)} = \phi[a_{i-r}^{(t-1)}, \; a_{i-r+1}^{(t-1)}, \ldots, a_{i+r}^{(t-1)}]. \tag{1.3}$$

This local rule leads to a global mapping

$$\Phi : \Sigma \rightarrow \Sigma \tag{1.4}$$

on complete cellular automaton configurations. Then in general

$$\Omega^{(t+1)} = \Phi\Omega^{(t)} \subseteq \Omega^{(t)}, \tag{1.5}$$

where

$$\Omega^{(t)} = \Phi^t \Sigma \tag{1.6}$$

is the set (ensemble) of configurations generated after t iterated applications of Φ (t time steps).

Formal languages consist of sets of words formed from strings of symbols in a finite alphabet S according to definite grammatical rules. Sets of cellular automaton configurations may thus be considered as formal languages, with each word in the language representing a cellular automaton configuration. Such infinite sets of configurations are then completely specified by finite sets of grammatical rules. (This

---

[2] The notation used here differs slightly from that of [2]. In particular, F in [2] is denoted here as ϕ.

descriptive use of formal grammars may be contrasted with the use of their transformation rules to define the dynamical evolution of developmental or L systems (e.g. [9]).)

Figure 1.1 gives typical examples of the evolution of cellular automata from disordered initial states according to various rules ϕ. Structure of varying complexity is seen to be formed. Four basic classes of behaviour are found in these and other cellular automata [2]. In order of increasing apparent complexity, qualitative characterizations of these classes are as follows:

1. Tends to a spatially homogeneous state.

2. Yields a sequence of simple stable or periodic structures.

3. Exhibits chaotic aperiodic behaviour.

4. Yields complicated localized structures, some propagating.

Approaches based on dynamical systems theory (e.g. [10, 11]) suggest some quantitative characterizations of these classes: the first three are analogous to the limit points, limits cycles and chaotic ("strange") attractors found in continuous dynamical systems. The fourth class exhibits more complex behaviour, and, as discussed below, is conjectured [2] to be capable of universal computation (e.g. [6, 7, 8]). The formal language theory approach discussed in this paper provides more precise and complete characterizations of the classes and their complexity.

The four classes of cellular automata generate distinctive patterns by evolution from finite initial configurations, as illustrated in Fig. 1.2:

1. Pattern disappears with time.

2. Pattern evolves to a fixed finite size.

3. Pattern grows indefinitely at a fixed rate.

4. Pattern grows and contracts with time.

The classes are also distinguished by the effects of small changes in initial configurations:

1. No change in final state.

2. Changes only in a region of finite size.

3. Changes over a region of ever-increasing size.

4. Irregular changes.

"Information" associated with the initial state thus propagates only a finite distance in classes 1 and 2, but may propagate an infinite distance in classes 3 and 4. In class 3, it typically propagates at a fixed positive speed.

Figure 1.1. Evolution of cellular automata with various typical local rules ϕ. The initial state is disordered; successive lines show configurations obtained at successive time steps. Four qualitative classes of behaviour are seen. (The first five rules shown have $k = 2$ and $r = 1$, and rule numbers 18, 22, 76, 90 and 128, respectively [1]. The last rule has $k = 2$, $r = 2$, and totalistic code number 20 [2].)

The grammar of a formal language gives rules for generating or recognizing the words in the language. An idealized computer (such as a Turing machine) may be constructed to implement these rules. Such a computer may be taken to consist of a "central processing unit" with a fixed finite number of internal states, together with a "memory" or "tape." Four types of formal language are conventionally identified, roughly characterized by the size of the memory in computers that implement them (e.g. [7]):

0. Unrestricted languages[3]: indefinitely large memory.

1. Context-sensitive languages: memory proportional to input word length.

2. Context-free languages: memory arranged in a stack, with a fixed number of elements available at a given time.

3. Regular languages: no memory.

These four types of languages (essentially) form a hierarchy, with type 0 the most general. Only type 0 languages require full universal computers; the other three types of language are associated with progressively simpler types of computer (linear-bounded automata, pushdown automata, and finite automata, respectively).

The grammatical rules for a formal language may be specified as "productions" which define transformations or rewriting rules for strings of symbols. In addition to the set S of "terminal" symbols s_i which appear directly in the words of the language, one introduces a set U of intermediate "non-terminal" symbols u_i. To generate words in the language, one begins with a particular non-terminal "start" symbol, then uses applicable productions in turn eventually to obtain strings containing only terminal symbols. The different types of languages involve productions of different kinds:

0. Arbitrary productions.

1. Productions $\alpha_1 \to \alpha_2$ for which $|\alpha_2| \geq |\alpha_1|$, where α_i is an arbitrary string of terminal and non-terminal symbols, and $|\alpha_i|$ is its length.

2. Productions of the form $u_i \to \alpha_j$ only (with a fixed bound on $|\alpha_j|$).

3. Productions of the form $u_i \to s_j u_k$ or $u_i \to s_j$ only.

Words in languages are recognized (or "parsed") by finding sequences of inverse productions that transform the words back to the start symbol.

The grammars for regular (type 3) languages may be specified by the finite state transition graphs for finite automata that recognize them. Each arc in such a graph carries a symbol s_i from the alphabet S. The nodes in the graph are labelled by non-terminal symbols, and connected according to the production rules of the grammar. Words in the language correspond to paths through the state transition graph. The (set) entropy of the language, defined as the exponential rate of increase in the number of words with length (see Sect. 3), is then given by the logarithm of the largest eigenvalue of the adjacency matrix for the state transition graph. This eigenvalue is always an algebraic integer.

The set of all possible sequences of zeroes and ones forms a trivial regular language, corresponding to a finite automaton with the state transition graph of Fig. 1.3a. Exclusion of all sequences with pairs of adjacent ones (so that any 1 must be followed by a 0) yields the regular language of Fig. 1.3b. The set of sequences in

---

[3] Also known as general, phrase-structure, and semi-Thue languages.

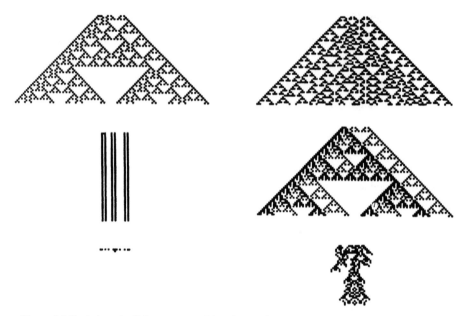

Figure 1.2. Evolution of cellular automata with various typical local rules from finite initial states. (The rules shown are the same as in Fig. 1.1.)

which, say, an even number of isolated ones appear between every 0110 block, again forms a regular language, now specified by the graph of Fig. 1.3c.

Regular expressions provide a convenient notation for regular languages. For example, $((0^*)(1^*))^*$ represents all possible sequences of zeroes and ones, corresponding to Fig. 1.3a. Here α^* denotes an arbitrary number of repetitions of the string α. With this notation, $(0^*(10)^*)^*$ represents Fig. 1.3b, and $(0(0^*)1(0^*)1)^*$ represents Fig. 1.3c.

Many regular grammars may in general yield the same regular language. However, it is always possible to find the simplest grammar for a given regular language (Myhill-Nerode theorem (e.g. [7])), whose corresponding finite automaton has the minimal number of states (nodes). This minimal number of states provides a measure of the "complexity" Ξ of the regular language. The regular languages of Fig. 1.3a–c are thus deemed to have progressively greater regular language complexities.

Section 2 shows that the sets of configurations $\Omega^{(t)}$ generated by any finite number of steps in the evolution of a cellular automaton form a regular language. For some cellular automata, the complexities of the regular languages obtained tend to a fixed limit after a few time steps, yielding a large time limiting set of configurations corresponding to a regular language. In general, it appears that the limit sets for all cellular automata that exhibit only class 1 or 2 behaviour are given by regular languages. For most class 3 and 4 cellular automata, however, the regular language complexities $\Xi^{(t)}$ of the sets $\Omega^{(t)}$ increase rapidly with time, presumably leading to non-regular language limit sets.

164

Sets of symbol sequences analogous to sets of cellular automaton configurations are obtained from the "symbolic dynamics" of continuous dynamical systems, in which the values of real number parameters are divided into discrete bins, labelled by symbols (e.g. [10, 11]). The simplest symbol sequences obtained in this way are "full shifts," corresponding to trivial regular languages Σ containing all possible sequences of symbols. More complicated systems yield finite complement languages, or "subshifts of finite type," in which a finite set of fixed blocks of symbols is excluded. "Sofic" systems, equivalent to general regular languages, have also been studied [12]. There is nevertheless evidence that, just as in cellular automata, regular languages are inadequate to describe the complete symbolic dynamics of even quite simple continuous dynamical systems.

Context-free (type 2) languages are generalizations of regular languages. Words in context-free languages may be viewed as sequences of terminal nodes (leaves) in trees constructed according to context-free grammatical rules. Each non-terminal symbol in the context-free grammar is taken to correspond to a type of tree node. The production rules for the non-terminal symbol then specify its possible descendents in the tree. For each word in the language, there corresponds such a "derivation" tree, rooted at the start symbol. (In most context-free languages, there are "ambiguous" words, obtained from multiple distinct derivation trees.) The syntax for most practical computer languages is supposed to be context-free. Each grammatical production rule corresponds to a subexpression with a particular structure (such as $u \bigcirc v$); the subexpressions may be arbitrarily nested [as in $((a \bigcirc (b \bigcirc c)) \bigcirc d) \bigcirc e$], corresponding to arbitrary derivation trees.

Regular languages correspond to context-free languages whose derivation trees consist only of a "trunk" sprouting a sequence of leaves, one at a time. An example of a context-free language not represented by any regular grammar is the sequence of strings of the form $0^n 10^n$ for any n. (Here, as elsewhere, α^n represents n-fold repetition of the string α.) A derivation tree for a word in this language is shown in Fig. 1.4. In general, the productions of any context-free language may in fact be

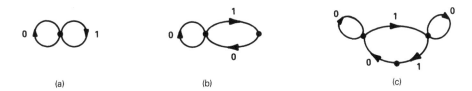

(a)   (b)   (c)

Figure 1.3a–c. State transition graphs for deterministic finite automata (DFA) corresponding to some regular languages: **a** the set of all possible sequences of zeroes and ones; **b** sequences in which 11 never occurs; **c** sequences in which an even number of isolated 1's appear between each 0110 block. Words in the languages correspond to sequences of symbols on arcs in paths through the DFA state transition graphs. The three DFA shown have successively larger numbers of states Ξ, and the sets of symbol sequences they represent may be considered to have successively larger "regular language complexities".

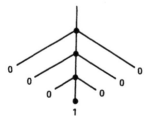

Figure 1.4. The derivation tree for a word in the context-free language consisting of sequences of the form $0^n 1 0^n$.

arranged so that all derivation trees are binary (Chomsky normal form)[4].

At each point in the generation of a word in a regular language, the next symbol depends only on the current finite automaton state, and not on any previous history. (Regular language words may thus be considered as Markov chains.) To generate words in a context-free language, however, one must maintain a "stack" (last-in first-out memory), which at each point represents the part of the derivation tree above the symbol (tree leaf) just generated. In this way, words in context-free languages may exhibit certain long-range correlations, as illustrated in Fig. 1.4. (In practical computer languages, these long-range correlations are typically manifest in the pairing of parentheses separated by many subexpressions.)

The production rules of a context-free grammar specify transformations for individual non-terminal symbols, independent of the "context" in which they appear. Context-sensitive grammars represent a generalization in which transformations for a particular symbol may depend on the strings of symbols that precede or follow it (its "context"). However, the transformation or production rule for any string α_1 is required to yield a longer (or equal length) string α_2. The set of all strings of the form $0^n 1^n 0^n$ for any n forms a context-sensitive language, not represented by any context-free or simpler language. The words in a context-sensitive language may be viewed as formed from sequences of terminal nodes in a directed graph. The graph is a derivation tree rooted at the start symbol, but with connections representing context sensitivities added. The requirement $|\alpha_2| \geq |\alpha_1|$ implies that there are progressively more nodes at each stage: the length of a word in context-sensitive language thus gives an upper bound on the number of nodes that occur at any stage in its derivation. A machine that recognizes words in a context-sensitive language by enumerating all applicable derivation graphs need therefore only have a memory as large as the words to be recognized.

Unrestricted (type 0) languages are associated with universal computers. A system is considered capable of "universal computation" if, with some particular input, it can simulate the behaviour of any other computational system[5]. A universal computer may thus be "programmed" to implement any finite algorithm. A universal

---

[4] Compare many implementations of the LISP programming language. Also, compare with models of multiparticle production cascade processes (e.g. [13]).

[5] Although there are some mathematically-defined operations which they cannot perform (as discussed below), it seems likely that the usual class of "universal computers" can simulate the behaviour of any physically-realizable system.

Turing machine has an infinite memory, and a central processing unit with a particular "instruction set." (The "simplest" known universal Turing machine has seven internal states, and a memory arranged as a line of sites, each having four possible values, and with one site accessible to the central processing unit at each time step (e.g. [8]).) Several quite different systems capable of universal computation have also been found. Among these are string manipulation systems which directly apply the production rules of type 0 languages; machines with one, infinite precision, arithmetic register; logic circuits analogous to those of practical digital electronic computers; and mathematical systems such as λ-calculus (general recursive functions). Some cellular automata have also been proved capable of universal computation. For example, a one-dimensional cellular automaton with $k = 18$ and $r = 1$ is equivalent to the simplest known universal Turing machine (e.g. [14]). (A two-dimensional cellular automata, the "Game of Life", with $k = 2$ and a nine site neighbourhood, has also been proved computationally universal (e.g. [15]).) It is conjectured that all cellular automata in the fourth class indicated above are in fact capable of universal computation [2].

There are many problems which can be stated in finite terms, but which are "undecidable" in a finite time, even for a universal computer[6]. An example is the "halting problem": to determine whether a particular computer will "halt" in a finite time, given particular input. The only way to predict the behaviour of some system S is to execute some procedure in a universal computer; but if, for example, S is itself a universal computer, then the procedure must reduce to a direct simulation, and can run no more than a finite amount faster than the evolution of S itself. The infinite time behaviour of S cannot therefore be determined in general in a finite time. For a cellular automaton, an analogue of the halting problem is to determine whether a particular finite initial configuration will ultimately evolve to the null configuration.

Any problem which depends on the results of infinite information processing may potentially be undecidable. However, when the information processing is sufficiently simple, there may be a finite "short-cut" procedure to determine the solution. For example, the information processing corresponding to the evolution of cellular automata with only class 1 or 2 behaviour appears to be sufficiently simple that their infinite time behaviour may be found by finite computation. Many problems concerning the infinite time behaviour of class 3 and 4 cellular automata may, however, be undecidable. For example, the entropies of the invariant sets for class 3 and 4 cellular automata may in general be non-computable numbers. This would be the case if the languages corresponding to these limit sets were of type 0 or 1.

It seems likely, in fact, that the consequences of infinite evolution in many dynamical systems may not be described in finite mathematical terms, so that many questions concerning their limiting behaviour are formally undecidable. Many features of the

[6] This is a form of Godel's theorem, in which the processes of mathematical proof are formalized in the operation of a computer.

167

behaviour of such systems may be determined effectively only by explicit simulation: no general predictions are possible.

Even for results that can in principle be obtained by finite computation there is a wide variation in the magnitude of time (or memory resources) required. Several classes of finite computations may be distinguished (e.g. [7]).

The first class (denoted P) consists of problems that can be solved by a deterministic procedure in a time given by some polynomial function of the size of their input. For example, finding the successor of a length n sequence in a cellular automaton takes (at most) a time linear in n, and is therefore a problem in the class P. Since most universal computers can simulate any other computer in a polynomial time, the times required on different computers usually differ at most by a polynomial transformation, and the set of problems in class P is defined almost independent of computer.

Nondeterministic polynomial time problems (NP) form a second class. Solutions to such problems may not necessarily be obtained in a polynomial time by a systematic procedure, but the correctness of a candidate solution, once guessed, can be tested in a polynomial time. Clearly $P \subseteq NP$, and there is considerable circumstantial evidence that $P \neq NP$. The problem of finding a pre-image for a length n sequence under cellular automaton evolution is in the class NP.

The problem classes P and NP are characterized by the times required for computations. One may also consider the class of problems PSPACE that require memory space given by a polynomial function of the size of the input, but may take an arbitrary time. There is again circumstantial evidence that $P \subset PSPACE$.

Just as there exist universal computers which, when given particular input, can simulate any other computer, so, analogously, there exist "NP-complete" (or "PSPACE-complete") problems which, with particular input, correspond to any NP (or PSPACE) problem of a particular size (e.g. [6, 7]). Many NP and PSPACE complete problems are known. An example of an NP-complete problem is "satisfiability": finding truth values for n variables which make a particular Boolean expression true. If $P \neq NP$ then there is essentially no faster method to solve this problem than to try the 2^n possible sets of values. (It appears that any method must at least require a time larger than any polynomial in n.) As discussed in Sect. 6, it is likely that the problem of finding pre-images for sequences in certain cellular automata, or of determining whether particular sequences are ever generated, is NP-complete. This would imply that no simple description exists even for some finite time properties of cellular automata: results may be found essentially only by explicit simulation of all possibilities.

2. Construction of Finite Time Sets

This section describes the construction of the set of configurations $\Omega^{(t)}$ generated after a finite number of time steps t of cellular automaton evolution, starting from the set

$\Omega^{(0)} = \Sigma$ of all possible configurations. It is shown that $\Omega^{(t)}$ may be represented as a regular language (cf. [2, 16]), and an explicit construction of the minimal grammar for this language is given. Section 3 describes some properties of such grammars, and Sect. 4 discusses their form for a variety of cellular automata.

To describe the construction we begin with a simple example. The procedure followed may be generalized directly.

Consider the construction of the set $\Omega^{(1)}$ generated by one time step in the evolution of the $k = 2, r = 1$ cellular automaton with a local rule ϕ given by ("rule number 76" [1])

$$111 \to 0, \quad 110 \to 1, \quad 101 \to 0, \quad 100 \to 0, \quad 011 \to 1,$$
$$010 \to 1, \quad 001 \to 0, \quad 000 \to 0. \tag{2.1}$$

The value $a_i^{(1)}$ of a site at position i in a configuration $A^{(1)} = \Phi A^{(0)} \in \Omega^{(1)}$ depends on a neighbourhood of three sites $\{a_{i-1}^{(0)}, a_i^{(0)}, a_{i+1}^{(0)}\}$ in the preceding configuration $A^{(0)} = \Omega^{(0)}$. The adjacent site $a_{i+1}^{(1)}$ depends on the overlapping neighbourhood $\{a_i^{(0)}, a_{i+1}^{(0)}, a_{i+2}^{(0)}\}$. The dependence of $a_{i+1}^{(1)}$ on $a_i^{(0)}$ associated with this two-site overlap in neighbourhoods may be represented by the graph g of Fig. 2.1 (analogous to a de Bruijn graph [17]). The nodes in the graph represent the overlaps $\{a_i^{(0)}, a_{i+1}^{(0)}\}$. These nodes are joined by directed arcs corresponding to three-site neighbourhoods. The local cellular automaton rule ϕ of Eq. (2.1) defines a transformation for each three-site neighbourhood, and thus associates a symbol with each arc of g. Each possible path through g corresponds to a particular initial configuration $A^{(0)}$. The successor $A^{(1)}$ of each initial configuration is given by the sequence of symbols associated with the

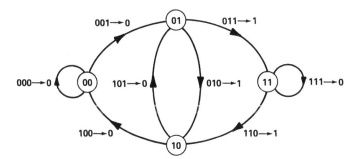

Figure 2.1. The state transition graph g for a non-deterministic finite automaton (NDFA) that generates configurations obtained after one time step in the evolution of the $k = 2, r = 1$ cellular automaton with rule number 76 [Eq. (2.1)]. Possible sequences of site values are represented by possible paths through the graph. The nodes in the graph are labelled by pairs of initial site values; the arcs then correspond to triples of initial site values. Each such triple is mapped under rule number 76 to a particular site value. The graph with arcs labelled by these site values corresponds to all possible configurations obtained after one time step. Note that the basic graph is the same for all $k = 2, r = 1$ cellular automata; only the images of the initial site value triples change from one rule to another.

arcs on the path. The sequences of symbols obtained by following all possible paths through g thus correspond to all possible configurations $A^{(1)}$ obtained after one time step in the evolution of the cellular automaton (2.1). The complete set $\Omega^{(1)}$ may thus be represented by the graph g. It is clear that not all possible sequences of 0's and 1's can appear in the configurations of $\Omega^{(1)}$. For example, no path in g can include the sequence 111, and thus no configuration in $\Omega^{(1)}$ can contain a block of sites 111.

The graph g of Fig. 2.1 may be considered as the state transition graph for a finite automaton which generates the formal language $\Omega^{(1)}$. Each node of g corresponds to a state of the finite automaton, and each arc to a transition in the finite automaton, or equivalently to a production rule in the grammar represented by the finite automaton. The set $\Omega^{(1)}$ thus forms a regular language. Labelling the states in g as u_0, u_1, u_2, u_3, the productions in the grammar for this language are:

$$u_0 \to 0u_0, \quad u_0 \to 0u_1, \quad u_1 \to 1u_2, \quad u_1 \to 1u_3, \quad u_2 \to 0u_0,$$
$$u_2 \to 0u_1, \quad u_3 \to 0u_3, \quad u_3 \to 1u_2. \tag{2.2}$$

This finite set of rules provides a complete specification of the infinite set $\Omega^{(1)}$.

Each path through g corresponds uniquely to a particular initial configuration $A^{(0)}$. But several different paths may yield the same successor configuration $A^{(1)}$. Each such path corresponds to a distinct inverse image of $A^{(1)}$ under Φ. Enumeration of paths in g shows, for example, that there are 5 distinct inverse images for the sequence 00 under the cellular automaton mapping (2.1), 5 also for 01 and 10, and 1 for 11.

The finite automaton g of Fig. 2.1 is not the only possible one that generates the language $\Omega^{(1)}$. An alternative finite automaton $\bar{g}$ is shown in Fig. 2.2, and may be considered "simpler" than g since it has fewer states. $\bar{g}$ is obtained from g by combining the 00 and 10 nodes, which are equivalent in that only paths carrying the same symbol sequences pass through these nodes. The complete set of symbol sequences generated by the possible paths through $\bar{g}$ is identical to that generated by possible paths through g.

The finite automata g and $\bar{g}$ are non-deterministic in the sense that multiple arcs carrying the same symbol emanate from some nodes, so that several distinct paths may generate the same word in the formal language. It is convenient for many purposes to find deterministic finite automata (DFA) equivalent to the non-deterministic finite automata (NDFA) g and $\bar{g}$. Such DFA may always be found by the standard "subset construction" (e.g. [6, 7]).

Consider for example the construction of a DFA G equivalent to the NDFA g of Fig. 2.1. Let ψ be the set of all possible subsets of the set of nodes $\{u_i\}$ (the power

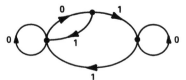

Figure 2.2. The state transition graph $\bar{g}$ for an alternative NDFA that generates the language $\Omega^{(1)}$ obtained after one time step of evolution according to rule 76. This NDFA is obtained by combining two equivalent states in the NDFA g of Fig. 2.1.

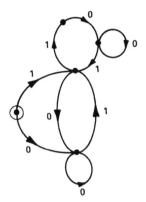

Figure 2.3. The state transition graph G for a deterministic finite automaton (DFA) obtained from the non-deterministic finite automaton of Fig. 2.1 by the subset construction [and represented by the productions of Eq. (2.3)]. Here and in other DFA graphs, the start node ψ_S is shown encircled. Words in the regular language $\Omega^{(1)}$ correspond to paths through G, starting at ψ_S.

set of $\{u_i\}$). There are $2^4 = 16$ elements ψ_i of ψ; each potentially corresponds to a state in G. The construction of G begins from the "start node" $\psi_S = \{u_0, u_1, u_2, u_3\}$. This node is joined by a 0 arc to the node $\{u_0, u_1, u_3\}$ corresponding to the set of NDFA states reached by a 0 arc according to (2.2) from any of the u_i in ψ_S. An analogous procedure is applied for each arc at each node in G. The resulting graph is shown in Fig. 2.3, and may be represented by the productions

$$\psi_S = \{u_0, u_1, u_2, u_3\} \to 0\{u_0, u_1, u_3\}, \quad \{u_0, u_1, u_2, u_3\} \to 1\{u_2, u_3\},$$
$$\{u_0, u_1, u_3\} \to 0\{u_0, u_1, u_3\}, \quad \{u_0, u_1, u_3\} \to 1\{u_2, u_3\},$$
$$\{u_2, u_3\} \to 0\{u_0, u_1, u_3\}, \quad \{u_2, u_3\} \to 1\{u_2\}, \qquad (2.3)$$
$$\{u_2\} \to 0\{u_0, u_1\}, \quad \{u_2\} \to 1\{\ \},$$
$$\{u_0, u_1\} \to 0\{u_0, u_1\}, \quad \{u_0, u_1\} \to 1\{u_2, u_3\}.$$

Notice that only 5 of the 16 possible ψ_i are reached by transitions from ψ_S. The production in Eq. (2.3) yielding the null set $\{\ \}$ (often denoted ε) signifies the absence of an arc carrying the symbol 1 emanating from the $\{u_2\}$ node.

The DFA G of Fig. 2.3 provides an alternative complete description of the language $\Omega^{(1)}$ represented by the NDFA g and $\bar{g}$ of Figs. 2.1 and 2.2. Possible sequences of symbols in words of $\Omega^{(1)}$ correspond to possible paths through G, starting at ψ_S. Consider the procedure for recognizing whether a sequence α can occur in $\Omega^{(1)}$. If α can occur, then it must correspond to a path through the NDFA g, starting at some node. The set of possible paths through g is represented by a single path through the DFA G. The start state ψ_S in G corresponds to the set of all possible states in g. As each symbol in the sequence α is scanned, the DFA G makes a transition to a state representing the set of states that g could reach at that point. The sequence α can thus occur in a word of $\Omega^{(1)}$ if and only if it corresponds to a path in G. The deterministic nature of G ensures that this path is unique.

Complete cellular automaton configurations consist of infinite sequences of symbols, and correspond to infinite paths in the DFA graph G. The possible words in $\Omega^{(1)}$ may thus be generated by following all possible paths through G.

Just as for the NDFA g, some of the states in the DFA G are equivalent, and may be combined. Two states are equivalent if and only if transitions from them with all possible symbols (here 0 or 1) lead to equivalent states. An equivalent DFA $\bar{G}$ shown in Fig. 2.4 may thus be obtained by representing each equivalence class of states in G by a single state. It may be shown that this DFA is the minimal one that recognizes the language $\Omega^{(1)}$ [18, 6, 7]. It is unique (up to state relabellings), and has fewer states than any equivalent DFA. Such a procedure yields the minimal form for any DFA; the analogous procedure for NDFA does not, however, necessarily yield a minimal form.

In most cases, the minimal DFA that generates all (two-way) infinite words of a regular language is the same as the minimal DFA constructed above that recognizes all finite (or one-way infinite) sequences of symbols in words of the language. In some cases, such as that of Fig. 2.5 (the set $\Omega^{(1)}$ for rule number 18), however, the latter DFA may contain additional "transient" subgraphs rooted at ψ_S, feeding into the main graph. The set of infinite paths through these transient subgraphs is typically a subset of the set of infinite paths in the main graph.

The minimal DFA $\bar{G}$ of Fig. 2.4 provides a simple description of the regular language $\Omega^{(1)}$. Regular expressions, mentioned in Sect. 1, provide a convenient notation for this and other regular languages. In terms of regular expressions,

$$\Omega^{(1)} = ((0^*)1(0 \vee 10)), \tag{2.4}$$

where infinite repetition to form each infinite word is understood. Here α^* represents an arbitrary number (possibly zero) of repetitions of the string α, and $\alpha_1 \vee \alpha_2$ stands for α_1 or α_2.

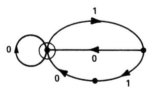

Figure 2.4. The state transition graph $\bar{G}$ for the minimal DFA that generates the regular language $\Omega^{(1)}$ obtained after one time step of evolution according to cellular automaton rule 76. The graph is obtained by combining equivalent nodes in the DFA G of Fig. 2.3. It has the smallest possible number of nodes.

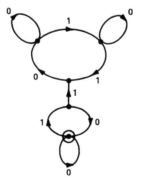

Figure 2.5. The state transition graph $\bar{G}$ for the minimal DFA corresponding to the regular language $\Omega^{(1)}$ obtained after one time step in the evolution of cellular automaton rule 18. This graph contains a "transient" subgraph rooted at the start state, feeding into the main graph. All symbol sequences occurring at any point in a word of $\Omega^{(1)}$ may be recognized as corresponding to paths through $\bar{G}$ beginning at the start state. Complete words in $\Omega^{(1)}$ may nevertheless be generated as possible infinite paths in $\bar{G}$, with the transient subgraph removed.

The example discussed so far generalizes immediately to show that the set $\Omega^{(t)}$ of configurations generated by t time steps of evolution according to any cellular automaton rule forms a regular language. Constructions analogous to those described above give grammars for these languages. The number of states in the initial NDFA g is in general k^{2rt}. (Two examples are shown in Fig. 2.6; graphs for successively larger values of rt may be obtained by a recursive construction [17].) The size of the DFA G obtained from g by the subset construction may be as large as $2^{k^{2rt}} - 1$, but is usually much smaller. (Note that the "reject" state { } is not counted in the size of the grammar.)

As an example, consider the language $\Omega^{(2)}$ generated by two time steps in the evolution of the cellular automaton (2.1). The original NDFA g which corresponds to this language has 16 states, and the DFA G obtained from it by the subset construction has nine states. Nevertheless, the resulting minimal DFA $\bar{G}$ has just three states, and is in fact identical to that found for $\Omega^{(1)}$ as shown in Fig. 2.4. Since $\bar{G}$ gives a complete (finite) specification of the languages $\Omega^{(t)}$, this implies that

$$\Omega^{(1)} = \Omega^{(2)} = \Phi\Omega^{(1)} \tag{2.5}$$

in this case. $\Omega^{(1)}$ is thus the limit set for the evolution of the cellular automaton of Eq. (2.1).

3. Properties of Finite Time Sets

This section discusses some properties of the regular language sets $\Omega^{(t)}$ generated by a finite number of steps of cellular automaton evolution, and constructed by the procedure of Sect. 2.

We consider as a sample set the 32 "legal" cellular automaton rules with $k = 2$ and $r = 1$. A rule ϕ is considered legal if it is symmetric, and maps the null configuration (with all site values 0) to itself. Each of the 256 possible $k = 2$, $r = 1$ cellular automaton rules is conveniently labelled by a "rule number," defined as the decimal equivalent of the sequence of binary digits $\phi[1, 1, 1]$, $\phi[1, 1, 0], \ldots, \phi[0, 0, 0]$ (analogous to Eq. (2.1)) [1].

Tables 3.1, 3.2 and 3.3 give some properties of the sets $\Omega^{(t)}$ generated by a few time steps in the evolution of the 32 legal $k = 2$, $r = 1$ cellular automata[7]. These properties are deduced from the minimal DFA which describe the $\Omega^{(t)}$, obtained according to the construction of Sect. 2.

The minimal DFA corresponding to the trivial language $\Omega^{(0)} = \Sigma$ illustrated in Fig. 1.1a has just one state. The minimal DFA corresponding to the minimal regular grammars for more complicated languages have progressively more states. The total number of states $\Xi^{(t)}$ in the minimal DFA that generates a set $\Omega^{(t)}$ provides a measure

---

[7] Requests for copies of the C language computer program used to obtain these and other results in this paper should be directed to the author.

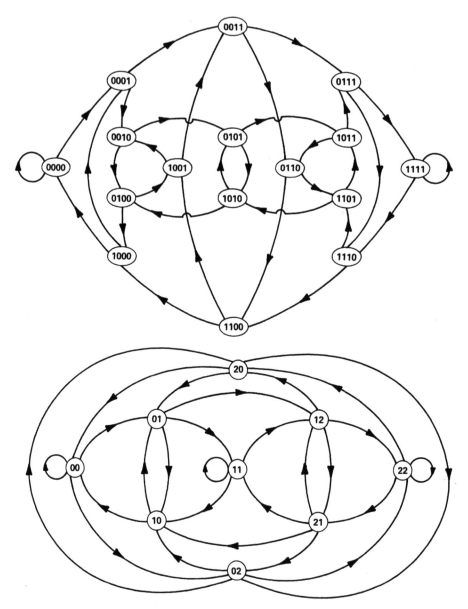

Figure 2.6. Non-deterministic finite automaton graphs (de Bruijn graphs) analogous to Fig. 2.1 for the cases $k = 2$, $r = 2$, and $k = 3$, $r = 1$.

of the "complexity" of the set $\Omega^{(t)}$, considered as a regular language. $\Xi^{(t)}$ gives the size of the shortest specification of the set $\Omega^{(t)}$ in terms of regular languages: this shortest specification becomes longer as the complexity of the set increases.

Table 3.1 gives the "regular language complexities" $\Xi^{(t)}$ for the sets $\Omega^{(t)}$ generated at the first few time steps in the evolution of the legal $k = 2$, $r = 1$ cellular automata.

174

Rule	$\Xi^{(0)}$	$\Xi^{(1)}$	$\Xi^{(2)}$	$\Xi^{(3)}$	$\Xi^{(4)}$
0	1 (2)	1 (1)	1 (1)	1 (1)	1 (1)
4	1 (2)	2 (3)	2 (3)	2 (3)	2 (3)
18	1 (2)	5 (9)	47 (91)	143 (270)	≥20 000
22	1 (2)	15 (29)	280 (551)	4506 (8963)	≥20 000
32	1 (2)	2 (3)	5 (7)	7 (9)	9 (11)
36	1 (2)	3 (5)	3 (4)	3 (4)	3 (4)
50	1 (2)	3 (5)	8 (14)	10 (17)	12 (20)
54	1 (2)	9 (16)	17 (32)	94 (179)	675 (1316)
72	1 (2)	5 (9)	5 (8)	5 (8)	5 (8)
76	1 (2)	3 (5)	3 (5)	3 (5)	3 (5)
90	1 (2)	1 (2)	1 (2)	1 (2)	1 (2)
94	1 (2)	15 (29)	230 (455)	3904 (7760)	≥20 000
104	1 (2)	15 (29)	265 (525)	2340 (4647)	1394 (2675)
108	1 (2)	9 (16)	11 (19)	11 (19)	11 (19)
122	1 (2)	15 (29)	179 (347)	5088 (9933)	≥20 000
126	1 (2)	3 (5)	13 (23)	107 (198)	2867 (5476)
128	1 (2)	4 (6)	6 (8)	8 (10)	10 (12)
132	1 (2)	5 (9)	7 (12)	9 (15)	11 (18)
146	1 (2)	15 (29)	92 (177)	1587 (3126)	≥20 000
150	1 (2)	1 (2)	1 (2)	1 (2)	1 (2)
160	1 (2)	9 (15)	16 (24)	25 (35)	36 (48)
164	1 (2)	15 (29)	116 (227)	667 (1310)	1214 (2363)
178	1 (2)	11 (20)	15 (26)	19 (32)	23 (38)
182	1 (2)	15 (29)	92 (177)	1587 (3126)	≥20 000
200	1 (2)	3 (5)	3 (5)	3 (5)	3 (5)
204	1 (2)	1 (2)	1 (2)	1 (2)	1 (2)
218	1 (2)	15 (29)	116 (227)	667 (1310)	1214 (2363)
222	1 (2)	5 (9)	7 (12)	9 (15)	11 (18)
232	1 (2)	11 (20)	15 (26)	19 (32)	23 (38)
236	1 (2)	3 (5)	3 (5)	3 (5)	3 (5)
250	1 (2)	9 (15)	16 (24)	25 (35)	36 (48)
254	1 (2)	4 (6)	6 (8)	8 (10)	10 (12)

Table 3.1. Numbers of nodes $\Xi^{(t)}$ (and arcs) in minimal deterministic finite automata (DFA) representing regular languages corresponding to sets of configurations $\Omega^{(t)}$ generated after t time steps in the evolution of legal $k = 2$, $r = 1$ cellular automata. Each configuration corresponds to a path through the DFA state transition graph. The construction of Sect. 2 yields the DFA with the minimal number of nodes (states) $\Xi^{(t)}$ that generates a given regular language $\Omega^{(t)}$. This DFA may be considered to give the shortest specification of $\Omega^{(t)}$ viewed as a regular language. Its size $\Xi^{(t)}$ measures the "complexity" of $\Omega^{(t)}$. The initial ($t = 0$) set of configurations include all possible sequences of zeroes and ones, and correspond to a trivial regular language. Cellular automata with only class 1 or 2 behaviour yield regular languages whose complexities become constant, or increase as polynomials in t. Cellular automata capable of class 3 or 4 behaviour usually lead to rapidly-increasing complexities. Bounds on these complexities are given when their exact calculation exceeded available computational resources. Some of the results in this table were obtained using the methods of [52] and [53].

Rule	$\chi^{(1)}(\lambda)$	λ_{max}
0	$1 - \lambda$	1.000
4	$-1 - \lambda + \lambda^2$	1.618
18	$(1 - \lambda - \lambda^2)(-1 + \lambda - 2\lambda^2 + \lambda^3)$	1.755
22	$\lambda(1 - \lambda)(2 - 2\lambda^2 + 6\lambda^3 - 3\lambda^4 - 5\lambda^5 + 10\lambda^6 - 5\lambda^7 - 3\lambda^8 + 6\lambda^9 - 2\lambda^{10} + 2\lambda^{11} - 3\lambda^{12} + \lambda^{13})$	1.917
32	$-1 - \lambda + \lambda^2$	1.618
36	$1 - \lambda + 2\lambda^2 - \lambda^3$	1.755
50	$1 + \lambda + \lambda^2 - \lambda^3$	1.839
54	$\lambda^3(1 + \lambda^2)(1 - \lambda + 2\lambda^3 - \lambda^4)$	1.867
72	$(1 + \lambda - \lambda^2)(-1 + \lambda - 2\lambda^2 + \lambda^3)$	1.755
76	$1 + \lambda + \lambda^2 - \lambda^3$	1.839
90	$2 - \lambda$	2.000
94	$-\lambda(2 - 2\lambda + 2\lambda^2 - \lambda^3 + 2\lambda^4 - 5\lambda^5 + 13\lambda^6 - 16\lambda^7 + 10\lambda^8 - 3\lambda^{10} - \lambda^{11} + 5\lambda^{12} - 4\lambda^{13} + \lambda^{14})$	1.883
104	$\lambda(1 - \lambda)(2 - 2\lambda^2 + 6\lambda^3 - 3\lambda^4 - 5\lambda^5 + 10\lambda^6 - 5\lambda^7 - 3\lambda^8 + 6\lambda^9 - 2\lambda^{10} + 2\lambda^{11} - 3\lambda^{12} + \lambda^{13})$	1.917
108	$\lambda^3(1 + \lambda^2)(1 - \lambda + 2\lambda^3 - \lambda^4)$	1.867
122	$-\lambda(2 - 2\lambda + 2\lambda^2 - \lambda^3 + 2\lambda^4 - 5\lambda^5 + 13\lambda^6 - 16\lambda^7 + 10\lambda^8 - 3\lambda^{10} - \lambda^{11} + 5\lambda^{12} - 4\lambda^{13} + \lambda^{14})$	1.883
126	$1 - \lambda + 2\lambda^2 - \lambda^3$	1.755
128	$(-1 - \lambda + \lambda^2)(1 - \lambda + \lambda^2)$	1.618
132	$1 - \lambda^2 + 2\lambda^4 - \lambda^5$	1.785
146	$-\lambda(-2 + 4\lambda - 6\lambda^2 + 4\lambda^3 + \lambda^4 - 7\lambda^5 + 12\lambda^6 - 13\lambda^7 + 9\lambda^8 - 4\lambda^9 + \lambda^{10} - 2\lambda^{11} + 5\lambda^{12} - 4\lambda^{13} + \lambda^{14})$	1.887
150	$2 - \lambda$	2.000
160	$(1 - \lambda^2 - \lambda^3)(1 - \lambda^2 + \lambda^3)(-1 + \lambda - 2\lambda^2 + \lambda^3)$	1.755
164	$-\lambda(2 - \lambda^2 - 2\lambda^4 + 5\lambda^5 - 9\lambda^6 + 14\lambda^7 - 9\lambda^8 + 2\lambda^9 - 6\lambda^{10} + 5\lambda^{11} + 3\lambda^{12} - 4\lambda^{13} + \lambda^{14})$	1.915
178	$\lambda(1 - \lambda^2 + \lambda^5)(1 - \lambda^2 + 2\lambda^4 - \lambda^5)$	1.785
182	$-\lambda(-2 + 4\lambda - 6\lambda^2 + 4\lambda^3 + \lambda^4 - 7\lambda^5 + 12\lambda^6 - 13\lambda^7 + 9\lambda^8 - 4\lambda^9 + \lambda^{10} - 2\lambda^{11} + 5\lambda^{12} - 4\lambda^{13} + \lambda^{14})$	1.887
200	$1 - \lambda + 2\lambda^2 - \lambda^3$	1.755
204	$2 - \lambda$	2.000
218	$-\lambda(2 - \lambda^2 - 2\lambda^4 + 5\lambda^5 - 9\lambda^6 + 14\lambda^7 - 9\lambda^8 + 2\lambda^9 - 6\lambda^{10} + 5\lambda^{11} + 3\lambda^{12} - 4\lambda^{13} + \lambda^{14})$	1.915
222	$1 - \lambda^2 + 2\lambda^4 - \lambda^5$	1.785
232	$\lambda(1 - \lambda^2 - \lambda^5)(-1 + \lambda^2 - 2\lambda^4 + \lambda^5)$	1.785
236	$1 - \lambda + 2\lambda^2 - \lambda^3$	1.755
250	$(1 - \lambda^2 - \lambda^3)(1 - \lambda^2 + \lambda^3)(-1 + \lambda - 2\lambda^2 + \lambda^3)$	1.755
254	$(-1 - \lambda + \lambda^2)(1 - \lambda + \lambda^2)$	1.618

Table 3.2. Characteristic polynomials $\chi^{(1)}(\lambda)$ for the adjacency matrices of state transition graphs for minimal DFA representing regular languages generated after one time step in the evolution of legal $k = 2$, $r = 1$ cellular automata. The nonzero roots of these polynomials determine the number of distinct symbol sequences that can appear in configurations generated by the cellular automaton evolution. The maximal root λ_{max} determines the limiting entropy of the sequences.

Rule	$L^{(1)}$	$L^{(2)}$	$L^{(3)}$	$L^{(4)}$
0	1*	−	−	−
4	2*	−	−	−
18	3	11	12	13
22	8	7	11	9
32	2*	4*	6*	8*
36	3*	2*	−	−
50	3*	5*	9*	11*
54	5	9	9	7
72	3	3*	−	−
76	3*	−	−	−
90	−	−	−	−
94	5	7	11	11
104	8	8	8	7
108	5	4*	−	−
122	5	7	8	10
126	3*	12	13	14
128	3*	5*	7*	9*
132	4*	5*	6*	7*
146	6	6	8	8
150	−	−	−	−
160	5*	7*	9*	11*
164	9	9	8	9
178	5*	6*	7*	8*
182	6	6	8	8
200	3*	−	−	−
204	−	−	−	−
218	9	9	8	9
222	4*	5*	6*	7*
232	5*	6*	7*	8*
236	3*	−	−	−
250	5*	7*	9*	11*
254	3*	5*	7*	9*

Table 3.3. The length $L^{(t)}$ of the shortest distinct blocks of site values newly-excluded after exactly t time steps in the evolution of legal $k = 2$, $r = 1$ cellular automata. The notation * indicates that the set of cellular automaton configurations $\Omega^{(t)}$ forms a finite complement language (finite number of distinct excluded blocks). The notation − signifies no new excluded blocks.

In all the cases given, $\Xi^{(t)}$ is seen to be non-decreasing with time. Cellular automata with only class 1 or 2 appear to give $\Xi^{(t)}$ which tend to constants after one or two time steps, or increase linearly or quadratically with time. Class 3 and 4 cellular automata usually give $\Xi^{(t)}$ which increase rapidly with time. In general,

$$1 \le \Xi^{(t)} \le 2^{k^{2rt}} - 1. \tag{3.1}$$

The upper bound is found to be attained in several cases for $t = 1$; for larger t, $\Xi^{(t)}$ appears to grow at most exponentially with t.

All possible sequences of symbols occur in the trivial language Σ. In more complicated regular languages, only some number $N(X)$ of the k^X possible sequences of X symbols may occur. Each sequence which occurs corresponds to a distinct path in the minimal DFA graph for the language. (Note that all distinct paths in a DFA correspond to different symbol sequences; this need not be the case in a NDFA graph.) The number of such paths is conveniently computed using a matrix representation for the DFA.

Consider as an example the set $\Omega^{(1)}$ obtained by one time step in the evolution of the cellular automaton (2.1). The minimal DFA graph $\bar{G}$ for this set is given in Fig. 2.4, and may be represented by the adjacency matrix

$$M = \begin{pmatrix} 1 & 1 & 0 \\ 1 & 0 & 1 \\ 1 & 0 & 0 \end{pmatrix}. \tag{3.2}$$

The elements of M^X give the numbers $N(X)$ of possible length X paths in $\bar{G}$. For lengths from 1 to 10 these numbers are 2, 4, 7, 13, 24, 44, 81, 149, 274, and 504. In general, at least for large X,

$$N(X) \simeq \text{Tr}[M^X] = \sum \lambda_i^X \sim \lambda_{max}^X, \tag{3.3}$$

where the λ_i are the eigenvalues of M, and λ_{max} is the largest of them. These eigenvalues are determined from the characteristic polynomial $\chi(\lambda)$ for the minimal DFA adjacency matrix, given in the case of Eq. (3.2) by[8]

$$\chi(\lambda) = 1 + \lambda + \lambda^2 - \lambda^3. \tag{3.4}$$

The largest (real) root of this characteristic polynomial (known as the "index" of the graph [19]) is given by the cubic algebraic integer

$$\lambda_{max} = [1 + \kappa + 4/\kappa] \simeq 1.83929,$$
$$\kappa = [(38 + \sqrt{1188})/2]^{1/3} \tag{3.5}$$

The set of infinite configurations $\Omega^{(t)}$ generated by cellular automaton evolution may be considered to form a Cantor set. The dimension of this Cantor set is given by

$$s = \lim_{X \to \infty} \frac{1}{X} \log_k N(X), \tag{3.6}$$

and is equal to the topological entropy of the shift mapping restricted to this set (e.g. [20]). For any regular language, this entropy is given according to Eqs. (3.2) by [21]

$$s = \log_k \lambda_{max}. \tag{3.7}$$

For the case of Eq. (3.2), the entropy is thus

$$s \simeq \log_2 1.83929 \simeq 0.87915. \tag{3.8}$$

---

[8] $1/\chi(\lambda)$ is related to the generating function for the sequence $N(X)$ (e.g. [19, Sect. 1.8]).

Table 3.2 gives the characteristic polynomials $\chi^{(1)}(\lambda)$ for the regular languages $\Omega^{(1)}$ obtained after one time step in the evolution of the 32 legal $k = 2, r = 1$ cellular automata, together with their largest real roots λ_{max}. All the nonzero roots of the $\chi(\lambda)$ appear in the expression (3.3) for $N(X)$, and are therefore the same for all possible DFA corresponding to a particular regular language. (They may thus be considered "topological invariants.") Additional powers of λ may appear in the characteristic polynomials obtained from non-minimal DFA.

The characteristic polynomials $\chi(\lambda)$ such as those in Table 3.2 obtained from regular languages are always monic (the term with the highest power of λ that appears in them always has unit coefficient). The largest roots λ_{max} of the $\chi(\lambda)$ for regular languages are thus always algebraic integers (e.g. [22])[9], so that the entropies for regular languages are always the logarithms of algebraic integers. The minimal polynomial with λ_{max} as a root has a degree not greater than the size $\Xi^{(t)}$ of the minimal DFA for a regular language $\Omega^{(t)}$. This bound is usually not reached, since the characteristic polynomial $\chi(\lambda)$ is usually reducible, as seen in Table 3.2. Notice that in many cases, $\chi(\lambda)$ has several factors with equal degrees. (The factorizations of the $\chi(\lambda)$ are related to the colouring properties of the corresponding graphs [19]. Note that graphs corresponding to minimal DFA always have trivial automorphism groups.) Factors (other than λ^n) with smaller degrees appear to be associated with transient subgraphs in the minimal DFA graph.

The entropy (3.6) characterizes the number of distinct symbol sequences generated by cellular automaton evolution, without regard to the probabilities with which they occur. One may also define a measure entropy (e.g. [20])

$$s_\mu = -\lim_{X \to \infty} \sum_{i=1}^{k^X} p_i \log_k p_i \tag{3.9}$$

in terms of the probabilities p_i for length X sequences. Starting from an initial ensemble in which all symbol sequences of a given length occur with equal probabilities, the probability for a sequence i after t time steps is given by

$$p_i = \xi_i / k^{X+2rt}, \tag{3.10}$$

where ξ_i is the number of (length $X + 2rt$) t-step preimages of the sequence i under the cellular automaton mapping Φ. This number is equal to the number of distinct paths through the NDFA graph analogous to g in Fig. 2.1 that yield the sequence i. It may also be computed from reduced NDFA graphs analogous to $\bar{g}$ of Fig. 2.2 by including a weight for each path, equal to the product of weights giving the number of unreduced nodes combined into each node on the path.

The set of configurations generated by cellular automaton evolution always contracts or remains unchanged with time, as implied by Eq. (1.5). The entropies

---

[9] The λ_{max} are always Perron numbers [23]. Any Perron number may be obtained from some regular language, and in fact also from some finite complement language [23].

associated with the sets $\Omega^{(t)}$ are therefore non-increasing with time. Class 1 cellular automata are characterized by (spatial) entropies that tend to zero with time [2]. Class 2, 3 and 4 cellular automata generate sets of configurations with nonzero limiting spatial entropy. (Class 2 cellular automata nevertheless yield patterns essentially periodic in time, with zero temporal entropy.)

Some cellular automata have the special property that

$$\Phi\Sigma = \Sigma, \tag{3.11}$$

so that all possible configurations can occur at any time in their evolution, and the entropies of the $\Omega^{(t)}$ are always equal to one. Such surjective cellular automaton rules may be recognized by the presence of all k possible outgoing arcs at each node in a DFA representing the grammar of the set $\Omega^{(1)}$ obtained after one time step in its evolution. The finite maximum size $2^{k^{2r}}$ for such a DFA, constructed as in Sect. 2, ensures that this procedure (cf. [24–27, 2]) for determining the surjectiveness of any cellular automaton rule is a finite one[10].

Since there are k outgoing arcs at each node in the original NDFA analogous to Fig. 2.1 for any cellular automaton rule, the rule is surjective if in all cases these arcs carry distinct symbols (so that the NDFA is in fact a DFA). This occurs whenever the local cellular automaton mapping ϕ is injective with respect to its first or last argument (as for additive rules [29, 16] such as 90, 150 or 204 in Tables 3.1–3.3). However, at least when $k > 2$ or $r > 1$, there exist surjective cellular automata for which this does not occur [25, 30]. Since all surjective cellular automata must yield the same trivial minimal DFA, it is possible that a reversal of the minimization and subset algorithms discussed in Sect. 2 could be used to generate all NDFA analogous to Fig. 2.1 that correspond to surjective rules.

Surjective cellular automata yield trivial regular languages, in which all possible blocks of symbols may appear. Some cellular automata generate the slightly more complicated "finite complement" regular languages, in which a finite set of distinct blocks are excluded. (Such languages are equivalent to "subshifts of finite type" (e.g. [10, 11]).) An example of a finite complement language, illustrated in Fig. 1.1b, consists of all sequences from which the block of sites 11 is absent. To construct the grammar for a finite complement language in which blocks of length b are excluded, first form a graph analogous to Fig. 2.1, but with sequences of length $b - 1$ at each node. Each arc then corresponds to a length b sequence, and may be labelled by the last symbol in the sequence. With this labelling, one arc carrying each of the k possible symbols emanates from each of the k^{b-1} nodes, so that the graph represents a DFA. Removing arcs corresponding to the excluded length b blocks then yields the graph for a DFA that recognizes the finite complement language with these blocks absent. Examples of the resulting graphs for two simple cases are shown in Fig. 3.1.

---

[10] The algorithm essentially involves testing whether a NDFA with k^{2r} states is equivalent to a NDFA that generates the trivial language Σ. This problem is known to be PSPACE-complete [28], and therefore presumably cannot be solved in a time polynomial in k^{2r}.

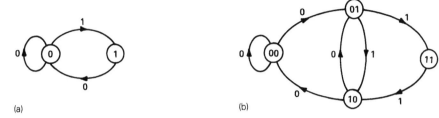

(a)   (b)

Figure 3.1 a and b. Non-deterministic finite automata (NDFA) corresponding to finite complement regular languages consisting of sequences of zeroes and ones in which **a** the block 11 is excluded, and **b** the block 111 is excluded. The graphs are constructed from analogues of Fig. 2.1 by dropping arcs corresponding to excluded blocks.

The minimal DFA for a finite complement language with a maximal distinct excluded block of length b has at most k^{b-1} states, and at least b states. An excluded block is considered "distinct" if it contains no excluded sub-blocks. (Hence, for example, in the language of Figs. 1.1a and 3.1a, the excluded block 11 is considered distinct, but 110, 111 and so on, are not.)

Any path through the minimal DFA graph for a regular language of length greater than $\Xi^{(t)}$ must contain a cycle, which retraverses some arcs. If no symbol sequence of length less than $\Xi^{(t)}$ is excluded, then no sequence of any length can therefore be excluded, and the corresponding language must be trivial. If some symbol sequences of length less than $\Xi^{(t)}$ are excluded, but no distinct sequences with lengths between $\Xi^{(t)}$ and $2\Xi^{(t)}$ are excluded, then no longer distinct sequences can be excluded, and the corresponding language must be a finite complement one. If further distinct excluded blocks with lengths between $\Xi^{(t)}$ and $2\Xi^{(t)}$ are found, then an infinite series of longer distinct excluded blocks must exist, and the language cannot be a finite complement one.

The language of Fig. 2.4, generated by the evolution of the $k = 2$, $r = 1$ cellular automaton with rule number 76, is a finite complement one, in which 111 is the only distinct excluded block. The language of Fig. 2.5, obtained after one time step in the evolution of rule number 18, is not a finite complement one. The block 111 is the shortest excluded in this case. But the distinct length 7 block 1101011 is also excluded, as are the two distinct length 8 blocks 11001011 and 11010011, three distinct length 9 blocks (11010100011, 110001011, 110010011), four distinct length 10 blocks, and so on.

The length $L^{(t)}$ of the shortest excluded block in a language $\Omega^{(t)}$ generated by cellular automaton evolution (denoted X_c in [2]) is in general given by the shortest distance from the start node in the corresponding DFA graph to an "incomplete" node, with less than k outgoing arcs. If the cellular automaton rule is not surjective, then

$$0 < L^{(t)} \le \Xi^{(t)}.$$
(3.12)

Whenever cellular automaton evolution is irreversible, the set of configurations $\Omega^{(t)}$ generated contracts with time, and progressively more distinct blocks are excluded. One may define $L^{(t)}$ to be the length of the shortest newly-excluded block at time step t in the evolution of a cellular automaton. The values of $L^{(t)}$ obtained in the first few time steps of evolution according to the 32 legal $k = 2$, $r = 1$ cellular automaton rules are given in Table 3.3. In most cases, $L^{(t)}$ is seen to increase with time, indicating that progressively finer subsets of Σ are excluded, and qualitatively reflecting the increase of $\Xi^{(t)}$. In general, however, $L^{(t)}$ need not increase monotonically with time. A length l block is excluded after t time steps if there is no initial length $l + 2rt$ block that evolves into it. A length l block is newly excluded at time step t if and only if no length $l + 2r$ blocks allowed at time step $t - 1$ evolve to it, but at least one length $l + 2r$ block newly excluded at time step $t - 1$ would evolve to it. The length $L^{(t)}$ of the shortest newly excluded block at time t is thus bounded by

$$L^{(t)} \geq L^{(t-1)} - 2r. \tag{3.13}$$

Table 3.3 includes several cases for which the lower bound is realized.

The sets of infinite symbol sequences $\Omega^{(t)}$ generated by cellular automaton evolution are characterized in part by the numbers and lengths of allowed and excluded finite blocks which appear in them. A further characterization may be given in terms of the number $\Pi(p)$ of infinite sequences with (spatial) period p that appear. This number is related to the number of distinct cycles in the minimal DFA graph for $\Omega^{(t)}$. Cycles are considered distinct if the sequences of symbols that appear in them are distinct. The enumeration of cycles thus requires knowledge of the arc labelling as well as connectivity of the DFA graph.

Just as the number of finite blocks $N(X)$ for all X may be summarized in the characteristic polynomial $\chi(\lambda)$, so also the number of periodic configurations $\Pi(p)$ may be summarized in the zeta function (e.g. [10, 11])

$$\zeta(\lambda) = \exp\left(\sum_{p=1}^{\infty} \Pi(p)\lambda^p / p\right). \tag{3.14}$$

For all regular languages $\zeta(\lambda)$ is a rational function of λ [31]. For the special case of finite complement languages,

$$\zeta(\lambda) = 1/\chi(\lambda). \tag{3.15}$$

A finite procedure may be given [32] to compute $\zeta(\lambda)$ for any regular language.

4. Evolution of Finite Time Sets

Tables 3.1–3.3 gave several properties of the sets of configurations generated by a finite number of steps in the evolution of legal $k = 2$, $r = 1$ cellular automata. This section discusses these results, identifies several types of behaviour, and considers analogies with classes of cellular automaton behaviour defined by dynamical systems theory means [2].

In the simplest cases, the set $\Omega^{(t)}$ generated by a cellular automaton evolves to a fixed form after a small number of time steps T (the case of surjective cellular automata, with $\Omega^{(t)} = \Sigma$ for all t, is considered separately). The minimal DFA corresponding to $\Omega^{(t)}$ for all $t \geq T$ are then identical, and the values of $\Xi^{(t)}$ and $\chi^{(t)}(\lambda)$ are thus constant. (Notice that $\Xi^{(t)} = \Xi^{(t+1)}$ does not necessarily imply $\Omega^{(t)} = \Omega^{(t+1)}$, as seen for rule 36 in Table 3.1.) In addition, for $t \geq T$, no more distinct blocks of sites are excluded. Such behaviour occurs in the trivial case of rule 0, under which all initial configurations are mapped to the null configuration after one time step. It also occurs for many other rules: one example is rule 76, discussed in Sects. 2 and 3. All the examples of this behaviour in Tables 3.1–3.3 have $T = 1$ (e.g. rule 76) or $T = 2$ (e.g. rule 108). In the trivial case of rule 0, only a single configuration (the null configuration) can appear when $t \geq T$. More complicated single configurations are sometimes generated, represented by minimal DFA consisting of a single cycle. In most cases (such as rule 76), however, $\Omega^{(T)}$ contains an infinite number of configurations. However, it appears that even in these cases, all configurations occur on finite cycles: each configuration is invariant under the cellular automaton mapping, or some finite iteration of it. (A result given in Sect. 5 then shows that the $\Omega^{(T)}$ must form finite complement languages in these cases.) This implies that changes in the initial state for such cellular automata propagate a distance of at most rT sites. A small initial change can thus ultimately affect a region no larger than $2rT$ sites. Such cellular automata must therefore exhibit class 1 or 2 behaviour [2].

For a second set of cellular automata, the form of the minimal DFA does not become fixed after a few time steps, but exhibits a simple growth with time, maintaining a fixed overall structure. The $L^{(t)}$ for such cellular automata typically increases linearly with time, and $\Xi^{(t)}$ increases as some polynomial function of t (linear or quadratic for legal $k = 2$, $r = 1$ rules). Rule 128 gives an example of this behaviour. Under this rule $111 \rightarrow 1$, but all other neighbourhoods map to 0. Any initial sequence of ones thus decreases steadily in length by one site on each side at each time step. After t time steps, any pair of ones must be separated by at least $2t + 1$ zeroes; all blocks of the form $10^j 1$ for $1 \leq j \leq 2t$ are thus excluded. The first few languages $\Omega^{(t)}$ in the sequence generated by successive time steps in the evolution of rule 128 are shown in Fig. 4.1. The minimal DFA are seen to maintain the same overall structure, but include a linearly increasing number of nodes at each time step. The characteristic polynomials corresponding to these DFA are given by

$$\chi^{(t)}(\lambda) = (1 - \lambda^t + \lambda^{t+1})(-1 - \lambda^t + \lambda^{t+1}), \tag{4.1}$$

yielding a set entropy which tends to zero at large times, roughly as $1/t$. Rule 160 provides another example in which the minimal DFA maintains the same overall structure, but increases in size with time. In this case, sequences of the form $1[(0 \vee 1)0]^j(0 \vee 1)1$ for all $j \leq t$, are excluded after t time steps, and the size $\Xi^{(t)}$ of

183

the corresponding minimal DFA grows quadratically with time.

Many cellular automata generate sets $\Omega^{(t)}$ whose corresponding minimal DFA become much more complicated at each successive time step, and appear to exhibit no simple overall structure.

Figure 4.2 shows the minimal DFA obtained after one and two time steps in the evolution of rule 126. No simple progression in the form of these minimal DFA is seen. $\Omega^{(1)}$ is a finite complement language, with only the block 010 excluded, yielding a characteristic polynomial

$$\chi^{(1)}(\lambda) = 1 - \lambda + 2\lambda^2 - \lambda^3, \tag{4.2}$$

giving $\lambda_{max} \simeq 1.7549$. After two time steps, an infinite sequence of distinct blocks is excluded, starting with the length 12 block 011101101110. The corresponding characteristic polynomial is

$$\begin{aligned}\chi^{(2)}(\lambda) = &-1 + \lambda - \lambda^2 + 2\lambda^3 - 4\lambda^4 + \lambda^5 + 3\lambda^6 - 5\lambda^7 \\ &+ 3\lambda^8 - 3\lambda^9 + 5\lambda^{10} - 6\lambda^{11} + 4\lambda^{12} - \lambda^{13},\end{aligned} \tag{4.3}$$

with $\lambda_{max} \simeq 1.7321$. The minimal DFA for $\Omega^{(3)}$ has 107 states, and the shortest newly-excluded block is 1011100011101 (length 13). $\Xi^{(t)}$ increases rapidly with time. After four time steps, the shortest newly-excluded blocks are 10111000011101, 10111000001110 and its reversal (length 14), and $\Xi^{(4)} = 2876$.

Figures 2.5 and 4.3 give the minimal DFA obtained after one and two time steps in the evolution of rule 18. A considerable increase in complication with time is again evident. After one time step, the shortest of an infinite number of distinct excluded blocks is 1101011 (length 7); after two time steps, the shortest newly-excluded block is 10011011001 (length 11); after three time steps, it is 110010010011 (length 12), and after four time steps it is 1001000010011 (length 13). In this case, as for rule 126, $L^{(t)}$ is found to increase monotonically over the range of times investigated. Progressively larger neighbourhoods of the start state are therefore left unchanged

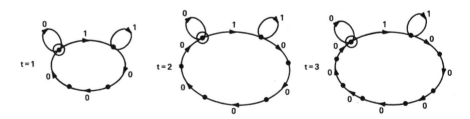

Figure 4.1. Minimal deterministic finite automata (DFA) corresponding to the regular languages $\Omega^{(t)}$ generated in the first few time steps of evolution according to cellular automaton rule 128. The DFA maintain the same structure, but increase in size with time. They correspond to finite complement languages, with all blocks of the form $10^j 1$ excluded for $1 \le j < 2t$.

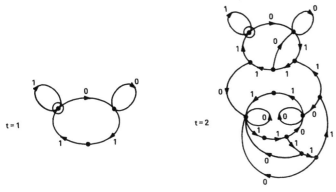

Figure 4.2. Minimal deterministic finite automata corresponding to the regular languages generated in the first two time steps of evolution according to the class 3 cellular automaton rule 126. A considerable increase in complexity with time is evident, characteristic of cellular automata which can exhibit class 3 behaviour.

in the corresponding minimal DFA. However, as discussed in Sect. 3, $L^{(t)}$ need not increase with time, but must in general only satisfy the inequality (3.13). Rule 22 provides an example in which $L^{(t)}$ decreases with time. The minimal DFA for $\Omega^{(1)}$ in this case is shown in Fig. 4.4; the shortest excluded blocks are 10101001 and 10010101 (length 8). After two time steps, the blocks 1110101 and 1010111 (length 7) are also excluded. The shortest newly-excluded blocks after three time steps are 01000010101, 01000110101, 10000010101 and their reversals (length 11). After four time steps, the shortest newly-excluded blocks are 010110011 and 110011010 (length 9), realizing the equality in (3.13).

Rule 126 provides an example in which the set generated after one time step is a finite complement language, but the sets generated at subsequent times are not. Rule 72 exhibits the opposite behaviour[11], as shown in Fig. 4.5. After one time step, it yields a set in which the infinite sequence of distinct blocks 111, 1101011, 11001011, ... are excluded (as in $\Omega^{(1)}$ for rule 18). After two time steps, however, the block 010 is also excluded. The exclusion of this single block implies exclusion of the infinite set of blocks excluded from $\Omega^{(1)}$. The resulting set thus corresponds to a finite complement language. In general, it can be shown that if a cellular automaton evolves to a finite complement language limit set, then it must do so in a finite number of time steps [34].

The sets $\Omega^{(t)}$ generated by most cellular automata never appear to become simpler with time. One exception is rule 72, in which the number of arcs in the minimal DFA for $\Omega^{(2)}$ is less than in that for $\Omega^{(1)}$. In most cases, the regular language complexity $\Xi^{(t)}$ appears to be non-decreasing with time. In fact, whenever the set of configurations generated continues to contract with time, a different regular language

---

[11] A more complicated example of this behaviour was given in [33].

185

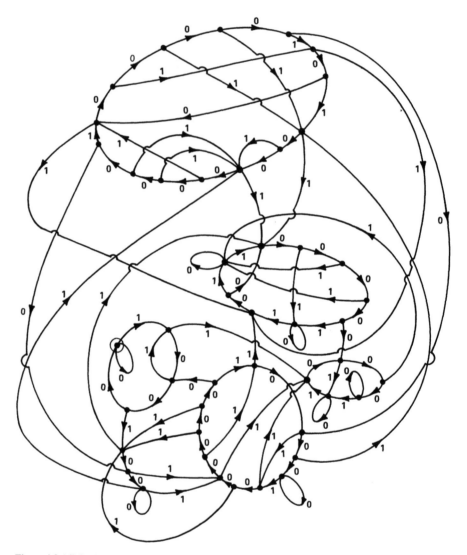

Figure 4.3. Minimal deterministic finite automata (DFA) corresponding to the regular language generated after two time steps of evolution according to the class 3 cellular automaton rule 18. The minimal DFA obtained for $t = 1$ is shown in Fig. 2.5. Rapidly-increasing complexity is again evident. The DFA illustrated here has 47 states.

must be obtained at each time step. Since there are a limited number of regular languages with complexities below any given value (certainly less than $2^{k\Xi^2}$), the complexity must on average increase at least slowly with time in this case.

Table 3.1 suggests that a definite set of cellular automata (including rules 18, 22 and 126) yield regular language complexities $\Xi^{(t)}$ that grow on average more rapidly than any polynomial in time (perhaps exponentially with time). Many of the cellular

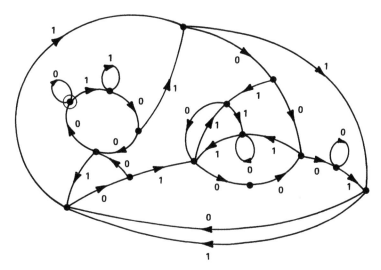

Figure 4.4. Minimal deterministic finite automaton (DFA) corresponding to the regular language $\Omega^{(1)}$ obtained after one time step in the evolution of the class 3 cellular automaton rule 22. The DFA has all 15 possible states. The shortest excluded block in $\Omega^{(1)}$ has length 8, and corresponds to the shortest path from the encircled start state to the one "incomplete" node in the DFA graph.

automata in this set generically exhibit class 3, chaotic, behaviour, suggesting that rapidly-increasing $\Xi^{(t)}$ are a signal for class 3 behaviour in cellular automata.

In a few cases, such as rule 94, $\Xi^{(t)}$ increases rapidly with time, but almost all initial configurations are found to give ultimately periodic behaviour. Nevertheless, special initial conditions (in this case, those in which successive pairs of sites have equal values) can yield chaotic behaviour. Since the set $\Omega^{(t)}$ includes all configurations that ever occur, it includes those that give chaotic behaviour, even though they occur with vanishingly small probability. Presumably these configurations would not affect a probabilistic grammar for the set $\Omega^{(t)}$ that included only nonzero probability configurations. But the $\Xi^{(t)}$ for the grammars discussed here appear to increase rapidly with time whenever any set of configurations in the cellular automaton yield class 3 behaviour.

Some exceptional cases are surjective class 3 cellular automata, such as the additive rules 90 and 150, in which every possible configuration can be generated at any time. The complexity of these and other cellular automata could perhaps be measured by constructing a grammar for the set of possible space-time patterns generated in their evolution. Such a grammar could presumably be characterized in terms of computers with memories arranged in a two-dimensional lattice (cf. [35])[12].

---

[12] This paper concentrates on one-dimensional cellular automata. Such cellular automata potentially correspond most directly with conventional formal languages. Two and higher dimensional cellular automata show some differences. For example the set of configurations obtained after a finite number of time steps in their evolution need not form a regular language and may in fact be nonrecursive [36, 51].

187

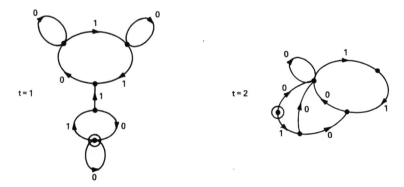

Figure 4.5. Minimal deterministic finite automata corresponding to the regular languages $\Omega^{(t)}$ generated in the first two time steps of evolution under rule 72. $\Omega^{(1)}$ is an infinite complement regular language, with the infinite sequence of distinct blocks 111, 1101011, 11001011, ... excluded. $\Omega^{(2)}$ is a finite complement language, with only the blocks 010 and 111 excluded.

The local rules ϕ for the 32 legal $k = 2$, $r = 1$ cellular automata of Tables 3.1–3.3 are all distinct. Yet in many cases sets of configurations with the same structure or properties are found to be generated. In some cases, there may exist bijective mappings which transform configurations evolving according to one cellular automaton rule into configurations evolving according to another rule. Several properties of the sets $\Omega^{(t)}$ are invariant under such mappings. One example is the set of non-zero roots of the characteristic polynomials $\chi^{(t)}(\lambda)$. While after one time step several of the cellular automata in Tables 3.1–3.3 yield the same sets of configurations $\Omega^{(1)}$, there are few examples of complete equivalence between pairs of cellular automaton rules. One simple example is rules 146 and 182, which are related by interchange of the roles of 0 and 1.

5. Some Invariant Sets

Section 2 showed that the set of configurations generated after a finite number of steps in the evolution of any cellular automaton forms a regular language. Sections 3 and 4 discussed some properties of such sets. This section and the next one consider the limiting sets of configurations generated after many time steps of cellular automaton evolution.

For all configurations A that appear in the limit set for a cellular automaton, there must exist some configuration A' such that $A = \Phi^t A'$ for any t. Any set of configurations invariant under the cellular automaton rule therefore appear in its limit set. This section considers some simple examples of invariant sets; Sect. 6 gives some comments on the complete structure of limit sets for cellular automata.

Periodic Sets

A simple class of invariant sets consist of configurations periodic with time under cellular automaton evolution. Such sets are found to form finite complement languages.

Consider the set of configurations that are stable (have temporal period 1) under a cellular automaton rule with $k = 2$ and $r = 1$. The set of such configurations is exactly those which contain only neighbourhoods $\{a_{i-1}, a_i, a_{i+1}\}$ for which

$$\phi[a_{i-1}, a_i, a_{i+1}] = a_i. \tag{5.1}$$

Only the finite set of distinct three-site blocks that violate (5.1) are forbidden, so that the complete set forms a finite-complement language, with a maximum distinct excluded block of length 3. A NDFA that generates the set of stable configurations is represented by a graph analogous to Fig. 3.1 in which only those arcs satisfying (5.1) are retained. The minimal grammar for this set is obtained by constructing the minimal equivalent DFA, as described in Sect. 2.

The procedure generalizes immediately to arbitrary cellular automaton rules, and to sets of configurations with any finite period (cf. [37]). The distinct excluded blocks in the finite complement languages corresponding to sets of configurations with period p have maximum length $2pr + 1$.

Figure 5.1 shows the minimal grammars for sets of configurations with various periods under the $k = 2$, $r = 1$ cellular automata with rule numbers 90, 18 and 22. The grammars are represented by graphs containing several disconnected pieces, each corresponding to a disjoint set of configurations.

Figure 5.1a suggests that only a finite number of configurations, all spatially periodic, are found with each temporal period in the surjective cellular automaton rule 90. For this and other surjective cellular automata whose local mappings ϕ are injective in their first and last arguments, the number of distinct configurations with any period p is always finite, and is exactly k^{hp}, where $\mathbf{h}$ is the invariant entropy of the cellular automaton mapping ($\mathbf{h} = 2$ for rule 90)[13]. This result follows from the fact that the complete space-time pattern generated by the evolution of such a cellular automaton is completely determined by any patch of site values with infinite temporal extent, but spatial width $\mathbf{h}$ (typically equal to $2r$). Moreover, any possible set of site values may occur in this patch. If the complete space-time pattern is to have period p, then so must the patch; but there are exactly k^{hp} possible patches with period p. (For large p, this result is as expected for any expansive homeomorphism (e.g. [10, 11]).)

In general, the sets of configurations with a particular periodicity under a cellular automaton rule are infinite, as illustrated for rules 18 and 22 in Figs. 5.1b and 5.1c.

---

[13] The actual configurations with particular periods may be found by methods analogous to those used in [29] for the complementary problem of determining the temporal periods of configurations with given spatial period.

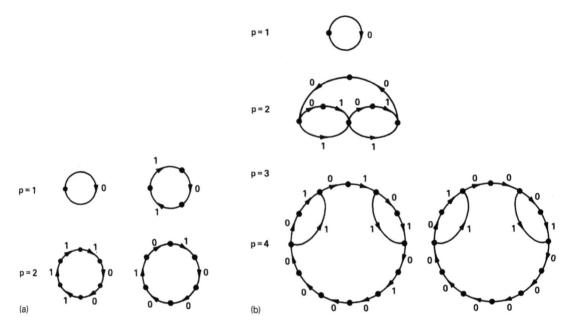

Figure 5.1a–c. Minimal deterministic finite automata corresponding to sets of configurations with (temporal) periods exactly p under cellular automaton rules **a** 90, **b** 18 and **c** 22.

Presumably there are sets of this kind with arbitrarily large periods. These infinite sets are nevertheless finite complement languages. For example, for the set of configurations with period two under rule 18, only the distinct blocks 111, 1011, 1101 and 10101 are excluded. It is common in class 3 cellular automata to find configurations with almost every possible period; for class 4 cellular automata, only some periods are typically found.

Periodic configurations form a small subset of all the configurations in the limit sets for cellular automata. Their entropy nevertheless provides a lower bound on the entropy of the complete limit sets. For rule 90, the set of periodic configurations has zero entropy, yet the complete limit contains all possible configurations, and thus has entropy 1. For rule 18, the period 2 set has entropy ≈ 0.4057 (given as the logarithm of the largest root of $\lambda^3 - \lambda - 1$), while the period 4 set has entropy ≈ 0.1824 ($\lambda^6 - \lambda - 1$). For rule number 22, the period 4 set has entropy ≈ 0.3219 ($\lambda^5 - \lambda^4 + \lambda^3 - \lambda^2 - 1$). Since irreversible cellular automaton mappings are contractive, the entropy of the set obtained after a finite number of time steps gives an upper bound on the entropy of the complete limit set. Using results from Table 3.3 one then finds

$$0.4057 \lesssim s_{[18]}^{(\infty)} \lesssim 0.8114,$$
$$0.1824 \lesssim s_{[22]}^{(\infty)} \lesssim 0.9390. \tag{5.2}$$

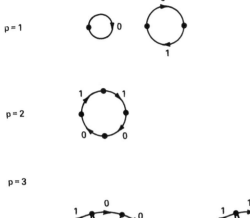

p = 1

p = 2

p = 3

p = 4

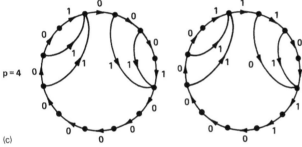

(c)

Figure 5.1 (continued).

Simulation Sets

The complete invariant sets for many cellular automata Φ are very complicated. Parts of these invariant sets may however have a simpler structure, and may consist of configurations for which Φ "simulates" a simpler cellular automaton rule. Thus for example stable configurations under Φ may be considered as those for which Φ "simulates" the identity mapping.

One class of configurations for which a cellular automaton rule Φ_1 may simulate a rule Φ_2 are those obtained by "blocking transformations." Each symbol in the possible configurations of Φ_2 is replaced by a length b_X block of symbols in Φ_1, and each time step in the evolution of Φ_2 is simulated by b_T time steps of evolution under Φ_1. Thus, for example, rule 18 simulates rule 90 under the $(b_X = 2, b_T = 2)$ blocking transformation $00 \to 0, 01 \to 1$ [1, 38, 5]. The evolution of an arbitrary configuration under rule 90 is thus simulated by the evolution under rule 18 of a configuration consisting of the digrams 00 and 01. But since rule 90 is surjective, all possible configurations correspond to an invariant set. Thus configurations containing only 00 and 01 digrams form an invariant set for rule 18. The entropy of these configurations is $1/2$, so that

$$0.5 \le s_{[18]}^{(\infty)} \le 0.8114. \tag{5.3a}$$

Rule 22 simulates rule 90 under the $(4, 4)$ blocking transformation $0000 \rightarrow 0, 0001 \rightarrow 1$, implying that

$$0.25 \leq s_{[22]}^{(\infty)} \lesssim 0.9390. \tag{5.3b}$$

A cellular automaton rule may simulate other rules with the same values of k and r under different blocking transformations (cf. the simulation network given in [5]). Some rules, apparently only surjective ones such as rule 90, simulate themselves, and thus correspond to fixed points of the blocking transformation. In other cases, one rule may simulate another under several distinct blocking transformations. For example, rule 18 simulates rule 90 under both $00 \rightarrow 0, 01 \rightarrow 1$, and $00 \rightarrow 0, 10 \rightarrow 1$, while rule 22 simulates rule 90 under any permutation of $0000 \rightarrow 0, 0001 \rightarrow 1$. One may consider the sets of blocks appearing in these blocking transformations to represent different "phases." An initial configuration then consists of several "domains," each of which contains blocks of one phase. The domains are separated by "walls." For rule 18, these walls appear to execute random walks, and annihilate in pairs, yielding progressively larger domains of a single phase [38]. The simulation of rule 90 by rule 18 may thus be considered "attractive" [3]. For rule 22, no such simple behaviour is observed.

Blocking transformations yield a particular class of configurations, corresponding to simple finite complement languages. Other classes of configurations, specified by more general grammars, may also yield simulations. (An example occurs for rule number 73, in which configurations containing only odd-length sequences of 0 and 1 sites simulate rule 90.) In addition, a set of configurations evolving under one rule may simulate an invariant set of configurations evolving under another rule.

6. Comments on Limiting Behaviour

Section 2 showed that after any finite number of time steps, the set of configurations $\Omega^{(t)}$ generated by any cellular automaton forms a regular language. Some cellular automata yield regular languages even in the infinite time limit; others appear to generate limit sets corresponding to more complicated formal languages. Cellular automata which exhibit different classes of overall behaviour appear to yield characteristically different limiting languages.

As discussed in Sect. 4, some cellular automata in Tables 3.1–3.3 yield regular languages which attain a fixed form after a few time steps. The limit sets for such cellular automata are thus regular languages. In fact, except for surjective rules, the limit sets found appear to contain only temporally periodic configurations, and are therefore finite complement languages. These cellular automata exhibit simple large time behaviour, characteristic of classes 1 and 2.

Rule 128 provides a more complicated example, discussed in Sect. 4. After t time steps, any pair of ones in configurations generated by this rule must be separated by at least $2t$ sites. The complete set of possible configurations forms a finite complement regular language, with a minimal DFA illustrated in Fig. 4.1 whose size $\Xi^{(t)}$ increases linearly with time. After many time steps, almost all initial configurations evolve to the null configuration. However, even after an arbitrarily long time, configurations containing just a single block of ones may still appear. A block of n ones, flanked by infinite sequences of zeroes, is generated after any number of time steps t from a block of $n + 2t$ ones. Such configurations therefore have exactly one predecessor under any number of time steps of the cellular automaton evolution. They thus appear in the limit set for rule 128, although if all initial configurations are given equal weight, they are generated with zero probability. Once generated, their evolution is never periodic. An increasing number of distinct blocks are excluded from the successive $\Omega^{(t)}$ obtained by evolution under rule 128. The set of configurations generated in the infinite time limit does not, therefore, correspond to a finite complement language. Nevertheless, the set does form a regular language, shown in Fig. 6.1. While the set contains an infinite number of configurations, its entropy vanishes, as given by the limit of Eq. (4.1).

Several rules given in Tables 3.1–3.3 exhibit behaviour similar to rule 128: they generate (finite complement) regular languages whose minimal grammars increase in size linearly or quadratically with time, but in the infinite time limit, yield regular language limit sets. These limit sets contain one or a few periodic configurations, together with an infinite number of aperiodic configurations, generated from a set of initial configurations of measure zero. The sets have zero entropy, and do not correspond to finite complement languages. (Only trivial finite complement languages can have zero entropy.) All the class 1 cellular automata (except for the trivial rule 0) in Tables 3.1–3.3 exhibit such limiting behaviour. The generation of limit sets corresponding to regular languages that are not finite complement languages appears to be a general feature of class 1 cellular automata.

Tables 3.1–3.3 suggest the result, discussed in Sect. 4, that cellular automata capable of class 3 or 4 behaviour give rise to sets of configurations represented by regular languages whose complexity increases rapidly with time. The limit sets for such cellular automata are therefore presumably not usually regular languages. If a finite description of them can be given, it must be in terms of more complicated formal languages.

Any language that can be described by a regular grammar must obey the regular language "pumping lemma" (e.g. [7]). This requires that it be possible to write all sufficiently long symbol sequences α appearing in the language in the form $\alpha_1 \alpha_2 \alpha_3$

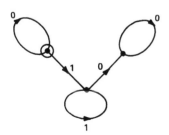

Figure 6.1. The deterministic finite automaton representing the regular language corresponding to the limit set for cellular automaton rule 128. This infinite complement regular language is obtained as the infinite time limit of the series of finite complement regular languages illustrated in Fig. 4.1. It contains an infinite number of configurations, but has zero limiting entropy.

so that for any n the symbol sequence $\alpha_1 \alpha_2^n \alpha_3$ also appears in the language. (This result follows from the fact that any sufficiently long sequence must correspond to a path containing a cycle in the DFA. This cycle may then be traversed any number of times, yielding arbitrarily repeated symbol sequences.) The sets generated after a finite number of time steps in cellular automaton evolution always obey this condition: arbitrary repetitions of the string α_2 are obtained by evolution from initial configurations containing arbitrarily-repeated sequences evolving to α_2.

It is possible to construct cellular automata for which the regular language pumping lemma fails in the large time limit, and which therefore yield non-regular language limit sets. In one class of examples [39, 34], there are pairs of localized structures which propagate with opposite velocities from point sources. After t time steps, such cellular automata generate configurations consisting roughly of repetitions of sequences

$$(10^j 20^j 1) \qquad j \leq t. \tag{6.1}$$

In the infinite time limit, arbitrarily long identical pairs of symbol sequences thus appear. The limit sets for such cellular automata are therefore not regular languages. Instead it appears that they correspond to context-free languages.

The pumping lemma for regular languages may be generalized to context-free languages. Any sufficiently long sequence in a context-free language must be of the form $\alpha_1 \alpha_2 \alpha_3 \alpha_4 \alpha_5$ such that $\alpha_1 \alpha_2^n \alpha_3 \alpha_4^n \alpha_5$ is also in the language for any n. The possibility for separated equal length identical substrings is a reflection of the non-local nature of context-free languages, manifest for example in the indefinitely large memory stacks required in machines to recognize them.

Limiting sets of configurations of the form (6.1) that violate the regular language pumping lemma nevertheless obey the context-free language pumping lemma, and thus correspond to context-free languages.

The correspondence between sets of infinite cellular automaton configurations and context-free languages is slightly more complicated than for regular languages. In all cases, the cellular automaton configurations correspond to infinite symbol sequences generated according to a formal grammar. For regular languages, it is also possible to construct finite automata which recognize words in the language, starting at any point. The necessity for a stack memory in the generation of context-free languages makes their recognition starting at any point in general impossible. Infinite configurations generated by context-free grammars must thus be viewed as concatenations of finite context-free language words. Only at the boundaries between these words is the stack memory for the machine generating the configuration empty, so that sequences of symbols may be recognized. Configurations generated by context-sensitive and more complicated grammars must be considered in an analogous way.

If the limit set for a cellular automaton is a context-free language, whose generation requires a computer with an indefinitely large stack memory, then one expects that the regular language sets obtained at successive finite time steps in the evolution of the cellular automaton would require progressively larger finite size stack memories. If the limiting context-free grammar contains say Q (non-terminal) productions, then there are $O(Q^t)$ possible stack configurations after t time steps, and the set of configurations obtained may be recognized by a finite automaton with about Q^t states. In addition, the context-free pumping lemma is satisfied for repetitions of substrings of length up to about t. Regular languages that approximate context-free languages for t time steps should have comparatively simple repetitive forms. The regular languages of Fig. 4.1 generated at finite times by rule 128 have roughly the expected form, but their limit is in fact a regular language. The absence of obvious patterns in the regular grammars such as Figs. 4.2–4.4 generated by typical class 3 cellular automata after even a few time steps suggests that the limiting languages in these cases are not context-free. They are presumably context-sensitive or unrestricted languages.

The entropies of regular languages are always logarithms of algebraic integers [as in Eq. (3.5)]. Context-free languages may, however, have entropies given by logarithms of general algebraic numbers (whose minimal polynomials are not necessarily monic). The enumeration of words in a formal language may be cast in algebraic terms by considering the sequence of words in the language as a formal power series satisfying equations corresponding to the production rules for the language (e.g. [40]). For the simple regular language $((0^*)10)$ (repetition understood) of Fig. 1.3b, with production rules

$$u_0 \rightarrow s_0 u_0, \quad u_0 \rightarrow s_1 u_1, \quad u_1 \rightarrow s_0 u_0, \tag{6.2}$$

(where the terminal symbols s_0 and s_1 represent 0 and 1 respectively), the corresponding equations are

$$u_0 = s_0 u_0 + s_1 u_1 + 1, \quad u_1 = s_0 u_0 + 1. \tag{6.3}$$

Solving for u_0 as the start symbol one obtains

$$u_0 = (s_1 + 1)/(1 - s_0 - s_1 s_0). \tag{6.4}$$

The expansion of this generating function (accounting for the non-commutative nature of symbol string concatenation) yields the sequence of possible words in the language. Replacing all terminal symbols in the generating function by a dummy variable x, the coefficient of x^n in its expansion gives the number of distinct symbol sequences of length n in the language. The asymptotic growth rate of this number, and thus the entropy of the language, are then determined by the smallest real root of the (monic) denominator polynomial. The generating function for any regular language is always a rational function of x. For a context-free language, however, the equations analogous to (3.8) are in general non-linear in the u_i. At least for unambiguous languages, the positions of the leading poles in the resulting generating functions obtained by solving these simultaneous polynomial equations are nevertheless algebraic numbers [41].

There is a finite procedure to find the minimal regular grammar that generates a given regular language, as described in Sect. 2. No finite procedure exists in general, however, to find the minimal context-free or other grammar corresponding to a more complicated language. The analogue of the regular language complexity is thus formally non-computable for context-free and more complicated languages. This is an example of the result that no finite procedure can in general determine the shortest input program that generates a particular output when run for an arbitrarily long time on some computer (e.g. [42]). Explicit testing of successively longer programs is inadequate, since the insolubility of the halting problem implies that no upper bound can in general be given for the time before the required sequence is generated. Particular simple cases of this problem are nevertheless soluble, so that, for example, the minimal grammars for regular languages are computable.

The entropies for regular and context-free languages may be computed by the finite procedures described above. The entropies for context-sensitive (type 1) and unrestricted (type 0) languages are, however, in general non-computable numbers [43]. Bounds on them may be given. But no finite procedure exists to calculate them to arbitrary precision. (They are in many respects analogous to the non-computable probabilities for universal computers to halt given random input.) If many class 3 and 4 cellular automata do indeed yield limit sets corresponding to context-sensitive or unrestricted languages, then the entropies of these sets are in general non-computable.

The discussion so far has concerned the generation of infinite configurations by cellular automaton evolution. One may also consider the evolution of initial configurations in which nonzero sites exist only in a finite region. Then for class 3 cellular automata with almost all initial states, the region of nonzero sites expands linearly with time. (Such expansion is guaranteed if, for example, $\phi[1, 0, \ldots, 0] = 1$ and so on.) For class 4 cellular automata, the region may expand and contract with time. One may characterize the structures generated by considering the set of finite sequences generated at any time by evolution from a set of finite initial configurations. For class 3 cellular automata, this set appears to be no more complicated than a context-sensitive language, while for class 4 cellular automata, it may be an unrestricted language. Notice that the set generated after a fixed finite number of time steps always corresponds to a regular language, just as for infinite configurations. (The regular grammar for these finite configurations consists of all paths with the relevant length that begin and end at the 00...0 node of the NDFA analogous to Fig. 2.1.)

Consider the language formed by the set of sequences of length n generated after any number of time steps in the evolution of a class 3 cellular automaton from all possible initial configurations with size n_0[14]. This language appears to be at most context-sensitive, since a word of length n in it can presumably be recognized in a finite time by a computer with a memory of size at most n. In its simplest form,

---

[14] This is analogous to but distinct from the problem of finding all initial configurations which ultimately evolve to a particular complete final configuration, such as the null configuration (cf. [2, 44, 45]).

the computer operates by testing configurations generated by evolution from all k^{n_0} possible initial states. Since the configurations expand steadily with time, the evolution of each configuration need be traced only until it is of size n; the required configuration of length n is either reached at that time, or will never be reached.

In a class 4 cellular automaton, evolution from an initial configuration of size n_0 may yield arbitrarily large configurations, but then ultimately contract to give a size n configuration. No upper bound on the time or memory space required to generate the size n configuration may therefore be given. The problem of determining whether a particular finite configuration is ever generated in the evolution of a class 4 cellular automaton from one of a finite set of initial configurations may therefore in general be formally undecidable. No finite computation can give all the structures of a particular size ultimately generated in the evolution of a class 4 cellular automaton.

The procedure for recognizing finite configurations generated by class 3 cellular automata, while finite in principle, may require large computational resources. Whenever the context-sensitive language corresponding to the set of finite configurations cannot be described by a context-free or simpler grammar, the problem of recognizing words in the language is PSPACE-complete with respect to the lengths of the words (e.g. [28]). It can thus presumably be performed essentially no more efficiently than by testing the structures generated by the evolution of each of the k^{n_0} possible finite initial configurations.

As well as considering the evolution of finite complete configurations, one may also consider the generation of finite sequences of symbols in the evolution of infinite configurations. Enumeration of sets of length n sequences that can and cannot occur provide partial characterizations of sets of infinite configurations. However, even for configurations generated at a finite time t, such enumeration in general requires large computational resources. A symbol sequence of length n appears only if at least one length $n_0 = n + 2rt$ initial block evolves to it after t time steps. A computation time polynomial in n and t suffices to determine whether a particular candidate initial block evolves to a required sequence. The problem of determining whether any such initial block exists is therefore in the class NP. One may expect that for many cellular automata, this problem is in fact NP-complete. (The procedure of Sect. 2 provides no short cut, since the construction of the required DFA is an exponential computational process.) It may therefore effectively be solved essentially only by explicit simulation of the evolution of all exponentially-many possible initial sequences.

In the limit of infinite time, the problem of determining whether a particular finite sequence is generated in the evolution of a cellular automata becomes in general undecidable. For a cellular automaton with only class 1 or 2 behaviour, the limit set always appears to correspond to a regular language, for which the problem is decidable. But for class 3 and 4 cellular automata, whose limit sets presumably correspond to more complicated formal languages, the problem may be undecidable. (The problem is in general in the undecidability class Π_1 [46]; the set of finite sequences that occur is thus recursively enumerable, but not necessarily recursive.)

Even when the general problem is undecidable, the appearance of particular finite sequences in the limit set for a cellular automaton may be decidable. The fraction of particular sequences whose appearance in the limit set is undecidable provides a measure of the degree of unpredictability or "computational achievement" of the cellular automaton evolution (presumably related to "logical depth" [47]).

7. Discussion

This paper has taken some preliminary steps in the application of computation theory to the global analysis of cellular automata. Cellular automata are viewed as computers, whose time evolution processes the information specified by their initial configurations. Many aspects of this information processing may be described in terms of computation theory. The intrinsic discreteness of cellular automata allows for immediate identifications with conventional computational systems; but the basic approach and many of the results obtained should be applicable to many other dynamical systems.

Self-organization in cellular automata involves the generation of distinguished sets of configurations with time. These sets are described as formal languages in computation theory terms. Each configuration corresponds to a word in a language, and is formed from a sequence of symbols according to definite grammatical rules. These grammatical rules provide a complete and succinct specification of the sets generated by the cellular automaton evolution.

Section 2 showed that, starting with all possible initial configurations, the sets generated by a finite number of time steps of cellular automaton evolution always correspond to regular formal languages. Such languages are recognized by finite automata. These finite automata are specified by finite state transition graphs; words in the languages correspond to all possible paths through these graphs. The (limiting) set entropies of such regular languages are then given as logarithms of the algebraic integers corresponding to the largest eigenvalues of the incidence matrices for their state transition graphs.

In general, several different finite automata or regular grammars may yield the same regular language. However, it is always possible to find a simplest finite automaton, or set of grammatical rules, which correspond to any particular regular language. This simplest finite automaton provides a canonical representation for sets generated by cellular automaton evolution, and its size (number of states) gives a measure of their "complexity." The larger the "regular language complexity" for a set of configurations, the more complicated is the minimal set of grammatical rules necessary to describe it as a regular language.

Section 4 suggests the general result that the regular language complexity is non-decreasing with time for all cellular automata. This result gives a quantitative characterization of progressive self-organization in cellular automata. It may give a first indication of a generalization of the second law of thermodynamics to irreversible systems.

Entropy may be estimated from experimental data by fitting parameters in simple models which reproduce the data. Extraction of regular language complexities from experimental data requires the identification of maximal (regular language) patterns in the data, or the construction of a minimal (finite automaton) model that generates the data. Given perfect data (and an upper bound on the regular language complexity), a direct method may be used (e.g. [48]). In practice, it will probably be convenient to construct stochastic finite automata which provide probabilistic reproductions of the available data (cf. estimates for the structure of Markovian sources (e.g. [49])).

Dynamical systems theory methods were used in [2] to identify four general classes of cellular automaton behaviour. Sections 4 and 6 suggested computation theory characterizations of these classes. The limit sets for cellular automata with only class 1 or 2 behaviour are regular languages. For most class 3 and 4 cellular automata, the regular language complexity increases steadily with time, so that the set of configurations obtained in the large time limit does not usually form a regular language. Instead (at least for appropriate finite size configurations) the limit sets for class 3 cellular automata appear to correspond to context-sensitive languages, while those for class 4 cellular automata correspond to general languages.

Regular languages are sufficiently simple that their properties may be determined by finite computational procedures. Properties of context-free and more complicated languages are, however, often not computable by finite means. Thus, for example, the minimal grammars for such languages (whose sizes would provide analogues of the regular language complexity) cannot in general be found by finite computations. Moreover, for context-sensitive and general languages, even quantities such as entropy are formally non-computable.

When cellular automaton evolution is viewed as computation, one may consider that the limiting properties of a cellular automaton are determined by an infinite computational process. One should not expect in general that the results of this infinite process can be summarized in finite mathematical terms. For sufficiently simple cellular automata, apparently those of classes 1 and 2, however, it is nevertheless possible to "short cut" the infinite processes of cellular automaton evolution, and to give a finite specification of their limiting properties. For most class 3 and 4 cellular automata, no such short cut appears possible: their behaviour may in general be determined by no procedure significantly faster than explicit simulation, and many of their limiting properties cannot be determined by any finite computational process. (Such non-computable limiting behaviour would be an immediate consequence of the universal computation capability conjectured for class 4 cellular automata, but does not depend on it.)

Non-computability and undecidability are common phenomena in the systems investigated in pure mathematics, logic and computation. But they have not been identified in the systems considered in theoretical physics. In many physical theories one can in fact imagine constructing complicated systems which behave, for example, as universal computers, and for which undecidable propositions may be formulated. Cellular automata (and other dynamical systems) may be considered as simple physical theories. This paper has suggested that in fact even simple, natural, questions concerning the limiting behaviour of cellular automata are often undecidable (except for very simple systems such as those corresponding to class 1 and 2 cellular automata). One may speculate that undecidability is common in all but the most trivial physical theories. Even simply-formulated problems in theoretical physics may be found to be provably insoluble.

Undecidability and non-computability are features of problems which attempt to summarize the consequences of infinite processes. Finite processes may always be carried out explicitly. For some particularly simple processes, the consequences of a large, but finite, number of steps may be deduced by a procedure involving only a small number of steps. But at least for many computational processes (e.g. [28]), it is believed that no such short cut exists: each step (or each possibility) must in fact be carried out explicitly. It was suggested that this phenomenon is common in cellular automata. One may speculate that it is widespread in physical systems. No simple theory or formula could ever be given for the overall behaviour of such systems: the consequences of their evolution could not be predicted, but could effectively be found only by direct simulation or observation.

Acknowledgements

I am grateful to A. Aho, C. Bennett, J. Conway, D. Hillis, L. Hurd, D. Lind, O. Martin, M. Mendes France, J. Milnor, A. Odlyzko, N. Packard, J. Reeds, and many others for discussions. A preliminary version of this paper was presented at a workshop on "Coding and Isomorphisms in Ergodic Theory," held at the Mathematical Sciences Research Institute, Berkeley (December 8–13, 1983). I thank M. Boyle, E. Coven, J. Franks, and many of the other participants for their comments. Some of the results given above were obtained using the computer mathematics system SMP [50].

References

1. Wolfram, S.: Statistical mechanics of cellular automata. Rev. Mod. Phys. **55**, 601 (1983).
2. Wolfram, S.: Universality and complexity in cellular automata. Physica **10D**, 1 (1984).
3. Packard, N. H.: Complexity of growing patterns in cellular automata, Institute for Advanced Study preprint (October 1983), and to be published in Dynamical behaviour of automata. Demongeot, J., Goles, E., Tchuente, M., (eds.). Academic Press (proceedings of a workshop held in Marseilles, September 1983).
4. Wolfram, S.: Cellular automata as models for complexity. Nature (to be published).
5. Wolfram, S.: Cellular automata. Los Alamos Science, Fall 1983 issue.
6. Beckman, F. S.: Mathematical foundations of programming. Reading, MA: Addison-Wesley 1980.

7. Hopcroft, J. E., Ullman, J. D.: Introduction to automata theory, languages, and computation. Reading, MA: Addison-Wesley 1979.

8. Minsky, M.: Computation: finite and infinite machines. Englewood Cliffs, NJ: Prentice-Hall 1967.

9. Rozenberg, G., Salomaa, A. (eds.): L systems. In: Lecture Notes in Computer Science, Vol. 15; Rozenberg, G., Salomaa, A.: The mathematical theory of L systems. New York: Academic Press 1980.

10. Guckenheimer, J., Holmes, P.: Nonlinear oscillations, dynamical systems, and bifurcations of vector fields. Berlin, Heidelberg, New York: Springer 1983.

11. Walters, P.: An introduction to ergodic theory. Berlin, Heidelberg, New York: Springer 1982.

12. Weiss, B.: Subshifts of finite type and sofic systems. Monat. Math. **17**, 462 (1973); Coven, E. M., Paul, M. E.: Sofic systems. Israel J. Math. **20** 165 (1975).

13. Field, R. D., Wolfram, S.: A QCD model for e^+e^- annihilation. Nucl. Phys. **B213**, 65 (1983).

14. Smith, A. R.: Simple computation-universal cellular spaces. J. ACM **18**, 331 (1971).

15. Berlekamp, E. R., Conway, J. H., Guy, R. K.: Winning ways for your mathematical plays. New York: Academic Press, Vol. 2, Chap. 25.

16. Lind, D.: Applications of ergodic theory and sofic systems to cellular automata. Physica **10D**, 36 (1984).

17. de Bruijn, N. G.: A combinatorial problem. Ned. Akad. Weten. Proc. **49**, 758 (1946); Good, I. J.: Normal recurring decimals. J. Lond. Math. Soc. **21**, 167 (1946).

18. Nerode, A.: Linear automaton transformations. Proc. AMS **9**, 541 (1958).

19. Cvetkovic, D., Doob, M., Sachs, H.: Spectra of graphs. New York: Academic Press 1980.

20. Billingsley, P.: Ergodic theory and information. New York: Wiley 1965.

21. Chomsky, N., Miller, G. A.: Finite state languages. Inform. Control **1**, 91 (1958).

22. Stewart, I. N., Tall, D. O.: Algebraic number theory. London: Chapman & Hall 1979.

23. Lind, D. A.: The entropies of topological Markov shifts and a related class of algebraic integers. Ergodic Theory and Dynamical Systems (to be published).

24. Milnor, J.: Unpublished notes (cited in [2]).

25. Hedlund, G. A.: Endomorphisms and automorphisms of the shift dynamical system. Math. Syst. Theor. **3**, 320 (1969); Hedlund, G. A.: Transformations commuting with the shift. In: Topological dynamics. Auslander, J., Gottschalk, W. H., (eds.). New York: Benjamin 1968.

26. Amoroso, S., Patt, Y. N.: Decision procedures for surjectivity and injectivity of parallel maps for tessellation structures. J. Comp. Sys. Sci. **6**, 448 (1972).

27. Nasu, M.: Local maps inducing surjective global maps of one-dimensional tessellation automata. Math. Syst. Theor. **11**, 327 (1978).

28. Garey, M. R., Johnson, D. S.: Computers and intractability: a guide to the theory of NP-completeness. San Francisco: Freeman 1979, Sect. A10.

29. Martin, O., Odlyzko, A. M., Wolfram, S.: Algebraic properties of cellular automata. Commun. Math. Phys. **93**, 219 (1984).

30. Hedlund, G.: Private communication.

31. Manning, A.: Axiom A diffeomorphisms have rational zeta functions. Bull. Lond. Math. Soc. **3**, 215 (1971); Coven, E., Paul, M.: Finite procedures for sofic systems. Monat. Math. **83**, 265 (1977).

32. Franks, J.: Private communication.

33. Coven, E.: Private communication.

34. Hurd, L.: Formal language characterizations of cellular automata limit sets (to be published).

35. Rosenfeld, A.: Picture languages. New York: Academic Press (1979).

201

36. Golze, U.: Differences between 1- and 2-dimensional cell spaces. In: Automata, Languages and Development, Lindenmayer, A., Rozenberg, G. (eds.). Amsterdam: North-Holland 1976; Yaku, T.: The constructibility of a configuration in a cellular automaton. J. Comput. System Sci. **7**, 481 (1983).

37. Grassberger, P.: Private communication.

38. Grassberger, P.: A new mechanism for deterministic diffusion. Phys. Rev. A, (to be published) Chaos and diffusion in deterministic cellular automata. Physica **10D**, 52 (1984).

39. Hillis, D., Hurd, L.: Private communications.

40. Salomaa, A., Soittola, M.: Automata-theoretic aspects of formal power series. Berlin, Heidelberg, New York: Springer 1978.

41. Kuich, W.: On the entropy of context-free languages. Inform. Cont. **16**, 173 (1970).

42. Chaitin, G.: Algorithmic information theory. IBM J. Res. Dev. **21**, 350 (1977).

43. Kaminger, F. P.: The non-computability of the channel capacity of context-sensitive languages. Inform. Cont. **17**, 175 (1970).

44. Smith, A. R.: Real-time language recognition by one-dimensional cellular automata. J. Comput. Syst. Sci. **6**, 233 (1972).

45. Sommerhalder, R., van Westrhenen, S. C.: Parallel language recognition in constant time by cellular automata. Acta Inform. **19**, 397 (1983).

46. Rogers, H.: Theory of recursive functions and effective computability. New York: McGraw-Hill 1967.

47. Bennett, C. H.: On the logical "depth" of sequences and their reducibilities to random sequences. Inform. Cont. (to be published).

48. Conway, J. H.: Regular algebra and finite machines. London: Chapman & Hall 1971.

49. Shannon, C. E.: Prediction and entropy of printed English. Bell Syst. Tech. J. **30**, 50 (1951).

50. Wolfram, S.: SMP reference manual. Computer Mathematics Group. Los Angeles: Inference Corporation 1983.

51. Packard, N. H., Wolfram, S.: Two dimensional cellular automata. Institute for Advanced Study preprint, May 1984.

52. Hopcroft, H.: An $n \log n$ algorithm for minimizing states in a finite automaton. In: Proceedings of the International Symposium on the Theory of Machines and Computations. New York: Academic Press 1971.

53. Hurd, L.: Private communication.

Undecidability and Intractability in Theoretical Physics

1985

*Physical processes are viewed as computations, and the difficulty of answering
questions about them is characterized in terms of the difficulty of performing the
corresponding computations. Cellular automata are used to provide explicit examples of various formally undecidable and computationally intractable problems. It
is suggested that such problems are common in physical models, and some other
potential examples are discussed.*

There is a close correspondence between physical processes and computations. On
one hand, theoretical models describe physical processes by computations that transform initial data according to algorithms representing physical laws. And on the
other hand, computers themselves are physical systems, obeying physical laws. This
paper explores some fundamental consequences of this correspondence.[1]

The behavior of a physical system may always be calculated by simulating explicitly each step in its evolution. Much of theoretical physics has, however, been
concerned with devising shorter methods of calculation that reproduce the outcome
without tracing each step. Such shortcuts can be made if the computations used in
the calculation are more sophisticated than those that the physical system can itself
perform. Any computations must, however, be carried out on a computer. But the
computer is itself an example of a physical system. And it can determine the outcome
of its own evolution only by explicitly following it through: No shortcut is possible.
Such computational irreducibility occurs whenever a physical system can act as a
computer. The behavior of the system can be found only by direct simulation or observation: No general predictive procedure is possible. Computational irreducibility
is common among the systems investigated in mathematics and computation theory.[2]
This paper suggests that it is also common in theoretical physics. Computational
reducibility may well be the exception rather than the rule: Most physical questions

Originally published in *Physical Review Letters*, volume 54, pages 735–738 (25 February 1985).

may be answerable only through irreducible amounts of computation. Those that concern idealized limits of infinite time, volume, or numerical precision can require arbitrarily long computations, and so be formally undecidable.

A diverse set of systems are known to be equivalent in their computational capabilities, in that particular forms of one system can emulate any of the others. Standard digital computers are one example of such "universal computers": With fixed intrinsic instructions, different initial states or programs can be devised to simulate different systems. Some other examples are Turing machines, string transformation systems, recursively defined functions, and Diophantine equations.[2] One expects in fact that universal computers are as powerful in their computational capabilities as any physically realizable system can be, so that they can simulate any physical system.[3] This is the case if in all physical systems there is a finite density of information, which can be transmitted only at a finite rate in a finite-dimensional space.[4] No physically implementable procedure could then shortcut a computationally irreducible process.

Different physically realizable universal computers appear to require the same order of magnitude times and information storage capacities to solve particular classes of finite problems.[5] One computer may be constructed so that in a single step it carries out the equivalent of two steps on another computer. However, when the amount of information n specifying an instance of a problem becomes large, different computers use resources that differ only by polynomials in n. One may then distinguish several classes of problems.[6] The first, denoted P, are those such as arithmetical ones taking a time polynomial in n. The second, denoted $PSPACE$, are those that can be solved with polynomial storage capacity, but may require exponential time, and so are in practice effectively intractable. Certain problems are "complete" with respect to $PSPACE$, so that particular instances of them correspond to arbitrary $PSPACE$ problems. Solutions to these problems mimic the operation of a universal computer with bounded storage capacity: A computer that solves $PSPACE$-complete problems for any n must be universal. Many mathematical problems are $PSPACE$-complete.[6] (An example is whether one can always win from a given position in chess.) And since there is no evidence to the contrary, it is widely conjectured that $PSPACE \neq P$, so that $PSPACE$-complete problems cannot be solved in polynomial time. A final class of problems, denoted NP, consist in identifying, among an exponentially large collection of objects, those with some particular, easily testable property. An example would be to find an n-digit integer that divides a given $2n$-digit number exactly. A particular candidate divisor, guessed nondeterministically, can be tested in polynomial time, but a systematic solution may require almost all $O(2^n)$ possible candidates to be tested. A computer that could follow arbitrarily many computational paths in parallel could solve such problems in polynomial time. For actual computers that allow only boundedly many paths, it is suspected that no general polynomial time solution is possible.[5] Nevertheless, in the infinite time limit, parallel paths are irrelevant, and a computer that solves NP-complete problems is equivalent to other universal computers.[6]

Figure 1. Seven examples of patterns generated by repeated application of various simple cellular automaton rules. The last four are probably computationally irreducible, and can be found only by direct simulation.

The structure of a system need not be complicated for its behavior to be highly complex, corresponding to a complicated computation. Computational irreducibility may thus be widespread even among systems with simple construction. Cellular automata (CA)[7] provide an example. A CA consists of a lattice of sites, each with k possible values, and each updated in time steps by a deterministic rule depending on a neighborhood of R sites. CA serve as discrete approximations to partial differential equations, and provide models for a wide variety of natural systems. Figure 1 shows typical examples of their behavior. Some rules give periodic patterns, and the outcome after many steps can be predicted without following each intermediate step. Many rules, however, give complex patterns for which no predictive procedure is evident. Some CA are in fact known to be capable of universal computation, so that their evolution must be computationally irreducible. The simplest cases proved have $k = 18$ and $R = 3$ in one dimension,[8] or $k = 2$ and $R = 5$ in two dimensions.[9] It is strongly suspected that "class-4" CA are generically capable of universal computation: There are such CA with $k = 3$, $R = 3$ and $k = 2$, $R = 5$ in one dimension.[10]

Computationally, irreducibility may occur in systems that are not full universal computers. For inability to perform, specific computations need not allow all computations to be shortcut. Though class-3 CA and other chaotic systems may not be universal computers, most of them are expected to be computationally irreducible, so that the solution of problems concerning their behavior requires irreducible amounts of computation.

As a first example consider finding the value of a site in a CA after t steps of evolution from a finite initial seed, as illustrated in Fig. 1. The problem is specified by giving the seed and the CA rule, together with the $\log t$ digits of t. In simple cases such as the first two shown in Fig. 1, it can be solved in the time $O(\log t)$ necessary to input this specification. However, the evolution of a universal computer CA for a polynomial in t steps can implement any computation of length t. As a consequence, its evolution is computationally irreducible, and its outcome found only by an explicit simulation with length $O(t)$: exponentially longer than for the first two in Fig. 1.

One may ask whether the pattern generated by evolution with a CA rule from a particular seed will grow forever, or will eventually die out.[11] If the evolution is computationally irreducible, then an arbitrarily long computation may be needed to answer this question. One may determine by explicit simulation whether the pattern dies out after any specified number of steps, but there is no upper bound on the time needed to find out its ultimate fate.[12] Simple criteria may be given for particular cases, but computational irreducibility implies that no shortcut is possible in general. The infinite-time limiting behavior is formally undecidable: No finite mathematical or computational process can reproduce the infinite CA evolution.

The fate of a pattern in a CA with a finite total number of sites N can always be determined in at most k^N steps. However, if the CA is a universal computer, then the problem is *PSPACE*-complete, and so presumably cannot be solved in a time polynomial in N.[13]

One may consider CA evolution not only from finite seeds, but also from initial states with all infinitely many sites chosen arbitrarily. The value $a^{(t)}$ of a site after many time steps t then in general depends on $2\lambda t \le Rt$ initial site values, where λ is the rate of information transmission (essentially Lyapunov exponent) in the CA.[9] In class-1 and -2 CA, information remains localized, so that $\lambda = 0$, and $a^{(t)}$ can be found by a length $O(\log t)$ computation. For class-3 and -4 CA, however, $\lambda > 0$, and $a^{(t)}$ requires an $O(t)$ computation.[14]

The global dynamics of CA are determined by the possible states reached in their evolution. To characterize such states one may ask whether a particular string of n site values can be generated after evolution for t steps from any (length $n+2\lambda t$) initial string. Since candidate initial strings can be tested in $O(t)$ time, this problem is in the class *NP*. When the CA is a universal computer, the problem is in general *NP*-complete, and can presumably be answered essentially only by testing all $O(k^{n+2\lambda t})$ candidate initial strings.[15] In the limit $t \to \infty$, it is in general undecidable whether particular strings can appear.[16] As a consequence, the entropy or dimension of the limiting set of CA configurations is in general not finitely computable.

Formal languages describe sets of states generated by CA.[17] The set that appears after t steps in the evolution of a one-dimensional CA forms a regular formal language: each possible state corresponds to a path through a graph with $\Xi^{(t)} < 2^{k^{Rt}}$ nodes. If, indeed, the length of computation to determine whether a string can occur increases exponentially with t for computationally irreducible CA, then the "regular language complexity" $\Xi^{(t)}$ should also increase exponentially, in agreement with empirical data on certain class-3 CA,[17] and reflecting the "irreducible computational work" achieved by their evolution.

Irreducible computations may be required not only to determine the outcome of evolution through time, but also to find possible arrangements of a system in space. For example, whether an $x \times x$ patch of site values occurs after just one step in a two-dimensional CA is in general *NP*-complete.[18] To determine whether there is any complete infinite configuration that satisfies a particular predicate (such as being invariant under the CA rule) is in general undecidable[18]: It is equivalent to finding the infinite-time behavior of a universal computer that lays down each row on the lattice in turn.

There are many physical systems in which it is known to be possible to construct universal computers. Apart from those modeled by CA, some examples are electric circuits, hard-sphere gases with obstructions, and networks of chemical reactions.[19] The evolution of these systems is in general computationally irreducible, and so suffers from undecidable and intractable problems. Nevertheless, the constructions used to find universal computers in these systems are arcane, and if computationally complex problems occurred only there, they would be rare. It is the thesis of this paper that such problems are in fact common.[20] Certainly there are many systems whose properties are in practice studied only by explicit simulation or exhaustive search: Few computational shortcuts (often stated in terms of invariant quantities) are known.

Many complex or chaotic dynamical systems are expected to be computationally irreducible, and their behavior effectively found only by explicit simulation. Just as it is undecidable whether a particular initial state in a CA leads to unbounded growth, to self-replication, or has some other outcome, so it may be undecidable whether a particular solution to a differential equation (studied say with symbolic dynamics) even enters a certain region of phase space, and whether, say, a certain n-body system is ultimately stable. Similarly, the existence of an attractor, say, with a dimension above some value, may be undecidable.

Computationally complex problems can arise in finding eigenvalues or extremal states in physical systems. The minimum energy conformation for a polymer is in general NP-complete with respect to its length.[21] Finding a configuration below a specified energy in a spin-glass with particular couplings is similarly NP-complete.[22] Whenever the stationary state of a physical system such as this can be found only by lengthy computation, the dynamic physical processes that lead to it must take a correspondingly long time.[5]

Global properties of some models for physical systems may be undecidable in the infinite-size limit (like those for two-dimensional CA). An example is whether a particular generalized Ising model (or stochastic multidimensional CA[23]) exhibits a phase transition.

Quantum and statistical mechanics involve sums over possibly infinite sets of configurations in systems. To derive finite formulas one must use finite specifications for these sets. But it may be undecidable whether two finite specifications yield equivalent configurations. So, for example, it is undecidable whether two finitely specified four-manifolds or solutions to the Einstein equations are equivalent (under coordinate reparametrization).[24] A theoretical model may be considered as a finite specification of the possible behavior of a system. One may ask for example whether the consequences of two models are identical in all circumstances, so that the models are equivalent. If the models involve computations more complicated than those that can be carried out by a computer with a fixed finite number of states (regular language), this question is in general undecidable. Similarly, it is undecidable what is the simplest such model that describes a given set of empirical data.[25]

This paper has suggested that many physical systems are computationally irreducible, so that their own evolution is effectively the most efficient procedure for determining their future. As a consequence, many questions about these systems can be answered only by very lengthy or potentially infinite computations. But some questions answerable by simpler computations may still be formulated.

This work was supported in part by the U. S. Office of Naval Research under Contract No. N00014-80-C-0657. I am grateful for discussions with many people, particularly C. Bennett, G. Chaitin, R. Feynman, E. Fredkin, D. Hillis, L. Hurd, J. Milnor, N. Packard, M. Perry, R. Shaw, K. Steiglitz, W. Thurston, and L. Yaffe.

1. For a more informal exposition see: S. Wolfram, Sci. Am. **251**, 188 (1984). A fuller treatment will be given elsewhere.

2. E.g., *The Undecidable: Basic Papers on Undecidable Propositions, Unsolvable Problems, and Computable Functions*, edited by M. Davis (Raven, New York, 1965), or J. Hopcroft and J. Ullman, *Introduction to Automata Theory, Languages, and Computations* (Addison-Wesley, Reading, Mass., 1979).

3. This is a physical form of the Church-Turing hypothesis. Mathematically conceivable systems of greater power can be obtained by including tables of answers to questions insoluble for these universal computers.

4. Real-number parameters in classical physics allow infinite information density. Nevertheless, even in classical physics, the finiteness of experimental arrangements and measurements, implemented as coarse graining in statistical mechanics, implies finite information input and output. In relativistic quantum field theory, finite density of information (or quantum states) is evident for free fields bounded in phase space [e.g., J. Bekenstein, Phys. Rev. D **30**, 1669 (1984)]. It is less clear for interacting fields, except if space-time is ultimately discrete [but cf. B. Simon, *Functional Integration and Quantum Physics* (Academic, New York, 1979), Sec. III.9]. A finite information transmission rate is implied by relativistic causality and the manifold structure of space-time.

5. It is just possible, however, that the parallelism of the path integral may allow quantum mechanical systems to solve any *NP* problem in polynomial time.

6. M. Garey and D. Johnson, *Computers and Intractability: A Guide to the Theory of NP-Completeness* (Freeman, San Francisco, 1979).

7. See S. Wolfram, Nature **311**, 419 (1984); *Cellular Automata*, edited by D. Farmer, T. Toffoli, and S. Wolfram, Physica **10D**, Nos. 1 and 2 (1984), and references therein.

8. A. R. Smith, J. Assoc. Comput. Mach. **18**, 331 (1971).

9. E. R. Banks, Massachusetts Institute of Technology Report No. TR-81, 1971 (unpublished). The "Game of Life," discussed in E. R. Berlekamp, J. H. Conway,

and R. K. Guy, *Winning Ways for Your Mathematical Plays* (Academic, New York, 1982), is an example with $k = 2$, $R = 9$. N. Margolus, Physica (Utrecht) **10D**, 81 (1984), gives a reversible example.

10. S. Wolfram, Physica (Utrecht) **10D**, 1 (1984), and to be published.

11. This is analogous to the problem of whether a computer run with particular input will ever reach a "halt" state.

12. The number of steps to check ("busy-beaver function") in general grows with the seed size faster than any finite formula can describe (Ref. 2).

13. Cf. C. Bennett, to be published.

14. Cf. B. Eckhardt, J. Ford, and F. Vivaldi, Physica (Utrecht) **13D**, 339 (1984).

15. The question is a generalization of whether there exists an assignment of values to sites such that the logical expression corresponding the t-step CA mapping is true (cf. V. Sewelson, private communication).

16. L. Hurd, to be published.

17. S. Wolfram, Commun. Math. Phys. **96**, 15 (1984).

18. N. Packard and S. Wolfram, to be published. The equivalent problem of covering a plane with a given set of tiles is considered in R. Robinson, Invent. Math. **12**, 177 (1971).

19. E.g., C. Bennett, Int. J. Theor. Phys. **21**, 905 (1982); E. Fredkin and T. Toffoli, Int. J. Theor. Phys. **21**, 219 (1982); A. Vergis, K. Steiglitz, and B. Dickinson, "The Complexity of Analog Computation" (unpublished).

20. Conventional computation theory primarily concerns possibilities, not probabilities. There are nevertheless some problems for which almost all instances are known to be of equivalent difficulty. But other problems are known to be much easier on average than in the worst case. In addition, for some *NP*-complete problems the density of candidate solutions close to the actual one is very large, so approximate solutions can easily be found [S. Kirkpatrick, C. Gelatt, and M. Vecchi, Science **220**, 671 (1983)].

21. Compare *Time Warps, String Edites, and Macromolecules*, edited by D. Sankoff and J. Kruskal (Addison-Wesley, Reading, Mass., 1983).

22. F. Barahona, J. Phys. A **13**, 3241 (1982).

23. E. Domany and W. Kinzel, Phys. Rev. Lett. **53**, 311 (1984).

24. See W. Haken, in *Word Problems*, edited by W. W. Boone, F. B. Cannonito, and R. C. Lyndon (North-Holland, Amsterdam, 1973).

25. G. Chaitin, Sci. Am. **232**, 47 (1975), and IBM J. Res. Dev. **21**, 350 (1977); R. Shaw, to be published.

Two-Dimensional Cellular Automata

1985

A largely phenomenological study of two-dimensional cellular automata is reported. Qualitative classes of behavior similar to those in one-dimensional cellular automata are found. Growth from simple seeds in two-dimensional cellular automata can produce patterns with complicated boundaries, characterized by a variety of growth dimensions. Evolution from disordered states can give domains with boundaries that execute effectively continuous motions. Some global properties of cellular automata can be described by entropies and Lyapunov exponents. Others are undecidable.

1. Introduction

Cellular automata are mathematical models for systems in which many simple components act together to produce complicated patterns of behavior. One-dimensional cellular automata have now been investigated in several ways (Ref. 1 and references therein). This paper presents an exploratory study of two-dimensional cellular automata.[1] The extension to two dimensions is significant for comparisons with many experimental results on pattern formation in physical systems. Immediate applications include dendritic crystal growth,[6] reaction-diffusion systems, and turbulent flow patterns. (The Navier–Stokes equations for fluid flow appear to admit turbulent solutions only in two or more dimensions.)

A cellular automaton consists of a regular lattice of sites. Each site takes on k possible values, and is updated in discrete time steps according to a rule ϕ that depends on the value of sites in some neighborhood around it. The value a_i of a site

---

Coauthored with Norman H. Packard. Originally published in *Journal of Statistical Physics*, volume 38, pages 901–946 (March 1985).

[1] Some aspects of two-dimensional cellular automata were discussed in Refs. 2 and 3, and mentioned in Ref. 4. Additive two-dimensional cellular automata were considered in Ref. 5.

Figure 1. Neighborhood structures considered for two-dimensional cellular automata. In the cellular automaton evolution, the value of the center cell is updated according to a rule that depends on the values of the shaded cells. Cellular automata with neighborhood (a) are termed "five-neighbor square"; those with neighborhood (b) are termed "nine-neighbor square." (These neighborhoods are sometimes referred to as the von Neumann and Moore neighborhoods, respectively.) Totalistic cellular automaton rules take the value of the center site to depend only on the sum of the values of the sites in the neighborhood. With outer totalistic rules, sites are updated according to their previous values, and the sum of the values of the other sites in the neighborhood. Triangular and hexagonal lattices are also possible, but are not used in the examples given here. Notice that five-neighbor square, triangular, and hexagonal cellular automaton rules may all be considered as special cases of general nine-neighbor square rules.

at position i in a one-dimensional cellular automata with a rule that depends only on nearest neighbors thus evolves according to

$$a_i^{(t+1)} = \phi[a_{i-1}^{(t)}, a_i^{(t)}, a_{i+1}^{(t)}] \tag{1.1}$$

There are several possible lattices and neighborhood structures for two-dimensional cellular automata. This paper considers primarily square lattices, with the two neighborhood structures illustrated in Fig. 1. A five-neighbor square cellular automaton then evolves in analogy with Eq. (1.1) according to

$$a_{i,j}^{(t+1)} = \phi[a_{i,j}^{(t)}, a_{i,j+1}^{(t)}, a_{i+1,j}^{(t)}, a_{i,j-1}^{(t)}, a_{i-1,j}^{(t)}] \tag{1.2}$$

Here we often consider the special class of totalistic rules, in which the value of a site depends only on the sum of the values in the neighborhood:

$$a_{i,j}^{(t+1)} = f[a_{i,j}^{(t)} + a_{i,j+1}^{(t)} + a_{i+1,j}^{(t)} + a_{i,j-1}^{(t)} + a_{i-1,j}^{(t)}] \tag{1.3}$$

These rules are conveniently specified by a code[7]

$$C = \sum_n f(n)k^n \tag{1.4}$$

We also consider outer totalistic rules, in which the value of a site depends separately on the sum of the values of sites in a neighborhood, and on the value of the site itself:

$$a_{i,j}^{(t+1)} = \tilde{f}(a_{i,j}^{(t)}, a_{i,j+1}^{(t)} + a_{i+1,j}^{(t)} + a_{i,j-1}^{(t)} + a_{i-1,j}^{(t)}) \tag{1.5}$$

Such rules are specified by a code

$$\tilde{C} = \sum_n \tilde{f}[a, n]k^{kn+a} \tag{1.6}$$

Rule type	5-neighbor square	9-neighbor square	Hexagonal
General	$2^{32} \simeq 4 \times 10^9$	$2^{512} \simeq 10^{154}$	$2^{128} \simeq 3 \times 10^{38}$
Rotationally symmetric	$2^{12} = 4096$	$2^{140} \simeq 10^{42}$	$2^{64} \simeq 2 \times 10^{19}$
Reflection symmetric	$2^{24} \simeq 2 \times 10^7$	$2^{288} \simeq 5 \times 10^{86}$	$2^{80} \simeq 10^{24},$
			$2^{74} \simeq 2 \times 10^{22}$
Completely symmetric	$2^{12} = 4096$	$2^{102} \simeq 5 \times 10^{30}$	$2^{28} \simeq 3 \times 10^8$
Outer totalistic	$2^{10} = 1024$	$2^{18} \simeq 3 \times 10^5$	$2^{14} = 16384$
Totalistic	$2^5 = 32$	$2^9 = 512$	$2^7 = 128$

Table 1. Numbers of possible rules of various kinds for cellular automata with two states per site, and neighborhoods of the form shown in Fig. 1. The two entries for reflectional symmetries of the hexagonal lattice refer to reflections across a cell and across a boundary, respectively. The number of quiescent rules (defined to leave the null configuration invariant) is always half the total number of rules of a given kind.

This paper considers two-dimensional cellular automata with values 0 or 1 at each site, corresponding to $k = 2$. Table 1 gives the number of possible rules of various kinds for such cellular automata. A notorious example of an outer totalistic nine-neighbor square cellular automaton is the "Game of Life,"[8] with a rule specified by code $\tilde{C} = 224$.

Despite the simplicity of their construction, cellular automata are found to be capable of very complicated behavior. Direct mathematical analysis is in general of little utility in elucidating their properties. One must at first resort to empirical means. This paper is a phenomenological study of typical two-dimensional cellular automata. Its approach is largely experimental in character: cellular automaton rules are selected and their evolution from various initial states is traced by direct simulation.[2] The emphasis is on generic properties. Typical initial states are chosen. Except for some restricted kinds of rules, Table 1 shows that the number of possible cellular automaton rules is far too great for each to be investigated explicitly. For the most part one must resort to random sampling, with the expectation that the rules so selected are typical. The phenomena identified by this experimental approach may then be investigated in detail using analytical approximations, and by conventional mathematical means. Generic properties are significant because they are independent of precise details of cellular automaton construction, and may be expected to be universal to a wide class of systems, including those that occur in nature.

---

[2] Several computer systems were used. The first was the special-purpose pipelined TTL machine built by the M.I.T. Information Mechanics group.[9] This machine updates all sites on a 256×256 square cellular automaton lattice 60 times per second. It is controlled by a microcomputer, with software written in FORTH. It allows for five- and nine-neighbor rules, with up to four effective values for each site. The second system was a software program running on the Ridge 32 computer. The kernel is written in assembly language; the top-level interface in the C programming language. A 128×128 cellular automaton lattice is typically updated about 7 times per second. Variants of the program, with kernels written in C and FORTRAN, were used on Sun Workstations, VAX, and Cray 1 computers. One-dimensional cellular automaton simulations were carried out with our CA cellular automaton simulation package, written in C, usually running on a Sun Workstation.

Empirical studies strongly suggest that the qualitative properties of one-dimensional cellular automata are largely independent of such features of their construction as the number of possible values for each site, and the size of the neighborhood. Four qualitative classes of behavior have been identified in one-dimensional cellular automata.[7] Starting from typical initial configurations, class-1 cellular automata evolve to homogeneous final states. Class-2 cellular automata yield separated periodic structures. Class-3 cellular automata exhibit chaotic behavior, and yield aperiodic patterns. Small changes in initial states usually lead to linearly increasing regions of change. Class-4 cellular automata exhibit complicated localized and propagating structures. Cellular automata may be considered as information-processing systems, their evolution performing some computation on the sequence of site values given as the initial state. It is conjectured that class-4 cellular automata are generically capable of universal computation, so that they can implement arbitrary information-processing procedures.

Dynamical systems theory methods may be used to investigate the global properties of cellular automata. One considers the set of configurations generated after some time from any possible initial configuration. Most cellular automaton mappings are irreversible (and not surjective), so that the set of configurations generated contracts with time. Class-1 cellular automata evolve from almost all initial states to a unique final state, analogous to a fixed point. Class-2 cellular automata evolve to collections of periodic structures, analogous to limit cycles. The contraction of the set of configurations generated by a cellular automaton is reflected in a decrease in its entropy or dimension. Starting from all possible initial configurations (corresponding to a set defined to have dimension one), class-3 cellular automata yield sets of configurations with smaller, but positive, dimensions. These sets are directly analogous to the chaotic (or "strange") attractors found in some continuous dynamical systems (e.g., Ref. 10).

Entropy or dimension gives only a coarse characterization of sets of cellular automaton configurations. Formal language theory (e.g., Ref. 11) provides a more complete and detailed characterization.[12] Configurations may be considered as words in a formal language; sets of configurations are specified by the grammatical rules of the language. The set of configurations generated after any finite number of time steps in the evolution of a one-dimensional cellular automaton can be shown to form a regular language: the possible configurations thus correspond to possible paths through a finite graph. For most class-3 and -4 cellular automata, the complexity of this graph grows rapidly with time, so that the limit set is presumably not a regular language (cf. Ref. 13).

This paper reports evidence that certain global properties of two-dimensional cellular automata are very similar to those of one-dimensional cellular automata. Many of the local phenomena found in two-dimensional cellular automata also have analogs in one dimension. However, there are a variety of phenomena that depend on the geometry of the two-dimensional lattice. Many of these phenomena involve complicated boundaries and interfaces, which have no direct analog in one dimension.

Section 2 discusses the evolution of two-dimensional cellular automata from simple "seeds," consisting of a few nonzero initial sites. Just as in one dimension, some cellular automata give regular and self-similar patterns; others yield complicated and apparently random patterns. A new feature in two dimensions is the generation of patterns with dendritic boundaries, much as observed in many natural systems. Most two-dimensional patterns generated by cellular automaton growth have a polytopic boundary that reflects the structure of the neighborhood in the cellular automaton rule (cf. Ref. 14). Some rules, however, yield slowly growing patterns that tend to a circular shape independent of the underlying cellular automaton lattice.

Section 3 considers evolution from typical disordered initial states. Some cellular automata evolve to stationary structures analogous to crystalline forms. The boundaries between domains of different phases may behave as if they carry a surface tension: positive surface tensions lead to large smooth-walled domains; negative surface tensions give rise to labyrinthine structures with highly convoluted walls. Other cellular automata yield chaotic, class-3, behavior. Small changes in their initial configurations lead to linearly increasing regions of change, usually circular or at least rounded.

Section 4 discusses some quantitative characterizations of the global properties of two-dimensional cellular automata. Many definitions are carried through directly from one dimension, but some results are rather different. In particular, the sets of configurations that can be generated after a finite number of time steps of cellular automaton evolution are no longer described by regular languages, and may in fact be nonrecursive. As a consequence, several global properties that are decidable for one-dimensional cellular automata become undecidable in two dimensions (cf. Ref. 15).

2. Evolution from Simple Seeds

This section discusses patterns formed by the evolution of cellular automata from simple seeds. The seeds consist of single nonzero sites, or small regions containing a few nonzero sites, in a background of zero sites. The growth of cellular automata from such initial conditions should provide models for a variety of physical and other phenomena. One example is crystal growth.[6] The cellular automaton lattice corresponds to the crystal lattice, with nonzero sites representing the presence of atoms or regions of the crystal. Different cellular automaton rules are found to yield both faceted (regular) and dendritic (snowflake-like) crystal structures. In other systems the seed may correspond to a small initial disturbance, which grows with time to produce a complicated structure. Such a phenomenon presumably occurs when fluid turbulence develops downstream from an obstruction or orifice.[3]

Figure 2 shows some typical examples of patterns generated by the evolution of two-dimensional cellular automata from initial states containing a single nonzero site.

---

[3] A cellular automaton approximation to the Euler equations is given in Ref. 16.

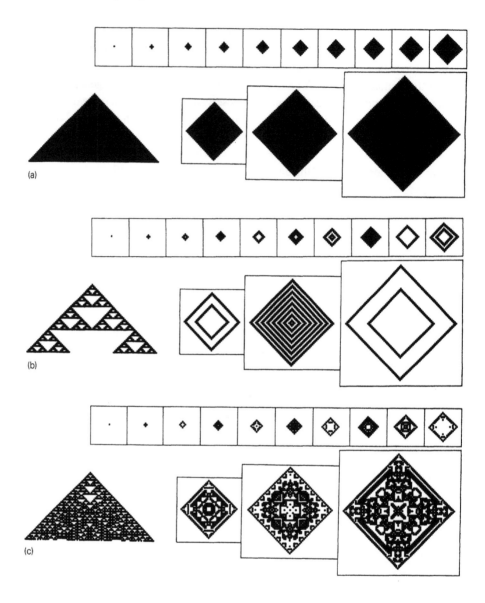

Figure 2. Examples of classes of patterns generated by evolution of two-dimensional cellular automata from a single-site seed. Each part corresponds to a different cellular automaton rule. All the rules shown are both rotation and reflection symmetric. For each rule, a sequence of frames shows the two-dimensional configurations generated by the cellular automaton evolution after the indicated number of time steps. Black squares represent sites with value 1; white squares sites with value 0. On the left is a space-time section showing the time evolution of the center horizontal line of sites in the two-dimensional lattice. Successive lines correspond to successive time steps. The cellular automaton rules shown are five-neighbor square outer totalistic, with codes (a) 1022, (b) 510, (c) 374, (d) 614 (sum modulo 2 rule), (e) 174, (f) 494.

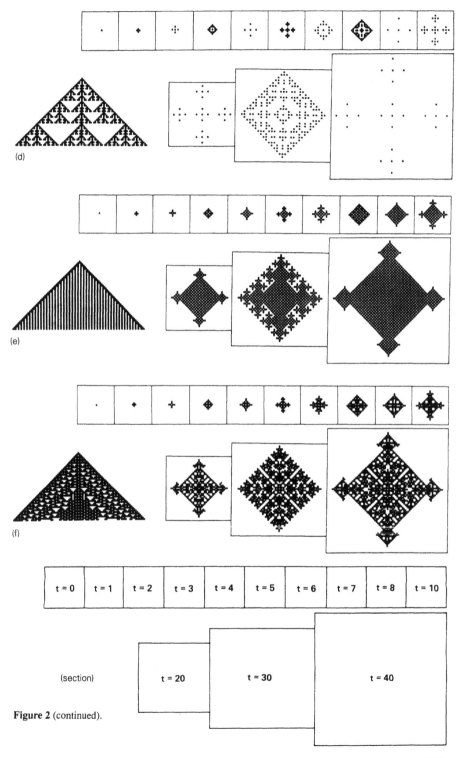

Figure 2 (continued).

In each case, the sequence of two-dimensional patterns formed is shown as a succession of "frames." A space-time "section" is also shown, giving the evolution of the center horizontal line in the two-dimensional lattice with time. Figure 3 shows a view of the complete three-dimensional structures generated. Figure 4 gives some examples of space-time sections generated by typical one-dimensional cellular automata.

With some cellular automaton rules, simple seeds always die out, leaving the null configuration, in which all sites have value zero. With other rules, all or part of the initial seed may remain invariant with time, yielding a fixed pattern, independent of time. With many cellular automaton rules, however, a growing pattern is produced.

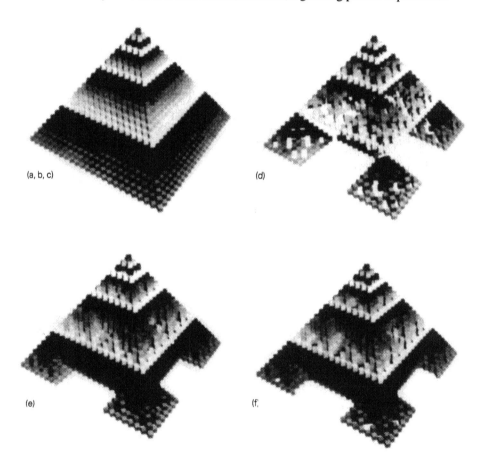

(a, b, c)

(d)

(e)

(f)

Figure 3. View of three-dimensional structures formed from the configurations generated in the first 24 time steps of the evolution of the two-dimensional cellular automata shown in Fig. 2. Rules (a), (b), and (c) all give rise to configurations with regular, faceted, boundaries. Rules (d), (e), and (f) yield dendritic patterns. In this and other three-dimensional views, the shading ranges periodically from light to dark when the number of time steps increases by a factor of two. The three-dimensional graphics here and in Figs. 10 and 14 is courtesy of M. Prueitt at Los Alamos National Laboratory.

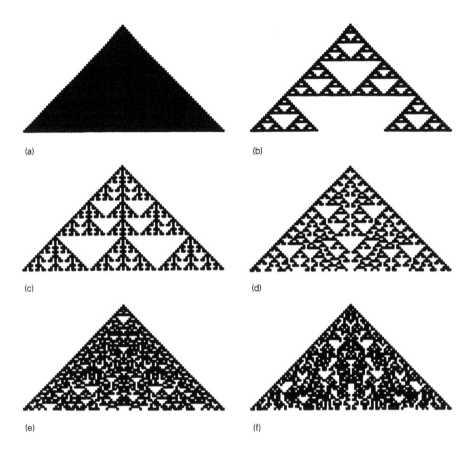

Figure 4. Examples of classes of patterns generated by evolution of one-dimensional cellular automata from a single-site seed. Successive time steps are shown on successive lines. Nonzero sites are shown black. The cellular automaton rules shown are totalistic nearest-neighbor ($r = 1$), with k possible values at each site: (a) $k = 2$, code 14, (b) $k = 2$, code 6, (c) $k = 2$, code 10, (d) $k = 3$, code 21, (e) $k = 3$, code 102, (f) $k = 3$, code 138. Irregular patterns are also generated by some $k = 2$, $r = 2$ rules (such as that with totalistic code 10), and by asymmetric $k = 2$, $r = 1$ rules (such as that with rule number 30).

Rule (a) in Figs. 2 and 3 is an example of the simple case in which the growing pattern is uniform. At each time step, a regular pattern with a fixed density of nonzero sites is produced. The boundary of the pattern consists of flat (linear) "facets," and traces out a pyramid in space-time, whose edges lie along the directions of maximal growth. Sections through this pyramid are analogous to the space-time pattern generated by the one-dimensional cellular automaton of Fig. 4(a).

Cellular automaton rule (b) in Figs. 2 and 3 yields a pattern whose boundary again has a simple faceted form, but whose interior is not uniform. Space-time sections through the pattern exhibit an asymptotically self-similar or fractal form: pieces of the pattern, when magnified, are indistinguishable from the whole. Figure 4(b) shows a one-dimensional cellular automaton that yields sections of the same form. The density of nonzero sites in these sections tends asymptotically to zero. The pattern of nonzero sites in the sections may be characterized by a Hausdorff or fractal dimension that is found by a simple geometrical construction to have value $\log_2 3 \simeq 1.59$.

Self-similar patterns are generated in cellular automata that are invariant under scale or blocking transformations.[17,18] Particular blocks of sites in a cellular automaton often evolve according to a fixed effective cellular automaton rule. The overall behavior of the cellular automaton is then left invariant by a replacement of each block with a single site and of the original cellular automaton rule by the effective rule. In some cases, the effective rule may be identical to the original rule. Then the patterns generated must be invariant under the blocking transformation, and are therefore self-similar. (All the rules so far found to have this property are additive.) In many cases, the effective rule obtained after several blocking transformations with particular blocks may be invariant under further blocking transformations. Then if the initial state contains only the appropriate blocks, the patterns generated must be self-similar, at least on sufficiently large length scales.

Cellular automaton (c) gives patterns that are not homogeneous, but appear to have a fixed nonzero asymptotic density. The patterns have a complex, and in some respects random, appearance. It is remarkable that simple rules, even starting from the simple initial conditions shown, can generate patterns of such complexity. It seems likely that the iteration of the cellular automaton rule is essentially the simplest procedure by which these patterns may be specified. The cellular automaton rule is thus "computationally irreducible" (cf. Ref. 19).

Cellular automata (a), (b), and (c) in Figs. 2 and 3 all yield patterns whose boundaries have a simple faceted form. Cellular automata (d), (e), and (f) give instead patterns with corrugated, dendritic, boundaries. Such complicated boundaries can have no analog in one-dimensional cellular automata: they are a first example of a qualitative phenomenon in cellular automata that requires two or more dimensions.

Cellular automaton (d) follows the simple additive rule that takes the value of each site to be the sum modulo two of the previous values of all sites in its five-site neighborhood. The space-time pattern generated by this rule has a fractal form. The fractal dimension of this pattern, and its analogs on d-dimensional lattices, is given by[4]: $\log_2\{d[(1+4/d)^{1/2}+1]\}$, or approximately 2.45 for $d = 2$. The average density of nonzero sites in the pattern tends to zero with time.

Rules (e) and (f) give patterns with nonzero asymptotic densities. The boundaries of the patterns obtained at most time steps are corrugated, and have fractal forms analogous to Koch curves. The patterns grow by producing "branches" along the four lattice directions. Each of these branches then in turn produces side branches, which themselves produce side branches, and so on. This recursive process yields a highly corrugated boundary. However, as the process continues, the side branches grow into each other, forming an essentially solid region. In fact, after each 2^j time steps the boundary takes on an essentially regular form. It is only between such times that a dendritic boundary is present.

Cellular automaton (e) is an example of a "solidification" rule,[6] in which any site, once it attains value one, never reverts to value zero. Such rules are of significance in studies of processes such as crystal growth. Notice that although the interior of the pattern takes on a fixed form with time, the possibility of a simple one-dimensional cellular automaton model for the boundary alone is precluded by nonlocal effects associated with interactions between different side branches.

The boundaries of the patterns generated by cellular automata (a), (b), and (c) expand with time, but maintain the same faceted form. So after a rescaling in linear dimensions by a factor of t, the boundaries take on a fixed form: the pattern obtained is a fixed point of the product of the cellular automaton mapping and the rescaling transformation (cf. Refs. 20 and 21). The boundaries of Figs. 2(d, e, f) and 3(d, e, f) continually change with time; a fixed limiting form after rescaling can be obtained only by considering a particular sequence of time steps, such as those of the form 2^j. The result depends critically on the sequence considered: some sequences yield dendritic limiting forms, while others yield faceted forms. The complete space-time patterns illustrated in Figs. 3(d, e, f) again approach a fixed limiting form after rescaling only when particular sequences of times are considered. It appears, however, that the forms obtained with different sequences have the same overall properties: they are asymptotically self-similar and have definite fractal dimensions.

The limiting structure of patterns generated by the growth of cellular automata from simple seeds can be characterized by various "growth dimensions." Two general types may be defined. The first, denoted generically D, depend on the overall space-time pattern. The second, denoted $\overline{D}$, depend only on the boundary of the pattern. The boundary may be defined as the set of sites that can be reached by some path on the lattice that begins at infinity and does not cross any nonzero sites. The boundary can thus be found by a simple recursive procedure (cf. Ref. 22). For rules that depend on more than nearest-neighboring sites, paths that pass within the range of the rule of any nonzero site are also excluded, and so no paths can enter any "pores" in the surface of the pattern.

Growth dimensions in general describe the logarithmic asymptotic scaling of the total sizes of patterns with their linear dimensions. For example, the spatial growth dimension D_x is defined in terms of the total number of sites n (interior and boundary) contained in patterns generated by a cellular automaton as a function of time t by the limit of $\log n / \log t$ as $t \to \infty$. Figure 5 shows the behavior of $\log n$ as a function of $\log t$ for the cellular automata of Figs. 2 and 3. For those with faceted boundaries, $D_x = \log n / \log t = 2$ for all sufficiently large t: the total size of the patterns scales as the square of the parameter t that determines their linear dimensions. When the boundaries can be dendritic, however, $\log n$ varies irregularly with $\log t$. In case (d), for example, $\log n$ depends on the number of nonzero digits in the binary decomposition of the integer t (cf. Ref. 4): $\log n / \log t$ is thus maximal when $t = 2^j - 1$, and is minimal when $t = 2^j$. One may define upper and lower spatial growth dimensions D_x^+ and D_x^- in terms of the upper and lower limits (lim *sup*

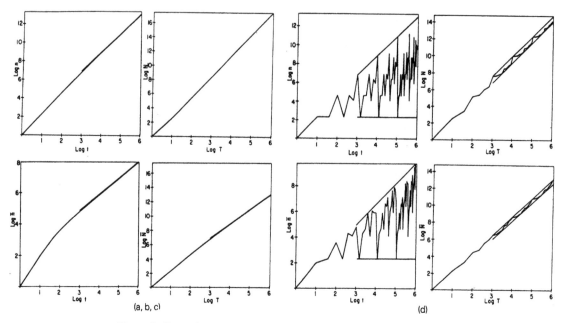

Figure 5. Sizes of structures generated by the two-dimensional cellular automata of Fig. 2 growing from single nonzero initial sites as a function of time. (Although the sizes are defined only at integer times, their successive values are shown joined by straight lines.) $\bar{n}$ gives the number of sites on the boundaries of patterns obtained at time t. n gives the total number of sites contained within these boundaries. $\bar{N}$ is the number of sites in the boundary (surface) of the complete three-dimensional space-time structures illustrated in Fig. 3 up to time T, and N is the number of

and lim *inf*) of $\log n / \log t$ as $t \to \infty$. For case (d), $D_x^+ = 2$, while $D_x^- = 0$. For cases (e) and (f), $\log n / \log t$ oscillates with time, achieving its maximal value at $t = 2^j - 1$, and its minimal value at or near $t = 3/2 \times 2^j$. However, in these cases numerical results suggest that the upper and lower growth dimensions are in fact equal, and in both cases have a value $\simeq 2$.

An alternative definition of the spatial growth dimension includes only nonzero sites in computing the total sizes of patterns generated by cellular automaton evolution. With this definition, the spatial growth dimension has no definite limit even for cellular automata such as that of case (b) which give patterns with faceted boundaries.

The spatial growth dimensions $\overline{D}_x$ for the boundaries of patterns generated by cellular automata are obtained from the limits of $\log \bar{n} / \log t$ at large t, where $\bar{n}$ gives the number of sites in the boundary at time t (cf. Ref. 23). Figure 5 shows the behavior of $\log \bar{n}$ with $\log t$ for the cellular automata of Figs. 2 and 3. For the faceted boundary cases (a), (b), and (c), $\overline{D}_x = 1$. In cases (d), (e), and (f), where dendritic boundaries occur, $\log \bar{n}$ varies irregularly with $\log t$. $\log \bar{n} / \log t$ is minimal when $t = 2^j$ and the boundary is faceted, and is maximal when the boundary is maximally dendritic, typically at $t = 2^j - 1$. No unique limit for $\overline{D}_x$ exists. In case (d), $\overline{D}_x^+ = 1.62 \pm 0.02$, while $\overline{D}_x^- = 0$. In case

222

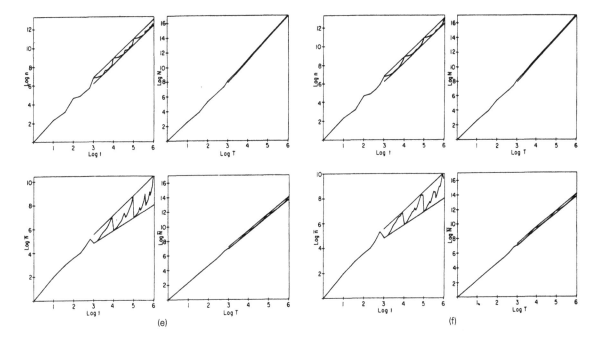

(e)

(f)

sites in their interior. The large-t limits of $\log n / \log t$ and so on give various growth dimensions for the structures. In cases (a), (b), and (c), structures with faceted boundaries are produced, and the growth dimensions have unique values. In cases (d), (e), and (f) the structures have dendritic boundaries, and the slopes of the bounding lines shown give upper (lim sup) and lower (lim inf) limits for the growth dimensions. In many of the cases shown, the numerical values of these upper and lower limits appear to coincide.

(e), $\overline{D}_x^+ = 1.65 \pm 0.02$ and $\overline{D}_x^- = 1$, while in case (f), $\overline{D}_x^+ = 1.53 \pm 0.02$ and $\overline{D}_x^- = 1$.

The limiting forms obtained after rescaling for the spatial patterns generated by the dendritic cellular automata (d), (e), and (f) depend on the sequences of time steps used in the limiting procedure, so that there are no unique values for their spatial growth dimensions. On the other hand, the overall forms of the complete space-time patterns generated by these cellular automata do have definite limits, so that the growth dimensions that characterize them have definite values. The total growth dimensions D and $\overline{D}^4$ may be defined as $\lim_{T \to \infty} \log N / \log T$ and $\lim_{T \to \infty} \log \overline{N} / \log T$, where N is the total number of sites contained in the space-time pattern generated up to time step T, and $\overline{N}$ is the number of sites in its boundary. [Notice that $N = \sum_{t=0}^{T} n(t)$.] Figure 5 shows the behavior of $\log N$ and $\log \overline{N}$ as a function of $\log T$ for the cellular automata of Figs. 2 and 3. Unique values of D and $\overline{D}$ are indeed found in all cases. Rules that give patterns with faceted boundaries have $D = 3$, $\overline{D} = 2$. The additive

----

[4] This quantity is referred to as the "growth rate dimension" in Ref. 20.

rule of case (d) gives $D = 2.36 \pm 0.02$, $\overline{D} = 2.19 \pm 0.02$. Cases (e) and (f) both give $D = 3$, $\overline{D} = 2.27 \pm 0.02$.

Growth dimensions may be defined in general by considering the intersection of the complete space-time pattern, or its boundary, with various families of hyperplanes. With fixed-time hyperplanes one obtains the spatial growth dimensions D_x and $\overline{D}_x$. Temporal growth dimensions $D_t^{(x)}$ and $\overline{D}_t^{(x)}$ are obtained by considering sections through the space-time pattern in spatial direction **x**. (The section typically includes the site of the original seed.) The total growth dimension may evidently be obtained as an appropriate average over temporal growth dimensions in different directions. (The average must be taken over pattern sizes n, and so requires exponentiation of the growth dimensions.) The values of the temporal growth dimensions for the patterns of Figs. 2 and 3 depend on their internal structure. Cases (a), (c), (e), and (f) have $D_t = 2$; case (b) has $D_t = \log_2 3 \simeq 1.59$, and case (d) has $D_t = \log_2(1 + \sqrt{5}) \simeq 1.69$. The temporal growth dimensions $\overline{D}_t^{(x)}$ for the boundaries of the patterns are equal to one for the faceted boundary cases. These dimensions vary with direction in cases with dendritic boundaries. They are equal to one in directions of maximal growth, but are larger in other directions.

In general the values of growth dimensions associated with particular hyperplanes are bounded by the topological dimensions of those hyperplanes. Empirical studies indicate that among all (symmetric) two-dimensional cellular automata, patterns with the form of case (c), characterized by $D = 3$, $\overline{D} = 2$, $D_t = 2$ are the most commonly generated. Fractal boundaries are comparatively common, but their growth dimensions $\overline{D}$ are usually quite close to the minimal value of two. Fractal sections with $D_t < 2$ are also comparatively common for five-neighbor rules, but become less common for nine-neighbor rules.

The rules for the the two-dimensional cellular automata shown in Figs. 2 and 3 are completely invariant under all the rotation and reflection symmetry transformations on their neighborhoods. Figure 6 shows patterns generated by cellular automaton rules with lower symmetries. These patterns are often complicated both in their boundaries and internal structure. Even though the patterns grow from completely symmetric initial states consisting of single nonzero sites, they exhibit definite directionalities and vorticities as a consequence of asymmetries in the rules. Asymmetric patterns may be obtained with symmetrical rules from asymmetric initial states containing several nonzero sites. For example, some rules should support periodic structures that propagate in particular directions with time. Other rules should yield spiral patterns with definite vorticities. Structures of these kinds are expected to be simpler in many $k > 2$ rules than for $k = 2$ rules (cf. Ref. 24) just as in one-dimensional cellular automata. Notice that spiral patterns in two-dimensional cellular automata have total growth dimensions $D = \overline{D} = 2$.

Figure 7 shows the evolution of various two-dimensional cellular automata from initial states containing both single nonzero sites, and small regions with a few nonzero sites. In most cases, the overall patterns generated after a sufficiently long

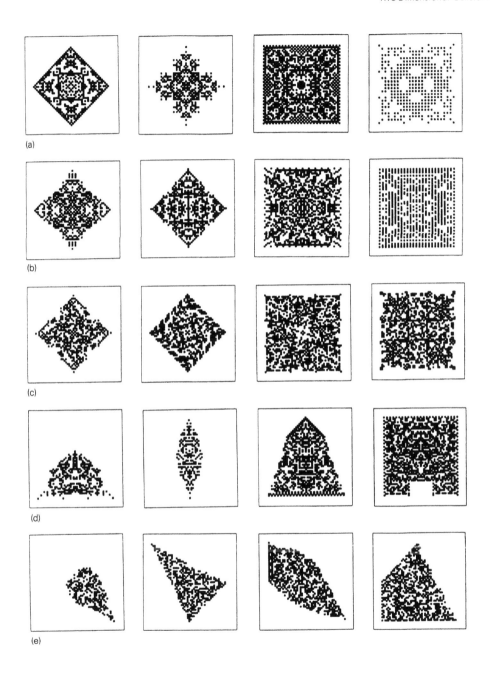

Figure 6. Examples of patterns generated by growth from single-site seeds for 24 time steps according to general nine-neighbor square rules, with symmetries: (a) all, (b) horizontal and vertical reflection, (c) rotation, (d) vertical reflection, (e) none.

(a)

(b)

(c)

(d)

(e)

(f)

(g)

(h)

(i)

minimal
seed
t = 44

disordered
region
t = 44

time are seen to be largely independent of the particular form of the initial state. In cases such as (c) and (e), features in the initial seed lead to specific dislocations in the final patterns. Nevertheless, deformations in the boundaries of the patterns usually occur only on length scales of order the size of the seed, and presumably become negligible in the infinite time limit. As a consequence, the growth dimensions for the resulting patterns are usually independent of the form of the initial seed (cf. Ref. 20 for additive rules).

There are nevertheless some cellular automaton rules for which slightly different seeds can lead to very different patterns. This phenomenon occurs when a cellular automaton whose configurations contain only certain blocks of site values satisfies an effective rule with special properties such as scale invariance. If the initial seed contains only these blocks, then the pattern generated follows the effective rule. However, if other blocks are present, a pattern of a different form may be generated. An example of this behavior for a one-dimensional cellular automaton is shown in Fig. 8. Patterns produced with one type of seed have temporal growth dimension $\log_2 3 \simeq 1.59$, while those with another type of seed have dimension 2.

Cellular automaton rules embody a finite maximum information propagation speed. This implies the existence of a "bounding surface" expanding at this finite speed. All nonzero sites generated by cellular automaton evolution from a localized seed must lie within this bounding surface. (The cellular automata considered here leave a background of zero sites invariant; such a background must be mapped to itself after at most k time steps with any cellular automaton rule.) Thus the pattern generated after t time steps by any cellular automaton is always bounded by the polytope (planar-faced surface) corresponding to the "unit cell" formed from the set of vectors specifying the displacements of sites in the neighborhood, magnified by a factor t in linear dimensions (cf. Ref. 14). Thus patterns generated by five-neighbor cellular automaton rules always lie within an expanding diamond-shaped region, while those with nine-neighbor rules may fill out a square region.

The actual minimal bounding surface for a particular cellular automaton rule often lies far inside the surface obtained by magnifying the unit cell. A sequence of better approximations to the bounding surface may be found as follows. First consider a set of sites representing the neighborhood for a cellular automaton rule. If the center site has value one at a particular time step, there could exist configurations for which all of the sites in the neighborhood would attain value one on the next time step. However, there may be some sites whose values cannot change from zero to one in a single time step with any configuration. Growth does not occur along directions corresponding to such sites. The polytope formed from sites in the neighborhood,

◀ **Figure 7.** Examples of patterns generated by evolution of two-dimensional cellular automata from minimal seeds and small disordered regions. In most cases, growth is initiated by a seed consisting of a single nonzero site; for some of the rules shown, a square of four nonzero sites is required. The cellular automaton rules shown are nine-neighbor square outer totalistic, with codes (a) 143954, (b) 50224, (c) five-neighbor 750, (d) 15822, (e) 699054, (f) 191044, (g) 11202, (h) 93737, (i) 85507.

Figure 8. Example of a one-dimensional cellular automaton in which space-time patterns with different temporal growth dimensions are obtained with different initial seeds. The cellular automaton has $k = 2$, $r = 1$, and rule number 218. With an initial state containing only the blocks 00 and 10, it behaves like the additive rule 90, and yields a self-similar space-time pattern with fractal dimension $\log_2 3$. But when the initial state contains 10 and 11 blocks, it behaves like rule 128, and yields a uniform space-time pattern.

excluding such sites, may be magnified by a factor t to yield a first approximation to the actual bounding surface for a cellular automaton rule. A better approximation is given by the polytope obtained after two time steps of cellular automaton evolution, magnified by a factor $t/2$.

The actual bounding surfaces for five-neighbor two-dimensional cellular automaton rules usually have their maximal diamond-shaped form. However, many nine-neighbor rules have a diamond-shaped form, rather than their maximal square form. Some nine-neighbor rules, such as those of Figs. 7(g) and 7(h) have octagonal bounding surfaces, while still others, such as those of Fig. 7(i) have dodecagonal bounding surfaces. The cellular automata rules with lower symmetries illustrated in Fig. 6 in many cases exhibit more complicated boundaries, with lower symmetries.

Patterns that maintain regular boundaries with time typically fill out their bounding surface at all times. Dendritic patterns, however, usually expand with the bounding surface only along a few axes. In other directions, they meet the bounding surface only at specific times, typically of the form 2^j. At other times, they lie within the bounding surface.

Dendritic boundaries seem to be associated with cellular automaton rules that exhibit "growth inhibition" (cf. Ref. 14). Growth inhibition occurs if there exist some a_i for which $\phi(a_1, \ldots, 0, \ldots, a_n) = 1$, but $\phi(a_1, \ldots, 1, \ldots, a_n) = 0$, or vice versa. Such behavior appears to be common in physical and other systems.

Figures 9 and 10 show examples of two-dimensional cellular automata that exhibit the comparatively rare phenomenon of slow, diffusive, growth from simple seeds. Figure 11 gives a one-dimensional cellular automaton with essentially analogous behavior.

The phenomenon is most easily discussed in the one-dimensional case. The pattern shown in Fig. 11 is such that it expands by one site at a particular time step only if the site on the boundary has value one. If the boundary site has one of its other three possible nonzero values, then on average, no expansion occurs. The cellular automaton rule is such that the boundary sites have values one through four with roughly equal frequencies. Thus the pattern expands on average at a speed of about $1/4$ sites per time step (on each side).

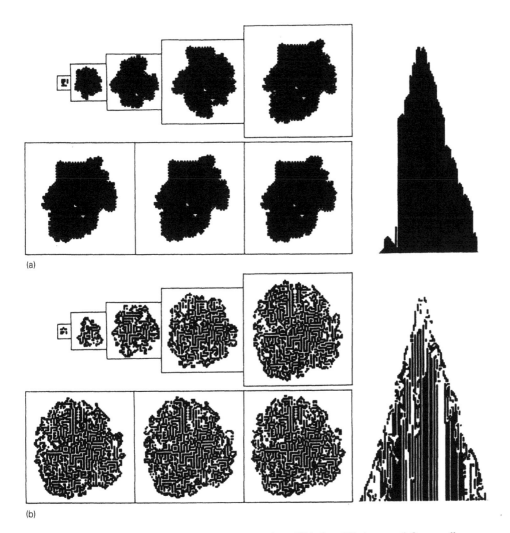

Figure 9. Examples of two-dimensional cellular automata that exhibit slow diffusive growth from small disordered regions. The cellular automaton rules shown are nine-neighbor square outer totalistic, with codes (a) 256746, (b) 736, (c) 291552.

The origin of diffusive growth is similar in the two-dimensional case. Growth occurs there only when some particular several-site structure appears on the boundary. For example, in the cellular automaton of Fig. 9(a), a linear interface propagates at maximal velocity. Deformations of the interface slow its propagation, and a maximally corrugated interface with a "battlement" form does not propagate at all. Since many boundary structures occur with roughly equal probabilities, the average growth rate is small. In the cases investigated, the growth rate is asymptotically constant, so that the growth dimensions have definite values. A remarkable feature is that the boundaries of the patterns produced do not follow the polytopic form

229

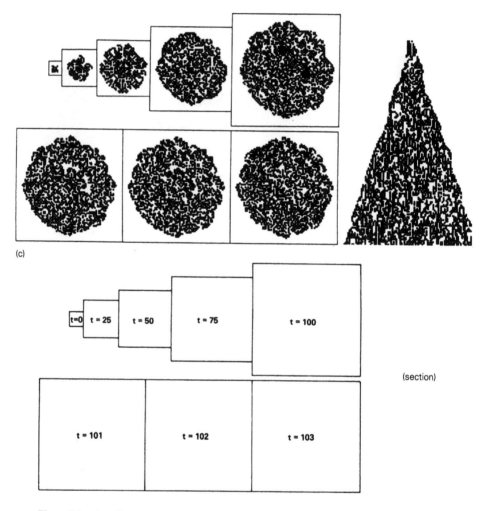

(c)

(section)

Figure 9 (continued).

suggested by the underlying lattice construction of the cellular automaton. Instead, in many cases, asymptotically circular patterns appear to be produced.

3. Evolution from Disordered Initial States

In this section, we discuss the evolution of cellular automata from disordered initial states, in which each site is randomly chosen to have value zero or one (usually with probability 1/2). Such disordered configurations are typical members of the set of all possible configurations. Patterns generated from them are thus typical of those obtained with any initial state. The presence of structure in these patterns is an indication of self-organization in the cellular automaton.

230

Figure 10. View of three-dimensional structure formed from the configurations generated in the first 24 time steps of evolution according to the two-dimensional cellular automaton rule of Fig. 9(a).

Figure 11. Example of a one-dimensional cellular automaton that exhibits slow growth. The rule shown is totalistic $k = 5$, $r = 1$, with code 985707700. All nonzero sites are shown black. The initial state contains a single site with value 3. Growth occurs when a site with value 1 appears on the boundary.

As mentioned in Section 1, four qualitative classes of behavior have been identified in the evolution of one-dimensional cellular automata from disordered initial states. Examples of these classes are shown in Fig. 12. Figure 13 shows the evolution of some typical two-dimensional cellular automata from disordered initial states. The same four qualitative classes of behavior may again be identified here. In fact, the space-time sections for two-dimensional cellular automata have a striking qualitative similarity to sections obtained from one-dimensional cellular automata, perhaps with some probabilistic noise added.

Just as in one dimension, some two-dimensional cellular automata evolve from almost all initial states to a unique homogeneous state, such as the null configuration. The final state for such class 1 cellular automata is usually reached after just a few time steps, but in some rare cases, there may be a long transient.

Figures 13(a) and 14(a) give an example of a two-dimensional cellular automaton with class-2 behavior. The disordered initial state evolves to a collection of separated simple structures, each stable or oscillatory with a small period. Each of these structures is a remnant of a particular feature in the initial state. The cellular automaton rule acts as a "filter" which preserves only certain features of the initial state. There is usually a simple pattern to the set of features preserved, and to the set of persistent structures produced. It should in fact be possible to devise cellular automaton rules that recognize particular sets of features, and to use such class-2 cellular automata for practical image processing tasks (cf. Ref. 25).

The patterns generated by evolution from several different disordered configurations according to a particular cellular automaton rule are almost always qualitatively similar. Yet in many cases the cellular automaton evolution is unstable, in that small changes in the initial state lead to increasing changes in the patterns generated with time. Figures 12 and 13 include difference patterns that illustrate the effect of changing the value of a single site in the initial state. For class-2 cellular automata, such a change affects only a finite region, and the difference pattern remains bounded with time. Information propagates only a finite distance in class-2 cellular automata, so that a particular region of the final state is determined from a bounded region in the initial state. For class-3 cellular automata, on the other hand, information generically propagates at a nonzero speed forever, and a small change in the initial state affects an ever-increasing region. The difference patterns for class-3 cellular automata thus grow without bound, usually at a constant rate.

The locally periodic patterns generated after many time steps by class-2 cellular automata such as in Fig. 13(a) consist of many separated structures located at

Figure 12. Examples of the evolution of one-dimensional cellular automata from disordered initial states. ▶ The difference patterns on the right show site values that change when a single initial site value is changed. All nonzero sites are shown black. The cellular automaton rules shown are totalistic nearest-neighbor ($r = 1$), with k possible values at each site: (a) $k = 2$, code 12, (b) $k = 5$, code 7530, (c) $k = 3$, code 681, (d) $k = 5$, code 3250, (e) $k = 2$, code 6, (f) $k = 3$, code 348, (g) $k = 3$, code 138, (h) $k = 3$, code 318, (i) $k = 3$, code 792.

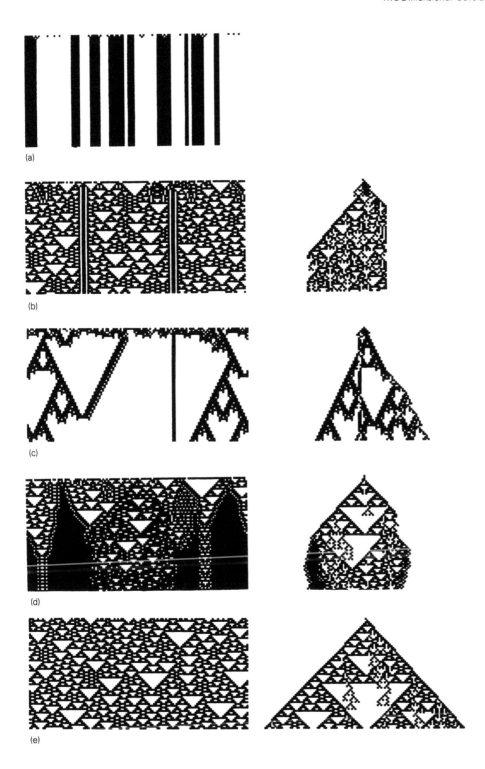

(a)

(b)

(c)

(d)

(e)

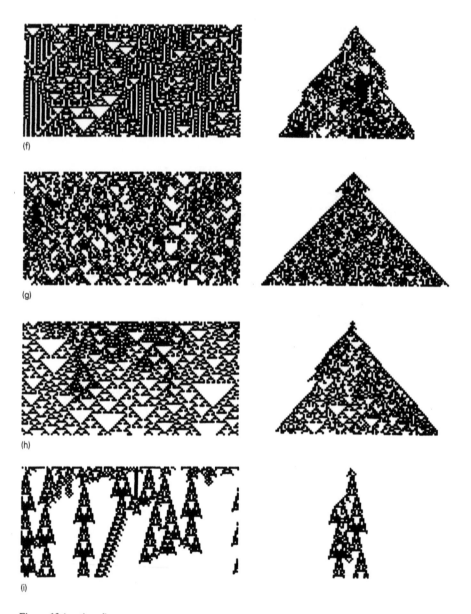

(f)

(g)

(h)

(i)

Figure 12 (continued).

essentially arbitrary positions. Figure 13(b) shows another form of class-2 cellular automaton. There are four basic "phases." Two phases have vertical stripes, with either on even or odd sites. The other two phases have horizontal stripes. Regions that take on forms corresponding to one of these phases are invariant under the cellular automaton rule. Starting from a typical disordered state, each region in the cellular automaton lattice evolves toward a particular phase. At large times, the cellular automaton thus "crystallizes" into a patchwork of "domains." The domains consist

234

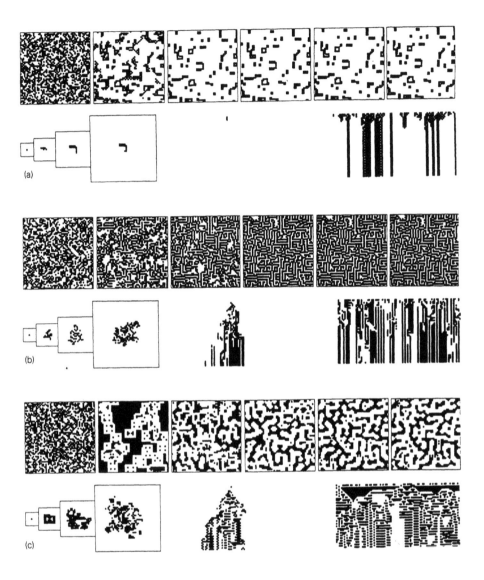

Figure 13. Examples of the evolution of two-dimensional cellular automata from disordered initial states. The cellular automaton rules shown are totalistic five-neighbor square with codes: (a) 24, (d) 510, (e) 52; and outer totalistic nine-neighbor with codes: (b) 736, (c) 196623, (f) 152822, (g) 143954, (h) 3276, (i) 224 (the "Game of Life").

of regions in particular phases. They are separated by domain walls. In the example of Fig. 13(b), these domain walls become essentially stationary after a finite time.

A change in a single initial site produces a difference pattern that ultimately spreads only along the domain walls. The spread continues only so long as each successive region on the domain wall contains only particular arrangements of site

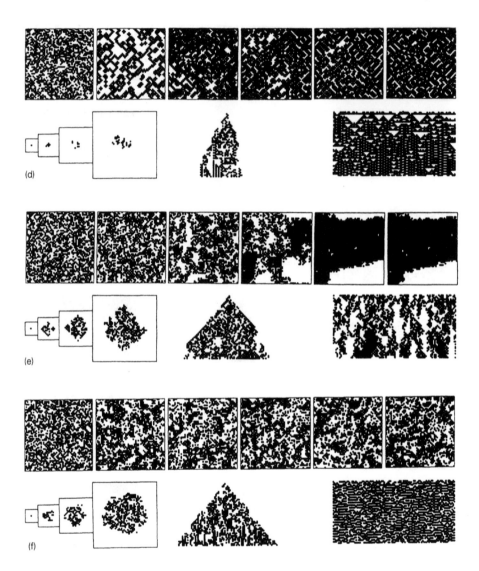

Figure 13 (continued).

values. The spread stops if a "pinning defect," corresponding to other arrangements of site values, is encountered. The arrangement of site values on the domain walls may in a first approximation be considered random. The difference pattern will thus spread forever only if the arrangements of site values necessary to support its propagation occur with a probability above the percolation threshold (e.g., Ref. 26), so that they form an infinite connected cluster with probability one.

Phases in cellular automata may in general be described by "order parameters" that specify the spatially periodic patterns of sites corresponding to each phase. The

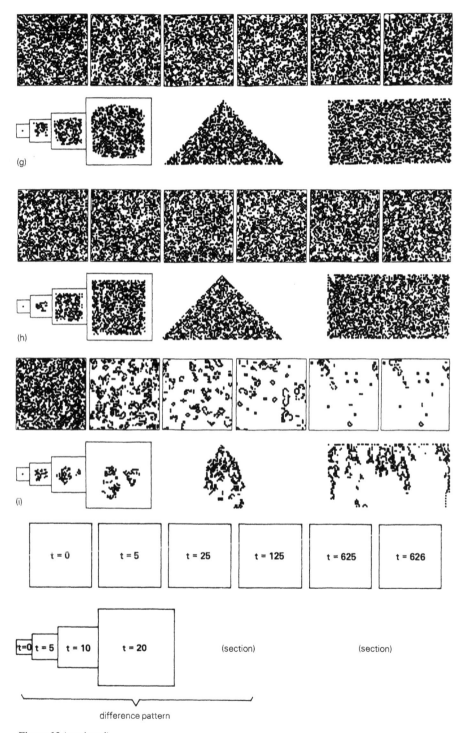

(g)

(h)

(i)

t = 0   t = 5   t = 25   t = 125   t = 625   t = 626

t=0 t = 5   t = 10   t = 20   (section)   (section)

difference pattern

Figure 13 (continued).

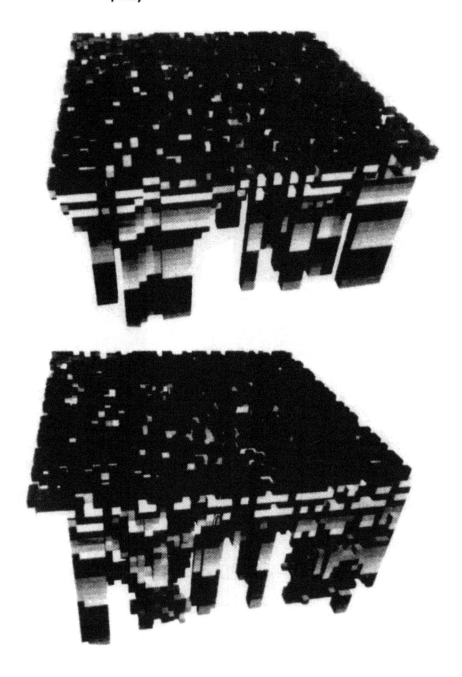

Figure 14. View of three-dimensional structures formed by configurations generated in the first 24 times of evolution from disordered initial states (in a finite region) according to the cellular automaton rules of Figs. 13(a) and 13(i).

238

size of domains generated by evolution from disordered initial states depends on the length of time before the domains become "frozen": slower relaxation leads to larger domains, as in annealing. A final state reached after any finite time can contain only finite size domains, and therefore cannot be a pure phase. States generated by two-dimensional cellular automata may contain "point" and "line" defects. Point defects are localized regions within domains. An example is the "L-shaped" region of zero sites in domains of the value one phase for the cellular automaton illustrated in Fig. 13(e). Line defects correspond to walls separating domains.

In the cellular automaton of Fig. 13(b), the domains become stationary after a few time steps. In the case of Fig. 13(e), however, the domains can continue to move forever, essentially by a diffusion process. Figure 12(d) shows a one-dimensional cellular automaton with domain walls that exhibit analogous behavior (cf. Refs. 27 and 17). In both cases, some domains become progressively larger with time, while others eventually disappear completely. The domain walls in Fig. 13(e) behave as if they carry a positive surface tension (cf. Ref. 28); the diffusion process responsible for their movement is biased to reduce the local curvature of the interface. A linear interface is stable under the cellular automaton rule, of Fig. 13(e). In addition, the heights of any protrusions or intrusions cannot increase with time. In general, they decay, often quite slowly, until they are of height at most one. Deformations of height one, analogous to surface waves, do not decay further, and are governed by a one-dimensional cellular automaton rule (with $k = 2$, $r = 1$, and rule number 150). At large times, therefore, a domain must either shrink to zero size, or must have walls with continually decreasing curvatures.

Figure 13(c) shows a two-dimensional cellular automaton with structures analogous to domains walls that carry a negative surface tension. More and more convoluted patterns are obtained with time. The resulting labyrinthine state is strongly reminiscent of behavior observed with ferrofluids or magnetic bubbles.[29]

Figures 13(f), 13(g), and 13(h) are examples of two-dimensional cellular automata that exhibit class-3 behavior. Chaotic aperiodic patterns are obtained at all times. Moreover, the difference patterns resulting from changes in single initial site values expand at a fixed rate forever. A remarkable feature is that in almost all cases (Fig. 13(h) is an exception), the expansion occurs at the same speed in all directions, resulting in an asymptotically circular difference pattern. For some rules, the expansion occurs at maximal speed; but often the speed is about 0.8 times the maximum. When the difference patterns are not exactly circular, they tend to have rounded corners. And even with asymmetrical rules, circular difference patterns are often obtained. A rough analog of this behavior is found in asymmetric one-dimensional cellular automata which generate symmetrical difference patterns. Such behavior is found to become increasingly common as k and r increase, or as the number of independent parameters in the rule ϕ increases.

An argument based on the central limit theorem suggests an explanation for the appearance of circular difference patterns in two-dimensional class-3 cellular automata. Consider the set of sites corresponding to the neighborhood for a cellular automaton rule. For each site, compute the probability that the value of that site changes after one time step of cellular automaton evolution when the value of the center site is changed, averaged over all possible arrangements of site values in the neighborhood. An approximation to the probability distribution of differences is then obtained as a multiple convolution of this kernel. (This approximation is effectively a linear one, analogous to Huygens' principle in optics.) The number of convolutions performed increases with time. If the number of neighborhood arrangements is sufficiently large, the kernel tends to be quite smooth. Convolutions of the kernel thus tend to a Gaussian form, independent of direction.

Some asymmetric class-3 cellular automata yield difference patterns that expand, say, in the horizontal direction, but contract in the vertical direction. At large times, such cellular automata produce patterns consisting of many independent horizontal lines, each behaving essentially as a one-dimensional class-3 cellular automaton.

Class-3 behavior is considerably the commonest among two-dimensional cellular automata, just as it is for one-dimensional cellular automata with large k and r. It appears that as the number of parameters or degrees of freedom in a cellular automaton rule increases, there is a higher probability for some degree of freedom to show chaotic behavior, leading to overall chaotic behavior.

Figure 12(i) shows an example of a class-4 one-dimensional cellular automaton. A characteristic feature of class-4 cellular automata is the existence of a complicated set of persistent structures, some of which propagate through space with time. Class-4 rules appear to occur with a frequency of a few per cent among all one-dimensional cellular automaton rules. Often one suspects that some degrees of freedom in a cellular automaton exhibit class-4 behavior, but they are masked by overall chaotic class-3 behavior.

Class-4 cellular automata appear to be much less common in two dimensions than in one dimension. Figures 13(i) and 13(b) show the evolution of a two-dimensional cellular automaton known as the "Game of Life."[8] Many persistent structures, some propagating, have been identified in this cellular automaton. It has in addition been shown that these structures can be combined to perform arbitrary information processing, so that the cellular automaton supports universal computation.[8] Starting from a disordered initial state, the density of propagating structures ("gliders") produced is about one per 2000 site region.

Except for a few simple variants on the Game of Life, no other definite class-4 two-dimensional cellular automata were found in a random sample of several thousand outer totalistic rules.[5] Some rules that appeared to be of class 2 were found to have long transients, characteristic of class-4 behavior, but no propagating structures were seen. Other rules seemed to exhibit some class-4 features, but they were overwhelmed by dominant class-3 behavior.

---

[5] A few examples of class-4 behavior were however found among general rules. Requests for copies of the relevant rule tables should be directed to the authors.

4. Global Properties

Section 3 discussed the typical behavior of cellular automata evolving from particular initial states. This section considers the global properties of cellular automata, determined by evolution from all possible initial states. Studies of the global properties of one-dimensional cellular automata have been made using methods both from dynamical systems theory[7] and from computation theory.[12] Here these studies are generalized to the case of two-dimensional cellular automata. For those based on dynamical systems theory the generalization is quite straightforward; but in the computation theory approach substantial additional complications occur. Whereas the sets of configurations generated after any finite number of steps in the evolution of one-dimensional cellular automata always correspond to regular formal languages,[12] the corresponding sets in two-dimensional cellular automata may be nonrecursive.[15]

Most cellular automaton rules are irreversible, so that several different initial states may evolve to the same final state. As a consequence, even starting from all possible initial states, only a subset of possible states may be generated with time. The properties of this set then determine the overall behavior of the cellular automaton, and the self-organization that occurs in it.

Entropy and dimension provide quantitative characterizations of the "sizes" of sets generated by cellular automaton evolution (e.g., Ref. 7). The spatial set entropy for a set of two-dimensional cellular automaton configurations is defined by considering a $X \times Y$ patch of sites. If the set contains all possible configurations, then all k^{XY} possible different arrangements of sites values must occur in the patch. In general $N(X, Y) \leq k^{XY}$ different arrangements will occur. Then the set entropy (or dimension) is defined as

$$s = \lim_{X,Y \to \infty} \frac{1}{XY} \log_k N(X, Y) \tag{4.1}$$

If the cellular automaton mapping is surjective, so that all possible configurations occur, then this entropy is equal to one. In general it decreases with time in the evolution of the cellular automaton.

Spatial set entropy characterizes the set of configurations that can possibly be generated in the evolution of a cellular automaton, regardless of their probabilities of occurrence. One may also define a spatial measure entropy in terms of the probabilities p_i for possible $X \times Y$ patches as

$$s_\mu = \lim_{X,Y \to \infty} \frac{-1}{XY} \sum_{i=1}^{k^{XY}} p_i \log_k p_i \tag{4.2}$$

The limiting value of s_μ at large times is typically nonzero for all but class-1 cellular automaton rules. Notice that in cases where domains with positive "surface tension" are formed, s_μ tends only very slowly to zero with time.

To find the spatial set entropy after, say, one time step in the evolution of a cellular automaton one must identify what configurations can be generated. In a one-dimensional cellular automaton, one can specify the set of configurations that can be generated in terms of rules that determine which sequences of site values can appear. These rules correspond to a regular formal grammar, and give the state transition graph for a finite state machine. The set of configurations that can be generated in a two-dimensional cellular automaton is more difficult to specify. In many circumstances in fact the occurrence of a particular patch of site values requires a global consistency that cannot be verified in general by any finite computation. As a consequence, many propositions concerning sets of configurations generated after even a finite number of steps in the evolution of two-(and higher-)dimensional cellular automata can be formally undecidable.

In a one-dimensional cellular automaton with a range-r rule, a particular sequence of X site values can be generated (reached) after one time step only if there exists some length $X + 2r$ sequence of initial site values that evolves to it. The locality of the cellular automaton rule ensures that in determining whether a length $X + 1$ sequence obtained by appending one new site can also be generated, it suffices to test only those length $X + 2r + 1$ predecessor configurations that differ in their last $2r + 1$ site values. In determining whether sequences of progressively greater lengths can be generated it suffices at each stage to record with which length $2r$ overlaps in the predecessor configuration a particular new site value can be appended. Since there are only k^{2r} possible sequences of site values in the overlaps, only a finite amount of information must be recorded, and a finite procedure can be given for determining whether any given sequence can be generated (cf. Ref. 12). Hence in particular there is a finite procedure to determine whether any given cellular automaton rule is surjective, so that all possible configurations can be reached in its evolution.[30]

In two-dimensional cellular automata there is no such simple iterative procedure for determining whether progressively larger patches of site values can be generated. An $X \times Y$ patch of site values is generated after one step in the evolution of a two-dimensional cellular automaton with a range-r rule if there exists some $(X+2r)(Y+2r)$ patch of initial site values that evolves to it. Progressively larger patches can be generated if appropriate progressively larger predecessor patches exist. The number of sites in the overlap between such progressively larger predecessor patches is not fixed, as in one-dimensional cellular automata, but instead grows essentially like the perimeter of the patch, $2r(X + Y + 2r)$. With this procedure, there is thus no upper bound on the amount of information that must be recorded to determine whether progressively larger patches can be generated. To find whether a patch of any particular size $X \times Y$ can be generated, it suffices to test all $k^{(X+2r)(Y+2r)}$ candidate predecessor patches. (As mentioned below, this is in fact an NP-complete problem, and therefore presumably cannot be solved in general in a time polynomial in the patch size.) However, questions concerning complete configurations can be answered only by considering arbitrarily large patches, and may require arbitrarily complex computations. As a consequence, there are global questions about configurations generated by two-dimensional cellular automata after a finite number of time steps that can posed, but cannot in general be answered by any finite computational process, and are therefore formally undecidable.[15]

Some examples of such undecidable questions about two-dimensional cellular automata are: (i) whether a particular complete (but finitely specified) configuration can be generated after one time step from any initial configuration; (ii) whether a particular cellular automaton rule is surjective, so that all possible configurations can be generated; (iii) whether the set of complete configurations generated after say one time step has a nonempty intersection with some recursive formal language such as a regular language, whose words can be recognized by a finite computation; (iv) whether there exist configurations that have a particular period in time (and are thus invariant under some number of iterations of the cellular automaton rule).

It seems that global questions about the finite time behavior of one-dimensional cellular automata are always decidable. Questions about their ultimate infinite time behavior may nevertheless be undecidable. To show this, one considers one-dimensional cellular automata whose evolution emulates that of a universal Turing machine. The successive arrangements of symbols on the Turing machine tape correspond to successive configurations of site values generated in the evolution of the cellular automaton. Undecidable questions such as halting for the Turing machines are then shown to be undecidable for the corresponding one-dimensional cellular automaton.[13]

In two-dimensional cellular automata, questions about global properties on infinite spatial scales can be undecidable even at finite times. This is proved[15] by considering the line-by-line construction of configurations. The rules used to obtain each successive line from the last can correspond to the rules for a universal Turing machine. The construction of the configuration can then be continued to infinity and completed only if this Turing machine does not halt with the input given, which is in general undecidable. Sets of configurations generated at finite times in the evolution of two-dimensional cellular automata can thus be nonrecursive.

Many global questions about two-dimensional cellular automata are closely analogous to geometrical questions associated with tilings of the plane. Consider for example the problem of finding configurations that remain invariant under a particular cellular automaton rule. All the neighborhoods in such configurations must be such that the values of their center sites are left unchanged by the cellular automaton rule. Each such neighborhood may be considered as a "tile." Complete invariant configurations are constructed from an array of tiles, with each adjacent pair of tiles subject to a consistency condition that the overlapping sites in the neighborhoods to which they correspond should agree. In a one-dimensional cellular automaton, the set of possible arrangements of tiles or configurations that satisfy the conditions can be enumerated immediately, and form a finite complement regular language (subshift of finite type).[12] In a two-dimensional cellular automaton, the problem of finding invariant configurations is equivalent to tiling the plane with a set of "dominoes" corresponding to the possible allowed neighborhoods, and subject to constraints that can be cast in the form of requiring adjacent pairs of edges to have complementary colors.[31] The problem of determining whether a particular set of dominoes can in fact be used to tile the plane is however known to be undecidable[32,33] (cf. Ref. 34). The problem of finding whether there exist invariant configurations under a particular two-dimensional cellular automaton rule is likewise undecidable.

If any infinite sequence can be constructed from some set of dominoes in one dimension, then it is clear that a spatially periodic sequence can be found. Hence if there are to be any configurations with a particular temporal period in a one-dimensional cellular automaton, then there must be spatially periodic configurations with this temporal period. (The maximum necessary spatial period for configurations with temporal period p is k^{2rp+1} [12]: the existence of such spatially periodic configurations can be viewed as a consequence of the pumping lemma (e.g., Ref. 11) for regular languages.) In two dimensions, however, there are sets of dominoes for which a tiling of the plane is possible, but the tiling cannot be spatially periodic.[32,33,35] In the examples known, it appears that the basic arrangement of tiles is always self-similar, so that it is almost periodic. In the simplest known examples, six square dominoes[33] or just two irregularly shaped dominoes[35] are required for this phenomenon to occur. (The simplest known example in three dimensions involves seven polyhedral "dominoes."[36])

The problem of whether a set of dominoes can tile a finite, say, $X \times X$ region of the plane is clearly decidable, but is NP complete.[37] The analogous problem of determining whether a particular patch can occur in an invariant configuration for a two-dimensional cellular automaton, or can in fact be generated by one time step of evolution from any initial state, is thus also NP complete. These problems can presumably be solved only by computations whose complication increases faster than a polynomial in X, and are essentially equivalent to explicit testing of all $O(k^{X^2})$ possible cases.

In addition to considering configurations of site values generated at a particular step in the evolution of a cellular automaton, one may also discuss sequences of site values obtained with time. In general one may consider the number of possible arrangements $N(\mathbf{v}_1, \ldots, \mathbf{v}_p)$ of site values in a space-time volume consisting of a parallelepiped with generator vectors $\mathbf{v}_i$. The set entropy may than be defined as the exponential rate of increase of N as the lengths of certain generators are taken to infinity (cf. Refs. 38 and 39):

$$s = \lim_{\alpha_1 \to \infty} \cdots \lim_{\alpha_p \to \infty} \frac{1}{\alpha_1 \cdots \alpha_{p'}} \log_k N(\alpha_1 \mathbf{v}_1, \alpha_2 \mathbf{v}_2, \ldots, \alpha_p \mathbf{v}_p) \qquad (4.3)$$

where the α_i are scalar parameters, and $p' \leq p \leq d$. These entropies are in fact functions of the unit p forms obtained as the exterior products of the generator vectors $\mathbf{v}_i$ considered as one-forms in space-time. Certain convergence properties of the limits in Eq. (4.3) can be proved from the fact that the number of arrangements $N(V)$ of site values in a volume V is submultiplicative, so that $N(V_1 \cup V_2) \leq N(V_1) N(V_2)$. A measure-theoretical analog of the set entropy (4.3) may be defined in correspondence with Eq. (4.2).

The spatial entropy (4.1) for two-dimensional cellular automata is obtained from the general definition (4.3) by choosing $p = 3$, $p' = 2$ and taking $\mathbf{v}_1$ and $\mathbf{v}_2$ to be orthogonal purely spacelike vectors along the two lattice directions. The generator vector $\mathbf{v}_3$ is taken to be in the positive time direction, but the number of arrangements N is independent of α_3 since a complete configuration at one time step determines all future configurations.

For a d-dimensional cellular automaton, there are critical values of p and p' such that entropies corresponding to higher or lower-dimensional parallelepipeds are zero or infinity. Entropies with exactly those critical values may be nonzero and bounded by quantities that depend on the cellular automaton neighborhood size.

Entropies are essentially determined by the correlations between values of sites at different space-time points. These correlations depend on the propagation of information in the cellular automaton. The difference patterns discussed in Section 3 provide measures of such information propagation. They can be considered as analogs of Green functions (cf. Ref. 4) which describe the change produced at some space-time point $\mathbf{x}'$ in a cellular automaton as a consequence changes at another point $\mathbf{x}$. The set-theoretical Green's function is defined to be nonzero whenever a change at $\mathbf{x}$ in any configuration could lead to a change at $\mathbf{x}'$. In the measure-theoretical Green's function the possible configurations are weighted with their probabilities. The maximum rate of information propagation is determined by the slope of the space-time ("light") cone within which the Green's function is nonzero. The slope corresponding to propagation in a particular spatial direction in say a two-dimensional cellular automaton gives the Lyapunov exponent in that direction for the cellular automaton evolution.[7,40] In most cases it appears that the space-time structure corresponding to the set of sites on which the Green's function is nonzero tends to a fixed form after rescaling at large times, so that the structure has a unique growth dimension, and the Lyapunov exponents have definite values. Exceptions may occur in rules where difference patterns spread along domain boundaries, typically producing asymptotically self-similar structures analogous to percolation clusters (e.g., Ref. 26).

The Green's functions describe not only how a change at some time affects site values at later times, but also how the value of a particular site is affected by the previous values of other sites. The backward light cone of a site contains all the sites whose values can affect it. (Notice that the backward light cone for a bijective rule in general has little relation with the forward light cone for the inverse rule.[41]) The values of all sites in a volume V are thus determined by the values of sites on a surface S that "absorbs" (covers) all the backward light cones of points in V. The number of possible configurations in V is then bounded from above by the number of possible configurations of the set of sites within one cellular automaton neighborhood of the surface S. The entropy associated with the volume V is then not greater than the entropy associated with the volume around S. By choosing various "absorbing surfaces" S, whose sizes are determined by the rates of information propagation in different directions, one can derive various inequalities between entropies.

Many entropies can be defined for cellular automata using Eq. (4.3). One significant class is those that are invariant under continuous invertible transformations on the space of cellular automaton configurations. Such entropies can be used to identify topologically inequivalent cellular automaton rules. For one-dimensional cellular automata, an invariant entropy may be defined by taking $p = 2$, $p' = 1$ in Eq. (4.3), and choosing v_1 in the positive time direction, and v_2 in the space direction. The entropy may be generalized by taking the v_i to be an arbitrary pair of orthogonal spacetime vectors (with v_1 having a positive time component).[38] The most direct generalization of these invariant entropies to two-dimensional cellular automata would have $p = 3$, $p' = 1$, and take the v_i to be an orthogonal triple of space-time vectors with v_1 having a positive time component. If v_1 were chosen purely timelike, then this entropy would have no dependence on spatial direction, and would correspond to the standard invariant entropy defined for the cellular automaton mapping. In general however, there is no upper bound on its value, and it is apparently infinite for most cellular automata that have positive Lyapunov exponents in more than one spatial direction. A finite entropy can nevertheless be constructed by choosing $p' = 2$. This entropy depends on the spatial (or in general space-time) vector $v_1 \times v_2$. To obtain an invariant entropy, one must perform some average over this vector (accounting for the fact that the entropy is a homogeneous function of degree one in the length of the vector). One possibility is to form the integral of the quantity (4.3) over those values of the vector for which the quantity is less than some constant (say, 1).

5. Discussion

This paper has presented an exploratory study of two-dimensional cellular automata. Much remains to be done, but a few conclusions can already be given.

A first approach to the study of cellular automaton behavior is statistical: one considers the average properties of evolution from typical initial configurations. Statistical studies of one-dimensional cellular automata have suggested that four basic qualitative classes of behavior can be identified. This paper has given analogs of these classes in two-dimensional cellular automata. One expects that the qualitative classification will also apply in three- and higher-dimensional systems.

Entropies and Lyapunov exponents are statistical quantities that measure the information content and rate of information transmission in cellular automata. Their definitions for one-dimensional cellular automata are closest to those used in smooth dynamical systems. But rather direct generalizations can nevertheless be found for two- and higher-dimensional cellular automata.

Beyond statistical properties, one may consider geometrical aspects of patterns generated by cellular automaton evolution. Even though the basic construction of a cellular automaton is discrete, its "macroscopic" behavior at large times and on large spatial scales may be a close approximation to that of a continuous system. In particular domains of correlated sites may be formed, with boundaries that at a large scale seem to show continuous motions and deformations. While some such phenomena do occur in one dimension, they are most significant in two and higher dimensions. Often their motion appears to be determined by attributes such as curvature, that have no analog in one dimension.

The structures generated by two- and higher-dimensional cellular automata evolving from simple seeds show many geometrical phenomena. The most significant is probably the formation of dendritic patterns, characterized by noninteger growth dimensions.

Statistical measurements provide one method for comparing cellular automaton models with experimental data. Geometrical properties provide another. The geometry of patterns formed by cellular automata may be compared directly with the geometry of patterns generated by natural systems.

Topology is another aspect of cellular automaton patterns. When domains or regions containing many correlated sites exist, one may approximate them as continuous structures, and consider their topology. For example, domains produced by cellular automaton evolution may exhibit topological defects that are stable under the cellular automaton rule. In two-dimensional cellular automata, only point and line defects occur. But in three dimensions, knotted line defects (e.g., Ref. 42) and other complicated topological forms are possible. The topology of the structures supported by a cellular automaton rule may be compared directly with the topology of structures that arise in natural systems (cf. Ref. 43).

Geometry and topology provide essentially local descriptions of the behavior of cellular automata. Computation theory potentially provides a more global characterization. One may classify the behavior and properties of cellular automata in terms of the nature of the computations required to reproduce them. Even in one dimension, there are cellular automata that can perform arbitrary computations, so that at least some of their properties can be reproduced only by direct simulation or observation, and their limiting behavior is formally undecidable. The range of properties for which undecidability can occur is much larger in two dimensions than in one dimension. In particular, properties that involve a limit of infinite spatial size, even at finite times, can be undecidable. As higher-dimensional cellular automata are considered, the degree of undecidability that can be encountered in studies of particular properties increases.

Acknowledgments

The research reported here made essential use of several computer systems other than our own. We are grateful to those who made the systems available, and helped us in using them. We thank the M.I.T. Information Mechanics Group (E. Fredkin, N. Margolus, T. Toffoli, and G. Vichniac) for the use of their special-purpose two-dimensional cellular automaton simulation system, and for their hospitality and assistance. We thank R. Shaw for writing the kernel of our software simulation system for two-dimensional cellular automata in Ridge assembly language. We thank the Theoretical Division and the Center for Nonlinear Studies at Los Alamos National Laboratory for hospitality during the final stages of this work. We thank M. Prueitt at Los Alamos for making the three-dimensional illustrations, and D. Umberger for help with some Cray-1 programming. We are grateful to those mentioned and to C. Bennett, J. Crutchfield, H. Hartman, L. Hurd, J. Milnor, S. Willson, and others for discussions.

This work was supported in part by the U.S. Office of Naval Research under Contract No. N00014-80-C-0657.

References

1. S. Wolfram, Cellular automata as models for complexity, *Nature* **311**:419 (1984).
2. T. Toffoli, Cellular automata mechanics, Ph.D. thesis and Technical Report 208, The Logic of Computers Group, University of Michigan (1977).
3. G. Vichniac, Simulating physics with cellular automata, *Physica* **10D**:96 (1984).
4. S. Wolfram, Statistical mechanics of cellular automata, *Rev. Mod. Phys.* **55**:601 (1983).
5. O. Martin, A. Odlyzko, and S. Wolfram, Algebraic properties of cellular automata, *Comm. Math. Phys.* **93**:219 (1984).
6. N. Packard, Cellular automaton models for dendritic crystal growth, Institute for Advanced Study, preprint (1985).
7. S. Wolfram, Universality and complexity in cellular automata, *Physica* **10D**:1 (1984).
8. E. R. Berlekamp, J. O. Conway, and R. K. Guy, *Winning Ways for Your Mathematical Plays*, Vol. 2 (Academic Press, New York, 1982), Chap. 25; M. Gardner, *Wheels, Life and Other Mathematical Amusements* (Freeman, San Francisco, 1983).
9. T. Toffoli, CAM: A high-performance cellular-automaton machine, *Physica* **10D**:195 (1984).
10. J. Guckenheimer and P. Holmes, *Nonlinear Oscillations, Dynamical Systems, and Bifurcations of Vector Fields* (Springer, New York, 1983).
11. J. E. Hopcroft and J. D. Ullman, *Introduction to Automata Theory, Languages, and Computation* (Addison-Wesley, Reading, Massachusetts, 1979).
12. S. Wolfram, Computation theory of cellular automata, *Commun. Math. Phys.* **96**:15 (1984).
13. L. Hurd, Formal language characterizations of cellular automaton limit sets, to be published.
14. S. Willson, On convergence of configurations, *Discrete Math.* **23**:279 (1978).
15. T. Yaku, The constructability of a configuration in a cellular automaton, *J. Comput. Syst. Sci.* **7**:481 (1973); U. Golze, Differences between 1- and 2-dimensional cell spaces, in *Automata, Languages, Development*, A. Lindenmayer and G. Rozenberg, eds. (North-Holland, Amsterdam, 1976).
16. J. Hardy, O. de Pazzis, and Y. Pomeau, Molecular dynamics of a classical lattice gas: transport properties and time correlation functions, *Phys. Rev.* **A13**:1949 (1976).
17. S. Wolfram, Cellular automata, *Los Alamos Science* (Fall 1983); Some recent results and questions about cellular automata, Institute for Advanced Study preprint (September 1983).
18. S. Wolfram, Twenty problems in the theory of cellular automata, *Physica Scripta* **T9**:170 (1985).
19. S. Wolfram, Computer software in science and mathematics, *Sci. Am.* **251**(3):188 (1984).
20. S. Willson, Comparing limit sets for certain increasing cellular automata, Mathematics Department, Iowa State University preprint (June 1984).

21. S. Willson, Growth rates and fractional dimensions in cellular automata, *Physica* **10D**:69 (1984).

22. N. Packard, Notes on a Go-playing program, unpublished (1984).

23. Y. Sawada, M. Matsushita, M. Yamazaki, and H. Kondo, Morphological phase transition measured by "surface kinetic dimension" of growing random patterns, *Phys. Scripta* (to be published).

24. J. M. Greenberg, B. D. Hassard, and S. P. Hastings, Pattern formation and periodic structures in systems modelled by reaction-diffusion equations, *Bull. Amer. Math. Soc.* **84**:1296 (1975); B. Madore and W. Freedman, Computer simulations of the Belousov-Zhabotinsky reaction, *Science* **222**:615 (1983).

25. K. Preston et al., Basics of cellular logic with some applications in medical image processing, *Proc. IEEE* **67**:826 (1979).

26. J. W. Essam, Percolation theory, *Rep. Prog. Phys.* **43**:833 (1980).

27. P. Grassberger, Chaos and diffusion in deterministic cellular automata, *Physica* **10D**:52 (1984).

28. G. Vichniac, Cellular automaton dynamics for interface motion and ordering, M.I.T. report, to appear.

29. R. Rosensweig, Fluid dynamics and science of magnetic liquids, *Adv. Electronics Electron Phys.* **48**:103 (1979); Magnetic fluids, *Sci. Am.* **247**(4):136 (1982).

30. G. A. Hedlund, Endomorphisms and automorphisms of the shift dynamical system, *Math. Syst. Theory* **3**:320 (1969); G. A. Hedlund, Transformations commuting with the shift, in *Topological dynamics*, J. Auslander and W. H. Gottschalk, eds. (Benjamin, New York, 1968); S. Amoroso and Y. N. Patt, Decision procedures for surjectivity and injectivity of parallel maps for tessellation structures, *J. Comp. Sys. Sci.* **6**:448 (1972); M. Nasu, Local maps inducing surjective global maps of one-dimensional tessellation automata, *Math. Syst. Theory* **11**:327 (1978).

31. H. Wang, Proving theorems by pattern recognition—II, *Bell Sys. Tech. J.* **40**:1 (1961).

32. R. Berger, Undecidability of the domino problem, *Mem. Am. Math. Soc.*, No. 66 (1966).

33. R. Robinson, Undecidability and nonperiodicity for tilings of the plane, *Inventiones Math.* **12**:177 (1971).

34. D. Ruelle, *Thermodynamic Formalism* (Addison-Wesley, Reading, Massachusetts, 1978), p. 68.

35. R. Penrose, Pentaplexity: a class of nonperiodic tilings of the plane, *Math. Intelligencer* **2**:32 (1979); M. Gardner, Extraordinary nonperiodic tiling that enriches the theory of tiles, *Sci. Am.* **236**(1):110 (1977); N. de Bruijn, Algebraic theory of Penrose's nonperiodic tilings of the plane, *Nederl. Akad. Wetensch. Indag. Math.* **43**:39 (1981).

36. P. Kramer, Non-periodic central space filling with icosahedral symmetry using copies of seven elementary cells, *Acta Crystallogr.* **A38**:257 (1982).

37. M. Garey and D. Johnson, *Computers and Intractability: A Guide to the Theory of NP-completeness*, (Freeman, San Francisco, 1979), p. 257.

38. J. Milnor, Entropy of cellular automaton-maps, Institute for Advanced Study preprint (May 1984).

39. J. Milnor, Directional entropies in higher dimensions, rough notes (September 1984).

40. N. Packard, Complexity of growing patterns in cellular automata, Institute for Advanced Study preprint (October 1983).

41. J. Milnor, Notes on surjective cellular automaton-maps, Institute for Advanced Study preprint (June 1984).

42. N. D. Mermin, The topological theory of defects in ordered media, *Rev. Mod. Phys.* **51**:591 (1979).

43. A. Winfree and E. Winfree, Organizing centers in a cellular excitable medium, Purdue University preprint (July 1984) and *Physica D* (to be published).

Origins of Randomness
in Physical Systems

1 9 8 5

Randomness and chaos in physical systems are usually ultimately attributed to external noise. But it is argued here that even without such random input, the intrinsic behavior of many nonlinear systems can be computationally so complicated as to seem random in all practical experiments. This effect is suggested as the basic origin of such phenomena as fluid turbulence.

There are many physical processes that seem random or chaotic. They appear to follow no definite rules, and to be governed merely by probabilities. But all fundamental physical laws, at least outside of quantum mechanics, are thought to be deterministic. So how, then, is apparent randomness produced?

One possibility is that its ultimate source is external noise, often from a heat bath. When the evolution of a system is unstable, so that perturbations grow, any randomness introduced through initial and boundary conditions is transmitted or amplified with time, and eventually affects many components of the system [1]. A simple example of this "homoplectic" behavior occurs in the shift mapping $x_t = 2x_{t-1}$ mod 1. The time sequence of bins, say, above and below $\frac{1}{2}$ visited by x_t is a direct transcription of the binary-digit sequence of the initial real number x_0 [2]. So if this digit sequence is random (as for most x_0 uniformly sampled in the unit interval) then so will the time sequence be; unpredictable behavior arises from a sensitive dependence on unknown features of initial conditions [3]. But if the initial condition is "simple," say a rational number with a periodic digit sequence, then no randomness appears.

There are, however, systems which can also generate apparent randomness internally, without external random input. Figure 1 shows an example, in which a cellular automaton evolving from a simple initial state produces a pattern so complicated that many features of it seem random. Like the shift map, this cellular automaton

Originally published in *Physical Review Letters*, volume 55, pages 449–452 (29 July 1985).

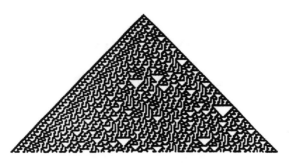

Figure 1. Pattern generated by cellular automaton evolution from a simple initial state. Site values 0 or 1 (represented by white or black, respectively) are updated at each step according to the rule $a_i' = a_{i-1} \oplus (a_i \vee a_{i+1})$ ($\oplus$ denotes addition modulo two, and $\vee$ Boolean disjunction). Despite the simplicity of its specification, many features of the pattern (such as the sequence of site values down the center column) appear random.

is homoplectic, and would yield random behavior given random input. But unlike the shift map, it can still produce random behavior even with simple input. Systems which generate randomness in this way will be called "autoplectic."

In developing a mathematical definition of autoplectic behavior, one must first discuss in what sense it is "random." Sequences are commonly considered random if no patterns can be discerned in them. But whether a pattern is found depends on how it is looked for. Different degrees of randomness can be defined in terms of the computational complexity of the procedures used.

The methods usually embodied in practical physics experiments are computationally quite simple [4,5]. They correspond to standard statistical tests for randomness [6], such as relative frequencies of blocks of elements (dimensions and entropies), correlations, and power spectra. (The mathematical properties of ergodicity and mixing are related to tests of this kind.) One characteristic of these tests is that the computation time they require increases asymptotically at most like a polynomial in the sequence length [7]. So if in fact no polynomial-time procedure can detect patterns in a sequence, then the sequence can be considered "effectively random" for practical purposes.

Any patterns that are identified in a sequence can be used to give a compressed specification for it. (Thus, for example, Morse coding compresses English text by exploiting the unequal frequencies of letters of the alphabet.) The length of the shortest specification measures the "information content" of a sequence with respect to a particular class of computations. (Standard Shannon information content for a stationary process [8] is associated with simple statistical computations of block frequencies.) Sequences are predictable only to the extent that they are longer than their shortest specification, and so contain information that can be recognized as "redundant" or "over-determined."

252

Sequences generated by chaotic physical systems often show some redundancy or determinism under simple statistical procedures. (This happens whenever measurements extract information faster than it can be transferred from other parts of the system [1].) But typically there remain compressed sequences in which no patterns are seen.

A sequence can, in general, be specified by giving an algorithm or computer program for constructing it. The length of the smallest possible program measures the "absolute" information content of the sequence [9]. For an "absolutely random" sequence the program must essentially give each element explicitly, and so be close in length to the sequence itself. But since no computation can increase the absolute information content of a closed system [except for $O(\log t)$ from input of "clock pulses"], physical processes presumably cannot generate absolute randomness [10]. However, the numbers of possible sequences and programs both increase exponentially with length, so that all but an exponentially small fraction of arbitrarily chosen sequences must be absolutely random. Nevertheless, it is usually undecidable what the smallest program for any particular sequence is, and thus whether the sequence is absolutely random. In general, each program of progressively greater length must be tried, and any one of them may run for an arbitrarily long time, so that the question of whether it ever generates the sequence may be formally undecidable.

Even if a sequence can ultimately be obtained from a small specification or program, and so is not absolutely random, it may nevertheless be effectively random if no feasible computation can recover the program [11]. The program can always be found by explicitly trying each possible one in turn [12]. But the total number of possible programs increases exponentially with length, and so such an exhaustive search would soon become infeasible. And if there is no better method the sequence must be effectively random.

In general, one may define the "effective information content" Θ of a sequence to be the length of the shortest specification for it that can be found by a feasible (say polynomial time) computation. A sequence can be considered "simple" if it has small Θ. Θ (often normalized by sequence length) provides a measure of "complexity," "effective randomness," or "computational unpredictability."

Increasing Θ can be considered the defining characteristic of autoplectic behavior. Examples such as Fig. 1 suggest that Θ can increase through polynomial-time processes. The rule and initial seed have a short specification, with small Θ. But one suspects that no polynomial time computation can recover this specification from the center vertical sequence produced, or can in fact detect any pattern in it [13]. The polynomial-time process of cellular automaton evolution thus increases Θ, and generates effective randomness. It is phenomena of this kind that are the basis for cryptography, in which one strives to produce effectively random sequences whose short "keys" cannot be found by any practical cryptanalysis [14].

The simplest mathematical and physical systems (such as the shift mapping) can be decomposed into essentially uncoupled components, and cannot increase Θ. Such systems are nevertheless often homoplectic, so that they transfer information, and with random input show random behavior. But when their input is simple (low Θ), their behavior is correspondingly simple, and is typically periodic. Of course, any system with a fixed finite total number of degrees of freedom (such as a finite cellular automaton) must eventually become periodic. But the phenomena considered here occur on time scales much shorter than such exponentially long recurrences.

Another class of systems widely investigated consists of those with linear couplings between components [such as a cellular automaton in which $a_i^{(t+1)} = (a_{i-1}^{(t)} + a_{i+1}^{(t)})$ mod 2]. Given random input, such systems can again yield random output, and are thus homoplectic. But even with simple input, they can produce sequences which pass some statistical tests of randomness. Examples are the standard linear congruence and linear-feedback shift-register (or finite additive cellular automaton [15]) systems used for pseudorandom number generation in practical computer programs [6,16].

Characteristic of such systems is the generation of self-similar patterns, containing sequences that are invariant under blocking or scaling transformations. These sequences are almost periodic, but may contain all possible blocks of elements with equal frequencies. They can be considered as the outputs of finite-state machines (generalized Markov processes) given the digits of the numerical positions of each element as input [17]. And although the sequences have certain statistical properties of randomness, their seeds can be found by comparatively simple polynomial-time procedures [18]. Such systems are thus not autoplectic (with respect to polynomial-time computations).

Many nonlinear mathematical systems seem, however, to be autoplectic, since they generate sequences in which no patterns have ever been found. An example is the sequence of leading digits in the fractional part of successive powers of $\frac{3}{2}$ [19] (which corresponds to a vertical column in a particular $k = 6$, $r = 1$ cellular automaton with a single site seed).

Despite extensive empirical evidence, almost nothing has, however, been proved about the randomness of such sequences. It is nevertheless possible to construct sequences that are strongly expected to be effectively random [20]. An example is the lowest-order bits of $x_t = x_{t-1}^2 \bmod (pq)$, where p and q are large primes [20]. The problem of deducing the initial seed x_0, or of substantially compressing this sequence, is equivalent to the problem of factoring large integers, which is widely conjectured to require more than polynomial time [21].

Standard statistical tests have also revealed no patterns in the digit sequences of transcendental numbers such as $\sqrt{2}$, e, and π [22] (or continued-fraction expansions of π or of most cubic irrational numbers). But the polynomial-time procedure of squaring and comparing with an integer does reveal the digits of, say, $\sqrt{2}$ as nonrandom [23]. Without knowing how the sequence was generated, however, such a very special "statistical test" (or program) can probably only be found by explicit enumeration of all exponentially many possible ones. And if a sequence passes all but perhaps exponentially few polynomial-time batteries of statistical tests, it should probably be considered effectively random in practice.

Within a set of homoplectic dynamical systems (such as class 3 or 4 cellular automata) capable of transmitting information, all but the simplest seem to support sophisticated information processing, and are thus expected to be autoplectic. In some cases (quite probably including Fig. 1 [24]) the evolution of the system represents a "complete" or "universal" computation, which, with appropriate initial conditions, can mimic any other (polynomial-time) computation [21]. If short specifications for sequences generated by any one such computation could in general be found in polynomial time, it would imply that all could, which is widely conjectured to be impossible. (Such problems are NP-complete [21].)

Many systems are expected to be computationally irreducible, so that the outcome of their evolution can be found essentially only by direct simulation, and no computational short cuts are possible [25]. To predict the future of these systems requires an almost complete knowledge of their current state. And it seems likely that this can be deduced from partial measurements only by essentially testing all exponentially many possibilities. The evolution of computationally irreducible systems should thus generically be autoplectic.

Autoplectic behavior is most clearly identified in discrete systems such as cellular automata. Continuous dynamical systems involve the idealization of real numbers on which infinite-precision arithmetic operations are performed. For systems such as iterated mappings of the interval there seems to be no robust notion of "simple" initial conditions. (The number of binary digits in images of, say, a dyadic rational grows like p^t, where p is the highest power of x in the map.) But in systems with many degrees of freedom, described for example by partial differential equations, autoplectism may be identified through discrete approximations.

Autoplectism is expected to be responsible for apparent randomness in many physical systems. Some features of turbulent fluid flow [26], say in a jet ejected from a nozzle, are undoubtedly determined by details of initial or boundary conditions. But when the flow continues to appear random far from the nozzle, one suspects that other sources of effective information are present. One possibility might be thermal fluctuations or external noise, amplified by homoplectic processes [1]. But viscous damping probably allows only sufficiently large-scale perturbations to affect large-scale features of the flow. (Apparently random behavior is found to be almost exactly repeatable in some carefully controlled experiments [27].) Thus, it seems more likely that the true origin of turbulence is an internal autoplectic process, somewhat like Fig. 1, operating on large-scale features of the flow. Numerical experiments certainly suggest that the Navier-Stokes equations can yield complicated behavior even with simple initial conditions [28]. Autoplectic processes may also be responsible for the widespread applicability of the second law of thermodynamics.

Many discussions have contributed to the material presented here; particularly those with C. Bennett, L. Blum, M. Blum, J. Crutchfield, P. Diaconis, D. Farmer, R. Feynman, U. Frisch, S. Goldwasser, D. Hillis, P. Hohenberg, E. Jen, R. Kraichnan, L. Levin, D. Lind, A. Meyer, S. Micali, J. Milnor, D. Mitchell, A. Odlyzko, N. Packard, I. Procaccia, H. Rose, and R. Shaw.

1. For example, R. Shaw, Z. Naturforsch. **36A**, 80 (1981), and in *Chaos and Order in Nature*, edited by H. Haken (Springer, New York, 1981).

2. An analogous cellular automaton [S. Wolfram, Nature (London) **311**, 419 (1984), and references therein] has evolution rule $a_i^{(t+1)} = a_{i+1}^{(t)}$, so that with time the value of a particular site is determined by the value of progressively more distant initial sites.

3. For example, *Order in Chaos*, edited by D. Campbell and H. Rose (North-Holland, Amsterdam, 1982). Many processes analyzed in dynamical systems theory admit "Markov partitions" under which they are directly equivalent to the shift mapping. But in some measurements (say of x_t with four bins) their deterministic nature may introduce simple regularities, and "deterministic chaos" may be said to occur. (This term would in fact probably be better reserved for the autoplectic processes to be described below.)

4. This is probably also true of at least the lower levels of human sensory processing [for example, D. Marr, *Vision* (Freeman, San Francisco, 1982); B. Julesz, Nature (London) **290**, 91 (1981)].

5. The validity of Monte Carlo simulations tests the random sequences that they use. But most stochastic physical processes are in fact insensitive to all but the simplest equidistribution and statistical independence properties. (Partial exceptions occur when long-range order is present.) And in general no polynomial-time simulation can reveal patterns in effectively random sequences.

6. For example, D. Knuth, *Seminumerical Algorithms* (Addison-Wesley, Reading, Mass., 1981).

7. Some sophisticated statistical procedures, typically involving the partitioning of high-dimensional spaces, seem to take exponential time. But most take close to linear time. It is possible that those used in practice can be characterized as needing $O(\log^p n)$ time on computers with $O(n^q)$ processors (and so be in the computational complexity class *NC*) [cf. N. Pippenger, *Proceedings of the Twentieth IEEE Symposium on Foundations of Computer Science* (IEEE, New York, 1979); J. Hoover and L. Ruzzo, unpublished.]

8. For example, R. Hamming, *Coding and Information Theory* (Prentice-Hall, Englewood Cliffs, 1980).

9. G. Chaitin, J. Assoc. Comput. Mach. **13**, 547 (1966), and **16**, 145 (1969), and Sci. Amer. **232**, No. 5, 47 (1975); A. N. Kolmogorov, Problems Inform.

Transmission **1**, 1 (1965); R. Solomonoff, Inform. and Control **7**, 1 (1964); L. Levin, Soviet Math. Dokl. **14**, 1413 (1973). Compare J. Ford, Phys. Today **33**, No. 4, 40 (1983). Note that the lengths of programs needed on different universal computers differ only by a constant, since each computer can simulate any other by means of a fixed "interpreter" program.

10. Quantum mechanics suggests that processes such as radioactive decay occur purely according to probabilities, and so could perhaps give absolutely random sequences. But complete quantum mechanical measurements are an idealization, in which information on a microscopic quantum event is spread through an infinite system. In finite systems, unmeasured quantum states are like unknown classical parameters, and can presumably produce no additional randomness. Suggestions of absolute randomness probably come only when classical and quantum models are mixed, as in the claim that quantum processes near black holes may lose information to space-time regions that are causally disconnected in the classical approximation.

11. In the cases now known, recognition of any pattern seems to involve essentially complete reconstruction of the original program, but this may not always be so (L. Levin, private communication).

12. In some cases, such as optimization or eigenvalue problems in the complexity class *NP* [e.g., M. Garey and D. Johnson, *Computers and Interactability: A Guide to the Theory of NP-Completeness* (Freeman, San Francisco, 1979)], even each individual test may take exponential time.

13. The sequence certainly passes the standard statistical tests of Ref. 6, and contains all possible subsequences up to length at least 12. It has also been proved that only at most one vertical sequence in the pattern of Fig. 1 can have a finite period [E. Jen, Los Alamos Report No. LA-UR-85-1218 (to be published)].

14. For example, D. E. R. Denning, *Cryptography and Data Security* (Addison-Wesley, Reading, Mass., 1982). Systems like Fig. 1 can, for example, be used for "stream ciphers" by adding each bit in the sequences produced with a particular seed to a bit in a plain-text message.

15. For example, O. Martin, A. Odlyzko, and S. Wolfram, Commun. Math. Phys. **93**, 219 (1984).

16. B. Jansson, *Random Number Generators* (Almqvist & Wiksells, Stockholm, 1966).

17. They are one-symbol-deletion tag sequences [A. Cobham, Math. Systems Theory **6**, 164 (1972)], and can be represented by generating functions algebraic over $GF(k)$ [G. Christol, T. Kamae, M. Mendes France, and G. Rauzy, Bull. Soc. Math. France **108**, 401 (1980); J.-M. Deshouillers, Seminar de Theorie des Nombres, Université de Bordeaux Exposé No. 5, 1979 (unpublished); M. Dekking, M. Mendes France, and A. van der Poorten, Math. Intelligencer **4**, 130,

173, 190 (1983)]. Their self-similarity is related to the pumping lemma for regular languages [e.g., J. Hopcroft and J. Ullman, *Introduction to Automata Theory, Languages and Computation* (Addison-Wesley, Reading, Mass., 1979)]. More complicated sequences associated with context-free formal languages can also recognized in polynomial time, but the recognition problem for context-sensitive ones is *P*-space complete.

18. For example, A. M. Frieze, R. Kannan, and J. C. Lagarias, in *Twenty-Fifth IEEE Symposium on Foundations of Computer Science* (IEEE, New York, 1984). The sequences also typically fail certain statistical randomness tests, such as multidimensional spectral tests (Ref. 6). They are nevertheless probably random with respect to all *NC* computations [J. Reif and J. Tygar, Harvard University Computation Laboratory Report No. TR-07-84 (to be published)].

19. For example, G. Choquet, C. R. Acad. Sci. (Paris), Sec. A **290**, 575 (1980); cf. J. Lagarias, Amer. Math. Monthly **92**, 3 (1985). (Note that with appropriate boundary conditions a finite-size version of this system is equivalent to a linear congruential pseudorandom number generator.)

20. A. Shamir, Lecture Notes in Computer Science, **62**, 544 (1981); S. Goldwasser and S. Micali, J. Comput. Sys. Sci. **28**, 270 (1984); M. Blum and S. Micali, SIAM J. Comput. **13**, 850 (1984); A. Yao, in *Twenty-Third IEEE Symposium on Foundations of Computer Science* (IEEE, New York, 1982); L. Blum, M. Blum, and M. Shub, in *Advances in Cryptology: Proceedings of CRYPTO-82*, edited by D. Chaum, R. Rivest, and A. T. Sherman (Plenum, New York, 1983); O. Goldreich, S. Goldwasser, and S. Micali, in *Twenty-Fifth IEEE Symposium on Foundations of Computer Science* (IEEE, New York, 1984).

21. For example, M. Garey and D. Johnson, Ref. 12.

22. For example, L. Kuipers and H. Niederreiter, *Uniform Distribution of Sequences* (Wiley, New York, 1974).

23. A polynomial-time procedure is also known for recognizing solutions to more complicated algebraic or trigonometric equations (R. Kannan, A. K. Lenstra, and L. Lovasz, Carnegie-Mellon University Technical Report No. CMU-CS-84-111).

24. Many localized structures have been found (D. Lind, private communication).

25. S. Wolfram, Phys. Rev. Lett. **54**, 735 (1985).

26. For example, U. Frisch, Phys. Scr. **T9**, 137 (1985).

27. G. Ahlers and R. W. Walden, Phys. Rev. Lett. **44**, 445 (1980).

28. For example, M. Brachet *et al.*, J. Fluid Mech. **130**, 411 (1983).

Thermodynamics and Hydrodynamics of Cellular Automata

1985

Simple cellular automata which seem to capture the essential features of thermo-dynamics and hydrodynamics are discussed. At a microscopic level, the cellular automata are discrete approximations to molecular dynamics, and show relaxation towards equilibrium. On a large scale, they behave like continuum fluids, and suggest efficient methods for hydrodynamic simulation.

Thermodynamics and hydrodynamics describe the overall behaviour of many systems, independent of the precise microscopic construction of each system. One can thus study thermodynamics and hydrodynamics using simple models, which are more amenable to efficient simulation, and potentially to mathematical analysis.

Cellular automata (CA) are discrete dynamical systems which give simple models for many complex physical processes [1]. This paper considers CA which can be viewed as discrete approximations to molecular dynamics. In the simplest case, each link in a regular spatial lattice carries at most one "particle" with unit velocity in each direction. At each time step, each particle moves one link; those arriving at a particular site then "scatter" according to a fixed set of rules. This discrete system is well-suited to simulation on digital computers. The state of each site is represented by a few bits, and follows simple logical rules. The rules are local, so that many sites can be updated in parallel. The simulations in this paper were performed on a Connection Machine Computer [2] which updates sites concurrently in each of 65536 Boolean processors [3].

In two dimensions, one can consider square and hexagonal (six links at 60°) lattices. On a square lattice [4], the only nontrivial local rule which conserves momentum and particle number takes isolated pairs of particles colliding head on to scatter in the orthogonal direction (no interaction in other cases). On a hexagonal

Coauthored with James B. Salem. Originally issued as a Thinking Machines Corporation technical report (November 1985).

lattice [5], such pairs may scatter in either of the other two directions, and the scattering may be affected by particles in the third direction. Four particles coming along two directions may also scatter in different directions. Finally, particles on three links separated by 120° may scatter along the other three links. At fixed boundaries, particles may either "bounce back" (yielding "no slip" on average), or reflect "specularly" through 120°.

On a microscopic scale, these rules are deterministic, reversible and discrete. But on a sufficiently large scale, a statistical description may apply, and the system may behave like a continuum fluid, with macroscopic quantities, such as hydrodynamic velocity, obtained by kinetic theory averages.

Figure 1 illustrates relaxation to "thermodynamic equilibrium". The system randomizes, and coarse-grained entropy increases. This macroscopic behaviour is robust, but microscopic details depend sensitively on initial conditions. Small perturbations (say of one particle) have microscopic effects over linearly-expanding regions [6]. Thus ensembles of "nearby" initial states usually evolve to contain widely-differing "typical" states. But in addition, individual "simply-specified" initial states can yield behaviour so complex as to seem random [7,8], as in figure 1. The dynamics thus "encrypts" the initial data; given only coarse-grained, partial, information, the initial simplicity cannot be recovered or recognized by computationally feasible procedures [7], and the behaviour is effectively irreversible.

Microscopic instability implies that predictions of detailed behaviour are impossible without ever more extensive knowledge of initial conditions. With complete knowledge (say from a simple specification), the behaviour can always be reproduced by explicit simulation. But if effective predictions are to be made, more efficient computational procedures should be found. The CA considered here can in fact act as universal computers [9]: with appropriate initial conditions, their evolution can implement any computation. Streams of particles corresponding to "wires" can meet in logical gates implemented by fixed obstructions or other streams. As a consequence, the evolution is computationally irreducible [10]; there is no general shortcut to explicit simulation. No simpler computation can reproduce all the possible phenomena.

Some overall statistical predictions can nevertheless be made. In isolation, the CA seem to relax to an equilibrium in which links are populated effectively randomly with a particular average particle density ρ and net velocity (as in figure 1). On length scales large compared to the mean free path λ, the system then behaves like a continuum fluid. The effective fluid pressure is $p = \rho/2$, giving a speed of sound $c = 1/\sqrt{2}$. Despite the microscopic anisotropy of the lattice, circular sound wavefronts are obtained from point sources (so long as their wavelength is larger than the mean free path) [11].

Assuming local equilibrium, the large-scale behaviour of the CA can be approximated by average rules for collections of particles, with particular average densities and velocities. The rules are like finite difference approximations to partial differential equations, whose form can be found by a standard Chapman-Enskog expansion

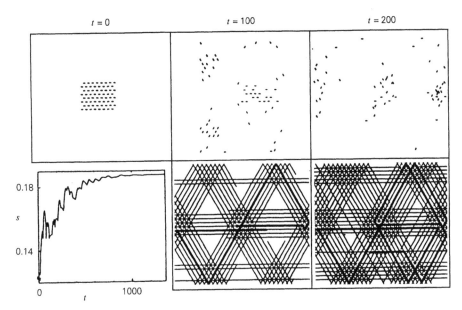

Figure 1. Relaxation to "thermodynamic equilibrium" in the hexagonal lattice cellular automaton (CA) described in the text. Discrete particles are initially in a simple array in the centre of a 32×32 site square box. The upper sequence shows the randomization of this pattern with time; the lower sequence shows the cells visited in the discrete phase space (one particle track is drawn thicker). The graph illustrates the resulting increase of coarse-grained entropy $\sum p_i \log_2 p_i$ calculated from particle densities in 32×32 regions of a 256×256 box.

[12] of microscopic particle distributions in terms of macroscopic quantities. The results are analogous to those for systems [13] in which particles occur with an arbitrary continuous density at each point in space, but have only a finite set of possible velocities corresponding to the links of the lattice. The hexagonal lattice CA is then found to follow exactly the standard Navier-Stokes equations [5,14]. As usual, the parameters in the Navier-Stokes equations depend on the microscopic structure of the system. Kinetic theory suggests a kinematic viscosity $v \simeq \lambda/2$ [15].

Figures 2 and 3 show hydrodynamic phenomena in the large scale behaviour of the hexagonal lattice CA. An overall flow U is obtained by maintaining a difference in the numbers of left- and right-moving particles at the boundaries. Since local equilibrium is rapidly reached from almost any state, the results are insensitive to the precise arrangement used. Random boundary fluxes imitate an infinite region; a regular pattern of incoming particles nevertheless also suffices, and reflecting or cyclic boundary conditions can be used on the top and bottom edges.

The hydrodynamics of the CA is much like a standard physical fluid [16]. For low Mach numbers $\mathrm{Ma} = U/c$, the fluid is approximately incompressible, and the flows show dynamical similarity, depending only on Reynolds number $\mathrm{Re} = UL/v\,(L \gg \lambda)$. The patterns obtained agree qualitatively with experiment [3]. At low Re, the flows are macroscopically stable; perturbations are dissipated into microscopic "heat".

261

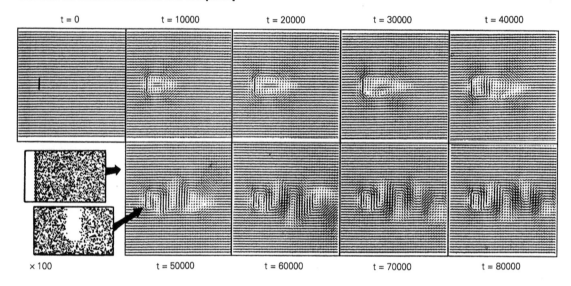

Figure 2. Time evolution of hydrodynamic flow around a plate in the CA of figure 1 on a 4096×4096 site lattice. Hydrodynamic velocities are obtained as indicated by averaging over 96×96 site regions. There is an average density of 0.3 particles per link (giving a total of 3×10^8 particles). An overall velocity $U = 0.1$ is maintained by introducing an excess of particles (here in a regular pattern) on the left hand boundary.

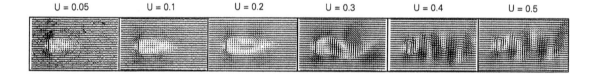

Figure 3. Hydrodynamic flows obtained after 10^5 time steps in the CA of figure 2, for various overall velocities U.

As Re increases, periodic vortex streets are at first produced, and then vortices are shed in an irregular, turbulent, fashion. Perturbations now affect details of the flow, though not its statistical properties. The macroscopic irregularity does not depend on microscopic randomness; it occurs even if microscopically simple (say spatially and temporally periodic) initial and boundary conditions are used, as illustrated in figure 2. As at the microscopic level, it seems that the evolution corresponds to a sufficiently complex computation that its results seem random [7].

The CA discussed here should serve as a basis for practical hydrodynamic simulations. They are simple to program, readily amenable to parallel processing, able to handle complex geometries easily [17], and presumably show no unphysical instabilities. (Generalization to three dimensions is straightforward in principle [18].)

Standard finite difference methods [19] consider discrete cells of fluid described by continuous parameters. These parameters are usually represented as digital numbers with say 64 bits of precision. Most of these bits are, however, probably irrelevant in determining observable features of flow. In the CA approach, all bits are of essentially equal importance, and the number of elementary operations performed is potentially closer to the irreducible limit.

The difficulty of computation in a particular case depends on the number of cells that must be used. Below a certain dissipation length scale $a \sim \mathrm{Re}^{-d/4}$ (in d dimensions), viscosity makes physical homogeneous turbulent fluids smooth [16]. In finite difference schemes, individual cells can represent fluid regions of this size. But complete calculations with the CA considered here probably require increasing numbers of cells in each region [20]. Approximate "turbulence models" involving fewer cells may however be devised.

Several further extensions of the CA scheme can be considered. First, on some or all of the lattice, basic units containing say n particles, rather than single particles, can be used. The properties of these units can be specified by digital numbers with $O(\log n)$ bits, but exact conservation laws can still be maintained. This scheme comes closer to adaptive grid finite difference methods [19], and potentially avoids detailed computation in featureless parts of flows.

A second, related, extension introduces discrete internal degrees of freedom for each particle. These could represent different particle types, directions of discrete vortices [19], or internal energy (giving variable temperature [21]).

This paper has given further evidence that simple cellular automata can reproduce the essential features of thermodynamic and hydrodynamic behaviour. These models make contact with results in dynamical systems theory and computation theory. They should also yield efficient practical simulations, particularly on parallel-processing computers.

Cellular automata can potentially reproduce behaviour conventionally described by partial differential equations in many other systems whose intrinsic dynamics involves many degrees of freedom with no large disparity in scales.

We are grateful to U. Frisch, B. Hasslacher, Y. Pomeau and T. Shimomura for sharing their unpublished results with us, and to N. Margolus, S. Omohundro, S. Orszag, N. Packard, R. Shaw, T. Toffoli, G. Vichniac and V. Yakhot for discussions. We thank many people at Thinking Machines Corporation for their help and encouragement. The work of S.W. was supported in part by the U.S. Office of Naval Research under contract number N00014-85-K-0045.

1. See for example S. Wolfram, "Cellular automata as models of complexity", Nature **311**, 419 (1984) where applications to thermodynamics and hydrodynamics were mentioned but not explored.

2. D. Hillis, *The Connection Machine* (MIT press, 1985). This application is discussed in S. Wolfram, "Scientific computation with the Connection Machine", Thinking Machines Corporation report (March 1985).

3. More detailed results of theory and simulation will be given in a forthcoming series of papers.

4. J. Hardy, Y. Pomeau and O. de Pazzis, "Time evolution of a two-dimensional model system. I. Invariant states and time correlation functions", J. Math. Phys. **14**, 1746 (1973); J. Hardy, O. de Pazzis and Y. Pomeau, "Molecular dynamics of a classical lattice gas: transport properties and time correlation functions", Phys. Rev. **A13**, 1949 (1976).

5. U. Frisch, B. Hasslacher and Y. Pomeau, "A lattice gas automaton for the Navier-Stokes equation", Los Alamos preprint LA-UR-85-3503.

6. The expansion rate gives the Lyapunov exponent as defined in N. Packard and S. Wolfram, "Two-dimensional cellular automata", J. Stat. Phys. **38**, 901 (1985). Note that the effect involves many particles, and does not arise from instability in the motion of single particles, as in the case of hard spheres with continuous position variables (e.g. O. Penrose, "Foundations of statistical mechanics", Rep. Prog. Phys. **42**, 129 (1979).)

7. S. Wolfram, "Origins of randomness in physical systems", Phys. Rev. Lett. **55**, 449 (1985); "Random sequence generation by cellular automata", Adv. Appl. Math. **7**, 123 (1986).

8. Simple patterns are obtained with very simple or symmetrical initial conditions. On a hexagonal lattice, the motion of an isolated particle in a rectangular box is described by a linear congruence relation, and is ergodic when the side lengths are not commensurate.

9. N. Margolus, "Physics-like models of computation", Physica **10D**, 81 (1984) shows this for some similar CA.

10. S. Wolfram, "Undecidability and intractability in theoretical physics", Phys. Rev. Lett. **54**, 735 (1985).

11. cf. T. Toffoli, "CAM: A high-performance cellular automaton machine", Physica **10D**, 195 (1984).

12. e.g. A. Sommerfeld, *Thermodynamics and statistical mechanics*, (Academic Press, 1955).

13. J. C. Maxwell, *Scientific Papers II*, (Cambridge University Press, 1890); J. Broadwell, "Shock structure in a simple discrete velocity gas", Phys. Fluids **7**, 1243 (1964); S. Harris, *The Boltzmann Equation*, (Holt, Reinhart and Winston, 1971); J. Hardy and Y. Pomeau, "Thermodynamics and hydrodynamics for a modeled fluid", J. Math. Phys. **13**, 1042 (1972); R. Gatignol, *Theorie cinetique des gaz a repartition discrete de vitesse*, (Springer, 1975).

14. On a square lattice, the total momentum in each row is separately conserved, and so cannot be convected by velocity in the orthogonal direction [4]. Symmetric

three particle collisions on a hexagonal lattice remove this spurious conservation law.

15. The symmetric rank four tensor which determines the nonlinear and viscous terms in the Navier-Stokes equations is isotropic for a hexagonal but not a square lattice (cf. [5]). Higher order coefficients are anisotropic in both cases. In two dimensions, there can be logarithmic corrections to the Newtonian fluid approximation: these can apparently be ignored on the length scales considered, but yield a formal divergence in the viscosity (cf. [4]).

16. e.g. D. J. Tritton, *Physical fluid dynamics*, (Van Nostrand, 1977).

17. They can also treat microscopic boundary effects beyond the hydrodynamic approximation.

18. Icosahedral symmetry yields isotropic fluid behaviour, and can be achieved with a quasilattice, or approximately by periodic lattices (cf. D. Levine et al., "Elasticity and dislocations in pentagonal and icosahedral quasicrystals", Phys. Rev. Lett. **54**, 1520 (1985); P. Bak, "Symmetry, stability, and elastic properties of icosahedral incommensurate crystals", Phys. Rev. **B32**, 5764 (1985)).

19. e.g. P. Roache, *Computational fluid dynamics*, (Hermosa, Albuquerque, 1976).

20. S. Orszag and V. Yakhot, "Reynolds number scaling of cellular automaton hydrodynamics", Princeton University Applied and Computational Math. report (November 1985).

21. In simple cases the resulting model is analogous to a deterministic microcanonical spin system (M. Creutz, "Deterministic Ising dynamics", Ann. Phys., to be published.)

Random Sequence Generation by Cellular Automata

1 9 8 6

A 1-dimensional cellular automaton which generates random sequences is discussed. Each site in the cellular automaton has value 0 or 1, and is updated in parallel according to the rule $a_i' = a_{i-1}$ XOR $(a_i$ OR $a_{i+1})$ $(a_i' = (a_{i-1} + a_i + a_{i+1} + a_i a_{i+1}) \bmod 2)$. Despite the simplicity of this rule, the time sequences of site values that it yields seem to be completely random. These sequences are analysed by a variety of empirical, combinatorial, statistical, dynamical systems theory and computation theory methods. An efficient random sequence generator based on them is suggested.

1. Random Sequence Generation

Sequences that seem random are needed for a wide variety of purposes. They are used for unbiased sampling in the Monte Carlo method, and to imitate stochastic natural processes. They are used in implementing randomized algorithms which require arbitrary choices. And their unpredictability is used in games of chance, and potentially in data encryption.

To generate a random sequence on a digital computer, one starts with a fixed length seed, then iteratively applies some transformation to it, progressively extracting as long as possible a random sequence (e.g., [1]). In general one considers a sequence "random" if no patterns can be recognized in it, no predictions can be made about it, and no simple description of it can be found (e.g., [2]). But if in fact the sequence can be generated by iteration of a definite transformation, then a simple description of it certainly does exist.[1] The sequence can nevertheless seem random if no computations

---

Originally published in *Advances in Applied Mathematics*, volume 7, pages 123–169 (June 1986).

[1] A stricter definition of randomness can be based on the non-existence of simple descriptions [3], rather than merely the difficulty of finding them. None of the sequences discussed here, nor many generally considered random, would qualify according to this definition.

done on it reveal this simple description. The original seed must be transformed in such a complicated way that the computations cannot recover it.

The degree of randomness of a sequence can be defined in terms of the classes of computations which cannot discern patterns in it. A sequence is "random enough" for application in a particular system if the computations that the system effectively performs are not sophisticated enough to be able to find patterns in the sequence. So, for example, a sequence might be random enough for Monte Carlo integration if the values it yields are distributed sufficiently uniformly. The existence say of particular correlations in the sequence might not be discerned in this calculation. Whenever a computation that uses a random sequence takes a bounded time, there is a limit to the degree of randomness that the sequence need have. Statistical tests of randomness emulate various simple computations encountered in practice, and check that statistical properties of the sequence agree with those predicted if every element occurred purely according to probabilities. It would be better if one could show in general that patterns could not be recognized in certain sequences by any computation whatsoever that, for example, takes less than a certain time. No such results can yet be proved, so one must for now rely on more circumstantial evidence for adequate degrees of randomness.

The fact that acceptably random sequences can indeed be generated efficiently by digital computers is a consequence of the fact that quite simple transformations, when iterated, can yield extremely complicated behaviour. Simple computations are able to produce sequences whose origins can apparently be deduced only by much more complex computations.

Most current practical random sequence generation computer programs are based on linear congruence relations (of the form $x' = ax + b \mod n$) (e.g., [1]), or linear feedback shift registers [4] (analogous to the linear cellular automata discussed below). The linearity and simplicity of these systems has made complete algebraic analyses possible and has allowed certain randomness properties to be proved [1, 4]. But it also leads to efficient algebraic algorithms for predicting the sequences (or deducing their seeds), and limits their degree of randomness.

An efficient random sequence generator should produce a sequence of length L in a time at most polynomial in L (and linear on most kinds of computers). It is always possible to deduce the seed (say of length s) for such a sequence by an exhaustive search which takes a time at most $O(2^s)$. But if in fact such an exponentially long computation were needed to find any pattern in the sequence, then the sequence would be random enough for almost any practical application (so long as it involved less than exponential time computations).

No such lower bounds on computational complexity are yet known. It is however often possible to show that one problem is computationally equivalent to a large class of others. So, for example, one could potentially show that the problem of deducing the seed for certain sequences was NP-complete [5]: special instances of the problem would then correspond to arbitrary problems in the class NP, and the problem would in general be as difficult as any in NP. (One should also show some form of uniform reducibility to ensure that the problem is difficult almost always, as well as in the worst case.) The class NP (nondeterministic polynomial time) includes many well-studied problems (such as integer factorization), which involve finding objects (such as prime factors) that satisfy polynomial-time-testable conditions, but for which no systematic polynomial time (P) algorithms have ever been discovered.

Random sequence generators have been constructed with the property that recognizing patterns in the sequences they produce is in principle equivalent to solving certain difficult number theoretical problems [2] (which are in the class NP, but are not NP-complete). An example is the sequence of least significant bits obtained by iterating the transformation $x' = x^2 \bmod (pq)$, where p and q are large primes (congruent to 3 modulo 4) [6]. Making predictions from this sequence is in principle equivalent to factoring the integer pq [6, 7].

There are in fact many standard mathematical processes which are simple to perform, yet produce sequences so complicated that they seem random. An example is taking square roots of integers. Despite the simplicity of its computation, no practical statistical procedures have revealed any regularity in say the digit sequence of $\sqrt{2}$ (e.g., [8]). (Not even its normality or equidistribution has however actually been proved.) An even simpler example is multiplication by $\frac{3}{2}$, say in base 6.[2] Starting with 1, one obtains the pattern shown in Fig. 1.1 The center vertical column of values, corresponding to the leading digit in the fractional part of $(\frac{3}{2})^n$, seems random [10]. (Though again not even its normality has actually been proved.) Given the complete number obtained at a particular stage, multiplication by $(\frac{2}{3})^n$ suffices to reproduce the original seed. But given only the center column, it seems difficult to deduce the seed.

Many physical processes also yield seemingly random behaviour. In some cases, the randomness can be attributed to the effects of external random input. Thus, for example, "analog" random sequence generators such as noise diodes work by sampling thermal fluctuations associated with a heat bath containing many components. Coin tossings and Roulette wheels produce outcomes that depend sensitively on initial velocities determined by complex systems with many components. It seems however that in all such cases, sequences extracted sufficiently quickly can depend on only a few components of the environment, and must eventually show definite correlations.

One suspects in fact that randomness in many physical systems (probably including turbulent fluids) arises not from external random input, but rather through intrinsic mathematical processes [11]. This paper discusses the generation of random sequences by simple procedures which seem to capture many features of this phenomenon. The investigations described may not only suggest practical methods

---

[2] This operation can be performed locally on a base 6 digit sequence, and so can be implemented as a cellular automaton. Given particular finite boundary conditions, it acts like a linear congruential sequence generator (e.g., [1]). But in an infinite region, its behaviour is more complicated, and is related to the so-called $3N + 1$ problem [9].

```
            1.
            1.3
            2.13
            3.213
            5.0213
           11.33213
           15.220213
           25.0303213
           41.34350213
          102.235433213
          133.3553520213
          222.25525003213
          333.425113050213
          522.3414514133213
         1203.53241532220213
         2005.521025203303213
         3012.5013420051350213
         4321.13223301152433213
        10501.5203513155510520213
        14132.5005451554442003213
        23221.13124155541030050213
        35031.5151025553134311133213
        54345.4544342551523445220213
       123542.42405342455054120303213
       205534.040122341244232004350213
       312523.1002035321103500105433213
       451204.4303055201435430142352021.3
      1115011.043442500235534323355003213
      1454314.405404130355523505254305021.3
      2423454.01231021355550544212344133213
      4035423.020443322555442403205402220213
     10055334.331105204255404005012303330321.3
     13125223.514442010425310011320435213502.13
     21512035.454103014042143015201055022433213
     32450055.42313432310323432500142433405202.13
     51113125.334523504435053511302340523120032.13
    114451512.224205441054422445133531204500050213
    154115450.340312401424034111522515011130113321.3
    253155413.53045100234005314550415431451315220213
    421555322.5141143035301215424402534541515503032.13
```

Figure 1.1. Successive powers of 3/2 in base 6. The leading digits in the fractional parts of these numbers form a sequence that seems random. The process of multiplication by 3/2 in base 6 corresponds to a $k = 6, r = 1$ cellular automaton rule.

for random sequence generation, but also provide further understanding of the nature and origins of randomness in physical processes.

2. Cellular Automata

A 1-dimensional cellular automaton [12, 13] consists of a line of sites with values a_i between 0 and $k - 1$. These values are updated in parallel (synchronously) in discrete time steps according to a fixed rule of the form

$$a_i' = \phi(a_{i-r}, a_{i-r+1}, \cdots, a_{i+r}). \tag{2.1}$$

Much of this paper is concerned with the study of a particular $k = 2$, $r = 1$ cellular automaton, described in Section 3.

For mathematical purposes, it is often convenient to consider cellular automata with an infinite number of sites. But practical implementations must contain a finite number of sites N. These are typically arranged in a circular register, so as to have

270

periodic boundary conditions, given in the $r = 1$ case by

$$a'_1 = \phi(a_N, a_1, a_2)$$
$$a'_N = \phi(a_{N-1}, a_N, a_1). \tag{2.2}$$

It is also possible to arrange the sites in a feedback shift register (cf. [4]), with boundary conditions

$$a'_1 = \phi(\phi(a_2, a_3, a_4), \phi(a_3, a_4, a_5), a_1),$$
$$a'_2 = \phi(\phi(a_3, a_4, a_5), a_1, a_2). \tag{2.3}$$

Cellular automata can be considered as discrete approximations to partial differential equations, and used as direct models for a wide variety of natural systems (e.g., [14]). They can also be considered as discrete dynamical systems corresponding to continuous mappings on the Cantor set (e.g., [15]). Finally they can be viewed as computational systems, whose evolution processes information contained in their initial configurations (e.g., [16]).

Despite the simplicity of their construction, cellular automata are found to be capable of diverse and complex behaviour. Figure 2.1 shows some patterns generated by evolution according to various cellular automaton rules, starting from typical disordered initial conditions. Four basic outcomes are seen [15]: (1) the pattern becomes homogeneous (fixed point), (2) the pattern degenerates into simple periodic structures (limit cycles), (3) the pattern is aperiodic, and appears chaotic, and (4) complicated localized structures are produced. The first two classes of cellular automata yield readily predictable behaviour, and show no seemingly random elements. But the third class gives rise to behaviour that is more complex. They can produce patterns whose features cannot readily be predicted in detail, and in fact often seem completely random. Such cellular automata can be used as models of randomness in nature. They can also be considered as abstract mathematical systems, and used for practical random sequence generation.

Figure 2.1 shows patterns produced by evolution according to various cellular automaton rules, starting from typical disordered initial conditions, in which the value of each site is randomly chosen to be zero or one. Figure 2.2 shows some patterns obtained instead by evolution from a very simple initial condition containing a single nonzero site. With such simple initial conditions, some class 3 cellular automata yield rather simple patterns, which are typically periodic or at least self similar (almost periodic). There are nevertheless class 3 cellular automata which yield complex patterns, even from simple initial states. Their evolution can intrinsically produce apparent randomness, without external input of random initial conditions. It is such "autoplectic" systems [11] which seem most promising for explaining randomness in nature, or for use as practical random sequence generation procedures.

Many class 3 cellular automata seem to perform very complicated transformations on their initial conditions. Their evolution thus corresponds to a complicated

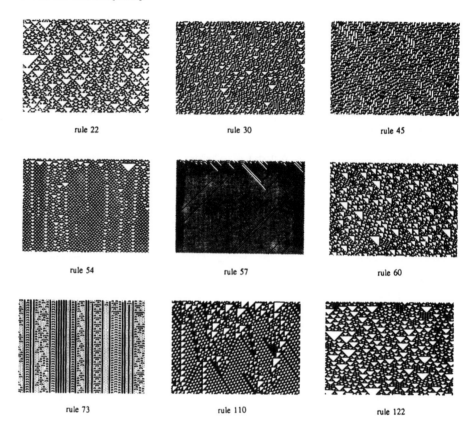

rule 22   rule 30   rule 45

rule 54   rule 57   rule 60

rule 73   rule 110   rule 122

Figure 2.1. Patterns generated by evolution of various $k = 2$, $r = 1$ cellular automata from disordered initial states. Successive lines give configurations obtained on successive time steps, with white and black squares representing sites with values 0 and 1 respectively. The coefficient of 2^i in the binary decomposition of each rule number gives the value of the function ϕ in Eq. (2.1) for the neighbourhood whose site values form the integer i (cf. [17]).

computation. But any predictions of the cellular automaton behaviour must also be obtained through computations. Effective predictions require computations that are more sophisticated than those corresponding to the cellular automaton evolution itself. One suspects however that the evolution of many class 3 cellular automata is in fact computationally as sophisticated as that of any (physically realizable) system can be [18, 19]. It is thus "computationally irreducible," and its outcome can effectively be found only by direct simulation or observation. There are no general computational shortcuts or finite mathematical formulae for it. As a consequence, many questions concerning infinite time or infinite size limits cannot be answered by bounded computations, and must be considered formally undecidable. In addition, questions about finite time or finite size behaviour, while ultimately computable, may be computationally intractable, and could require, for example, exponential time computations.

Most class 3 cellular automata are expected to be computationally irreducible. A few rules however have special simplifying features which make predictions and analysis possible. One class of such rules are those for which the function ϕ is linear (modulo k) in the a_{i+j}. Such cellular automata are analogous to linear feedback shift registers [4]. An example with $k = 2$ is

$$a_i' = (a_{i-1} + a_i) \bmod 2 = (a_{i-1} \text{ XOR } a_i), \qquad (2.4)$$

where XOR stands for exclusive disjunction (this is rule number 60 in the scheme of [17]). Linear cellular automata satisfy a superposition principle, which implies that patterns generated with arbitrary initial states can be obtained as appropriate superpositions of the self-similar pattern produced with a single nonzero initial site (as illustrated in Fig. 2.2). As a result, it is possible to give a complete algebraic description of the behaviour of the system [20], and to deduce the outcome of its evolution by a much reduced computation.

Most class 3 cellular automata are however nonlinear. No general methods to predict their behaviour have been found, and from their likely computational irreducibility one expects that no such methods even in principle exist. In studying such systems one must therefore to a large extent forsake conventional mathematical techniques and instead rely on empirical and experimental mathematical results.

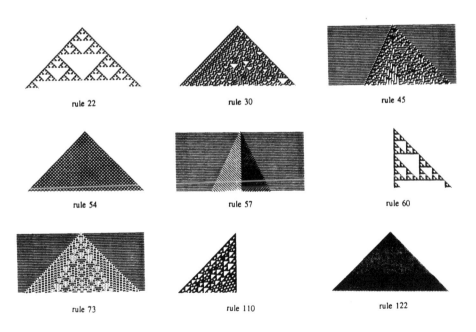

rule 22   rule 30   rule 45

rule 54   rule 57   rule 60

rule 73   rule 110   rule 122

Figure 2.2. Patterns generated by evolution of various $k = 2$, $r = 1$ cellular automata from an initial state containing a single nonzero site. Complex patterns are seen to be produced even with such simple initial conditions.

273

3. A Random Sequence Generator

There are a total of $2^{2^3} = 256$ cellular automaton rules that depend on three sites, each with two possible values ($k = 2$, $r = 1$). Among these are several linear rules similar to that of Eq. (2.4). But the two rules that seem best as random sequence generators are nonlinear, and are given by

$$a_i' = a_{i-1} \text{ XOR } (a_i \text{ OR } a_{i+1}) \tag{3.1a}$$

or, equivalently,

$$a_i' = (a_{i-1} + a_i + a_{i+1} + a_i a_{i+1}) \bmod 2 \tag{3.1b}$$

(rule number 30 [17]; equivalent to rule 86 under reflection), and

$$a_i' = a_{i-1} \text{ XOR } (a_i \text{ OR } (\text{NOT } a_{i+1})) \tag{3.2a}$$

or

$$a_i' = (1 + a_{i-1} + a_{i+1} + a_i a_{i+1}) \bmod 2 \tag{3.2b}$$

(rule 45; reflection equivalent to rule 75). Here XOR stands for exclusive disjunction (addition modulo two); OR for inclusive disjunction (Boolean addition), and NOT for negation. The patterns obtained by evolution from a single nonzero site with each of these rules were shown in Fig. 2.2. It is indeed remarkable that such complexity can arise in systems of such simple construction. A first indication of their potential for random sequence generation is the apparent randomness of the center vertical column of values in the patterns of Fig. 2.2.

This paper concentrates on the cellular automaton of Eq. (3.1). The methods used carry over directly to the cellular automaton of Eq. (3.2), but some of the results obtained in this case are slightly less favourable for random sequence generation.

The cellular automaton rule (3.1) is essentially nonlinear. Nevertheless, its dependence on a_{i-1} is in fact linear. This feature (termed "left permutivity" in [21], and also studied in [22]) is the basis for many of its properties. In the form (3.1), the rule gives the new value a_i' of a site in terms of the old values a_{i-1}, a_i and a_{i+1}. But the linear dependence on a_{i-1} allows the rule to be rewritten as

$$a_{i-1} = a_i' \text{ XOR } (a_i \text{ OR } a_{i+1}), \tag{3.3}$$

giving a_{i-1} in terms of a_i', a_i and a_{i+1}. This relation implies that the spacetime patterns shown, for example, in Figs. 2.1 and 2.2 can be found not only by direct time evolution according to (3.1) from a given initial configuration, but also by extending spatially according to (3.3), starting with the temporal sequence of values of two adjacent sites.

Random sequences are obtained from (3.1) by sampling the values that a particular site attains as a function of time. In practical implementations, a finite number of sites are considered, and are typically arranged in a circular register. Given almost any initial "seed" configuration for the sites in the register, a long and seemingly random sequence can apparently be obtained. This paper discusses several approaches to the analysis of the cellular automaton (3.1) and the sequences it produces. While little can rigourously be proved, the overwhelming weight of evidence is that the sequences indeed have a high degree of randomness.

4. Global Properties

This section considers the behaviour of the cellular automaton (3.1) starting from all possible initial states. The basic approach is to count the possible sequences and patterns that can occur, and to characterize them using methods from dynamical systems theory (e.g., [23]). The next section discusses the behaviour obtained by evolution from particular initial configurations. For purposes of simplicity, this section concentrates on the infinite size limit; Section 9 considers finite size effects.

Figure 4.1 shows a spacetime pattern produced by evolution according to (3.1) starting from a typical disordered initial state. While definite structure is evident, one may suspect that a single line of sites at any angle in the pattern can have an arbitrary sequence of values. Below we shall show that this is in fact the case: given an appropriate initial condition, any sequence can be generated in an infinite cellular automaton with the rule (3.1).

Figure 4.1. Pattern produced by evolution according to the cellular automaton rule (3.1) from a typical disordered initial state.

The rule (3.1) can be considered as a mapping from one (say infinite) cellular automaton configuration to another. An important property of this mapping is that it is surjective or onto. Any configuration A can thus always be obtained as the image of some configuration A^-, according to $A = \phi A^-$. A possible configuration A^- (not necessarily unique) can be found by starting with a candidate pair of site values, then extending to the left using Eq. (3.3). So if all possible initial configurations are considered, then any configuration can be generated at any time step. Thus with appropriate initial conditions, any spatial sequence of site values can be produced.

Every length X spatial sequence of site values that occurs is determined by a length $X + 2$ sequence on the previous time step. The surjectivity of the rule (3.1) implies that such a predecessor exists for any length X sequence. But Eq. (3.3) also implies that there are exactly four predecessors for any sequence. Given values a_i, a_{i-1}, and so on, in one sequence, the values a_{i+1}^- and a_i^- in its predecessor can be chosen in all the four possible ways; in each case the remaining a_{i-j}^- are then uniquely determined by Eq. (3.3). Thus starting from an ensemble that contains all possible (infinite) cellular automaton configurations with equal probabilities, each configuration will be generated with equal probability throughout the evolution of the cellular automaton, and so every possible spatial sequence of a particular length will occur with equal frequency.

One may also consider sequences of values attained by a single site as a function of time. Starting from an initial ensemble which contains all configurations with equal probabilities, all such sequences again occur with equal frequencies. For, given any temporal sequence, iteration of Eq. (3.3) yields an equal number of initial configurations which evolve to it. The same is true for sequences of site values on lines at any angle in the spacetime pattern.

Entropies provide characterizations of the number of possible sequences that occur. First, let the number of distinct length n blocks in these sequences be $N(n)$, and let the ith such sequence appear with probability p_i. Then the topological entropy of the sequence is given by (e.g., [15])

$$s = \lim_{n \to \infty} \frac{1}{n} \log_2 N(n), \tag{4.1}$$

and the measure entropy by

$$s_\mu = \lim_{n \to \infty} \frac{-1}{n} \sum_i^{2^n} p_i \log_2 p_i. \tag{4.2}$$

If the cellular automaton configurations are considered as elements of a Cantor set, then these entropies give respectively the Hausdorff (strictly Kolmogorov) and measure dimensions of this set. If the sequences are considered as "messages," then the entropies give respectively their capacity and Shannon information content.

For the cellular automaton of Eq. (3.1), all possible sequences occur with equal probabilities (given an equal probability initial ensemble) so both entropies are maximal:

$$s_\mu = s = 1. \tag{4.3}$$

Any reduction in entropy would reveal redundancy in the sequences, and would imply a lack of randomness. Equation (4.3) is thus a necessary (though not sufficient) condition for randomness. (It is related to statistical test A of Section 10 and Appendix A.)

Although Eq. (4.3) implies that all possible sequences of values for single sites can occur along any spacetime direction, the deterministic nature of the cellular automaton rule (3.1) implies that only certain spacetime patches of values can occur. In fact, all the site values in a particular patch are completely determined by the values that appear on its upper, left and right boundaries. Once these boundaries are specified, the values of remaining sites in the patch are redundant, and can be found simply by applying (3.1) and (3.3).

In general the degree of redundancy in such spacetime patterns can be characterized by the invariant topological and measure entropies for the cellular automaton mapping, given by (e.g., [15, 24])

$$\mathbf{h} = \lim_{X \to \infty} \lim_{T \to \infty} \frac{1}{T} \log_2 N(X, T) \tag{4.4}$$

and

$$\mathbf{h}_\mu = \lim_{X \to \infty} \lim_{T \to \infty} \frac{-1}{T} \sum_{i=1}^{2^{XT}} p_i \log_2 p_i, \tag{4.5}$$

where $N(X, T)$ gives the total number of distinct $X \times T$ spacetime patches of site values that occur, and the p_i give their probabilities.

It is clear from the locality of the rule (3.1) that

$$\mathbf{h}_\mu \le \mathbf{h} \le 2. \tag{4.6}$$

A calculation based on the method of [25] in fact shows that[3]

$$\mathbf{h}_\mu \lesssim 1.20. \tag{4.7}$$

Hence a knowledge of the time sequences of values of about 1.2 sites suffice in principle to determine the values of all other sites. In practice however the function which gives the initial configuration in terms of these temporal sequences seems rapidly to become intractably complicated, as discussed in Section 7.

---

[3] Recent results [45] suggest in fact that $\mathbf{h}_\mu \simeq 1 + T^{-(0.6\pm0.1)}$, yielding a final value of 1.

5. Stability Properties

Section 4 considered properties of possible patterns generated by evolution with the cellular automaton rule of Eq. (3.1), starting from all possible initial configurations. This section considers the change in the patterns produced by small perturbations in the initial state. Figure 5.1 shows the differences resulting from reversal of a single site value in a typical disordered initial configuration. The region affected increases in size with time, reflecting the instability of the patterns generated.

This instability implies that information on localized changes eventually propagates throughout the cellular automaton. The rates of information transmission to the left and right are determined by the slopes of the difference pattern in Fig. 5.1. These in turn give left and right Lyapunov exponents λ_L and λ_R for the cellular automaton evolution [15, 26]. (The sequence of site values in a configuration, starting from a particular point, can be represented as a real number. Linear growth of the difference pattern in Fig. 5.1 then implies exponential divergence of the numbers representing nearby configurations.)

The form of the cellular automaton rule (3.1) immediately implies that

$$\lambda_R = 1. \tag{5.1}$$

For consider a configuration in which the difference pattern has reached site −1. Whatever the current values of sites 0 and 1, the XOR in (3.1) leads to a change in the new value of site 0. The value (5.1) is the maximum allowed by the locality of the rule (3.1).

Empirical measurements suggest that the left-hand side of the difference pattern expands at an asymptotically linear rate, with a slope [45]

$$\lambda_L = (0.2428 \pm 0.0003). \tag{5.2}$$

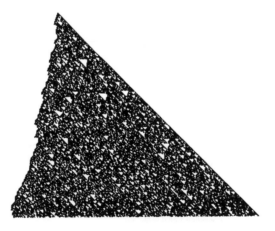

Figure 5.1. Differences in patterns produced by evolution according to the cellular automaton rule of Eq. (3.1) from two typical disordered states which differ by reversal of the centre site value. The growth of the region of differences reflects the instability of the cellular automaton evolution.

A simple statistical estimate for λ_L can be given. Consider a pair of configurations for which the front of the difference pattern has reached site 0. As a first approximation, one may assume that the motion of this front depends only on the neighbouring values a_{-1} and a_{+1}, where, by construction, a_{-1} is the same for the two configurations. When $a_{-1} = 0$, the front advances (left) by one site, independent of the values of the a_1. When $a_{-1} = 1$, the front remains stationary if the a_{+1} for the two configurations are equal, and retreats by one site if they are unequal. If possible sets of site values occurred with equal probabilities, the front should thus follow a biased random walk, advancing at average speed $1/4$. In practice, however, Fig. 5.1 shows that the front can retreat by many sites in a single time step. This occurs when the cellular automaton rule yields the same image for multiple site value sequences, as for say 10100 and 11001. Such phenomena make the probabilities for different difference patterns unequal, and invalidate this purely statistical approach discussed. (The values of λ_L obtained in this approach by considering the effects of between 1 and 5 sites on the right are 0.25, 0.1875, 0.15625, 0.140625 and 0.134766.)

The result (5.2) gives the average speed of the left-hand side of the difference pattern. As the random walk interpretation suggests, however, one can choose initial configurations for which a single site change leads to differences which expand at speed 1 on the left. In general, one can construct the analog of a Green's function, giving the probability that a site at a particular position and time will be affected by an initial perturbation. This function is nonzero within a "light cone" with edges expanding at speed 1. It appears to be uniform on the right-hand side. But on the left-hand side, it appears to be determined by a diffusion equation which gives the average behaviour of the biased random walk. The difference pattern can thus extend beyond the line given by Eq. (5.2), but with an exponentially damped probability.

Lyapunov exponents measure the rate of information transmission in cellular automata, and provide upper bounds on entropies, which measure the information content of patterns generated by cellular automaton evolution. For surjective cellular automata it can be shown, for example, that [15]

$$\mathbb{h}_\mu \le (\lambda_L + \lambda_R), \tag{5.3}$$

consistent with Eq. (4.6) and (5.2). The existence of positive Lyapunov exponents is a characteristic feature of class 3 cellular automata.

The difference pattern of Fig. 5.1, and the related Green's function, measure the effect of initial perturbations on the values of individual sites. In studying random sequence generation, one must also consider the effect of such perturbations on time sequences of site values, say of length T. These sequences are always completely determined from the initial values of $2T + 1$ sites. But not all these initial values necessarily affect the time sequences. A change in any of the $T + 1$ left-hand initial sites necessarily leads to a change in at least one element of the time sequence. But some changes in the T right-hand initial sites have no effect on any element of the time sequence. It seems that the probability for a particular initial site to affect the time sequence decreases exponentially with distance to the right. The average number of sites on the right which affect the time sequence is found to be approximately $0.26 + 0.19T$. Thus the total number of initial sites on which a length T time sequence depends is on average approximately $1.91 + 1.19T$. This result is presumably related to the entropy (4.6).

6. Particular Initial States

Sections 4 and 5 have discussed some properties of the patterns produced by evolution according to Eq. (3.1) from generic initial conditions. This section considers evolution from particular special initial configurations.

Figure 6.1 shows on two scales the pattern produced by evolution from a configuration containing a single nonzero site. (This could be considered a difference pattern for the special time-invariant state in which all sites have value zero.) Remarkable complexity is evident.

There are however some definite regularities. For example, diagonal sequences of sites on the left-hand side of the pattern are periodic, with small periods. In general, the value of a site at a depth N from the edge of the pattern depends only on sites at depths N or less; all the other sites on which it could depend always have value 0 because of the initial conditions given. As a consequence, the sites down to depth N are independent of those deeper in the pattern, and in fact follow a shifted version of the cellular automaton rule (3.1), with boundary conditions that constrain two sites at one end to have value zero. Since such a finite cellular automaton has a total of 2^N possible states, any time sequence of values in it must have a period of at most 2^N. The corresponding diagonal sequences in the pattern of Fig. 6.1 must therefore also have periods not greater than 2^N.

Table 6.1 gives the actual periods of diagonal sequences found at various depths on the left- and right-hand sides of the pattern in Fig. 6.1. These are compared with those for the self-similar pattern shown in Fig. 2.2 generated by evolution according to the linear cellular automaton rule (2.4).

The short periods on the left-hand side of the pattern in Fig. 6.1 are related to the high degree of irreversibility in the effective cellular automaton rule for diagonal sequences in this case [27]. Starting with any possible initial configuration, this cellular automaton always yields cycles with period 2^j. The maximum value of j increases very slowly with N, yielding maximum cycle lengths which increase in jumps, on average slower than linearly with N. (Between the N values at which the maximum cycle length increases, a single additional cycle of maximal length seems to be added each time N increases by one. The total number of cycle states thus increases at most quadratically with N, implying an increasing degree of irreversibility.) The actual sequences that occur near the left-hand boundary of the pattern in Fig. 6.1 correspond to a particular set of those possible in this effective cellular automaton. In

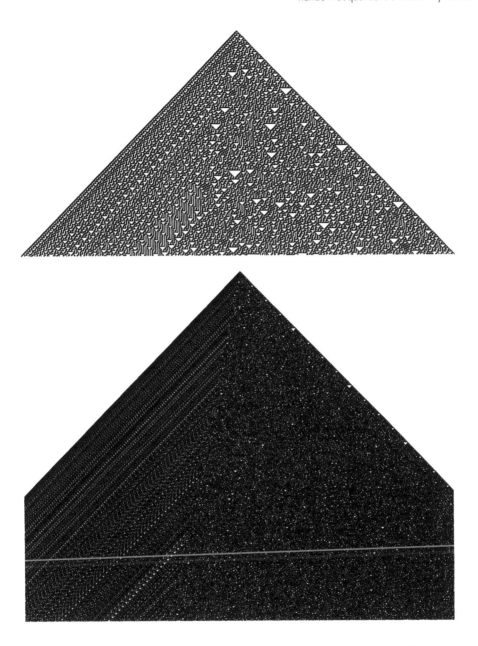

Figure 6.1. Patterns generated by evolution for 250 and 2000 generations, respectively, according to the cellular automaton rule (3.1) from an initial state containing a single nonzero site. (The second pattern was obtained by Jim Salem using a prototype Connection Machine computer.)

Depth	CA30 π_R	π_L	CA60 π_R
0	1	1	1
1	2	1	2
2	2	1	4
3	4	2	4
4	8	1	8
5	8	2	8
6	16	2	8
7	32	1	8
8	32	4	16
9	64	1	16
10	64	4	16
11	64	4	16
12	64	4	16
13	64	4	16
14	64	4	16
15	128	4	16
16	256	4	32
32		8	64
64		4	128
128		8	256
256		8	512
512		16	1024
1024		16	2048

Table 6.1. Period lengths for diagonal sequences in patterns generated by evolution from a single nonzero site according to the cellular automaton rules of Eqs. (3.1) and (2.4). π_R and π_L signify respectively periods for diagonal sequences on the right and left of the patterns, at the specified depth. (The entries left blank were not found.)

a first approximation, they can be considered uniformly distributed among possible N-site configurations, and their periods increase very slowly with N.

The effective rule for the right-hand side diagonal pattern in Fig. 6.1 is a shifted version of Eq. (3.1)

$$a_i' = a_i \text{ XOR } (a_{i+1} \text{ OR } a_{i+2}), \tag{6.1a}$$

with boundary conditions

$$a_{N-1}' = a_{N-1} \text{ XOR } a_N,$$
$$a_N' = a_N. \tag{6.1b}$$

This system is exactly reversible: all of its 2^N possible configurations have unique predecessors. All the configurations thus lie on cycles, and again the cycles have periods of the form 2^j. Figure 6.2 shows the lengths of longest cycles as a function of N. These lengths increase roughly exponentially with N; a least squares fit to the data of Fig. 6.2 yields

$$\log_2 \Pi_N \simeq 0.5(N + 1). \tag{6.2}$$

This length is small compared to the total number of states 2^N; few states in fact lie on

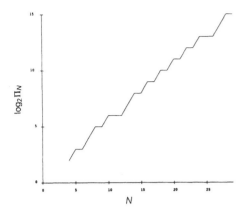

Figure 6.2. Maximal period lengths Π_N for the effective cellular automaton which gives the right-hand diagonal sequences in Fig. 6.1 down to depth N. Points plotted at integer N are joined for pictorial purposes.

such longest cycles. Nevertheless, the periods of the right-hand diagonal sequences in Fig. 6.1 do seem to increase roughly exponentially with depth, as suggested by Table 6.1.

The boundary in Fig. 6.1 between regular behaviour on the left and irregular behaviour on the right seems to be asymptotically linear, and to move to the left with speed 0.25. A statistical argument for this result can be given in analogy with that for Eq. (5.2). Each site at depth d on the left-hand side of the pattern could in principle be affected by sites down to depth d arbitrarily far up in the pattern. In practice, however, it is unaffected by changes in sites outside a cone whose boundary propagates at speed $\lambda_L \simeq 0.25$. Thus the irregularity on the right spreads to the left only at this speed.

While diagonal sequences at angles ± 1 in Fig. 6.1 must ultimately become periodic, sequences closer to the vertical need not. In fact, no periodicity has been found in any such sequences. The center vertical (i.e., temporal) sequence has, for example, been tested up to length $2^{19} \simeq 5 \times 10^5$, and no periodicity is seen. One can prove in fact that only one such vertical sequence (obtained from any initial state containing a finite number of nonzero sites) can possibly be periodic [22]. For if two sequences were both periodic, then it would follow that all sequences to their right must also be, which would lead to a contradiction at the edge of the pattern.

Not only has no periodicity been detected in the center vertical sequence of Fig. 6.1; the sequence has also passed all other statistical tests of randomness applied to it, as discussed in Section 10.

While individual sequences seem random, there are local regularities in the overall pattern of Fig. 6.1. Examples are the triangular regions of zero sites. Such regularities are associated with invariants of the cellular automaton rule.

The particular configuration in which all sites have value 0 is invariant under the cellular automaton rule of Eq. (3.1). As a consequence, any string of zeroes that appears can be corrupted only by effects that propagate in from its ends. Thus each string of zeroes that is produced leads to a uniform triangular region.

Table 6.2 and Fig. 6.3 give other configurations which are periodic under the rule (3.1). (They can be considered as invariant under iterations of the rule.) Again,

Period	Element
1	0
	01
3	000011111001
4	0000001
	0000111
	0010011
	0111111

Table 6.2. Configurations periodic under the cellular automaton mapping (3.1) consist of infinite repetitions of the elements given. Notice that the four elements given for period four correspond simply to different phases in a cycle. The patterns generated by these periodic configurations are shown in Fig. 6.3.

any string that contains just the sequences in these configurations can be corrupted only through end effects, and leads to a regular region in spacetime patterns generated by Eq. (3.1).

In general, there is a finite set of configurations with any particular period p under a permutive cellular automaton rule such as (3.1). The configurations may be found by starting with a candidate length $2p$ string, then testing whether this and the string it yields through Eq. (3.3) on the left are in fact invariant under ϕ^p. The string to be tested need never be longer than 2^{2p}, since such a string can contain all possible length $2p$ strings. Thus the periodic configurations consist of repetitions of blocks containing 2^{2p} or less site values. (For an arbitrary cellular automaton rule, the set of invariant configurations forms a finite complement language which contains in general an infinite number of sequences with the constraint that certain blocks are excluded [16].)

The pattern in Fig. 6.1 can be considered the effect of a single site "defect" in the periodic pattern resulting from a configuration with all sites 0. Figure 6.4 shows difference patterns produced by single site defects in the other periodic configurations of Table 6.2 and Fig. 6.3.

The periodic configurations of Table 6.2 and Fig. 6.3 can be viewed as special states in which the cellular automaton of Eq. (3.1) behaves just like the identity rule. Concatenations of other blocks could simulate other cellular automata: one block might correspond to a value 0 site, and another to a value 1 site in the effective cellular automaton. Some cellular automata (such as that of Eq. (2.4)) simulate themselves

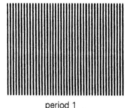
period 1

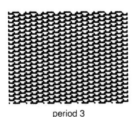
period 3

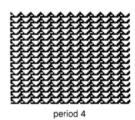
period 4

Figure 6.3. Periodic patterns for the cellular automaton rule of Eq. (3.1). The form of these patterns is given in Table 6.2.

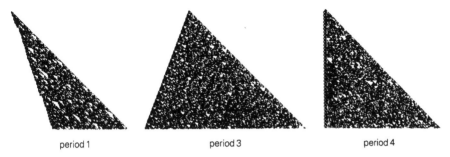

period 1   period 3   period 4

Figure 6.4. Patterns produced by evolution according to the cellular automaton rule (3.1) by single site initial defects in the periodic patterns of Fig. 6.2 and Table 6.2.

under such "blocking transformations," and thus evolve to self-similar patterns. The cellular automata of Eqs. (3.1) and (3.2) are unique among $k = 2$, $r = 1$ rules in simulating no other rules, at least with blocks of length up to eight [14].

7. Functional Properties

Cellular automaton rules such as (3.1) can be considered as functions ϕ which map three Boolean values to one. Iterations of these rules for say t steps correspond to functions of $2t + 1$ Boolean values. The complexity of these functions reflects the intrinsic complexity of the cellular automaton evolution.

The complexity of a Boolean function can be characterized by the number of logic gates that would be needed to evaluate it with a particular kind of circuit, or the number of terms that it would have in a particular symbolic representation. Explicit evolution according to the cellular automaton rule (3.1) corresponds to a circuit with $O(t^2)$ components and depth t. But for purposes of comparison, it is convenient to consider fixed depth representations. One such representation is disjunctive normal form (DNF), in which the function is written as a disjunction of conjunctions. A two-level circuit can be constructed in direct correspondence with this form (as programmable logic arrays often are).

For the function of Eq. (3.1), the DNF is

$$\phi(a_{-1}, a_0, a_1) = (\overline{a_{-1}}a_0) + (a_{-1}\overline{a_0}\,\overline{a_1}) + (\overline{a_{-1}}a_1), \tag{7.1}$$

where + stands for OR, concatenation for AND, and bar for NOT. Notice that by using in addition an XOR operation, Eq. (3.1) itself gives a shorter form for this function.

The general problem of finding the absolute shortest representation for an arbitrary Boolean function, even in DNF, is NP-complete (e.g., [5]), and so presumably requires an exponential time computation. But a definite approximation can be found in terms of "prime implicants" (e.g., [28]). A Boolean function of n variables can be considered as a colouring of the Boolean n-cube. Prime implicants give the hyperplanes (with different dimensions) in the n-cube which must be superimposed

t	CA30		CA60
	P.I.	Min.	P.I./Min.
1	3	3	2
2	9	7	2
3	23	17	8
4	76	41	2
5	185	105	8
6	666	272	8

Table 7.1. Number of terms in disjunctive normal form Boolean expressions corresponding to iterations of the mappings (3.1) (CA30) and (2.4) (CA60). P.I. gives the number of prime implicants; min. the number of terms obtained by [29]. (The two numbers are the equal in the case of Eq. (2.4).)

to obtain the region with value 1. Each prime implicant can thus be used as a term in a DNF for the function. The number of prime implicants required gives a measure of the total number of "holes" in the colouring of the n-cube, and thus of the complexity of the function.

The minimal DNF obtained with prime implicants for the function corresponding to two iterations of the cellular automaton mapping (3.1) is

$$\phi^2(a_{-2}, a_{-1}, a_0, a_1, a_2) = (\overline{a_{-2}}\, \overline{a_{-1}}\, \overline{a_0}\, a_1 \overline{a_2}) + (\overline{a_{-2}} a_{-1} a_0 a_1\, \overline{a_2})$$
$$+ (a_{-2}\overline{a_{-1}} a_0 a_1\, \overline{a_2}) + (a_{-2} a_{-1} a_0 \overline{a_1}\, \overline{a_2})$$
$$+ (a_{-2}\overline{a_{-1}}\, \overline{a_1}\, \overline{a_2}) + (\overline{a_{-2}}\, \overline{a_{-1}}\, \overline{a_0}\, a_2) \tag{7.2}$$
$$+ (a_{-2}\overline{a_{-1}} a_0 a_2) + (\overline{a_{-2}} a_{-1} a_0 a_2) + (a_{-2} a_{-1}\, \overline{a_0}).$$

Table 7.1 gives the number of prime implicants for successive iterations of the mapping (3.1). These results are plotted in Fig. 7.1. For arbitrary Boolean functions of $2t + 1$ variables, the number of prime implicants could increase like 4^t. In practice, however, a least squares fit to the data of Table 7.1 suggests growth like $4^{0.77t}$.

Various efficient methods are known to find DNF that are somewhat simpler than those obtained using prime implicants. With one such method [28, 29], the DNF of Eq. (7.2) can be reduced to

$$\phi^2(a_{-2}, a_{-1}, a_0, a_1, a_2) = (\overline{a_{-2}}\, \overline{a_{-1}}\, \overline{a_0}\, a_1) + (\overline{a_{-2}}\, a_{-1} a_0 a_1)$$
$$+ (\overline{a_{-2}}\, \overline{a_{-1}}\, \overline{a_0}\, a_2) + (\overline{a_{-2}}\, a_{-1} a_0 a_2) \tag{7.3}$$
$$+ (a_{-2} \overline{a_1}\, \overline{a_2}) + (a_{-2} \overline{a_{-1}}\, a_0) + (a_{-2} a_{-1}\overline{a_0}).$$

The sizes of the minimal DNF obtained by this method for iterations of Eq. (3.1) are shown in Table 7.1 and Fig. 7.1. They are seen to grow more slowly than those obtained with prime implicants; the data given are however again fit by exponential growth like $4^{0.65t}$.

Table 7.1 and Fig. 7.1 also give the size of the minimal DNF for iterations of the linear cellular automaton mapping (2.4). This number remains much smaller, apparently increasing like $2^{2\#_1(t)-1} < t^2$, where $\#_1(t)$ gives the number of ones in the binary representation for the integer t (cf. [30]).

The rapid increase in the size of the minimal DNF found for iterations of Eq. (3.1) indicates the increasing computational complexity of determining the result of evolution according to (3.1), and supports the conjecture of its computational irreducibility.

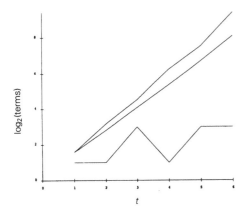

Figure 7.1. Number of terms in disjunctive normal form Boolean expressions for t step iterations of the mappings (3.1) and (2.4). The upper curve gives the number of prime implicants for iterations of Eq. (3.1). The next curve gives the minimal number of terms obtained in this case using [29]. The lowest curve gives the minimal number of terms for the linear cellular automaton mapping (2.4).

(Note however that even the parity function cannot be computed by any DNF, or in general fixed-depth, circuit of polynomial size [31].)

Equation (7.3) gives the function which determines the value of a single site after two iterations of the cellular automaton rule (3.1). One can also construct a function which gives the length t sequence of values of a particular site attained through time by evolution from a given length $2t + 1$ initial sequence. The minimal DNF representation for this function is found (using [29]) to grow in size approximately as $2^{1.36t}$.

The results of Table 7.1 and Fig. 7.1 concern the difficulty of finding the outcome of cellular automaton evolution according to Eq. (3.1) from a given initial state. One may also consider the problem of deducing the initial state from time sequences of site values produced in the evolution. Given say t steps in the time sequence of values for two adjacent sites, the initial configuration up to t sites to the left can be deduced directly by iteration of Eq. (3.3). The combinatorial results of Section 4 indicate in fact that only about 1.2 such temporal sequences should on average be required. And in principle from a single sufficiently long temporal sequence, it should be possible to deduce a complete initial configuration for a finite cellular automaton. In practice, however, the necessary computation seems to become increasingly intractable as the size of the system increases.

Given a particular temporal sequence, say at position 0, Eq. (3.3) uniquely determines the values of all sites in a triangle to the left as a function of values in the temporal sequence at position 1. The number of values in the position 1 temporal sequence on which a given site depends varies with the form of the position 0 sequence [32]. For example, if the position 0 sequence consists solely of ones, then the whole triangle of sites is completely determined, entirely independent of the position 1 sequence. Table 7.2 gives some results from considering the dependence of the site value a_{-t} at position $-t$ (the apex of the triangle) on the position 1 sequence, for all 2^t possible position 0 sequences. The number of values in the position 1 sequence on which a_{-t} depends seems to be roughly Poisson distributed, with a mean that grows like $0.4t$, as shown in Fig. 7.2. This is consistent with the combinatorial result (4.6).

n	$\langle \text{Var.} \rangle$	$\langle \text{P.I.} \rangle$	Max. P.I.
2	0.5	0.75	1
3	1	1.125	2
4	1.375	1.375	3
5	1.125	1.219	3
6	2.281	2.719	12
7	2.828	3.539	17
8	3.164	4.105	26
9	3.699		
10	4.254		

Table 7.2. Properties of Boolean expressions for left-most initial site values deduced from length n time sequences, obtained by evolution according to Eq. (3.1). The average number of variables appearing in the Boolean expressions is given, together with the number of prime implicants in the disjunctive normal form for the expression. The maximum number of variables which can appear is always $n - 1$. (Results for $n \geq 9$ were obtained by Carl Feynman using a Symbolics 3600 LISP machine. The entries left blank were not found.)

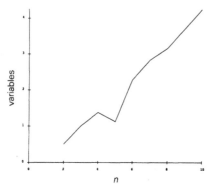

Figure 7.2. Average number of additional site values necessary to "back-track" and determine uniquely the initial site value a_{-n} given the sequence of values a_0 for n subsequent time steps.

Table 7.2 also gives some properties of the prime implicant forms for a_{-t}. It is clear that the complexity of the function that determines a_{-t} from temporal sequences grows with t, probably at an increasingly rapid rate. Again this suggests that the problem of deducing the initial sequence for evolution according to Eq. (3.1), while combinatorially possible, is computational complex.

By comparison, the corresponding problem for evolution according to the linear rule (2.4) is quite straightforward. For each possible position 0 sequence, there are only two possible forms for the dependence of a_{-t} on the position 1 sequence, and each of them involves exactly $2^{\#_1(t-1)}$ prime implicants. This simplicity can be viewed as a consequence of the algebraic structure associated with this system.

8. Computation Theoretical Properties

The discussion of the previous section can be considered as giving a characterization of the computational complexity of iterations of the cellular automaton mapping (3.1) in a particular simple model of computation. The results obtained suggest that at least in this model, there is no shortcut method for finding the outcome of the evolution: the computations required are no less than for an explicit simulation of each time step. As discussed above, one suspects in fact that the evolution is in general computationally irreducible, so that no possible computation could find its outcome more efficiently than by direct simulation.

This would be the case if the cellular automaton of Eq. (3.1) could act as an efficient universal computer (e.g., [33]), so that with an appropriate initial state, its evolution could mimic any possible computation. In particular, it could be that the problem of finding the value of a particular site after t steps (given say a simply-specified initial state, as in Fig. 6.1) must take a time polynomial in t on any computer. (Direct simulation takes $O(t^2)$ time on a serial-processing computer, and $O(t)$ time with $O(t)$ parallel processors.) For a linear cellular automaton such as that of Eq. (2.4), this problem can be solved in a time polynomial in $\log(t)$; but for the cellular automaton of Eq. (3.1) it quite probably cannot [18].

In addition to studying cellular automaton evolution from given initial configurations, one may consider the problem of deducing configurations of the cellular automaton from partial information such as temporal sequences. In particular, one may study the computational complexity of finding the seed for a cellular automaton in a finite region from the temporal sequences it generates.

There are 2^N possible seeds for a size N cellular automaton, and one can always find which ones produce a particular sequence by trying each of them in turn. Such a procedure would however rapidly become impractical. The results in Section 7 suggest a slightly more efficient method. If it were possible to find two adjacent temporal sequences, then the seed could be found easily using Eq. (3.3). Given only one temporal sequence, however, some elements of the seed are initially undetermined. Nevertheless, in a finite size system, say with periodic boundary conditions, one can derive many distinct equations for a single site value. The site value can then be deduced by solving the resulting system of simultaneous Boolean equations. The equations will however typically involve many variables. As discussed in Section 7, the number of variables seems to be Poisson-distributed with a mean around $0.4N$.

The general problem of solving a Boolean equation in n variables is NP-complete (e.g., [5]), and so presumably cannot be solved in a time polynomial in n. In addition, it seems likely that the average time to solve an arbitrary Boolean equation is correspondingly long. To relate the problem of deducing the seed discussed above to this would however require a demonstration that the Boolean equations generated were in a sense uniformly distributed over all possibilities. Out of all 2^{2^n} n-variable equations, the problem here typically involves $O(2^n)$, but these seem to have no special simplifying features. At least with the method discussed above, it is thus conceivable that the problem of deducing the seed is equivalent to the general problem of solving Boolean equations, which is NP-complete.

9. Finite Size Behaviour

Much of the discussion above has concerned the behaviour of the cellular automaton (3.1) in the idealized limit of an infinite lattice of sites. But practical implementations must use finite size registers, and certain global properties can depend on the size and boundary conditions chosen.

The total number of possible states in a size N cellular automaton is 2^N. Evolution between these states can be represented by a finite state transition diagram. Figure 9.1 gives some examples of such diagrams for the cellular automaton of Eq. (3.1) with periodic boundary conditions, as in Eq. (2.2). Table 9.1 summarizes some of their properties. The results are seen to depend not only on the magnitude of N, but also presumably on its number theoretical properties.

Each state transition diagram contains a set of cycles, fed by trees representing transients. The cycles may be considered as "attractors" to which states in their "basins of attraction" irreversibly evolve.

There are many regularities in the structure of the state transition diagrams obtained from Eq. (3.1). The evolution is thus not well-approximated by a random mapping between 2^N states.

N	Cycles	Frac. longest	Cyc. frac.	(Transient)
4	$1 \times 8, 3 \times 1$	0.75	0.69	0.5
5	$1 \times 5, 1 \times 1$	0.94	0.19	4.3
6	3×1	1.00	0.05	3.3
7	$1 \times 63, 7 \times 4, 1 \times 1$	0.60	0.72	0.4
8	$1 \times 40, 1 \times 8, 3 \times 1$	0.88	0.20	3.1
9	$1 \times 171, 1 \times 72, 1 \times 1$	0.81	0.48	1.1
10	$2 \times 15, 1 \times 5, 3 \times 1$	0.82	0.04	14.8
11	$1 \times 154, 11 \times 17, 1 \times 1$	0.76	0.17	3.3
12	$4 \times 102, 1 \times 8, 4 \times 3, 3 \times 1$	0.93	0.11	4.4
13	$1 \times 832, 1 \times 260, 1 \times 247, 1 \times 91, 1 \times 1$	0.32	0.17	2.2
14	$1 \times 1428, 2 \times 133, 1 \times 112, 2 \times 84, 1 \times 63, 1 \times 14, 3 \times 1$	0.84	0.13	2.7
15	$1 \times 1455, 5 \times 30, 5 \times 9, 15 \times 7, 4 \times 5, 1 \times 1$	0.93	0.05	5.7
16	$1 \times 6016, 1 \times 4144, 3 \times 40, 1 \times 8, 3 \times 1$	0.50	0.16	
17	$1 \times 10846, 1 \times 1632, 1 \times 867, 1 \times 306, 1 \times 136, 1 \times 17, 1 \times 1$	0.96	0.11	

Table 9.1. Properties of state transition diagrams for the cellular automaton rule of Eq. (3.1) in a circular register of size N. The multiplicity and length of each cycle is given, followed by the fraction of initial states which evolve to a longest cycle (size of attractor basin), the total fraction of all 2^N states which lie on cycles, and the average length of transient before a cycle is reached in evolution from an arbitrary initial state. (Results for $N \geq 16$ were obtained by Holly Peck.)

A first observation is that most configurations have unique predecessors under the mapping (3.1) (as mentioned for infinite lattices in Section 4), so there is little branching in the state transition diagram. In fact, it can be shown [32] that a configuration has a unique predecessor unless it contains a pair of value zero sites separated by a sequence of $3n + 1$ value one sites (with $n \geq 0$), or unless N is divisible by 3, and all sites have value one. In the former case, the configuration has exactly zero or two predecessors; in the latter case, it has three. The numbers of configurations with zero and two predecessors are equal when N is not divisible by

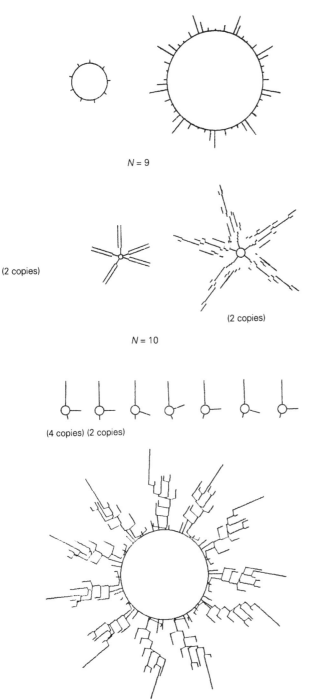

$N = 9$

(2 copies)

$N = 10$

(2 copies)

(4 copies) (2 copies)

$N = 11$

Figure 9.1. State transition diagrams for configurations of cellular automata evolving according to Eq. (3.1) in circular registers of size N. Each node represents one of the 2^N possible length N configurations, and is joined by an arc to its successor under the cellular automaton mapping. Transients corresponding to trees in the graph are seen ultimately to evolve to periodic cycles. Some properties of these state transition diagrams are given in Table 9.1. (Graphics by Steve Strassmann.)

3; there are two more with zero predecessors when $3|N$. For large N, the number of configurations with zero or two predecessors behaves as [32] κ^N, where $\kappa \simeq 1.696$ is the real root of $4\kappa^3 - 2\kappa^2 - 1 = 0$. Since the total number of configurations grows like 2^N, the fraction of nodes in the state transition diagram that are branch points thus tends exponentially to zero.

A second observation is that there are often many identical parts in the state transition diagrams of Table 9.1 and Fig. 9.1. This is largely a consequence of shift invariance. States in a cellular automaton with periodic boundary conditions that are related by shifts (translations) evolve equivalently. Thus, for example, there are often several identical cycles, related by shifts in their configurations. In addition, the periods of the cycles are often divisible by N or its factors, since they contain several sequences of configurations related by shifts. The transient trees that feed each of these sequences are then identical.

The evolution of a finite cellular automaton with periodic boundary conditions is equivalent to the evolution of an infinite cellular automaton with a periodic initial configuration. Thus the results on cycle length distributions in Table 9.1 can be considered as inverse to those in Table 6.2 on configurations with given temporal periods. Cycles of lengths corresponding to these temporal periods occur whenever N is divisible by the spatial periods of these configurations. Such short cycles are absent if N has none of these factors.

For large N, the state transition diagrams for Eq. (3.1) appear to be increasingly dominated by a single cycle. This cycle is longer than the others, and its basin of attraction is large enough that most arbitrarily chosen initial states evolve to it. The low degree of branching in the transient trees implies that the points reached from arbitrary initial states should be roughly uniformly distributed around the cycle.

The shorter cycles in Table 9.1 can be considered as related to subsets of states invariant under the cellular automaton rule. With N even, for example, configurations which consist of two identical length $N/2$ subsequences can evolve only to configurations of the same type. Once such a configuration has been reached, the evolution is "trapped" within this subset of configurations, and must yield shorter cycles. (This phenemonon also occurs for cellular automata with essentially trivial rules, such as the shift mapping $a_i' = a_j$. All states are on cycles in this case. The different cycles correspond to the possible "necklaces" with N beads of two kinds, which are inequivalent under shifts or rotations. These necklaces in turn correspond to cyclotomic polynomials; there are $\sum_{d|N} \phi(d) 2^{N/d}$ of them, where ϕ the Euler totient function (e.g., [4]).) In general, there may exist subsets of states with certain special symmetry properties that are preserved by the cellular automaton rule. Initial states with particular, symmetrical, forms can be expected to have these properties, and thus to be trapped in subsets of state space, and to yield short cycles. For example, with $N = 36$, a configuration containing a single nonzero site evolves to a length 2844 cycle, while most initial configurations evolve to the longest cycle, with 2237472 states.

In the infinite size limit, patterns such as that of Fig. 6.1 generated by the cellular automaton of Eq. (3.1) never become periodic. But with a total of N sites, a cycle must occur after 2^N or less steps. Table 9.2 and Fig. 9.2 give the actual maximal cycle lengths Π_N found. A roughly exponential increase of Π_N with N is seen, and a least squares fit to the data of Table 9.2 yields

$$\log_2 \Pi_N \simeq 0.61(N + 1). \tag{9.1}$$

Note that if the state transition diagram corresponded to an entirely random mapping between the 2^N cellular automaton states, then cycles of average length $2^{N/2}$ would be expected [34]. The cycles actually obtained are significantly longer. The exponent in Eq. (9.1) may be related to the entropy (4.6) as a result of the expansivity or instability of the mapping discussed in Section 5.

If there were very short cycles, then the sequences produced by the cellular automaton would readily be predictable. So if in fact no such prediction can be made by any polynomial time computation, the length of the cycles that occur should in general increase asymptotically faster than polynomial in N (cf. [2]). This behaviour is supported by Eq. (9.1).

If indeed the evolution of cellular automata such as (3.1) is computationally irreducible, then a complex computation may always be required to determine for example the lengths of cycles that appear. For in this case, there can effectively be no better way to find the succession of states that occur, except by explicit application of the rule (3.1). One expects in fact that the problem of finding say whether two configurations lie on the same cycle is PSPACE-complete, and so presumably cannot be solved in a time polynomial in N, but rather essentially requires a direct simulation of the cellular automaton evolution. (Note that if the lengths of the cycles studied are $O(2^M)$, where both 2^{N-M} and 2^M are large, then parallel processing is essentially of no avail in this problem.)

While the determination of cycle lengths and structures may be computationally intractable for cellular automata such as (3.1), it should be much easier for linear

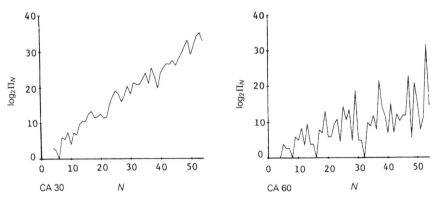

Figure 9.2. Maximal cycle lengths Π_N for the cellular automaton of Eqs. (3.1) (CA30) and (2.4) (CA60) in circular registers of size N.

	CA30		CA60	
N	Π_N	$\log_2 \Pi_N$	Π_N	$\log_2 \Pi_N$
4	8	3.0	1	0.0
5	5	2.3	15	3.9
6	1	0.0	6	2.6
7	63	6.0	7	2.8
8	40	5.3	1	0.0
9	171	7.4	63	6.0
10	15	3.9	30	4.9
11	154	7.3	341	8.4
12	102	6.7	12	3.6
13	832	9.7	819	9.7
14	1428	10.5	14	3.8
15	1455	10.5	15	3.9
16	6016	12.6	1	0.0
17	10845	13.4	255	8.0
18	2844	11.5	126	7.0
19	3705	11.9	9709	13.2
20	6150	12.6	60	5.9
21	2793	11.4	63	6.0
22	3256	11.7	682	9.4
23	38249	15.2	2047	11.0
24	185040	17.5	24	4.6
25	588425	19.2	25575	14.6
26	312156	18.3	1638	10.7
27	67554	16.0	13797	13.7
28	249165	17.9	28	4.8
29	1466066	20.5	475107	18.9
30	306120	18.2	30	4.9
31	2841150	21.4	31	5.0
32	2002272	20.9	1	0.0
33	2038476	21.0	1023	10.0
34	5656002	22.4	510	9.0
35	18480630	24.1	4095	12.0
36	2237472	21.1	252	8.0
37	49276415	25.6	3233097	21.6
38	9329228	23.2	19418	14.2
39	961272	19.9	4095	12.0
40	19211080	24.2	120	6.9
41	51151354	25.6	41943	15.4
42	109603410	26.7	126	7.0
43	93537212	26.5	5461	12.4
44	192218312	27.5	1364	10.4
45	75864495	26.2	4095	12.0
46	261598274	28.0	4094	12.0
47	811284813	29.6	8388607	23.0
48	3035918676	31.5	48	5.6
49	9937383652	33.2	2097151	21.0
50	593487780	29.1	51150	15.6
51	3625711023	31.8	255	8.0
52	20653434880	34.3	3276	11.7
53	40114679273	35.2	3556769739	31.7
54	7551779562	32.8	27594	14.8

Table 9.2. Maximum cycle lengths Π_N found for the cellular automata of Eqs. (3.1) (CA30) and (2.4) (CA60) in circular registers of size N. In the former case, a selection of seeds, including single nonzero sites, were used. In the latter case, maximal length cycles are always obtained with single nonzero site seeds. The results are plotted in Fig. 9.2. (Results for $N \geq 32$ were obtained by Holly Peck and Tsutomu Shimomura with an assembly-language program on a Celerity C-1200 computer.)

cases such as (2.4). From the algebraic theory of these systems it is possible to show for example that the maximal cycle length Π_N satisfies [20]

$$\Pi_N | 2^{\mathrm{ord}_N(2)} - 1, \tag{9.2}$$

where $n|m$ states that the integer n exactly divides m. Here $\mathrm{ord}_N(k)$ is the multiplicative order function, equal to the minimum integer j such that $k^j = 1 \bmod N$. This function divides the totient function $\phi(N)$ (equal to the number of integers less than N which are relatively prime to N), which is maximal for prime N. Table 9.2 and Fig. 9.2 give the actual maximal periods found in this case. Equation (9.2) rarely holds as an equality, and the Π_N found are usually much shorter than the corresponding ones for the nonlinear rule (3.1).

The cycle structures of finite cellular automata depend in detail on the boundary conditions chosen. Table 9.3 gives the maximal cycle lengths found for rules (3.1)

N	CA30		CA60	
	Π_N	$\log_2 \Pi_N$	Π_N	$\log_2 \Pi_N$
4	5	2.3	15	3.9
5	2	1.0	21	4.4
6	7	2.8	21	4.4
7	4	2.0	127	7.0
8	17	4.1	63	6.0
9	65	6.0	73	6.2
10	6	2.6	889	9.8
11	57	5.8	1533	10.6
12	50	5.6	1085	10.1
13	118	6.9	7905	12.9
14	185	7.5	11811	13.5
15	257	8.0	32767	15.0
16	481	8.9	255	8.0
17	907	9.8	273	8.1
18	1681	10.7	253921	18.0
19	707	9.5	413385	18.7
20	2679	11.4	761763	19.5
21	5630	12.5	5461	12.4
22	1368	10.4	4194303	22.0
23	31241	14.9	2088705	21.0
24	3567	11.8	2097151	21.0
25	60503	15.9	2192337	21.1
26	4752	12.2	22995	14.5
27	46519	15.5	41943035	25.3
28	35569	15.1	17895697	24.1
29	207197	17.7		
30	149899	17.2		
31	482717	18.9		

Table 9.3. Maximum cycle lengths Π_N found for the cellular automata of Eqs. (3.1) (CA30) and (2.4) (CA60) in shift registers of size N (with boundary conditions given by Eq. (2.3)).

and (2.4) with shift register boundary conditions. The results differ substantially from those with periodic boundary conditions given in Table 9.2. One notable feature is the presence of length $2^N - 1$ cycles in the linear cellular automaton (2.4) for certain N. These correspond to maximal length linear feedback shift registers, and can be identified by a direct algebraic procedure [4].

Other boundary conditions may also be considered. Among them are twisted ones, in which the sites a_1 and a_N are negated in Eq. (2.2). The maximum cycle lengths found with such boundary conditions seem typically shorter than in the purely periodic case.

One may in addition consider boundary conditions in which the boundary site values are fixed, rather than being periodically identified. Section 6 (particularly Fig. 6.2) gave some examples of results with such boundary conditions. Different cycles are obtained in different cases; all those investigated nevertheless give maximal cycle lengths shorter than those of Table 9.2 found with periodic boundary conditions.

What has been discussed so far are cycles in complete finite cellular automaton configurations. But in obtaining random sequences one samples single sites. The sequences found could potentially have periods which were sub-multiples of the periods for the complete configuration. For permutive rules such as (3.1) (or (2.4)) this cannot, however, occur.

The state transition diagrams summarized in Table 9.1 give the number of complete N-site configurations that can occur at various stages in the evolution of the cellular automaton (3.1). One may also consider the number of single site temporal sequences that can occur. Table 9.4 gives the fraction of the 2^L possible length L temporal sequences that are actually generated from any of the 2^N possible initial states in a size N cellular automaton evolving according to Eq. (3.1) (with periodic boundary conditions). The results are plotted in Fig. 9.3. Whenever $N \gtrsim L + 2$, all possible sequences seem to be generated. They appear with roughly equal frequencies.

10. Statistical Properties

The sequences generated by the cellular automaton of Eq. (3.1) may be considered effectively random if no feasible procedure can identify a pattern in them, or allow their behaviour to be predicted. Even though it may not be possible to prove that no such procedure can exist, circumstantial evidence can be accumulated by trying various statistical procedures and finding that they reveal no regularities. The basic approach is to compare statistical results on sequences generated by (3.1) with those calculated for sequences whose elements occur purely according to probabilities.

To establish the validity of (3.1) as a general-purpose random sequence generator, one should apply a variety of statistical procedures, related to various different kinds of calculations. The choice of tests is necessarily as ad hoc as the choice of calculations done. Appendix A lists those used here. (But see also [35].) Some can be considered related to Monte Carlo simulations of physical and other systems. Others

L	3	4	5	6	7	8	9	10	11	12	13	14	15
3	0.500	1.000	1.000	1.000	1.000	1.000	1.000	1.000	1.000	1.000	1.000	1.000	1.000
4	0.250	0.625	0.875	0.938	1.000	1.000	1.000	1.000	1.000	1.000	1.000	1.000	1.000
5	0.125	0.313	0.656	0.844	1.000	1.000	1.000	1.000	1.000	1.000	1.000	1.000	1.000
6	0.063	0.156	0.344	0.594	0.906	1.000	1.000	1.000	1.000	1.000	1.000	1.000	1.000
7	0.031	0.078	0.180	0.352	0.609	0.891	1.000	1.000	1.000	1.000	1.000	1.000	1.000
8	0.016	0.039	0.094	0.188	0.328	0.633	0.949	0.992	1.000	1.000	1.000	1.000	1.000
9	0.008	0.020	0.047	0.094	0.168	0.361	0.668	0.895	0.996	1.000	1.000	1.000	1.000
10	0.004	0.010	0.023	0.047	0.085	0.195	0.386	0.644	0.917	0.989	1.000	1.000	1.000
11	0.002	0.005	0.012	0.023	0.042	0.102	0.204	0.377	0.666	0.897	0.995	1.000	1.000
12	0.001	0.002	0.006	0.012	0.021	0.052	0.105	0.204	0.387	0.651	0.911	0.995	1.000
13	0.000	0.001	0.003	0.006	0.011	0.026	0.054	0.105	0.209	0.385	0.669	0.913	0.995
14	0.000	0.001	0.001	0.003	0.005	0.013	0.027	0.053	0.109	0.209	0.397	0.671	0.906
15	0.000	0.000	0.001	0.001	0.003	0.007	0.013	0.027	0.055	0.109	0.215	0.399	0.668

Table 9.4. Fraction of length L temporal sequences generated from all possible seeds by evolution according to Eq. (3.1) in a length N circular register. Results for successive values of N are given in successive columns. The results are plotted in Fig. 9.3.

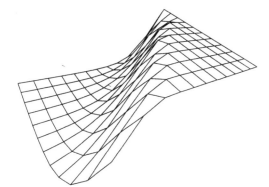

Figure 9.3. Fraction of length L sequences obtained by evolution from all possible seeds according to Eq. (3.1) in a size N circular register. The three-dimensional view is from the point $N = L = 20$, with elevation 2.

to statistical analyses that would be done on data from various kinds of measurements. While quite ad hoc, the tests seem to be sensitive, and reasonably independent.

As an example, consider the "equidistribution" or "frequency" test. If a sequence of zeroes and ones is to be random, the digits zero and one must occur in it with equal frequency. In general, in fact, all 2^n possible length n blocks of digits must also occur with equal frequency. (The measure entropy of (4.2) is maximal exactly when such equidistribution occurs.) However, in a finite sample of length m, there are expected to be statistical fluctuations, which lead to slightly different numbers of zeroes and ones. (The value of entropy deduced from a finite sample is thus almost always not maximal, even if it would be maximal were the sequence to be continued forever.) As a consequence, one can never definitively conclude by studying a finite sample that the complete sequence is not random. One can however calculate the probability that a truly random sequence would have the properties seen in the finite sample.

To do this (e.g., [36]), one evaluates χ^2, defined in terms of the observed and expected frequencies p_0 and p_e as

$$\chi^2 = \sum_{1}^{v} (p_0 - p_e)^2 / p_e. \tag{10.1}$$

Here v gives the number of degrees of freedom, or number of distinct objects whose frequencies are included in the sum. If blocks of length n are studied then $v = 2^n$. Now one must find the probability that a value of χ^2 larger than that observed would occur for a random sequence. This "confidence interval" is obtained immediately from the integral of the χ^2 distribution (e.g., [36]).

If the confidence interval is very close to zero or one, then the observed χ^2 is unlikely to be produced from a random sequence, and one may infer that the observed sequence is not random. Of course, if say a total of k tests are done, it is to be expected that the confidence interval for at least one of them will be less than $1/k$. Evidence for nonrandomness in a sequence must come from an excess of confidence interval values close to zero or one, over and above the number expected for a uniform distribution.

Table 10.1 gives results from the statistical tests described in Appendix A for sequences generated by the cellular automaton (3.1) in a finite circular register. Except when the sample sequence is comparable in length to the period of the system, as given by Table 9.2, no significant deviations from randomness are found.

	CA30 $N = 17$ $L = 8k$	CA30 $N = 17$ $L = 64k$	CA30 $N = 23$ $L = 64k$	CA30 $N = 29$ $L = 64k$	CA30 $N = 37$ $L = 64k$	CA30 $N = 49$ $L = 64k$
A	**0.0039**	**1.0000**	**0.0456**	0.7375	0.3852	0.8003
B	**0.0171**	**0.9944**	0.3391	0.4888	0.1010	0.1494
C	0.4164	0.4783	0.7256	0.4847	0.4083	0.9407
D	0.3227	**0.9998**	0.1506	0.1434	0.1678	0.6074
E	0.4576	0.4484	0.6790	0.8492	0.5414	0.7991
F	0.4306	0.8644	0.8751	0.5590	0.6681	0.6606
G	0.2942	**0.9944**	0.1232	0.7359	0.4448	0.6961

Table 10.1. Results of the statistical tests described in Appendix A for sequences of length L ($k = 1024$) generated by the cellular automaton of Eq. (3.1) (rule number 30) in circular registers of length N. In each case, the seed used consists of a single nonzero site. The numbers given are the probabilities (confidence intervals) for statistical averages of truly random sequences to exceed those of the sequences analysed. The numbers should be uniformly distributed between 0 and 1 if the sequences analysed are indeed truly random. Results below 0.05 and above 0.95 are shown in bold type. Accumulations close to 0 or 1 suggest deviations from randomness. Such accumulations are seen in this case only when the period of the cellular automaton is comparable to the length of the sequence sampled. (The statistical test programs used here were written in C by Don Mitchell.)

	CA60 $N = 29$ $L = 64k$	LFSR $N = 17$ $L = 64k$	LFSR $'N = 29$ $L = 64k$	LCG $N = 32$ $L = 64k$	$\sqrt{2}$ $L = 51906k$	e $L = 9501k$	π $L = 26755k$
A	**1.0000**	**0.0390**	**0.9998**	**0.0167**	0.6255	0.5505	0.1441
B	**1.0000**	**0.9773**	0.4378	0.0841	0.0801	0.4556	**0.9525**
C	**1.0000**	0.2654	**1.0000**	0.1676	0.0582	0.8615	0.2799
D	**1.0000**	0.8797	0.8400	0.8322	0.8553	0.7605	**0.9986**
E	0.9256	**1.0000**	0.9435	0.5850	0.6363	0.6890	**0.0049**
F	**0.9998**	**1.0000**	**0.9674**	0.9248	0.8499	0.7031	0.1297
G	**1.0000**	**0.9790**	0.3476	0.3137	0.8465	0.4086	0.5473

Table 10.2. Results of statistical tests for sequences generated by various procedures. CA60 is the linear cellular automaton rule of Eq. (2.4), in a size N circular register. LFSR is a linear feedback shift register of length N with period $2^N - 1$. For $N = 17$ the shift register taps are at positions 14 and 17; for $N = 29$ they are at positions 27 and 29. For CA60 and LFSR seeds consisting of a single nonzero site were used. LCG is the linear congruential generator $x' = (1103515245x + 12345)$ mod 2^{31} (used, for example, in many implementations of the UNIX operating system). The seed $x = 1$ was used. The behaviour of CA60, LFSR and LCG are illustrated in Fig. 11.1. $\sqrt{2}$, e and π are the binary digit sequences of the square root of two, the exponential constant, and pi, respectively. (These digit sequences were obtained by R. W. Gosper using a Symbolics 3600 LISP machine.)

Table 10.2 gives statistical results for sequences generated by other procedures. Those obtained from linear feedback shift registers, while provably random in some respects (e.g., [4]), are revealed as significantly nonrandom by several of the tests used here. Many sequences obtained from linear congruential generators are also found to be significantly nonrandom with respect to these tests. No regularities are detected in the digit sequence of $\sqrt{2}$ (and other surds tried) (cf. [37]). There is, however, some possible evidence for nonrandomness in the digit sequences of e and π (cf. [38]). (This will be explored elsewhere.)

Table 10.3 gives statistical results for temporal sequences in the pattern of Figure 6.1 obtained by evolution according to Eq. (3.1) from a single nonzero initial site on an infinite lattice. Once again, no significant deviations from randomness are seen.

If deviations from randomness were detected by some statistical procedure, then this procedure could be used to make statistical predictions about the sequence. In addition, it could be used to obtain a compressed representation for the sequence, and would thus demonstrate that the sequence did not have maximal information content. The fact that deviations from randomness have not been found by any of the statistical procedures considered lends strong support to the belief that sequences produced by Eq. (3.1) with large N are indeed random for practical purposes.

	$i = 0$ $L = 8k$	$i = 0$ $L = 64k$	$i = 0$ $L = 512k$	$i = 1$ $L = 512k$	$i = -1$ $L = 512k$	$i = 32$ $L = 512k$	$i = -32$ $L = 512k$
A	0.1536	0.2234	0.6453	0.8629	0.8630	0.8733	0.2677
B	0.5996	0.0637	0.4891	0.7639	0.8343	0.2525	0.1751
C	0.6448	0.6538	0.5443	0.5887	0.4000	0.8271	0.8815
D	0.5921	0.2643	**0.0051**	**0.0105**	0.7030	0.4550	0.7832
E	0.1358	0.1348	0.6631	0.8430	0.7498	0.1264	0.8353
F	0.2622	0.1957	0.9385	0.4324	0.9009	0.4736	0.8022
G	0.4542	0.8773	0.6658	0.1080	0.7169	0.7744	0.2364

Table 10.3. Results of statistical tests for vertical sequences at position i in the pattern of Fig. 6.1 generated by evolution according to Eq. (3.1) from a single nonzero initial site on an infinite lattice. Leading zeroes in each sequence were truncated. (The sequences were obtained by Jim Salem using a prototype Connection Machine computer.)

11. Practical Implementation

The simplicity and intrinsic parallelism of the cellular automaton rule (3.1) makes possible efficient implementation on many kinds of computers.

On a serial-processing computer, each site could be updated in turn according to (3.1). But in practice, site values can be represented by single bits in say a 32-bit word, and updated in parallel using standard word-wise Boolean operations. (Additional bit-wise operations are often needed for boundary conditions.)

On a synchronous parallel-processing computer, different sites or groups of sites in the cellular automaton can be assigned to different processors. They can then be updated independently (though synchronously), using the same instructions, and with only local communications.

Very efficient hardware implementations of (3.1) should also be possible. For short registers, explicit circuitry can be included for each site. And for long registers, a pipelined approach analogous to a feedback shift register can be used (cf. [39]).

The evidence presented above suggests that the cellular automaton of Eq. (3.1) can serve as a practical random sequence generator. The most appropriate detailed choices of parameters depend on the application intended. The most obvious constraint is one of cycle length. To obtain a cycle length larger than $2^{32} \simeq 4 \times 10^9$, Table 9.2 shows that a circular register of length $N = 49$ can be used. Cycle lengths tend to increase with N, but Table 9.2 shows some irregularities. Thus it is not clear, for example, how large N need be to obtain a cycle length larger than $2^{64} \simeq 10^{19}$. But based on Eq. (9.1), a value $N = 127$ should certainly suffice.

Random sequences can be obtained by sampling the sequence of values of a particular site in a register updated according to Eq. (3.1). The theoretical and statistical studies described above support the contention that such sequences show

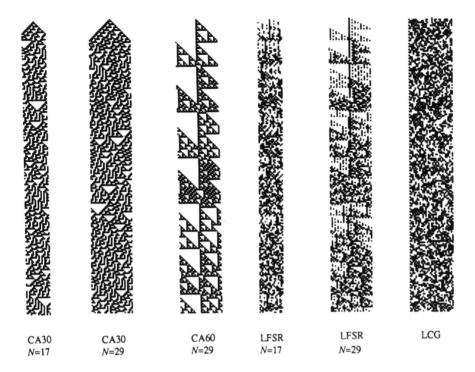

CA30	CA30	CA60	LFSR	LFSR	LCG
N=17	N=29	N=29	N=17	N=29	

Figure 11.1. Patterns obtained by various procedures in registers of size N. CA30 stands for the cellular automaton of Eq. (3.1), with periodic boundary conditions. CA60 is the linear cellular automaton of Eq. (2.4), again with periodic boundary conditions. LFSR is a linear feedback shift register with size N and period $2^N - 1$. For $N = 17$ the taps are at positions 14 and 17; for $N = 29$, they are at positions 27 and 29. LCG is a linear congruential sequence generator, operating on the 32-bit integers whose binary digit sequences are given. The seed in all cases consists of a single nonzero bit in the center of the register. Statistical properties of the sequences produced are given in Tables 10.1 and 10.2.

no regularities. For some critical applications, it may be best however, to sample site values only say on alternate time steps. While this method generates a sequence more slowly, it should foil prediction procedures along the lines discussed in Section 7.

Sequences could potentially be obtained more quickly by extracting the values of several sites in the register at each time step. But Eq. (4.6) implies that some statistical correlations must exist between these values. The correlations are probably minimized if the sites sampled are equally spaced around the register. Nevertheless, in some applications where only a low degree of randomness is needed, it may even be satisfactory to use all site values in the register. (An example appears to be approximation of partial differential equations, where randomness can be used to emulate additional low-order digits.)

The random sequences obtained from Eq. (3.1) have an equal fraction of 0 and 1. Many applications, however, involve random binary choices with unequal probabilities. There is nevertheless a simple algorithm [40] to obtain digits with arbitrary probabilities. First write the probability p for outcome 1 as a binary number. Then generate a random binary sequence s with a length equal to this number. The output is obtained by an iterative procedure. Begin with a "current result" of 1. Then, starting from the least significant digit in p, successively find a new result by combining the old result with the corresponding digit of s, using a function AND or OR, depending on whether the digit in p is 0 or 1, respectively. The final result thus obtained is equal to 1 with probability exactly p.

Configurations in two length N registers with slightly different seeds should become progressively less correlated under the action (3.1) as a result of the instability discussed in Section 5. The characteristic time for this process is governed by Eqs. (5.1) and (5.2), and should be $\simeq 0.8N$. Thus, if several sequences are to be generated with seeds that differ only slightly (obtained for example from addresses of computer elements), then (3.1) should be applied at least $O(N)$ times to the seeds before beginning to extract random sequences.

One may compare the scheme for random sequence generation described here with the linear methods now in common use (e.g., [1]). Figure 11.1 shows patterns produced by these various schemes. The primary feature of linear schemes is that they can be analysed by algebraic methods. As a consequence, certain randomness properties can be proved for the sequences they generate, and cases that give long cycles can be identified. But the simplicity in structure which underlies this analysis also limits the degree of randomness that such schemes can produce. The nonlinear scheme described here is not readily amenable to complete analysis, and no significant limits on the degree of randomness it yields are known. But on the other hand, no conventional mathematical proofs for particular randomness properties can be given, and it must be investigated by largely empirical methods.

12. Alternative Schemes

The cellular automaton of Eq. (3.1) is one of the simplest that seems good for random sequence generation. But other cellular automata may also be considered, and some potentially have certain advantages.

Among $k = 2$, $r = 1$ cellular automata, Eq. (3.2) is the only other serious contender. No direct equivalence between this rule and that of Eq. (3.1) is known, but their properties are very similar. Equation (3.2) gives however [45]

$$\lambda_L = (0.1724 \pm 0.0004), \tag{12.1}$$

slightly smaller than the corresponding result (5.2) for Eq. (3.1). In addition, it gives a slightly smaller invariant entropy $\mathbf{h}_\mu$. It seems to have no advantages over (3.1).

Cellular automata with $k > 2$ or $R > 3$ may also be studied. (Here R is defined as the total number of sites in the neighbourhood for the rule.) Any class 3 (chaotic) cellular automaton rule can be considered a candidate random sequence generator.

ϕ	ϕ^{-1}
$k = 2, R = 4$	
1kng	1kng
1s5k	1s5k
1hmc	1hmc
1j4s	1j4s
$k = 2, R = 5$	
3nh1vo0	3nh1vo0
3ug5vo0	3ug5vo0
39gtvo0	f20nv1jogtvo0

Table 12.1. Bijective cellular automata rules with k possible values for each site and depending on strictly R previous site values. The rules given are "totally quiescent," so that $\phi(a, a, \ldots, a) = a$ for all a. The rules are specified by giving the values of ϕ as digits in a binary number indexed by a number formed from the arguments of ϕ. The binary number is then stated in base 32, with letters of the alphabet representing successive digits greater than 9. Leading zeroes are not truncated. Long specifications correspond to rules with larger values of R.

Autoplectic rules which produce complex patterns even from simple initial conditions are probably best. Some of these rules have larger Lyapunov exponents and invariant entropies than Eq. (3.1), but they are also more difficult to compute. In addition, many rules that seem to produce chaotic overall patterns nevertheless yield sequences that show definite regularities, resulting, for example, in non-maximal temporal entropies. Permutive chaotic rules avoid such problems, but are very similar in character to the rule of Eq. (3.1), and so potentially share any of its possible deficiencies.

One possibility is to consider bijective cellular automaton rules, which are invertible, so that each configuration has both a unique successor in time, and a unique predecessor. The state transition diagrams for such cellular automata in finite regions with periodic boundary conditions can contain only cycles, and no transients. But only a very small fraction of all cellular automaton rules are bijective, and very few of those that are exhibit chaotic behaviour. Table 12.1 gives some non-trivial bijective

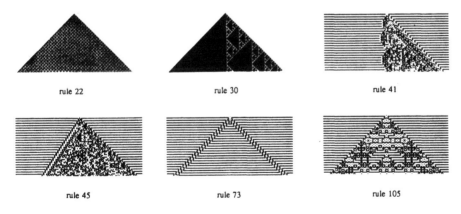

Figure 12.1. Patterns generated by various bijective (reversible) $k = 2$, $r = 1$ cellular automata with rules of the form (12.2).

cellular automaton rules with $k = 2$ and $R \leq 5$ (cf. [41]). None of those with $R \leq 4$ are chaotic.

With larger effective k, it is nevertheless possible to construct chaotic bijective rules explicitly. One method [42] yields cellular automaton rules that are most easily stated in terms of dependence on second-to-last as well as immediately preceding site values:

$$a_i^{(t)} = \phi(a_{i-r}^{(t-1)}, \cdots, a_{i+r}^{(t-1)}) \text{ XOR } a_i^{(t-2)}. \tag{12.2}$$

Such rules may be stated in the standard form (2.1) by considering sites with k^2 possible values. Some examples of patterns generated by rules of the form (12.2) are shown in Fig. 12.1. The rules are bijective, so that all states lie on cycles. However, there are often many distinct cycles, each quite short, making the system unsuitable for random sequence generation.

13. Discussion

This paper has used methods from several disciplines to study the behaviour of the nonlinear cellular automaton of Eq. (3.1). Despite the simplicity of its construction, all the approaches taken support the conjecture that its behaviour is so complicated as to seem random for practical purposes. It is remarkable that such a simple system can give rise to such complexity. But it is in keeping with the observation that mathematical systems with few axioms, or computers with few intrinsic instructions, can lead to essentially arbitrary complexity. And it seems likely that the mathematical mechanisms at work are also responsible for much of the randomness and chaos seen in nature.

The simplicity of Eq. (3.1) makes it amenable to highly efficient practical implementation. And the analyses carried out here suggest that the sequences it produces have a high degree of randomness. In fact, if any regularity could be found in these sequences, it would probably have substantial consequences for studies of many complex and seemingly random phenomena.

Appendix A:
Statistical Procedures

This Appendix describes the statistical randomness testing procedures used in Section 10. The procedures are mostly taken from [1], although their numbering has been changed slightly. The basic method in each case is to compare an observed distribution with that calculated for a purely probabilistic sequence.

The sequences studied consist of strings of binary bits. In many of the tests, these bits are grouped into blocks: either length 8 (non-overlapping) bytes, or length 4 (non-overlapping) nybbles. The possible bit sequences in these blocks can be represented by integer "values" between 0 and 255 or 16, respectively.

A. Block Frequency Distribution. Each of the 2^n possible n-blocks should occur with equal frequency. ($n = 8$ is used.)

B. Gap Length Distribution. The lengths of runs of n-blocks whose values are all greater than i_2 or less than i_1 should follow a binomial distribution. ($n = 8$, $i_1 = 100$, $i_2 = 200$ are used; runs longer than 16 blocks are lumped together.)

C. Distinct Blocks Distribution. The frequencies with which p out of q successive m-blocks are distinct should follow a definite distribution. ($m = 4$, $q = 4$ are used.)

D. Block Accumulation Distribution. The number of successive n-blocks necessary for all possible m-blocks to appear in order as their first m elements should follow a definite distribution. ($n = 8$, $m = 3$ are used; numbers greater than 40 are lumped together.)

E. Permutation Frequency Distribution. The values of q successive n-blocks should occur in all $q!$ possible orderings with equal frequency. ($n = 8$, $q = 5$ are used.)

F. Monotone Sequence Length Distribution. The lengths of sequences in which successive n-blocks have monotonically increasing values should follow a definite distribution. ($n = 8$ is used; lengths greater than 6 are lumped together; elements immediately following each run are discarded to make successive runs statistically independent.)

G. Maxima Distribution. The maximum values of n-blocks in sequences of q n-blocks should follow a power law distribution. ($n = 8$, $q = 8$ are used.)

Acknowledgments

Many people have contributed in various ways to the material presented here. For specific suggestions I thank: Persi Diaconis, Carl Feynman, Richard Feynman, Shafi Goldwasser, Peter Grassberger, Erica Jen, and John Milnor.

For discussions I thank: Lenore Blum, Manuel Blum, Whit Diffie, Rolf Fiebrich, Danny Hillis, Doug Lind, Silvio Micali, Marvin Minsky, Andrew Odlyzko, Steve Omohundro, Norman Packard, and Jim Reeds.

For help with computational matters I thank: Keira Bromberg, Bill Gosper, Don Mitchell, Bruce Nemnich, Holly Peck, Jim Salem, Tsutomu Shimomura, Steve Strassmann, and Don Webber.

The computer mathematics system SMP [43] was used for some of the calculations. I thank the Science Office of Sun Microsystems for the loan of a SUN workstation on which most of the graphics and many of the calculations were done. And finally I thank Thinking Machines Corporation for the use of a prototype Connection Machine computer [44], without which much more about the cellular automaton of Eq. (3.1) would still be unknown.

Note added in proof. Eq. (3.1) can also be used to generate efficiently a key sequence for stream encryption [46].

References

1. D. Knuth, "Seminumerical Algorithms," Addison-Wesley, Reading, Mass., 1981.

2. A. Shamir, "On the generation of cryptographically strong pseudorandom sequences," Lecture Notes in Computer Science Vol. 62, p. 544, Springer-Verlag, New York/Berlin, 1981; S. Goldwasser and S. Micali, Probabilistic encryption, *J. Comput. System Sci.* **28** (1984), 270; M. Blum and S. Micali, How to generate cryptographically strong sequences of pseudorandom bits, *SIAM J. Comput.* **13** (1984), 850; A. Yao, Theory and applications of trapdoor functions, in "Proc. 23rd IEEE Symp. on Foundations of Computer Science," 1982.

3. G. Chaitin, On the length of programs for computing finite binary sequences, I, II, *J. Assoc. Comput. Mach.* **13** (1966), 547; **16** (1969), 145, Randomness and mathematical proof, *Sci. Amer.* **232**, No. 5 (1975), 47; A. N. Kolmogorov, Three approaches to the concept of "the amount of information," *Problems Inform. Transmission* **1**, 1 (1965); R. Solomonoff, A formal theory of inductive inference, *Inform. Control* **7** (1964), 1; P. Martin-Lof, The definition of random sequences, *Inform. Control* **9** (1966), 602; L. Levin, On the notion of a random sequence, *Soviet Math. Dokl.* **14** (1973), 1413.

4. S. W. Golomb, "Shift Register Sequences," Holden-Day, San Francisco, 1967.

5. M. Garey and D. Johnson, "Computers and Intractability: A Guide to the Theory of NP-Completeness," W. H. Freeman, San Francisco, 1979.

6. L. Blum, M. Blum, and M. Shub, Comparison of two pseudorandom number generators, in "Advances in Cryptology: Proc. of CRYPTO-82" (D. Chaum, R. Rivest, and A. T. Sherman, Eds.), Plenum, New York, 1983.

7. W. Alexi, B. Chor, O. Goldreich, and C. Schnorr, RSA/Rabin bits are $\frac{1}{2} + 1/\text{poly}(\log N)$ secure, in "Proc. Found. Comput. Sci.," (1984); U. Vazirani and V. Vazirani, Efficient and secure pseudorandom number generation, in "Proc. Found. Comput. Sci.," (1984).

8. L. Kuipers and H. Niederreiter, "Uniform Distribution of Sequences," Wiley, New York, 1974.

9. J. Lagarias, The $3x + 1$ problem and its generalizations, *Amer. Math. Monthly* **92** (1985), 3.

10. K. Mahler, An unsolved problem on the powers of $\frac{3}{2}$, *Proc. Austral. Math. Soc.* **8** (1968), 313; G. Choquet, Repartition des nombres $k(\frac{3}{2})^n$; mesures et ensembles associes, *C. R. Acad. Sci. Paris A* **290** (1980), 575.

11. S. Wolfram, Origins of randomness in physical systems, *Phys. Rev. Lett.* **55** (1985), 449.

12. S. Wolfram, Cellular automata as models of complexity, *Nature* **311** (1984), 419.

13. D. Farmer, T. Toffoli, and S. Wolfram, (Eds.), Cellular automata, *Physica D* **10** Nos. 1, 2, (1984).

14. S. Wolfram, Cellular automata and condensed matter physics, in Proc. NATO Advanced Study Institute on Scaling phenomena in disordered systems, April 1985.

15. S. Wolfram, Universality and complexity in cellular automata, *Physica D* **10** (1984), 1.

16. S. Wolfram, Computation theory of cellular automata, *Commun. Math. Phys.* **96** (1984), 15.

17. S. Wolfram, Statistical mechanics of cellular automata, *Rev. Modern Phys.* **55** (1983), 601.

18. S. Wolfram, Undecidability and intractability in theoretical physics, *Phys. Rev. Lett.* **54** (1985), 735.

19. S. Wolfram, Computer software in science and mathematics, *Sci. Amer.* **251** September 1984.

20. O. Martin, A. Odlyzko, and S. Wolfram, Algebraic properties of cellular automata, *Comm. Math. Phys.* **93** (1984), 219.

21. J. Milnor, Notes on surjective cellular automaton-maps, Institute for Advanced Study preprint, June 1984.

22. E. Jen, "Global Properties of Cellular Automata," Los Alamos report LA-UR-85-1218, 1985; *J. Stat. Phys.*, in press.

23. J. Guckenheimer and P. Holmes, "Nonlinear Oscillations, Dynamical Systems, and Bifurcations of Vector Fields," Springer-Verlag, New York/Berlin, 1983.

24. J. Milnor, Entropy of cellular automaton-maps, Institute for Advanced Study preprint, May 1984; Directional entropies of cellular automaton maps, Institute for Advanced Study preprint, October 1984.

25. Ya. Sinai, An answer to a question by J. Milnor, *Comment. Math. Helv.* **60** (1985), 173.

26. N. Packard, Complexity of growing patterns in cellular automata, *in* "Dynamical systems and cellular automata," (J. Demongeot, E. Goles, and M. Tchuente, Eds.), Academic Press, New York, 1985.

27. R. Feynman, private communication.

28. R. Brayton, G. Hachtel, C. McMullen, and A. Sangiovanni-Vincentelli, "Logic Minimization Algorithms for VLSI Synthesis," Kluwer, Boston, 1984.

29. R. Rudell, "*Espresso* software program," Computer Science Dept., University of California, Berkeley, 1985.

30. S. Wolfram, Geometry of binomial coefficients, *Amer. Math. Monthly* **91** (1984), 566.

31. M. Furst, J. Saxe, and M. Sipser, Parity, circuits, and the polynomial-time hierarchy, *Math Systems Theory* **17** (1984), 13.

32. C. Feynman and R. Feynman, private communication.

33. M. Minsky, "Computation: Finite and Infinite Machines," Prentice-Hall, Englewood Cliffs, N.J., 1967.

34. B. Harris, Probability distributions related to random mappings, *Ann. Math. Statist.* **31** (1960), 1045.

35. G. Marsaglia, A current view of random number generators, *in* "Proc. Computer Sci. and Statistics, 16th Sympos. on the Interface," Atlanta, March 1984.

36. G. W. Snedecor and W. G. Cochran, "Statistical Methods," Iowa State Univ. Press, Ames, 1967.

37. W. Beyer, N. Metropolis and J. R. Neergaard, Statistical study of digits of some square roots of integers in various bases, *Math. Comput.* **24** (1970), 455.

38. S. Wagon, Is π normal?, *Math. Intelligencer* **7** (1985), 65.

39. T. Toffoli, CAM: A high-performance cellular-automaton machine, *Physica D* **10** (1984), 195; K. Steiglitz and R. Morita, A multi-processor cellular automaton chip, *in* "Proc. 1985 IEEE International Conf. on Acoustics, Speech, and Signal Processing," March 1985.

40. J. Salem, Thinking Machines Corporation report, to be published.

41. G. Hedlund, Endomorphisms and automorphisms of the shift dynamical system, *Math. Systems Theory* **3** (1969), 320; G. Hedlund, private communication.

42. N. Margolus, Physics-like models of computation, *Physica D* **10** (1984), 81.

43. S. Wolfram, "SMP Reference Manual," Computer Mathematics Group, Inference Corporation, Los Angeles, 1983.

44. D. Hillis, "The Connection Machine," MIT Press, Cambridge, Mass., 1985.

45. P. Grassberger, "Towards a quantitative theory of self-generated complexity," Wuppertal preprint (1986).

46. S. Wolfram, Cryptography with cellular automata, *in* "Proc. CRYPTO 85," August 1985.

Approaches to
Complexity Engineering

1986

Principles for designing complex systems with specified forms of behaviour are discussed. Multiple scale cellular automata are suggested as dissipative dynamical systems suitable for tasks such as pattern recognition. Fundamental aspects of the engineering of such systems are characterized using computation theory, and some practical procedures are discussed.

The capabilities of the brain and many other biological systems go far beyond those of any artificial systems so far constructed by conventional engineering means. There is however extensive evidence that at a functional level, the basic components of such complex natural systems are quite simple, and could for example be emulated with a variety of technologies. But how a large number of these components can act together to perform complex tasks is not yet known. There are probably some rather general principles which govern such overall behaviour, and allow it to be moulded to achieve particular goals. If these principles could be found and applied, they would make new forms of engineering possible. This paper discusses some approaches to such forms of engineering with complex systems. The emphasis is on general concepts and analogies. But some of the specific systems discussed should nevertheless be amenable to implementation and detailed analysis.

In conventional engineering or computer programming, systems are built to achieve their goals by following strict plans, which specify the detailed behaviour of each of their component parts. Their overall behaviour must always be simple enough that complete prediction and often also analysis is possible. Thus for example motion in conventional mechanical engineering devices is usually constrained simply to be periodic. And in conventional computer programming, each step consists of a single operation on a small number of data elements. In both of these cases, much more com-

Originally published in *Physica D*, volume 22, pages 385–399 (October 1986).

plex behaviour could be obtained from the basic components, whether mechanical or logical, but the principles necessary to make use of such behaviour are not yet known.

Nature provides many examples of systems whose basic components are simple, but whose overall behaviour is extremely complex. Mathematical models such as cellular automata (e.g. [1]) seem to capture many essential features of such systems, and provide some understanding of the basic mechanisms by which complexity is produced for example in turbulent fluid flow. But now one must use this understanding to design systems whose complex behaviour can be controlled and directed to particular tasks. From complex systems science, one must now develop complex systems engineering.

Complexity in natural systems typically arises from the collective effect of a very large number of components. It is often essentially impossible to predict the detailed behaviour of any one particular component, or in fact the precise behaviour of the complete system. But the system as a whole may nevertheless show definite overall behaviour, and this behaviour usually has several important features.

Perhaps most important, it is robust, and is typically unaffected by perturbations or failures of individual components. Thus for example a change in the detailed initial conditions for a system usually has little or no effect on the overall outcome of its evolution (although it may have a large effect on the detailed behaviour of some individual elements). The visual system in the brain, for example, can recognize objects even though there are distortions or imperfections in the input image. Its operation is also presumably unaffected by the failure of a few neurons. In sharp contrast, however, typical computer programs require explicit account to be taken of each possible form of input. In addition, failure of any one element usually leads to catastrophic failure of the whole program.

Dissipation, in one of many forms, is a key principle which lies behind much of the robustness seen in natural systems. Through dissipation, only a few features in the behaviour of a system survive with time, and others are damped away. Dissipation is often used to obtain reliable behaviour in mechanical engineering systems. Many different initial motions can for example be dissipated away through viscous damping which brings particular components to rest. Such behaviour is typically represented by a differential equation whose solution tends to a fixed point at large times, independent of its initial conditions. Any information on the particular initial conditions is thus destroyed by the irreversible evolution of the system.

In more complicated systems, there may be several fixed points, reached from different sets of initial conditions. This is the case for an idealized ball rolling on a landscape, with dissipation in the form of friction. Starting at any initial point, the ball is "attracted" towards one of the local height minima in the landscape, and eventually comes to rest there. The set of initial positions from which the ball goes to a particular such fixed point can be considered as the "basin of attraction" for that fixed point. Each basin of attraction is bounded by a "watershed" which typically lies along a ridge in the landscape. Dissipation destroys information on details of initial conditions, but preserves the knowledge of which basin of attraction they were in. The evolution of the system can be viewed as dividing its inputs into various "categories", corresponding to different basins of attraction. This operation is the essence of many forms of pattern recognition: despite small changes, one recognizes that a particular input is in a particular category, or matches a particular pattern. In the example of a ball rolling on a landscape, the categories correspond to different regions of initial positions. Small changes in input correspond to small changes in initial position.

The state of the system just discussed is given by the continuous variables representing the position of the ball. More familiar examples of pattern recognition arise in discrete or digital systems, such as those used for image processing. An image might be represented by a 256×256 array of cells, each black or white. Then a simple image processing (or "image restoration") operation would be to replace any isolated black cell by a white cell. In this way certain single cell errors in the images can be removed (or "damped out"), and classes of images differing just by such errors can be recognized as equivalent (e.g. [3]). The process can be considered to have attractors corresponding to the possible images without such errors. Clearly there are many of these attractors, each with a particular basin of attraction. But in contrast to the example with continuous variables above, there is no obvious measure of "distance" on the space of images, which could be used to determine which basin of attraction a particular image is in. Rather the category of an image is best determined by explicit application of the image processing operation.

Length n sequences of bits can be considered as corners of an n-dimensional unit hypercube. The Hamming distance between two sequences can then be defined as the number of edges of the hypercube that must be traversed to get from one to the other, or, equivalently, the total number of bits that differ between them. It is possible using algebraic methods to devise transformations with basins of attraction corresponding to spheres which enclose all points at a Hamming distance of at most say two bits from a given point [4]. This allows error-correcting codes to be devised in which definite messages can be reconstructed even though they may contain say up to two erroneous bits.

The transformations used in error-correcting codes are specially constructed to have basins of attraction with very simple forms. Most dissipative systems, however, yield much more complicated basins of attraction, which cannot for example be described by simple scalar quantities such as distances. The form of these basins of attraction determines what kinds of perturbations are damped out, and thus what classes of inputs can be recognized as equivalent.

As a first example, consider various idealizations of the system discussed above consisting of a ball rolling with friction on a landscape, now assumed one dimensional. In the approximation of a point ball, this is equivalent to a particle moving with damping in a one-dimensional potential. The attractors for the system are again fixed points corresponding to minima of the potential. But the basins of attraction depend substantially on the exact dynamics assumed. In the case of very large friction, the particle satisfies a differential equation in which velocity is proportional to force, and force is given by the gradient of the potential. With zero initial velocity, the basins of attraction in this case have a simple form, separated by boundaries at the positions of maxima in the potential. In a more realistic model, with finite friction and the inertia of the ball included, the system becomes similar to a Roulette wheel. And in this case it is known that the outcome is a sensitive function of the precise initial conditions. As a consequence, the basins of attraction corresponding for example to different holes around the wheel must have a complicated, interdigitated, form (cf. [5]).

Complicated basin boundaries can also be obtained with simpler equations of motion. As one example, one can take time to be discrete, and assume that the potential has the form of a polynomial, so that the differential equation of motion is approximated by an iterated polynomial mapping. The sequence of positions found from this mapping may overshoot the minimum, and for some values of parameters may in fact never converge to it. The region of initial conditions which evolves to a particular attractor may therefore be complicated. In the case of the complex iterated mapping $z \rightarrow z^2 + c$, the boundary of the basin of attraction (say for the attractor $z = \infty$) is a Julia set, and has a very complicated fractal form (e.g. [6]).

The essentials of the problem of finding basins of attraction already arise in the problem of determining what set of inputs to a function of discrete variables yields a particular output. This problem is known in general to be computationally very difficult. In fact, the satisfiability problem of determining which if any assignments of truth values to n variables in a Boolean expression make the whole expression true is NP-complete, and can presumably be solved in general essentially only by explicitly testing all 2^n possible assignments (e.g. [7]). For some functions with a simple, perhaps algebraic, structure, an efficient inversion procedure to find appropriate inputs may exist. But in general no simple mathematical formula can describe the pattern of inputs: they will simply seem random (cf. [8]).

Many realistic examples of this problem are found in cellular automata. Cellular automata consist of a lattice of sites with discrete values updated in discrete steps according to a fixed local rule. The image processing operation mentioned above can be considered as a single step in the evolution of a simple two-dimensional cellular automaton (cf. [9]). Other cellular automata show much more complicated behaviour, and it seems in fact that with appropriate rules they capture the essential features of many complex systems in nature (e.g. [1]). The basic problems of complexity engineering thus presumably already arise in cellular automata.

Most cellular automata are dissipative, or irreversible, so that after many steps, they evolve to attractors which contain only a subset of their states. In some cellular automata (usually identified as classes 1 and 2), these attractors are fixed points (or limit cycles), and small changes in initial conditions are usually damped out [10]. Other cellular automata (classes 3 and 4), however, never settle down to a fixed state with time, but instead continue to show complicated, chaotic, behaviour. Such cellular automata are unstable, so that most initial perturbations grow with time to

312

affect the detailed configuration of an ever-increasing number of sites. The statistical properties of the behaviour produced are nevertheless robust, and are unaffected by such perturbations.

It can be shown that the set of fixed points of a one-dimensional cellular automaton consists simply of all those configurations in which particular blocks of site values do not appear [11]. This set forms a (finite complement) regular language, and can be represented by the set of possible paths through a certain labelled directed graph [11]. Even when they are not fixed points, the set of states that can occur after say t time steps in the evolution of a one-dimensional cellular automaton in fact also forms a regular language (though not necessarily a finite complement one). In addition, the basin of attraction, or in general the set of all states which evolve after t steps to a given one, can be represented as a regular language. For class 1 and 2 cellular automata, the size of the minimal graph for this language stays bounded, or at most increases like a polynomial with t. For class 3 and 4 cellular automata, however, the size of the graph often increases apparently exponentially with t, so that it becomes increasingly difficult to describe the basin of attraction. The general problem of determining which states evolve to a particular one after t steps is in fact a generalization of the satisfiability problem for logical functions mentioned above, and is thus NP complete. The basin of attraction in the worst case can thus presumably be found only by explicit testing of essentially all $O(2^t)$ possible initial configurations (cf. [12]). Its form will again often be so complicated as to seem random. For two-dimensional cellular automata, it is already an NP-complete problem just to find fixed points (say to determine which $n \times n$ blocks of sites with specified boundaries are invariant under the cellular automaton rule) [13].

It is typical of complex systems that to reproduce their behaviour requires extensive computation. This is a consequence of the fact that the evolution of the systems themselves typically corresponds to a sophisticated computation. In fact, the evolution of many complex systems is probably computationally irreducible: it can be found essentially only by direct simulation, and cannot be predicted by any short-cut procedure [12, 14]. Such computational irreducibility is a necessary consequence of the efficient use of computational resources in a system. Any computational reducibility is a sign of inefficiency, since it implies that some other system can determine the outcome more efficiently.

Many systems in nature may well be computationally irreducible, so that no general predictions can be made about their behaviour. But if a system is to be used for engineering, it must be possible to determine in advance at least some aspects of its behaviour. Conventional engineering requires detailed specification of the precise behaviour of each component in a system. To make use of complex systems in engineering, one must relax this constraint, and instead require only some general or approximate specification of overall behaviour.

One goal is to design systems which have particular attractors. For the example of an inertialess ball rolling with friction on a landscape, this is quite straightforward (cf. [15]). In one dimension, the height of the landscape at position x could be given by the polynomial $\prod_i (x - x_i)^2$, where the x_i are the desired minima, or attractors. This polynomial is explicitly constructed to yield certain attractors in the dynamics. However, it implies a particular structure for the basins of attraction. If the attractors are close to equally spaced, or are sufficiently far apart, then the boundaries of the basins of attraction for successive attractors will be roughly half way between them. Notice, however, that as the parameters of the landscape polynomial are changed, the structure of the attractors and basins of attraction obtained can change discontinuously, as described by catastrophe theory.

For a more complex system, such as a cellular automaton, it is more difficult to obtain a particular set of attractors. One approach is to construct cellular automaton rules which leave particular sequences invariant [16]. If these sequences are say of length L, and are arbitrarily chosen, then it may be necessary to use a cellular automaton rule which involves a neighbourhood of up to $L - 1$ sites. The necessary rule is straightforward to construct, but takes up to 2^{L-1} bits to specify.

Many kinds of complex systems can be considered as bases for engineering. Conventional engineering suggests some principles to follow. The most important is the principle of modularity. The components of a system should be arranged in some form of hierarchy. Components higher on the hierarchy should provide overall control for sets of components lower on the hierarchy, which can be treated as single units or modules. This principle is crucial to software engineering, where the modules are typically subroutines. It is also manifest in biology in the existence of organs and definite body parts, apparently mirrored by subroutine-like constructs in the genetic code.

An important aspect of modularity is the abstraction it makes possible. Once the construction of a particular module has been completed, the module can be treated as a single object, and only its overall behaviour need be considered, wherever the module appears. Modularity thus divides the problem of constructing or analysing a system into many levels, potentially making each level manageable.

Modularity is used in essentially all of the systems to be discussed below. In most cases, there are just two levels: controlling (master) and controlled (slave) components. The components on these two levels usually change on different time scales. The controlling components change at most slowly, and are often fixed once a system say with a particular set of attractors has been obtained. The controlled components change rapidly, processing input data according to dynamical rules determined by the controlling components. Such separation of time scales is common in many natural and artificial systems. In biology, for example, phenotypes of organisms grow by fast processes, but are determined by genotypes which seem to change only slowly with time. In software engineering, computer memory is divided into a part for "programs", which are supposed to remain fixed or change only slowly, and another part for intermediate data, which changes rapidly.

Multiple scale cellular automata provide simple but quite general examples of such hierarchical systems. An ordinary cellular automaton consists of a lattice of sites, with each site having say k possible values, updated according to the same definite rule. A two-scale cellular automaton can be considered to consist of two lattices, with site values changing on different characteristic time scales. The values of the sites on the "slow" lattice control the rules used at the corresponding sites on the "fast" lattice. With q possible values for the slow lattice sites, there is an array of q possible rules for each site on the fast lattice. (Such a two-scale cellular automaton could always be emulated by specially chosen configurations in an ordinary cellular automaton with at most qk possible values at each site.)

If the sites on the slow lattice are fixed, then a two-scale cellular automaton acts like a dynamic random field spin system (e.g. [17]), or a spin glass (e.g. [18]) (cf. [19]). Examples of patterns generated by cellular automata of this kind are shown in figure 1. If instead the "slow" lattice sites change rapidly, and take on essentially random values, perhaps as a result of following a chaotic cellular automaton rule, then the evolution of the fast lattice is like that of a stochastic cellular automaton, or a directed percolation system (e.g. [21]).

With dissipative dynamics, the evolution of the fast lattice in a two-scale cellular automaton yields attractors. The form of these attractors is determined by the control configuration on the slow lattice. By choosing different slow lattice configurations, it is thus possible to engineer particular attractor structures.

In a typical case, a two-scale cellular automaton might be engineered to recognize inputs in different categories. Each category would be represented by a fixed point in the fast lattice dynamics. The system could then be arranged in several ways. Assume that the input is a one-dimensional symbol sequence (such as a text string). Then one possibility would be to consider a one-dimensional cellular automaton whose fixed points correspond to symbol sequences characteristic of each category. But if the required fixed points are arbitrarily chosen, only a few of them can be obtained with a single slow configuration. If the cellular automaton has N sites, then each fixed point of the fast lattice is specified by $N \log_2 k$ bits. A configuration of the slow lattice involves only $N \log_2 q$ bits. As a consequence, the number of arbitrarily-chosen fixed points that can be specified is just $\log q / \log k$, a result independent of N. (More fixed points may potentially be specified if there is redundancy between their symbol sequences.)

It is usually not necessary, however, to give all $N \log_2 k$ bits of a fixed point to specify the form of the attractor for a particular category. The number of bits actually needed presumably increases with the number of categories. It is common to find a small number of possible categories or responses to a wide variety of input data. The responses can then for example be represented by the values of a small number of sites on the fast lattice of a two-scale cellular automaton. The input data can be used to give initial values for a larger number of sites, possibly a different set. (In an analogy with the nervous system, some sites might receive input from afferent nerves while others, typically smaller in number, might generate output for efferent nerves.)

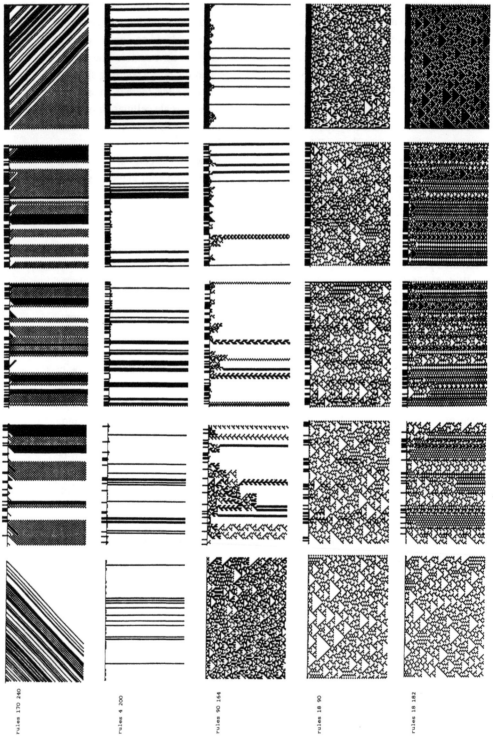

rules 170 240

rules 4 200

rules 90 164

rules 18 90

rules 18 182

A second possibility is to consider a two-dimensional two-scale cellular automaton, in which the input is specified along a line, and the dynamics of the fast lattice transfers information only in a direction orthogonal to this line [22] (cf. [23]). This arrangement is functionally equivalent to a one-dimensional two-scale cellular automaton in which the slow lattice configuration changes at each time step. In its two-dimensional form, the arrangement is very similar to a systolic array [24], or in fact to a multistage generalization of standard modular logic circuits. In an $N \times M$ system of this kind, a single slow lattice configuration can specify $M \log q / \log k$ length N fixed points in the fast configuration (cf. [2]).

In the approaches just discussed, input is given as an initial condition for the fast lattice. An alternative possibility is that the input could be given on the slow lattice, and could remain throughout the evolution of the fast lattice. The input might for example then specify boundary conditions for evolution on a two-dimensional fast lattice. Output could be obtained from the final configuration of the fast lattice. However, there will often be several different attractors for the fast lattice dynamics even given boundary conditions from a particular slow lattice configuration. Which attractor is reached will typically depend on the initial conditions for the fast lattice, which are not specified in this approach. With appropriate dynamics, however, it is nevertheless possible to obtain almost unique attractors: one approach is to add probabilistic elements or noise to the fast lattice dynamics so as to make it ergodic, with a unique invariant measure corresponding to a definite "phase" [25].

Cellular automata are arranged to be as simple as possible in their basic microscopic construction. They are discrete in space and time. Their sites are all identical, and are arranged on a regular lattice. The sites have a finite set of possible values, which are updated synchronously according to identical deterministic rules that depend on a few local neighbours. But despite this microscopic simplicity, the overall macroscopic behaviour of cellular automata can be highly complex. On a large scale, cellular automata can for example show continuum features [13, 26], randomness [8], and effective long-range interactions [27]. Some cellular automata are even known to be universal computers [28], and so can presumably simulate any possible form of behaviour. Arbitrary complexity can thus arise in cellular automata. But for engineering purposes, it may be better to consider basic models that are more sophisticated than cellular automata, and in which additional complexity is included from the outset (cf. [2]). Multiple scale cellular automata incorporate modularity, and need not be homogeneous. Further generalizations can also be considered, though one suspects that in the end none of them will turn out to be crucial.

◀ **Figure 1.** Patterns generated by two-scale cellular automata with $k = 2$, $q = 2$ and $r = 1$. The configuration of the slow lattice is fixed in each case, and is shown at the top. The rule used at a particular site on the fast lattice is chosen from the two rules given according to the value of the corresponding site on the slow lattice. (The rule numbers are as defined in ref. [20].)

First, cellular automaton dynamics is local: it involves no long-range connections which can transmit information over a large distance in one step. This allows (one or two-dimensional) cellular automata to be implemented directly in the simple planar geometries appropriate, for example, for very large-scale integrated circuits. Long range electronic signals are usually carried by wires which cross in the third dimension to form a complicated network. (Optical communications may also be possible.) Such an arrangement is difficult to implement technologically. When dynamically-changing connections are required, therefore, more homogeneous switching networks are used, as in computerized telephone exchanges, or the Connection Machine computer [29]. Such networks are typically connected like cellular automata, though often in three (and sometimes more) dimensions.

Some natural systems nevertheless seem to incorporate intrinsic long-range connections. Chemical reaction networks are one example: reaction pathways can give almost arbitrary connectivity in the abstract space of possible chemical species [30, 31, 32]. Another example is the brain, where nerves can carry signals over long distances. In many parts of the brain, the pattern of connectivity chosen seems to involve many short-range connections, together with a few long-range ones, like motorways (freeways) or trunk lines [33]. It is always possible to simulate an arbitrary arrangement of long range connections through sequences of short range connections; but the existence of a few intrinsic long range connections may make large classes of such simulations much more efficient [33].

Many computational algorithms seem to involve arbitrary exchange of data. Thus for example, in the fast Fourier transform, elements are combined according to a shuffle-exchange graph (e.g. [29]). Such algorithms can always be implemented by a sequence of local operations. But they seem to be most easily conceived without reference to the dynamics of data transfer. Indeed, computers and programming languages have traditionally been constructed to enforce the idealization that any piece of data is available in a fixed time (notions such as registers and pipelining go slightly beyond this). Conventional computational complexity theory also follows this idealization (e.g. [34]). But in developing systems that come closer to actual physical constraints, one must go beyond this idealization. Several classes of algorithms are emerging that can be implemented efficiently and naturally with local communications (e.g. [35]). A one-dimensional cellular automaton ("iterative array") can be used for integer multiplication [36, 37] and sorting [38]. Two dimensional cellular automata ("systolic arrays") can perform a variety of matrix manipulation operations [24].

Although the basic rules for cellular automata are local, they are usually applied in synchrony, as if controlled by a global clock. A generalization would allow asynchrony, so that different sites could be updated at different times (e.g. [39]). Only a few sites might, for example, be updated at each time step. This typically yields more gradual transitions from one cellular automaton configuration to another, and can prevent certain instabilities. Asynchronous updating makes it more difficult for information to propagate through the cellular automaton, and thus tends to prevent initial perturbations from spreading. As a result, the evolution is more irreversible and dissipative. Fixed point and limit cycle (class 1 and 2) behaviour therefore becomes more common.

For implementation and analysis, it is often convenient to maintain a regular updating schedule. One possibility is to alternate between updates of even and odd-numbered sites (e.g. [40]). "New" rather than "old" values for the nearest neighbours of a particular cell are then effectively used. This procedure is analogous to the implicit, rather than explicit, method for updating site values in finite difference approximations to partial differential equations (e.g. [41]), where it is known to lead to better convergence in certain cases. The scheme also yields, for example, systematic relaxation to thermodynamic equilibrium in a cellular automaton version of the microcanonical Ising model [42]: simultaneous updating of all sites would allow undamped oscillations in this case [40].

One can also consider systems in which sites are updated in a random order, perhaps one at a time. Such systems can often be analysed using "mean field theory", by assuming that the behaviour of each individual component is random, with a particular average (cf. [2]). Statistical predictions can then often be made from iterations of maps involving single real variables. As a result, monotonic approach to fixed points is more easily established.

Random asynchronous updating nevertheless makes detailed analysis more difficult. Standard computational procedures usually require definite ordering of operations, which can be regained in this case only at some cost (cf. [43]).

Rather than introducing randomness into the updating scheme, one can instead include it directly in the basic cellular automaton rule. The evolution of such stochastic cellular automata can be analogous to the steps in a Monte Carlo simulation of a spin system at nonzero temperature [44]. Randomness typically prevents the system from being trapped in metastable states, and can therefore accelerate the approach to equilibrium.

In most practical implementations, however, supposedly random sequences must be obtained from simple algorithms (e.g. [37]). Chaotic cellular automata can produce sequences with a high degree of randomness [8], presumably making explicit insertion of external randomness unnecessary.

Another important simplifying feature of cellular automata is the assumption of discrete states. This feature is convenient for implementation by digital electronic circuits. But many natural systems seem to involve continuously-variable parameters. There are usually components, such as molecules or vesicles of neurotransmitter, that behave as discrete on certain levels. But very large numbers of these components can act in bulk, so that for example only their total concentration is significant, and this can be considered as an essentially continuous variable. In some systems, such bulk quantities have simple behaviour, described say by partial differential equations. But the overall behaviour of many cellular automata and other systems can be sufficiently

319

complex that no such bulk or average description is adequate. Instead the evolution of each individual component must be followed explicitly (cf. [12]).

For engineering purposes, it may nevertheless sometimes be convenient to consider systems which involve essentially continuous parameters. Such systems can for example support cumulative small incremental changes. In a cellular automaton, the states of n elements are typically represented by $O(n)$ bits of information. But bulk quantities can be more efficiently encoded as digital numbers, with only $O(\log n)$ bits. There may be some situations in which data is best packaged in this way, and manipulated say with arithmetic operations.

Systems whose evolution can be described in terms of arithmetic operations on numbers can potentially be analysed using a variety of standard mathematical techniques. This is particularly so when the evolution obeys a linear superposition principle, so that the complete behaviour can be built up from a simple superposition of elementary pieces. Such linear systems often admit extensive algebraic analysis (cf. [45]), so that their behaviour is usually too simple to show the complexity required.

Having selected a basic system, the problem of engineering consists in designing or programming it to perform particular tasks. The conventional approach is systematically to devise a detailed step-by-step plan. But such a direct constructive approach cannot make the most efficient use of a complex system.

Logic circuit design provides an example (e.g. [46]). The task to be performed is the computation of a Boolean function with n inputs specified by a truth table. In a typical case, the basic system is a programmable logic array (PLA): a two-level circuit which implements disjunctive normal form (DNF) Boolean expressions, consisting of disjunctions (ORs) of conjunctions (ANDs) of input variables (possibly negated) [24, 47]. The direct approach would be to construct a circuit which explicitly tests for each of the 2^n cases in the truth table. The resulting circuit would contain $O(n2^n)$ gates. Thus for example, the majority function, which yields 1 if two or more of its three inputs a_i are one would be represented by the logical circuit corresponding to $a_1 a_2 a_3 + a_1 a_2 \overline{a_3} + a_1 \overline{a_2} a_3 + \overline{a_1} a_2 a_3$, where multiplication denotes AND, addition OR, and bar NOT. (This function can be viewed as the $k = 2$, $r = 1$ cellular automaton rule number 232 [20].)

Much smaller circuits are, however, often sufficient. But direct constructive techniques are not usually appropriate for finding them. Instead one uses methods that manipulate the structure of circuits, without direct regard to the meaning of the Boolean functions they represent. Many methods start by extracting prime implicants [46, 47]. Logical functions of n variables can be considered as colourings of the Boolean n-cube. Prime implicants represent this colouring by decomposing it into pieces along hyperplanes with different dimensionalities. Each prime implicant corresponds to a single conjunction of input variables: a circuit for the original Boolean function can be formed from a disjunction of these conjunctions. This

320

circuit is typically much smaller than the one obtained by direct construction. (For the majority function mentioned above, it is $a_1 a_2 + a_1 a_3 + a_2 a_3$.) And while it performs the same task, it is usually no longer possible to give an explicit step-by-step "explanation" of its operation.

A variety of algebraic and heuristic techniques are used for further simplification of DNF Boolean expressions [47]. But it is in general very difficult to find the absolutely minimal expression for any particular function. In principle, one could just enumerate all possible progressively more complicated expressions or circuits, and find the first one which reproduces the required function. But the number of possible circuits grows exponentially with the number of gates, so such an exhaustive search rapidly becomes entirely infeasible. It can be shown in fact that the problem of finding the absolute minimal expression is NP hard, suggesting that there can never be a general procedure for it that takes only polynomial time [7]. Exhaustive search is thus effectively the only possible exact method of solution.

Circuits with a still smaller number of gates can in principle be constructed by allowing more than two levels of logic. (Some such circuits for the majority function are discussed in ref. [48].) But the difficulty of finding the necessary circuits increases rapidly as more general forms are allowed, and as the absolute minimum circuit is approached. A similar phenomenon is observed with many complex systems: finding optimal designs becomes rapidly more difficult as the efficiency of the designs increases (cf. [49]).

In most cases, however, it is not necessary to find the absolutely minimal circuit: any sufficiently simple circuit will suffice. As a result, one can consider methods that find only approximately minimal circuits.

Most approximation techniques are basically iterative: they start from one circuit, then successively make changes which preserve the functionality of the circuit, but modify its structure. The purpose is to find minima in the circuit size or "cost" ("fitness") function over the space of possible circuits. The effectiveness of different techniques depends on the form of the circuit size "landscape".

If the landscape was like a smooth bowl, then the global minimum could be found by starting at any point, and systematically descending in the direction of the local gradient vector. But in most cases the landscape is presumably more complicated. It could for example be essentially flat, except for one narrow hole containing the minimum (like a golf course). In such a case, no simple iterative procedure could find the minimum.

Another possibility, probably common in practice, is that the landscape has a form reminiscent of real topographical landscapes, with a complicated pattern of peaks and valleys of many different sizes. Such a landscape might well have a self similar or fractal form: features seen at different magnifications could be related by simple scalings. Straightforward gradient descent would always get stuck in local minima on such a landscape, and cannot be used to find a global minimum (just as water forms localized lakes on a topographical landscape). Instead one should

321

use a procedure which deals first with large-scale features, then progressively treats smaller and smaller scale details.

Simulated annealing is an example of such a technique [50]. It is based on the gradient descent method, but with stochastic noise added. The noise level is initially large, so that all but the largest scale features of the landscape are smeared out. A minimum is found at this level. Then the noise level ("temperature") is reduced, so that smaller scale features become relevant, and the minimum is progressively refined. The optimal temperature variation ("annealing schedule") is probably determined by the fractal dimension of the landscape.

In actual implementations of the simulated annealing technique, the noise will not be truly random, but will instead be generated by some definite, and typically quite simple, procedure. As a consequence, the whole simulated annealing computation can be considered entirely deterministic. And since the landscape is probably quite random, it is possible that simple deterministic perturbations of paths may suffice (cf. [51]).

In the simulated annealing approach, each individual "move" might consist of a transformation involving say two logic gates. An alternative procedure is first to find the minimal circuit made from "modules" containing many gates, and then to consider rearranging progressively smaller submodules. The hierarchical nature of this deterministic procedure can again mirror the hierarchical form of the landscape.

The two approaches just discussed involve iterative improvement of a single solution. One can also consider approaches in which many candidate solutions are treated in parallel. Biological evolution apparently uses one such approach. It generates a tree of different genotypes, and tests the "fitness" of each branch in parallel. Unfit branches die off. But branches that fare well have many offspring, each with a genotype different by a small random perturbation ("genetic algorithm" [52]). These offspring are then in turn tested, and can themselves produce further offspring. As a result, a search is effectively conducted along many paths at once, with a higher density of paths in regions with greater fitness. (This is analogous to decision tree searching with, say, $\alpha\beta$-pruning [53].) Random perturbations in the paths at each generation may prevent getting stuck in local minima, but on a fractal landscape of the type discussed above, this procedure seems less efficient than one based on consideration of progressively finer details.

In the simplest iterative procedures, the possible changes made to candidate solutions are chosen from a fixed set. But one can also imagine modifying the set of possible changes dynamically [54] (cf. [55]). To do this, one must parametrize the possible changes, and in turn search the space of possibilities for optimal solutions.

The issues discussed for logic circuit design also arise in engineering complex systems such as two-scale cellular automata. A typical problem in this case is to find a configuration for the slow lattice that yields particular fixed points for evolution on the fast lattice. With simple linear rules, for example, a constructive algebraic solution to this problem can be given. But for arbitrary rules, the problem is in general NP hard. An exact solution can thus presumably be found only by exhaustive search. Approximation procedures must therefore again be used.

The general problem is to find designs or arrangements of complex systems that behave in specified ways. The behaviour sought usually corresponds to a comparatively simple, usually polynomial time, computation. But to find exactly the necessary design may require a computation that effectively tests exponentially many possibilities. Since the correctness of each possibility can be tested in polynomial time, the problem of finding an appropriate design is in the computational complexity class NP (non-deterministic polynomial time). But in many cases, the problem is in fact NP complete (or at least NP hard). Special instances of the problem thus correspond to arbitrary problems in NP; any general solution could thus be applied to all problems in NP.

There are many NP complete problems, all equivalent in the computational difficulty of their exact solution [7]. Examples are satisfiability (finding an assignment of truth values to variables which makes a Boolean expression true), Hamilton circuits (finding a path through a graph that visits each arc exactly once), and spin glass energy minima (finding the minimum energy configuration in a spin glass model). In no case is an algorithm known which takes polynomial time, and systematically yields the exact solution.

Many approximate algorithms are nevertheless known. And while the difficulty of finding exact solutions to the different problems is equivalent, the ease of approximation differs considerably. (A separate consideration is what fraction of the instances of a problem are difficult to solve with a particular algorithm. Some number theoretical problems, for example, have the property that all their instances are of essentially equivalent difficulty [56].) Presumably the "landscapes" for different problems fall into several classes. There is already some evidence that the landscapes for spin glass energy and the "travelling salesman" problem have a hierarchical or ultrametric, and thus fractal, form [57]. This may explain why the simulated annealing method is comparatively effective in these cases.

Even though their explicit forms cannot be found, it could be that particular, say statistical, features of solutions to NP problems could easily be predicted. Certainly any solution must be distinguished by the P operation used to test its validity. But at least for some class of NP hard problems, one suspects that solutions will appear random according to all standard statistical procedures. Despite the "selection" process used to find them, this would imply that their statistical properties would be typical of the ensemble of all possibilities (cf. [58]).

There are many potential applications for complex systems engineering. The most immediate ones are in pattern recognition. The basic problem is to take a wide variety of inputs, say versions of spoken words, and to recognize to which category or written word they correspond (e.g. [59]).

In general, there could be an arbitrary mapping from input to output, so that each particular case would have to be specified explicitly. But in practice the number of possible inputs is far too large for this to be feasible, and redundancy in the inputs must be used. One must effectively make some form of model for the inputs, which can be used, for example, to delineate categories that yield the same output. The kinds of models that are most appropriate depend on the regularities that exist in the input data. In human language and various other everyday forms of input, there seem for example to be regularities such as power laws for frequencies (e.g. [60]).

In a simple case, one might take inputs within one category to differ only by particular kinds of distortions or errors. Thus in studies of DNA sequences, changes associated with substitution, deletion, insertion, or transposition of elements are usually considered [61]. (These changes have small effects on the spatial structure of the molecule, which determines many of its functions.) A typical problem of pattern recognition is to determine the category of a particular input, regardless of such changes.

Several approaches are conventionally used (e.g. [59]).

One approach is template matching. Each category is defined by a fixed "template". Then inputs are successively compared with each template, and the quality of match is determined, typically by statistical means. The input is assigned to the category with the best match.

A second approach is feature extraction. A fixed set of "features" is defined. The presence or absence of each feature in a particular input is then determined, often by template-matching techniques. The category of the input is found from the set of features it contains, typically according to a fixed table.

In both these approaches, the pattern recognition procedure must be specially designed to deal with each particular set of categories considered. Templates or features that are sufficiently orthogonal must be constructed, and small changes in the behaviour required may necessitate large changes in the arrangement used.

It would be more satisfactory to have generic systems which would take simple specifications of categories, and recognize inputs using "reasonable" boundaries between the categories. Dissipative dynamical systems with this capability can potentially be constructed. Different categories would be specified as fixed points (or other attractors). Then the dynamics of the system would determine the forms of the basins of attraction. Any input within a particular basin would be led to the appropriate fixed point. In general, however, different inputs would take varying numbers of steps to reach a fixed point. Conventional pattern recognition schemes typically take a fixed time, independent of input. But more flexible schemes presumably require variable times.

It should be realized, however, that such schemes implicitly make definite models for the input data. It is by no means clear that the dynamics of such systems yield basin structures appropriate for particular data. The basins are typically complicated and difficult to specify. There will usually be no simple distance measure or metric,

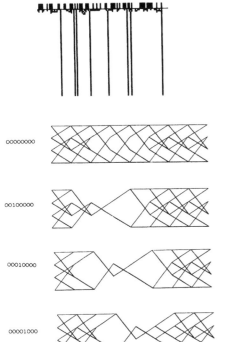

Figure 2. Representation of the basins of attraction for fixed points in a length 8 two-scale cellular automaton with $q = 2$ and rules 36 and 72. The configurations in each basin correspond to possible paths traversing each graph from left to right. Descending segments represent value one, ascending segments value zero.

analogous to the quality of template matches, which determines the basin for a particular input from the fixed point to which it is "closest". Figure 2 shows a representation of the basins of attraction in a two-scale cellular automaton. No simple metric is evident.

While the detailed behaviour of a system may be difficult to specify, it may be possible to find a high-level phenomenological description of some overall features, perhaps along the lines conventional in psychology, or in the symbolic approach to artificial intelligence (e.g. [62]). One can imagine, for example, proximity relations for attractors analogous to semantic networks (e.g. [62]). This high level description might have the same kind of relation to the underlying dynamics as phenomenological descriptions such as vortex streets have to the basic equations of fluid flow.

To perform a particular pattern recognition task, one must design a system with the appropriate attractor structure. If, for example, categories are to be represented by certain specified fixed points, the system must be constructed to have these fixed points. In a two-scale cellular automaton with $k = 2$ and $q = 2$, a single fixed point on the whole lattice can potentially be produced by an appropriate choice of the slow configuration. Arbitrary fixed points can be obtained in this way only with particular pairs of rules. (The rules that take all configurations to zero, and all configurations to one, provide a trivial example.) But even in this case, it is common for several different slow configurations to yield the same required fixed point, but to give very different basin structures. Often spurious additional fixed points are also produced. It is not yet clear how best to obtain only the exact fixed points required.

It would be best to devise a scheme for "learning by example" (cf. [63]). In the simplest case, the fixed points would be configurations corresponding to "typical" members of the required categories. In a more sophisticated case, many input and output pairs would be presented, and an iterative algorithm would be used to design an attractor structure to represent them. In a multiple scale cellular automaton, such an algorithm might typically make "small" incremental changes of a few sites in the slow configuration. Again such a procedure involves inferences about new inputs, and requires a definite model.

Acknowledgements

I am grateful for discussions with many people, including Danny Hillis, David Johnson, Stuart Kauffman, Alan Lapedes, Marvin Minsky, Steve Omohundro, Norman Packard, Terry Sejnowski, Rob Shaw, and Gerry Tesauro.

References

1. S. Wolfram, "Cellular automata as models of complexity", Nature 311 (1984) 419.
2. J. Hopfield, "Neural networks and physical systems with emergent collective computational abilities", Proc. Natl. Acad. Sci. 79 (1982) 2554.
3. W. Green, *Digital image processing*, Van Nostrand (1983).
4. R. Hamming, *Coding and information theory*, Prentice-Hall (1980).
5. V. Vulovic and R. Prange, "Is the toss of a true coin really random?", Maryland preprint (1985).
6. S. McDonald, C. Grebogi, E. Ott and J. Yorke, "Fractal basin boundaries", Physica 17D (1985) 125.
7. M. Garey and D. Johnson, *Computers and intractability: a guide to the theory of NP-completeness*, Freeman (1979).
8. S. Wolfram, "Random sequence generation by cellular automata", Adv. Applied Math. 7, 123–169 (1986).
9. K. Preston and M. Duff, *Modern cellular automata*, Plenum (1984).
10. S. Wolfram, "Universality and complexity in cellular automata", Physica 10D (1984) 1.
11. S. Wolfram, "Computation theory of cellular automata", Commun. Math. Phys. 96 (1984) 15.
12. S. Wolfram, "Undecidability and intractability in theoretical physics", Phys. Rev. Lett. 54 (1985) 735.
13. N. Packard and S. Wolfram, "Two-dimensional cellular automata", J. Stat. Phys. 38 (1985) 901.
14. S. Wolfram, "Computer software in science and mathematics", Sci. Amer. (September 1984).
15. R. Sverdlove, "Inverse problems for dynamical systems in the plane", in *Dynamical systems*, A. R. Bednarek and L. Cesari (eds.), Academic Press (1977).
16. E. Jen, "Invariant strings and pattern-recognizing properties of one-dimensional cellular automata", J. Stat. Phys., to be published; Los Alamos preprint LA-UR-85-2896.
17. J. Villain, "The random field Ising model", in *Scaling phenomena and disordered systems*, NATO ASI, Geilo, Norway (April 1985).
18. Proc. Heidelberg Colloq. on Spin Glasses, Heidelberg (June 1983); K. H. Fischer, Phys. Status Solidi 116 (1983) 357.

19. G. Vichniac, P. Tamayo and H. Hartman, "Annealed and quenched inhomogeneous cellular automata", J. Stat. Phys., to be published.

20. S. Wolfram, "Statistical mechanics of cellular automata", Rev. Mod. Phys. 55 (1983) 601.

21. W. Kinzel, "Phase transitions of cellular automata", Z. Phys. B58 (1985) 229.

22. T. Hogg and B. Huberman, "Parallel computing structures capable of flexible associations and recognition of fuzzy inputs", J. Stat. Phys. 41 (1985) 115.

23. T. Sejnowski and C. R. Rosenberg, "NETtalk: A parallel network that learns to read aloud", Physica D, to be published (Johns Hopkins Elec. Eng. and Comput. Sci. Tech. Report 86-01).

24. C. Mead and L. Conway, *An introduction to VLSI systems*, Addison-Wesley (1980).

25. D. Ackley, G. Hinton and T. Sejnowski, "A learning algorithm for Boltzmann machines", Cognitive Sci. 9 (1985) 147.

26. U. Frisch, B. Hasslacher and Y. Pomeau, "A lattice gas automaton for the Navier-Stokes equation", Los Alamos preprint LA-UR-85-3503; J. Salem and S. Wolfram, "Thermodynamics and hydrodynamics with cellular automata", IAS preprint (November 1985).

27. S. Wolfram, "Glider gun guidelines", report distributed through Computer Recreations section of Scientific American; J. Park, K. Steiglitz and W. Thurston, "Soliton-like behaviour in cellular automata", Princeton University Computer Science Dept. report (1985).

28. A. Smith, "Simple computation-universal cellular spaces", J. ACM 18 (1971) 331; E. R. Berlekamp, J. H. Conway and R. K. Guy, *Winning ways for your mathematical plays*, Academic Press (1982).

29. D. Hillis, *The Connection Machine*, MIT press (1985).

30. S. Kauffman, "Metabolic stability and epigenesis in randomly constructed genetic nets", J. Theoret. Biol. 22 (1969) 437; "Autocatalytic sets of proteins", J. Theor. Biol. (in press).

31. A. Gelfand and C. Walker, "Network modelling techniques: from small scale properties to large scale systems", University of Connecticut report (1982).

32. E. Goles Chacc, "Comportement dynamique de reseaux d'automates", Grenoble University report (1985).

33. C. Leiserson, "Fat trees: universal networks for hardware efficient supercomputing", IEEE Trans. Comput. C-36 (1985) 892.

34. J. Hopcroft and J. Ullman, *Introduction to automata theory, languages and computation*, Addison-Wesley (1979).

35. S. Omohundro, "Connection Machine algorithms primer", Thinking Machines Corporation (Cambridge, Mass.) report in preparation.

36. A. J. Atrubin, "A one-dimensional real-time iterative multiplier", IEEE Trans. Comput. EC-14 (1965) 394.

37. D. Knuth, *Seminumerical algorithms*, Addison-Wesley (1981).

38. H. Nishio, "Real time sorting of binary numbers by 1-dimensional cellular automata", Kyoto University report (1981).

39. T. E. Ingerson and R. L. Buvel, "Structure in asynchronous cellular automata", Physica 10D (1984) 59.

40. G. Vichniac, "Simulating physics with cellular automata", Physica 10D (1984) 96.

41. C. Gerald, *Applied numerical analysis*, Addison-Wesley (1978).

42. M. Creutz, "Deterministic Ising dynamics", Ann. Phys. (in press).

43. A. Grasselli, "Synchronization of cellular arrays: the firing squad problem in two dimensions", Info. & Control 28 (1975) 113.

44. E. Domany and W. Kinzel, "Equivalence of cellular automata to Ising models and directed percolation", Phys. Rev. Lett. 53 (1984) 311.

45. M. Minsky and S. Papert, *Perceptrons*, MIT press (1969).

46. Z. Kohavi, *Switching and finite automata theory*, McGraw-Hill (1970).

47. R. Brayton, G. Hachtel, C. McMullen and A. Sangiovanni-Vincentelli, *Logic minimization algorithms for VLSI synthesis*, Kluwer (1984).

48. L. Valiant, "Short monotone formulae for the majority function", Harvard University report TR-01-84 (1983).

49. M. Conrad, "On design principles for a molecular computer", Commun. ACM 28 (1985) 464.

50. S. Kirkpatrick, C. Gelatt and M. Vecchi, "Optimization by simulated annealing", Science 220 (1983) 671.

51. J. Hopfield and D. Tank, "Neural computation of decisions in optimization problems", Biol. Cybern. 52 (1985) 141.

52. J. Holland, "Genetic algorithms and adaptation", Tech. Rep. #34, Univ. Michigan (1981).

53. A. Barr and E. Feigenbaum, *The handbook of artificial intelligence*, HeurisTech Press (1983), vol. 1.

54. J. Holland, "Escaping brittleness: the possibilities of general purpose learning algorithms applied to parallel rule-based systems", University of Michigan report.

55. D. Lenat, "Computer software for intelligent systems", Scientific American (September 1984).

56. M. Blum and S. Micali, "How to generate cryptographically strong sequences of pseudo-random bits", SIAM J. Comput. 13 (1984) 850.

57. S. Kirkpatrick and G. Toulouse, "Configuration space analysis of travelling salesman problems", J. Physique 46 (1985) 1277.

58. S. Kauffman, "Self-organization, selection, adaptation and its limits: a new pattern of inference in evolution and development", in *Evolution at a crossroads*, D. J. Depew and B. H. Weber (eds.), MIT press (1985).

59. C. J. D. M. Verhagen *et al.*, "Progress report on pattern recognition", Rep. Prog. Phys. 43 (1980) 785; B. Batchelor (ed.), *Pattern recognition*, Plenum (1978).

60. B. Mandelbrot, *The fractal geometry of nature*, Freeman (1982).

61. D. Sankoff and J. Kruskal (eds.), *Time warps, string edits, and macromolecules: the theory and practice of sequence comparison*, Addison-Wesley (1983).

62. M. Minsky, *Society of mind*, in press.

63. L. Valiant, "A theory of the learnable", Commun. ACM 27 (1984) 1134.

Minimal Cellular Automaton Approximations to Continuum Systems

1986

1. Introduction

The basic components of cellular automata are discrete. But at least in some cases the aggregrate behaviour of large numbers of these components can be effectively continuous. As a result, it is possible to use cellular automata as models of continuum systems, such as fluids.

The mathematical origins of continuum behaviour in cellular automata are much the same as they are for many physical systems. A gas, for example, consists of many discrete molecules. Nevertheless, on a large scale, it can be described as a fluid.

Several conditions are necessary for the overall behaviour of a system with discrete elements to seem continuous.

First, continuum behaviour must be associated with some kind of extensive quantity. Such a quantity must be additive, and must be conserved in the dynamical evolution of the system. In a gas, one example of such a quantity is particle number. Other examples are energy and momentum.

A continuum system such as a fluid has the feature that its state can be described (locally) by just a few extensive quantities. To describe the precise microscopic state of a real gas one must, of course, specify the precise configuration of molecules. But it is believed that unless the gas is highly rarefied, this precise configuration is irrelevant to the macroscopic behaviour of the gas. Only the values of the few, averaged, extensive quantities are significant, so that a fluid approximation can be used.

The basis for this belief is embodied in the Second Law of thermodynamics. It seems that almost regardless of the initial microscopic configuration, collisions rapidly tend to randomize the configuration of gas molecules, so that at least for macroscopic purposes, it suffices to specify merely the values of certain average quantities.

Originally presented at Cellular Automata '86 (June 1986).

329

The true basis for this phenomenon has never been very clear. Some descriptions of it can be given in terms of the apparent increase of coarse-grained entropy. But no fundamental derivation has ever been given. The investigation of cellular automaton models seems likely to provide some new insights.

If microscopic randomization is assumed, then overall continuum behaviour can be derived using statistical mechanics. Based on master or transport equations, one can find partial differential equations satisfied by the densities of the extensive quantities conserved by the cellular automaton evolution.

Thus for example there has been much recent work on cellular automata which reproduce the Navier-Stokes equations for viscous fluid flow (see various other CA '86 posters).

Statistical mechanics, and the continuum equations derived from it, provide a considerably reduced description of the system. There may in fact be many systems with different detailed microscopic dynamics, which nevertheless yield identical large-scale statistical or continuum behaviour. Thus, for example, the Navier-Stokes equations describe the aggregate behaviour of fluids such as air and water with very different microscopic constitutions.

Given generic macroscopic behaviour, it is important for both theoretical and practical purposes to try and find the simplest microscopic dynamics which can reproduce the macroscopic behaviour. One may, for example, seek the simplest cellular automaton rule which reproduces a particular form of continuum behaviour. ("Simplest" can be defined for example as requiring minimum storage space and minimum number of logical operations to implement.)

Specific rules which reproduce given macroscopic behaviour can conceivably be produced by explicit construction. Different elements of the rules can for example be arranged to mimic particular forms of particle collisions, and so on. The result of such a procedure will be some rule with the desired behaviour. But it will most likely not be the simplest such rule. Finding the simplest rule is in general a difficult optimization problem.

It is in some respects akin to problems such as logic circuit design in which a device with a particular form of overall behaviour must be constructed with the minimum number of circuit elements. Such problems have recently increasingly been tackled by iterative or adaptive procedures. Some dynamics in the space of possible circuits is defined, and the optimization process consists in applying this dynamics with certain constraints imposed.

Thus one can consider finding minimal cellular automaton rules by various iterative and adaptive procedures.

Such methods are examples of a general approach to computer programming and other design problems which one expects will become increasingly common. At present, most systems are designed in a step-by-step fashion, with their complete progression of states foreseen in detail by the designer. But more efficient designs may potentially be found by a more "goal-oriented" approach. Having specified the constraints, a definite adaptive or iterative procedure traverses the space of possible designs, seeking the one which optimizes some measure of success. The result will typically be a more efficient "computer-generated" design, whose operation cannot necessarily be "understood" in an explicit step-by-step fashion.

This poster considers as an example the problem of finding the simplest cellular automaton rule which reproduces the one-dimensional diffusion equation.

The potential interest of these investigations is severalfold.

1. They may provide practical methods for solving problems related to continuum systems (and these methods may be compared in detail with existing methods).

2. They provide examples of systems which exhibit the basic phenomena of thermodynamics, and should allow further elucidation of the foundations of thermodynamics.

3. They give examples of the procedure of "adaptive programming".

2. The Approach

The diffusion equation can be derived by considering the behaviour of the aggregate density of a large number of particles, each of which executes a random walk. The random walk may result from collisions with other particles of the same kind (as in self diffusion), or from interactions with some separate stochastic background.

The overall statistical behaviour of random walks is well known to be highly insensitive to the precise details of the walk. Thus for example walks whose steps are constrained to lie on various discrete lattices give in the large scale limit the same statistical behaviour as walks whose steps have no constraints.

By constructing a cellular automaton rule which involves various discrete particles, whose total number is conserved, one should thus be able to reproduce the diffusion equation.

A crucial issue, which relates to the foundations of thermodynamics, is the degree of randomness which is produced by a cellular automaton, or which, for that matter, is really necessary to reproduce macroscopic diffusion phenomena.

Nevertheless, following the approach discussed in the introduction, one is concerned not merely with finding some cellular automaton rule which reproduces diffusion, but rather with finding the simplest or optimal one. One must delineate a class of rules capable of reproducing diffusion, and then search within these to find the optimal one.

The conservation laws necessary for macroscopic diffusion turn out to be quite straightforward to ensure in a class of cellular automata. The capability for randomness generation cannot, it seems, be guaranteed directly by the structure of the rule, but must rather be deduced by studying the explicit behaviour of the system.

Diffusion requires that a scalar quantity (which in some cases can be identified as a particle number) is additively conserved.

331

In the simplest cellular automata, one considers rules which specify the new value of a single site in terms of the values of a neighbourhood of sites around it on the previous time step. In most such rules, no additive quantities can be conserved. In addition, such rules are usually highly irreversible, so that they evolve towards attractors which contain only a subset of the possible states. The accessibility of only a subset of states makes an adequate degree of randomness less likely. It does however necessarily preclude diffusion equation behaviour; the various statistical mechanical tools used in derivations can still be applied, but now not to all possible states, but only to those on the attractor.

There are several methods for constructing classes of cellular automata whose evolution satisfies certain conservation laws. (See Y. Pomeau "Invariant in cellular automata", J. Phys. A17 (1984) L415 and N. Margolus "Physics-like models of computation", Physica 10D (1984) 81, both reprinted in *Theory and Applications of Cellular Automata* (edited by S. Wolfram).) The method used here involves considering cellular automata which map one block of sites into another block of the same size.

In the simplest case, one considers a one-dimensional cellular automaton which maps pairs of binary site values to other pairs of binary values. The dynamics is chosen to be such that the boundaries of the pairs are taken to be at even and at odd sites on alternate time steps.

Figure 2.1 shows patterns generated by all the $4^4 = 256$ possible cellular automata of this kind. A variety of phenomena are observed.

Most of the cellular automata of this class show neither additive conservation laws nor reversibility. But unlike cellular automata whose rules are constructed in the usual way, the conditions for conservation and reversibility in these blocked cellular automata are comparatively simple to state.

The condition for reversibility is simply that the mapping from one set of blocks to another be a permutation (so that this mapping is invertible). (There are 24 such rules in the set shown in figure 2.1.)

The condition for additive conservation laws is that for some values v_0 and v_1 the quantity $v_0 N_0 + v_1 N_1$ be conserved, where N_i is the number of sites with value i in each possible block.

Table 2.1 gives the possible rules which satisfy this condition. Two are reversible; two are not. Inspection of figure 2.1 shows that in none of the cases is sufficient randomness generated.

As a result, one must conclude that two possible values at each site ($k = 2$) and block size 2 ($b = 2$) are not sufficient to yield diffusion equation behaviour.

Figure 2.1. Patterns generated by evolution from disordered initial states according to all possible one-dimensional $k = 2$, $b = 2$ blocked cellular automaton rules. These rules have 2 possible values at each site. They are updated by mapping each block of two adjacent sites on to another block of two sites. On one "half step", blocks which begin on even-numbered sites are updated; on the other "half step", blocks beginning at odd-numbered sites are updated. The rules are numbered as follows. The output blocks for each of the possible input blocks 11, 10, 01 and 00 are written down in order. Then each output block is converted to a base 4 digit. The resulting base 4 number is then quoted in base 10.

332

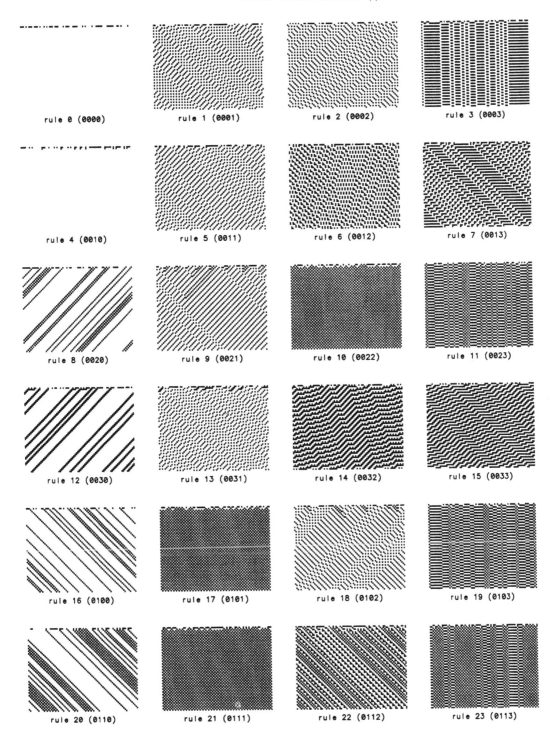

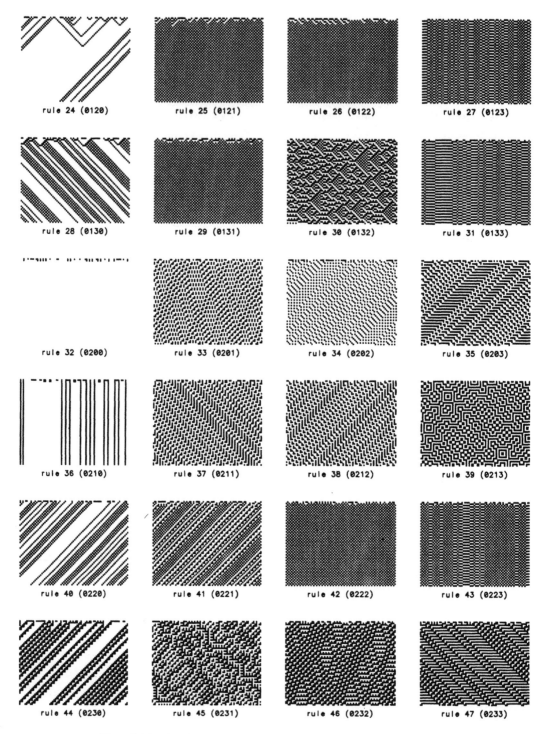

rule 24 (0120)   rule 25 (0121)   rule 26 (0122)   rule 27 (0123)

rule 28 (0130)   rule 29 (0131)   rule 30 (0132)   rule 31 (0133)

rule 32 (0200)   rule 33 (0201)   rule 34 (0202)   rule 35 (0203)

rule 36 (0210)   rule 37 (0211)   rule 38 (0212)   rule 39 (0213)

rule 40 (0220)   rule 41 (0221)   rule 42 (0222)   rule 43 (0223)

rule 44 (0230)   rule 45 (0231)   rule 46 (0232)   rule 47 (0233)

Figure 2.1 (continued).

334

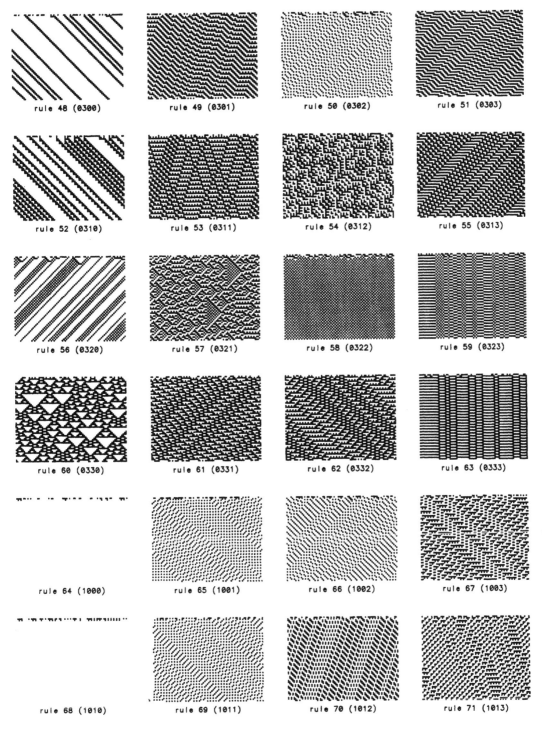

Figure 2.1 (continued).

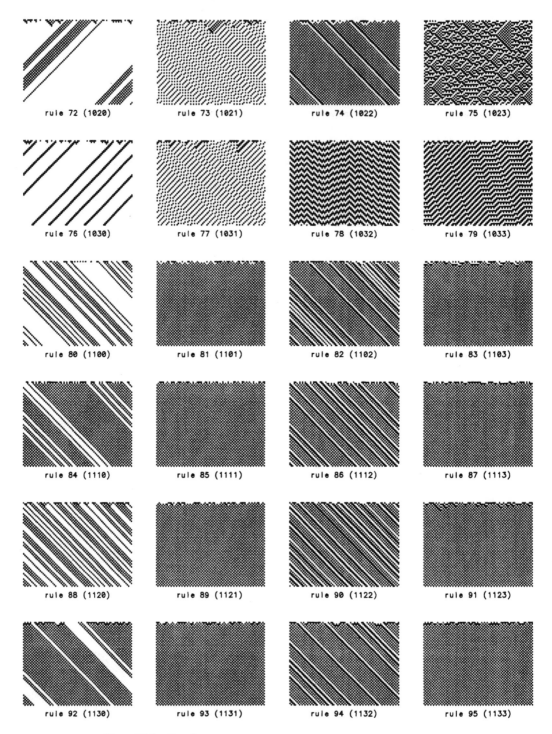

Figure 2.1 (continued).

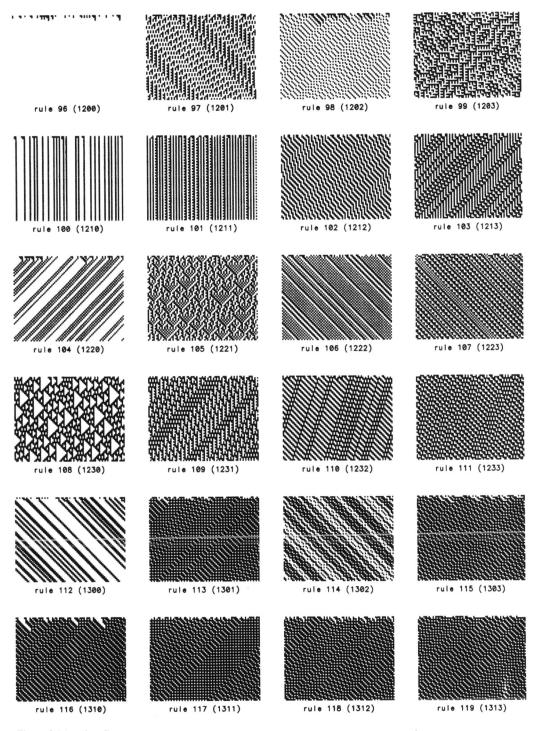

Figure 2.1 (continued).

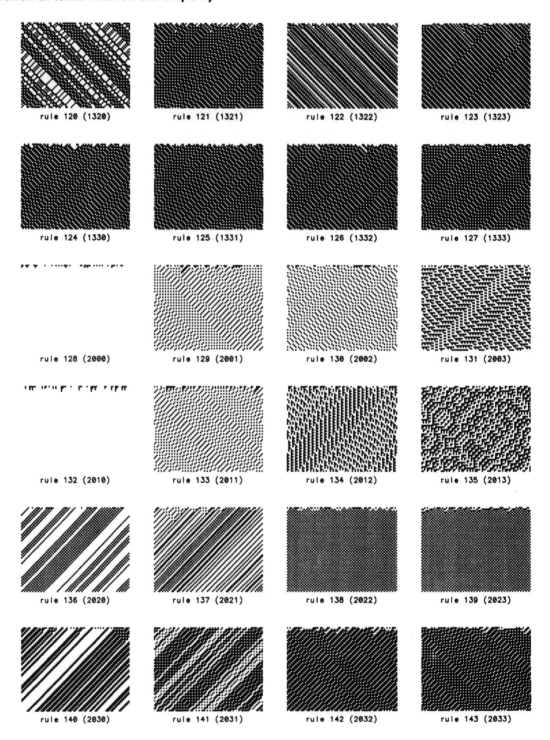

rule 120 (1320) rule 121 (1321) rule 122 (1322) rule 123 (1323)

rule 124 (1330) rule 125 (1331) rule 126 (1332) rule 127 (1333)

rule 128 (2000) rule 129 (2001) rule 130 (2002) rule 131 (2003)

rule 132 (2010) rule 133 (2011) rule 134 (2012) rule 135 (2013)

rule 136 (2020) rule 137 (2021) rule 138 (2022) rule 139 (2023)

rule 140 (2030) rule 141 (2031) rule 142 (2032) rule 143 (2033)

Figure 2.1 (continued).

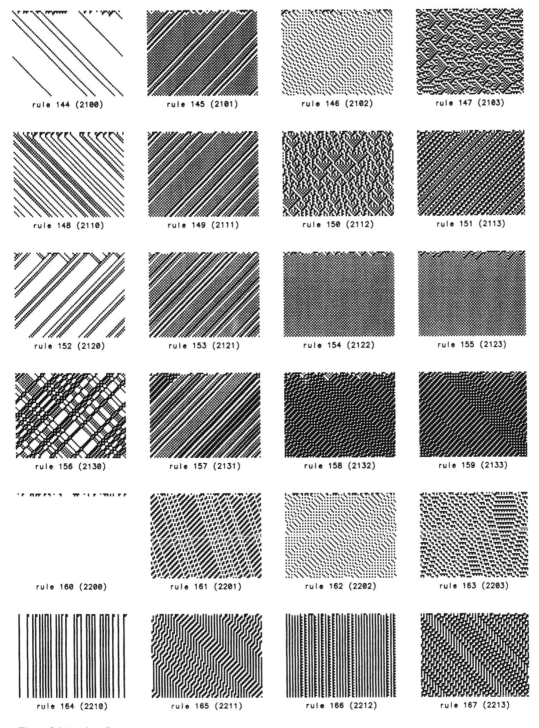

Figure 2.1 (continued).

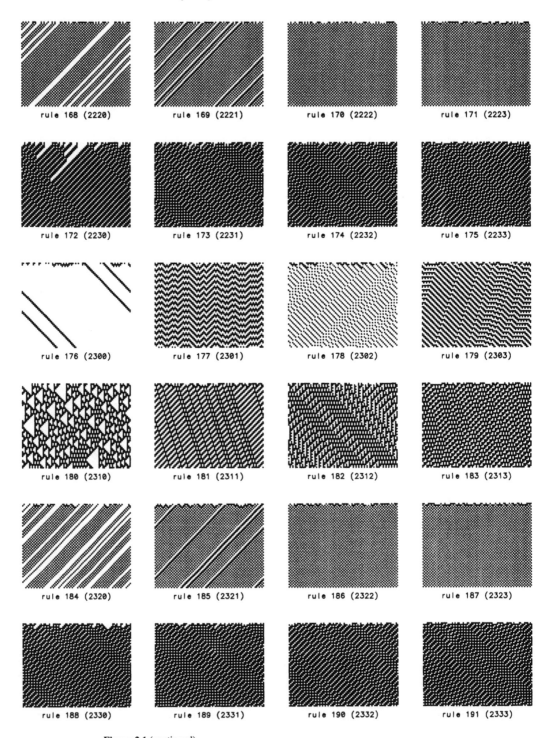

rule 168 (2220)   rule 169 (2221)   rule 170 (2222)   rule 171 (2223)

rule 172 (2230)   rule 173 (2231)   rule 174 (2232)   rule 175 (2233)

rule 176 (2300)   rule 177 (2301)   rule 178 (2302)   rule 179 (2303)

rule 180 (2310)   rule 181 (2311)   rule 182 (2312)   rule 183 (2313)

rule 184 (2320)   rule 185 (2321)   rule 186 (2322)   rule 187 (2323)

rule 188 (2330)   rule 189 (2331)   rule 190 (2332)   rule 191 (2333)

Figure 2.1 (continued).

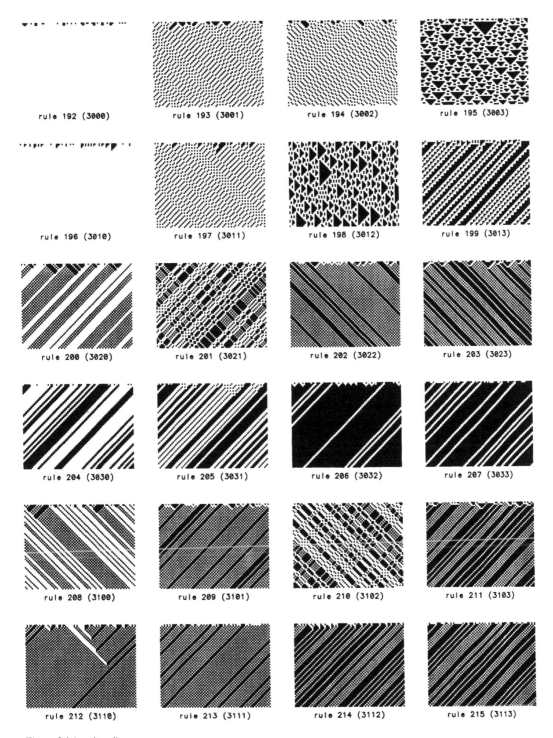

Figure 2.1 (continued).

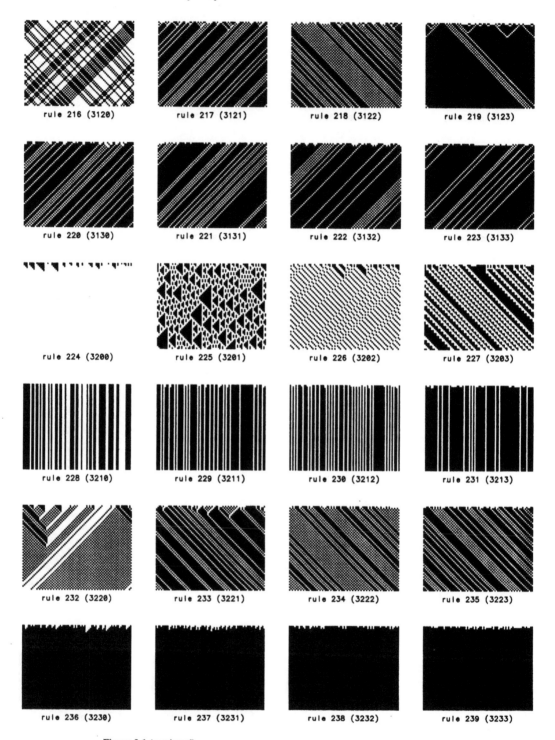

rule 216 (3120)   rule 217 (3121)   rule 218 (3122)   rule 219 (3123)

rule 220 (3130)   rule 221 (3131)   rule 222 (3132)   rule 223 (3133)

rule 224 (3200)   rule 225 (3201)   rule 226 (3202)   rule 227 (3203)

rule 228 (3210)   rule 229 (3211)   rule 230 (3212)   rule 231 (3213)

rule 232 (3220)   rule 233 (3221)   rule 234 (3222)   rule 235 (3223)

rule 236 (3230)   rule 237 (3231)   rule 238 (3232)   rule 239 (3233)

Figure 2.1 (continued).

342

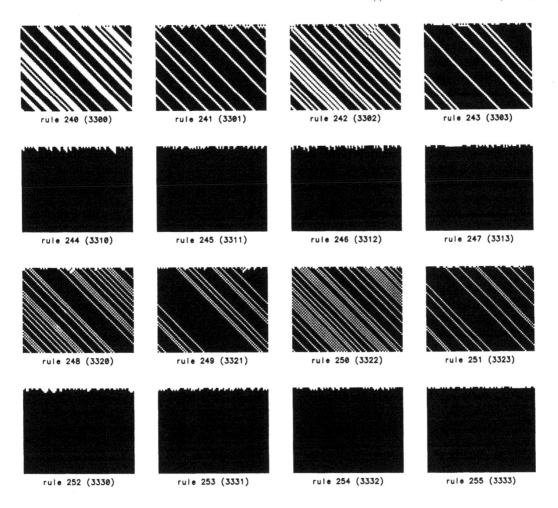

Figure 2.1 (continued).

$n_0 + n_1$ constant	216	invertible
	228	identity
	232	
	212	
$n_0 + n_1$ mod 2 constant	27	invertible
	39	invertible
	23	
	43	

Table 2.1. $k = 2$, $b = 2$ block cellular automaton rules as illustrated in figure 2.1, with certain conservation laws relating to the total numbers n_i of sites with values i.

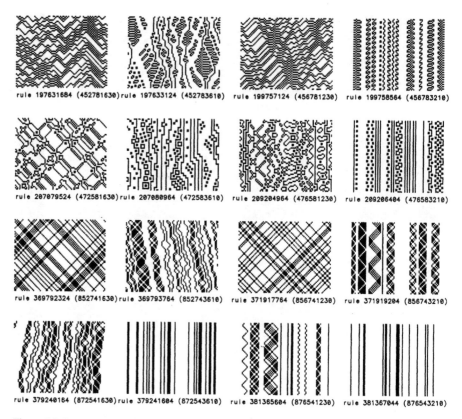

Figure 2.2. Patterns generated by all $k = 3$, $b = 2$ rules which are reversible and conserve the number of binary bits in each configuration. These rules are candidates for simulation of the one-dimensional diffusion equation.

It turns out that $k = 2$, $b = 3$ is also not sufficient.

As a result, one must consider $k = 3$, $b = 2$ rules. (A rough estimate of the "complexity" of rules can be obtained from the number of bits necessary, without compression, to specify their complete rule tables. This number is given by $\log_2\left((k^b)^{(k^b)}\right)$. It is slightly larger for $k = 3$, $b = 2$ than for $k = 2$, $b = 3$.)

With $k = 3$, there is slightly more freedom in the definition of additive quantities. One might, for example, consider adding the numerical values of sites. It turns out, again, that the set of rules with this quantity conserved is too highly constrained to allow a sufficient degree of randomness generation.

An alternative class of rules are those which conserve not the sum of the numerical values of sites, but the total number of binary bits contained in these values. There are 16 possible rules which satisfy this condition, and are reversible. Patterns generated by them are illustrated in figure 2.2.

Some of these rules obviously do not show sufficient randomness to yield diffusion behaviour. But others require more sophisticated analysis.

344

3. Randomization and Thermodynamics

It is observed that many systems, starting from almost any state, evolve rapidly to states which seem for practical purposes random. The sense in which the states are random is that their properties (say, statistical ones) are typical of the ensemble of all possible states. Several explanations and conditions for such randomness have been given. No complete understanding yet exists.

A common approach is based on ergodicity. Only reversible systems can be ergodic. The condition for ergodicity is that starting from any initial state, the evolution of the system eventually visits all possible states. The state transition diagram for the system thus consists of a single large cycle. If a system is ergodic, then at least after a sufficiently long time, it must evolve to an arbitrary, and thus "typical" state. In practice, however, the maximum period of time necessary to reach arbitrary states is usually astronomically large (it is typically exponential in the system size, and comparable to the recurrence time). Evolution for practical times reaches only some small subset of possible states.

What must now be explained is why these states seem random.

This is a subtle issue. There are always special choices of initial conditions for which the states reached are far from random. For example, one could choose initial conditions which are obtained from some orderly state by time reversal of the dynamics for some number of steps. These initial conditions would yield evolution which would not show degradation to randomness: rather it would suddenly yield orderly behaviour, seemingly violating the Second Law of thermodynamics.

One approach often taken is to consider the dependence of the evolution on small changes in initial conditions. It is supposed that the initial conditions cannot be determined precisely, so that in practice, measurements or experimental preparations can be guaranteed to yield only one of an ensemble of states, which differ slightly. The effects of small changes in initial conditions can be seen quite clearly in cellular automata.

One considers the evolution of a cellular automaton from two states, which differ say by a change in the value of a single site. The pattern of differences between states produced as a function of time shows the effect of this small initial perturbation. Figure 3.1 shows such difference patterns for the rules of figure 2.2. In some cases, initial changes remain localized; the evolution in such cases may be considered "stable". (Notice that in a reversible cellular automaton, the effects of changes in initial conditions can never die out completely, because information on the initial state must be preserved.) In other cases ("class 3" cellular automata), small initial changes are progressively amplified by the evolution. Change of the value of one site can ultimately affect the values of sites an arbitrary distance away. The patterns produced by such cellular automata can thus be considered unstable with respect to arbitrarily small perturbations.

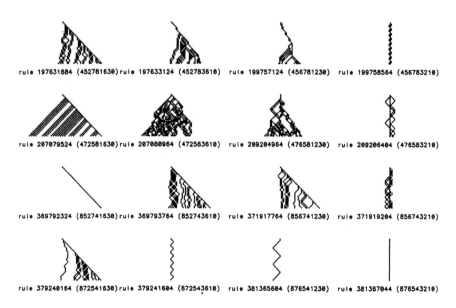

rule 197631684 (452781630) rule 197633124 (452783610) rule 199757124 (456781230) rule 199758564 (456783210)

rule 207079524 (472581630) rule 207080964 (472583610) rule 209204964 (476581230) rule 209206404 (476583210)

rule 369792324 (852741630) rule 369793764 (852743610) rule 371917764 (856741230) rule 371919204 (856743210)

rule 379240164 (872541630) rule 379241604 (872543610) rule 381365604 (876541230) rule 381367044 (876543210)

Figure 3.1. Difference patterns for the rules of figure 2.2. The patterns show the evolution of the difference between two random configurations which initially differ just by a change in a single bit. Some rules are seen to be stable under such perturbations; for other rules, the effect of these changes grows with time. The rate of growth gives the Lyapunov exponent of dynamical systems theory. Such instability leads to a sensitive dependence on the initial conditions for the evolution.

This phenomenon is central to much of what has been studied in the theory of chaotic dynamical systems. It implies that with incomplete knowledge of initial conditions, a time must ultimately come at which the results of measurements can no longer be predicted, because they depend on unknown features of the initial conditions.

It is not however guaranteed that the system will at this point be random. Its randomness depends on the randomness of the unknown features of the initial conditions. It is by no means clear in fact that in actual experiments, these features are indeed adequately random. Certainly one can consider cases in which for example only a few cellular automaton sites are nonzero, and all sites beyond some point are zero. In this case, randomness in final configurations cannot be directly attributed to random unknown data in the initial conditions.

A further, related, problem is the exact definition of "apparently random" states. A sequence or configuration is commonly considered "random" if no pattern can be discerned in it, so that no procedure can be used to predict additional elements of it, or to compress the information associated with it. The meaning of randomness depends on the kinds of pattern recognition which are considered.

If one starts with an orderly initial state, all states generated with time can be specified by giving this state, and the number of steps required to generate them. Such

346

a specification will usually represent a substantial compression in the information associated with the state. Yet despite the possibility for such compression, many aspects of the state may still seem random. Although compression is possible, it may not be revealed by the kinds of statistical procedures commonly used to analyse the states.

Figure 3.2 shows examples of some simple $k = 2$, $r = 1$ cellular automata which illustrate this phenomenon. In each case, a simple initial condition is chosen, consisting of a single nonzero site. With these initial conditions, some cellular automata yield simple patterns, and sequences of sites in these patterns are for example periodic. Other cellular automata yield slightly more complicated, self similar, patterns. But here again sequences of site values are almost periodic, and are readily predictable. Some cellular automata, however, can yield apparently random sequences even starting from these simple initial conditions. The two simplest examples (found by explicit search) are rules 30 and 45. In both cases, the sequences generated seem random according to all standard statistical tests (see S. Wolfram, "Random sequence generation by cellular automata", Adv. Applied Math. 7 (1986) 123 and in *Theory and Applications of Cellular Automata*). Figure 3.3 shows a more detailed example of evolution according to rule 30.

The phenomenon observed in this case occurs in other mathematical systems. Even though a simple specification for π, for example, can be given, its digit sequence, once generated, seems random for all practical purposes. The fractional parts of successive powers of 3/2, which can be generated by a $k = 6$, $r = 1$ cellular automaton, provide another example. In all cases, what is observed is that a sequence which can be generated easily can be hard to invert or compress. This phenomenon is the basis for the possibility of pseudorandom number generation or cryptography. Given a short seed or key as an initial specification, there are algorithms (such as that of figure 3.3) which yield long sequences from which the simplicity of the initial conditions is not apparent. The dynamics of the evolution has effectively "encrypted" the initial data to the point where it cannot be recovered by any simple computation.

Computation theory provides a characterization of this phenomenon. The process of generating a sequence is in the polynomial time class P. But the process of recognizing the origins of the sequence is in the class NP of non-deterministic polynomial time computations. It seems that P ≠ NP so that there exist at least some cases in which the problem of recognition cannot be solved by a polynomial time computation.

It is clear that an exhaustive search through all possible initial conditions would reveal whether any "simple" one yielded a particular sequence. But the number of such possible initial conditions is exponentially large, so that such a search could take an exponentially long time. As a result, it would rapidly become infeasible.

The standard statistical tests of randomness applied to physical systems are computationally quite simple. As a result, they are unable to detect regularities that require say exponential time computations to recognize. Thus if a system "encrypts" its initial data to the same degree as that of say figure 3.3 does, it will yield behaviour that appears random for practical purposes.

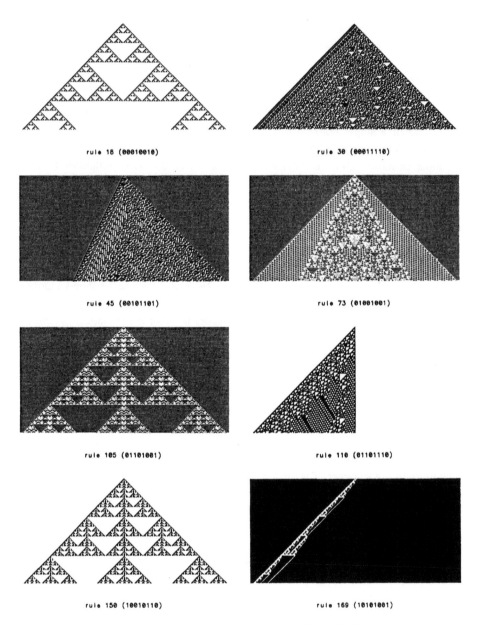

rule 18 (00010010)

rule 30 (00011110)

rule 45 (00101101)

rule 73 (01001001)

rule 105 (01101001)

rule 110 (01101110)

rule 150 (10010110)

rule 169 (10101001)

Figure 3.2. Examples of patterns generated by various simple $k = 2$, $r = 1$ cellular automata, evolving from a single nonzero initial site. Some rules are seen to give comparatively simple patterns, while other rules give patterns which seem in many respects random. The generation of randomness in this way may well be the source of thermodynamic behaviour in many systems. It is necessary for the reproduction of continuum phenomena such as diffusion.

Figure 3.3. The pattern generated by $k = 2$, $r = 1$ rule 30 starting from a single site initial seed. This rule has the form

$$a'_i = (a_{i-1} + a_i + a_{i+1} + a_i a_{i+1}) \bmod 2.$$

Despite the simplicity of this rule, the patterns it generates are so complicated as to seem in many respects random. Thus for example the centre column in this picture seems random for at least a million sites according to standard statistical randomness tests. This rule is probably the simplest cellular automaton which generates random behaviour in this way. It was found by an explicit search over all possible rules.

I believe that most of the randomization associated with thermodynamic behaviour is of the mathematical type illustrated in figure 3.3. Even though the initial conditions are simple, the system encrypts them to the point where no feasible measurements or computations can recover them.

4. The Winning Rule

The phenomenon of randomization from simple initial conditions occurs in some but not all of the candidate diffusion equation cellular automata of figure 2.2. Figure 4.1

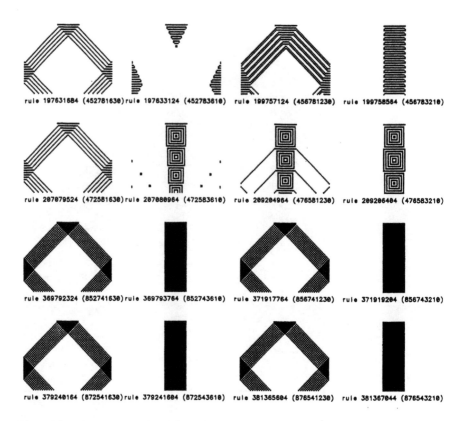

Figure 4.1. Patterns generated by the $k = 3$, $b = 2$ blocked cellular automata of figure 2.2, starting from a simple initial condition. The cellular automata are shown on a size 80 lattice with periodic boundary conditions. The initial condition consists of a block of 20 sites with value 1 in the centre of the system. Most of the rules are seen to give rise to simple periodic patterns.

shows evolution from a simple initial condition for all of these rules. Only one rule, and its (2,1) conjugate, show randomization in this case. Figure 4.2 shows the longer time evolution of this rule, on a size 80 with periodic boundary conditions. Regularities are still seen, but many features seem random.

The degree of randomness generated by this rule can be tested by applying certain statistical procedures. A simple one is the computation of coarse-grained entropy. Figure 4.3 shows the coarse-grained entropy for the system. It is seen to tend rapidly to a maximum value, as expected for an apparently random system.

Table 4.1 gives the block transformations for this rule. Interpretations in terms of particles and so on can be given. But it is noteworthy that making the rule "increasingly mixing" by including transitions for various other blocks does not yield an increase in the randomness of the overall behaviour. In fact, as figures 2.2 and 4.1 show, such "additional mixing" usually leads to simpler overall behaviour.

350

Figure 4.2. Longer sequences generated by one rule from figure 4.1 which seems to generate randomness from simple initial conditions. The patterns on this page were made on a size 80 lattice, with a size 20 initial block. The patterns on the next page were made with a size 21 initial block. The degradation of orderly initial conditions into apparent randomness is clearly visible in these pictures.

351

Figure 4.2 (continued).

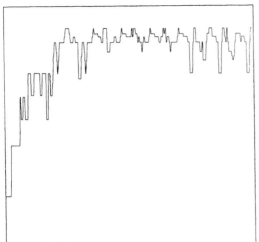

Figure 4.3. Time development of the coarse-grained entropy associated with the process of figure 4.2. The density of bits was computed in 10 bins across the system. Then the densities p_i found were combined to give the entropy

$$s = -\sum_i p_i \log p_i.$$

This coarse-grained entropy is seen to tend to a maximum, as expected from the Second Law of thermodynamics.

The large scale average density of bits in evolution according to the rule of table 4.1 should satisfy a diffusion equation. Figure 4.4 shows the microscopic dynamics of this rule for the cases of low and high bit density. At low bit densities, the rule exhibits particle dynamic phenomena, as might be seen in a rarefied gas. At high bit densities, however, it acts like a dense gas, and defects or particles executing apparently random walks can be seen.

$22 \rightarrow \mathbf{11}$	**Table 4.1.** Block transitions which define the $k = 3$, $b = 2$ rule which reproduces the
$21 \rightarrow 21$	diffusion equation. Blocks which change under the rule are shown in bold. The rule is
$20 \rightarrow \mathbf{02}$	applied on alternate time steps to even and odd blocks in the one-dimensional cellular
$12 \rightarrow 12$	automaton configuration. The rule is arranged to be reversible, so that each block has
$\mathbf{11} \rightarrow \mathbf{22}$	a unique predecessor as well as a unique successor under the time evolution. It is also
$10 \rightarrow 10$	bit conserving, so that the total number of binary bits in each block is invariant under
$\mathbf{02} \rightarrow \mathbf{20}$	these transitions.
$01 \rightarrow 01$	
$00 \rightarrow 00$	

The microscopic configurations of this system are highly sensitive to small changes in initial conditions. Figure 4.5 shows the pattern of differences associated with the change in single initial site value. The pattern of differences is seen to expand at a fixed "speed of sound".

The overall average behaviour of this system however obeys the diffusion equation, and so is insensitive to small changes. This phenomenon is just the same as occurs in real gases.

The cellular automaton of table 4.1 can be considered as a system which contains particles executing random walks. What is perhaps remarkable about it is that the randomness necessary to produce appropriate average behaviour in these walks is generated intrinsically by the system, apparently at a low computational cost.

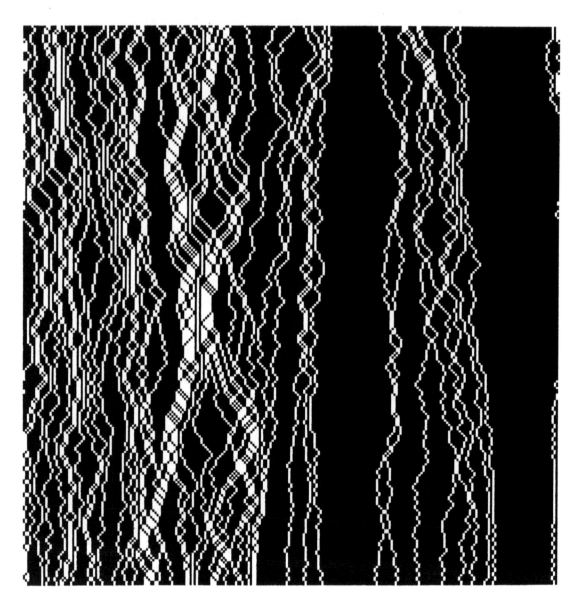

Figure 4.4. Microscopic diffusion at two densities in the minimal cellular automaton approximation to the one-dimensional diffusion equation.

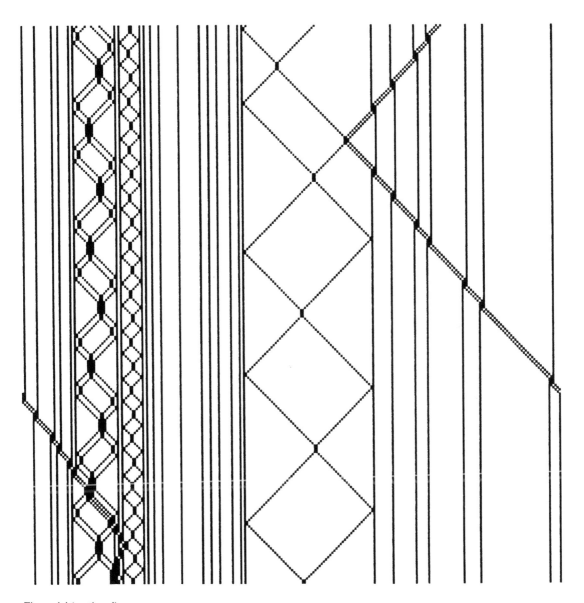

Figure 4.4 (continued).

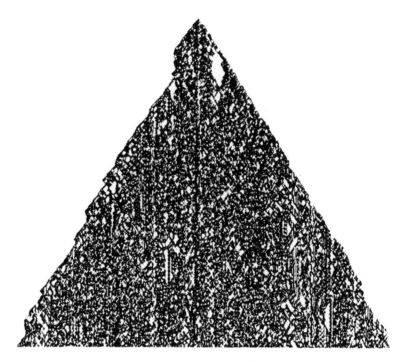

Figure 4.5. Difference pattern for the rule of figure 4.2. This shows the bits which change as a result of a change in a single initial bit in a random initial configuration.

One can consider this system as a random sequence generator. The effectiveness of the system as a model for diffusion is related to its effectiveness as a random sequence generator.

One issue is what the global behaviour of the system on a finite lattice is. Since the system is reversible, all states lie on cycles. Table 4.2 gives the multiplicities and sizes of these cycles for various lattice sizes.

The lattice sizes so far investigated are not large enough to determine whether the maximum cycle time for the system does indeed increase exponentially with its size.

The exact sets of cycles that occur for particular lattice sizes depend on the number-theoretical properties of the lattice size. It is clear that the system is not ergodic, since there are often many distinct cycles. Some of these cycles may however be largely spurious. For example, when the lattice size is not prime, there are classes of initial states whose site values show a periodicity which is some divisor of the lattice size. Such classes of states must lie on distinct cycles.

For complete randomization to occur, the system should have no conservation laws other than that of total bit number. The presence of multiple cycles implies that some other conservation laws may exist. However, no simple invariant quantities seem likely to be associated with these additional conservation laws.

---

size 7

total number: 1918; number distinct lengths: 6

$3 \times 6, 2 \times 5, 8 \times 4, 24 \times 3, 174 \times 2, 1707 \times 1$

---

size 9

total number: 17135; number distinct lengths: 11

$1 \times 12, 1 \times 10, 1 \times 9, 7 \times 8, 12 \times 7, 19 \times 6, 31 \times 5, 93 \times 4, 182 \times 3, 1537 \times 2, 15251 \times 1$

---

size 11

total number: 24219; number distinct lengths: 143

$4 \times 816, 4 \times 672, 4 \times 654, 12 \times 547, 4 \times 540, 4 \times 372, 8 \times 366, 4 \times 354, 12 \times 349, 4 \times 342, 6 \times 330, 12 \times 315$

$4 \times 312, 2 \times 270, 12 \times 264, 2 \times 246, 12 \times 244, 8 \times 240, 12 \times 239, 8 \times 234, 8 \times 225, 8 \times 222, 24 \times 220, 24 \times 219$

$12 \times 194, 8 \times 183, 12 \times 179, 14 \times 174, 12 \times 169, 36 \times 168, 4 \times 162, 12 \times 161, 8 \times 159, 12 \times 153, 18 \times 150$

$12 \times 149, 36 \times 148, 12 \times 143, 12 \times 141, 24 \times 139, 10 \times 138, 12 \times 137, 12 \times 135, 6 \times 130, 36 \times 126, 36 \times 124$

$12 \times 121, 40 \times 120, 12 \times 119, 12 \times 117, 24 \times 115, 28 \times 114, 12 \times 110, 12 \times 109, 48 \times 108, 24 \times 103, 36 \times 102$

$28 \times 96, 12 \times 95, 48 \times 94, 48 \times 93, 24 \times 91, 46 \times 90, 48 \times 89, 12 \times 86, 12 \times 85, 16 \times 84, 12 \times 83, 24 \times 81, 24 \times 80$

$12 \times 79, 26 \times 78, 12 \times 77, 24 \times 75, 24 \times 74, 60 \times 73, 64 \times 72, 24 \times 71, 32 \times 69, 72 \times 68, 12 \times 67, 84 \times 66, 60 \times 65$

$24 \times 64, 64 \times 63, 168 \times 60, 36 \times 59, 32 \times 57, 48 \times 56, 48 \times 55, 168 \times 54, 12 \times 52, 24 \times 51, 12 \times 50, 108 \times 49$

$164 \times 48, 84 \times 47, 132 \times 46, 124 \times 45, 108 \times 44, 168 \times 43, 266 \times 42, 132 \times 41, 72 \times 40, 96 \times 39, 156 \times 38$

$96 \times 37, 136 \times 36, 132 \times 35, 108 \times 34, 116 \times 33, 72 \times 32, 84 \times 31, 218 \times 30, 60 \times 29, 84 \times 28, 156 \times 27, 258 \times 26$

$192 \times 25, 140 \times 24, 168 \times 23, 162 \times 22, 408 \times 21, 222 \times 20, 360 \times 19, 304 \times 18, 372 \times 17, 492 \times 16, 1018 \times 15$

$498 \times 14, 528 \times 13, 576 \times 12, 546 \times 11, 612 \times 10, 1415 \times 9, 1710 \times 8, 1194 \times 7, 1740 \times 6, 495 \times 5, 1248 \times 4$

$1725 \times 3, 1460 \times 2, 1190 \times 1$

Table 4.2. Cycles in finite size systems evolving according to the rule of table 4.1. In each case, the cellular automaton is taken to have periodic boundary conditions. The multiplicities and sizes of all cycles are given. For width 11, the cycles have lengths which contain all primes up to 149, excluding 101, 107, 113, 127 and 131. An ergodic system would have just one cycle.

5. Discussion

This poster has illustrated a simple cellular automaton rule which exhibits continuum average behaviour in the large scale limit. It is possible to construct similar rules in two dimensions, and to give various other kinds of continuum behaviour (see several CA '86 posters).

In each case, one may compare the cellular automaton rules with traditional approaches to emulating these continuum systems on digital computers. In the conventional approach, one starts from partial differential equations, then makes discrete numerical approximations to them. These approximations involve considering a discrete lattice of points. But unlike in cellular automata, each of these lattice points carries a continuous variable which represents the precise value of a continuum quantity, such as particle density, at that point. In actual computer implementations, the continuous variable is represented by a floating-point number, say 64 bits in length. The number is updated in a sequence of time steps, according to a discrete rule. The rule in general involves arithmetic operations, which cannot be carried out precisely on the finite precision number. As a result, low-order bits of the number are truncated.

Numerical analysis has studied in detail the propagation of such round-off errors, and has suggested schemes which minimize their effects.

In a cellular automaton, the values of variables such as particle density are stored in a distributed fashion. It is necessary to average over a region of the system to find the values of such macroscopic variables. Each bit which contributes to this average is however treated according to a precise deterministic rule, and each bit is equally important. Nevertheless, the need for averaging introduces $1/\sqrt{N}$ fluctuations in the values of measured quantities. For some systems, such as turbulent ones, only statistical averages are expected to be reproducible. But in others, such as the diffusion equation, the need for averaging represents a limit on accuracy.

One can imagine a hybrid of cellular automaton and numerical analysis schemes. Consider the case of the diffusion equation. On a lattice of sites, one stores values which consist of sequences of bits. The high-order bits are encoded digitally, so that n bits can represent 2^n possible numbers of particles. The low-order bits are however encoded in unary, and correspond to individual particles. The update scheme can conserve the total number of particles.

Viewed as a numerical analysis procedure, the dynamics of the low-order bits represents a dynamics of round-off errors. Instead of systematically truncating the numbers, their low-order bits are modified according to dynamics which yields effectively random behaviour. The result is similar to random round-off, but includes a precise particle conservation law.

By adjusting the number of unary and digital bits, one can determine the tradeoffs between cellular automaton and numerical analysis approaches.

One of the major issues in numerical analysis is convergence. This is very difficult to prove for all but the simplest equations and the simplest schemes. But in cellular automata, the analogue of convergence is the process of coming to "thermodynamic" equilibrium. Thus the problem of "convergence" is related to a fundamental problem of physics.

Cellular Automaton Fluids: Basic Theory

1986

Continuum equations are derived for the large-scale behavior of a class of cellular automaton models for fluids. The cellular automata are discrete analogues of molecular dynamics, in which particles with discrete velocities populate the links of a fixed array of sites. Kinetic equations for microscopic particle distributions are constructed. Hydrodynamic equations are then derived using the Chapman-Enskog expansion. Slightly modified Navier-Stokes equations are obtained in two and three dimensions with certain lattices. Viscosities and other transport coefficients are calculated using the Boltzmann transport equation approximation. Some corrections to the equations of motion for cellular automaton fluids beyond the Navier-Stokes order are given.

1. Introduction

Cellular automata (e.g., Refs. 1 and 2) are arrays of discrete cells with discrete values. Yet sufficiently large cellular automata often show seemingly continuous macroscopic behavior (e.g., Refs. 1 and 3). They can thus potentially serve as models for continuum systems, such as fluids. Their underlying discreteness, however, makes them particularly suitable for digital computer simulation and for certain forms of mathematical analysis.

On a microscopic level, physical fluids also consist of discrete particles. But on a large scale, they, too, seem continuous, and can be described by the partial differential equations of hydrodynamics (e.g., Ref. 4). The form of these equations is in fact quite insensitive to microscopic details. Changes in molecular interaction laws can affect parameters such as viscosity, but do not alter the basic form of the macroscopic equations. As a result, the overall behavior of fluids can be found without accurately reproducing the details of microscopic molecular dynamics.

Originally published in *Journal of Statistical Physics*, volume 45, pages 471–526 (November 1986).

This paper is the first in a series which considers models of fluids based on cellular automata whose microscopic rules give discrete approximations to molecular dynamics.[1] The paper uses methods from kinetic theory to show that the macroscopic behavior of certain cellular automata corresponds to the standard Navier-Stokes equations for fluid flow. The next paper in the series[16] describes computer experiments on such cellular automata, including simulations of hydrodynamic phenomena.

Figure 1 shows an example of the structure of a cellular automaton fluid model. Cells in an array are connected by links carrying a bounded number of discrete "particles." The particles move in steps and "scatter" according to a fixed set of deterministic rules. In most cases, the rules are chosen so that quantities such as particle number and momentum are conserved in each collision. Macroscopic variations of such conserved quantities can then be described by continuum equations.

Particle configurations on a microscopic scale are rapidly randomized by collisions, so that a local equilibrium is attained, described by a few statistical average quantities. (The details of this process will be discussed in a later paper.) A master equation can then be constructed to describe the evolution of average particle densities as a result of motion and collisions. Assuming slow variations with position and time, one can then write these particle densities as an expansion in terms of macroscopic quantities such as momentum density. The evolution of these quantities is determined by the original master equation. To the appropriate order in the expansion, certain cellular automaton models yield exactly the usual Navier-Stokes equations for hydrodynamics.

The form of such macroscopic equations is in fact largely determined simply by symmetry properties of the underlying cellular automaton. Thus, for example, the structure of the nonlinear and viscous terms in the Navier-Stokes equations depends on the possible rank three and four tensors allowed by the symmetry of the cellular automaton array. In two dimensions, a square lattice of particle velocities gives anisotropic forms for these terms.[6] A hexagonal lattice, however, has sufficient symmetry to ensure isotropy.[7] In three dimensions, icosahedral symmetry would guarantee isotropy, but no crystallographic lattice with such a high degree of symmetry exists. Various structures involving links beyond nearest neighbors on the lattice can instead be used.

Although the overall form of the macroscopic equations can be established by quite general arguments, the specific coefficients which appear in them depend on details of the underlying model. In most cases, such transport coefficients are found from explicit simulations. But, by using a Boltzmann approximation to the master equation, it is possible to obtain some exact results for such coefficients, potentially valid in the low-density limit.

---

[1] This work has many precursors. A discrete model of exactly the kind considered here was discussed in Ref. 6. A version on a hexagonal lattice was introduced in Ref. 7, and further studied in Refs. 8, 9. Related models in which particles have a discrete set of possible velocities, but can have continuously variable positions and densities, were considered much earlier.[10-14] Detailed derivations of hydrodynamic behavior do not, however, appear to have been given even in these cases (see, however, e.g., Ref. 15).

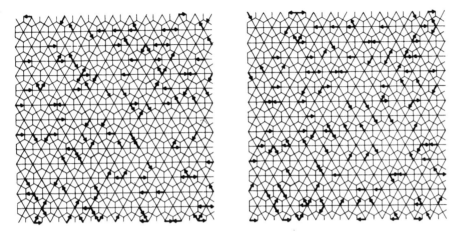

Figure 1. Two successive microscopic configurations in the typical cellular automaton fluid model discussed in Section 2. Each arrow represents a discrete "particle" on a link of the hexagonal grid. Continuum behavior is obtained from averages over large numbers of particles.

This paper is organized as follows. Section 2 describes the derivation of kinetic and hydrodynamic equations for a particular sample cellular automaton fluid model. Section 3 generalizes these results and discusses the basic symmetry conditions necessary to obtain standard hydrodynamic behavior. Section 4 then uses the Boltzmann equation approximation to investigate microscopic behavior and obtain results for transport coefficients. Section 5 discusses a few extensions of the model. The Appendix gives an SMP program[17] used to find macroscopic equations for cellular automaton fluids.

2. Macroscopic Equations for a Sample Model

2.1. Structure of the Model

The model[7] is based on a regular two-dimensional lattice of hexagonal cells, as illustrated in Fig. 1. The site at the center of each cell is connected to its six neighbors by links corresponding to the unit vectors $\mathbf{e}_1$ through $\mathbf{e}_6$ given by

$$\mathbf{e}_a = (\cos(2\pi a/6),\ \sin(2\pi a/6)) \tag{2.1.1}$$

At each time step, zero or one particles lie on each directed link. Assuming unit time steps and unit particle masses, the velocity and momentum of each particle is given simply by its link vector $\mathbf{e}_a$. In this model, therefore, all particles have equal kinetic energy, and have zero potential energy.

The configuration of particles evolves in a sequence of discrete time steps. At each step, every particle first moves by a displacement equal to its velocity $\mathbf{e}_a$. Then the particles on the six links at each site are rearranged according to a definite set of rules. The rules are chosen to conserve the number and total momentum of the particles. In a typical case, pairs of particles meeting head on might scatter through 60°, as would triples of particles 120° apart. The rules may also rearrange other configurations, such as triples of particles meeting asymmetrically. Such features are important in determining parameters such as viscosity, but do not affect the form of the macroscopic equations derived in this section.

To imitate standard physical processes, the collision rules are usually chosen to be microscopically reversible. There is therefore a unique predecessor, as well as a unique successor, for each microscopic particle configuration. The rules for collisions in each cell thus correspond to a simple permutation of the possible particle arrangements. Often the rules are self-inverse. But in any case, the evolution of a complete particle configuration can be reversed by applying inverse collision rules at each site.

The discrete nature of the cellular automaton model makes such precise reversal in principle possible. But the rapid randomization of microscopic particle configurations implies that very complete knowledge of the current configuration is needed. With only partial information, the evolution may be effectively irreversible.[8,19]

2.2. Basis for Kinetic Theory

Cellular automaton rules specify the precise deterministic evolution of microscopic configurations. But if continuum behavior is seen, an approximate macroscopic description must also be possible. Such a description will typically be a statistical one, specifying not, for example, the exact configuration of particles, but merely the probabilities with which different configurations appear.

A common approach is to consider ensembles in which each possible microscopic configuration occurs with a particular probability (e.g., Ref. 18). The reversibility of the microscopic dynamics ensures that the total probability for all configurations in the ensemble must remain constant with time. The probabilities for individual configurations may, however, change, as described formally by the Liouville equation.

An ensemble is in "equilibrium" if the probabilities for configurations in it do not change with time. This is the case for an ensemble in which all possible configurations occur with equal probability. For cellular automata with collision rules that conserve momentum and particle number, the subsets of this ensemble that contain only those configurations with particular total values of the conserved quantities also correspond to equilibrium ensembles.

If the collision rules effectively conserved absolutely no other quantities, then momentum and particle number would uniquely specify an equilibrium ensemble. This would be the case if the system were ergodic, so that starting from any initial configuration, the system would eventually visit all other microscopic configurations with the same values of the conserved quantities. The time required would, however, inevitably be exponentially long, making this largely irrelevant for practical purposes.

A more useful criterion is that starting from a wide range of initial ensembles, the system evolves rapidly to ensembles whose statistical properties are determined solely from the values of conserved quantities. In this case, one could assume for statistical purposes that the ensemble reached contains all configurations with these values of the conserved quantities, and that the configurations occur with equal probabilities. This assumption then allows for the immediate construction of kinetic equations that give the average rates for processes in the cellular automaton.

The actual evolution of a cellular automaton does not involve an ensemble of configurations, but rather a single, specific configuration. Statistical results may nevertheless be applicable if the behavior of this single configuration is in some sense "typical" of the ensemble.

This phenomenon is in fact the basis for statistical mechanics in many different systems. One assumes that appropriate space or time averages of an individual configuration agree with averages obtained from an ensemble of different configurations. This assumption has never been firmly established in most practical cases; cellular automata may in fact be some of the clearest systems in which to investigate it.

The assumption relies on the rapid randomization of microscopic configurations, and is closely related to the second law of thermodynamics. At least when statistical or coarse-grained measurements are made, configurations must seem progressively more random, and must, for example, show increasing entropies. Initially ordered configurations must evolve to apparent disorder.

The reversibility of the microscopic dynamics nevertheless implies that ordered initial configurations can always in principle be reconstructed from a complete knowledge of these apparently disordered states. But just as in pseudorandom sequence generators or cryptographic systems, the evolution may correspond to a sufficiently complex transformation that any regularities in the initial conditions cannot readily be discerned. One suspects in fact that no feasibly simple computation can discover such regularities from typical coarse-grained measurements.[19,20] As a result, the configurations of the system seem random, at least with respect to standard statistical procedures.

While most configurations may show progressive randomization, some special configurations may evolve quite differently. Configurations obtained by computing time-reversed evolution from ordered states will, for example, evolve back to ordered states. Nevertheless, one suspects that the systematic construction of such "antithermodynamic" states must again require detailed computations of a complexity beyond that corresponding to standard macroscopic experimental arrangements.

Randomization requires that no additional conserved quantities are present. For some simple choices of collision rules, spurious conservation laws can nevertheless be present, as discussed in Section 4.5. For most of the collision rules considered in this paper, however, rapid microscopic randomization does seem to occur.

As a result, one may use a statistical ensemble description. Equilibrium ensembles in which no statistical correlations are present should provide adequate approximations for many macroscopic properties. At a microscopic level, however, the deterministic dynamics does lead to correlations in the detailed configurations of particles.[2] Such correlations are crucial in determining local properties of the system. Different levels of approximation to macroscopic behavior are obtained by ignoring correlations of different orders.

Transport and hydrodynamic phenomena involve systems whose properties are not uniform in space and time. The uniform equilibrium ensembles discussed above cannot provide exact descriptions of such systems. Nevertheless, so long as macroscopic properties vary slowly enough, collisions should maintain approximate local equilibrium, and should make approximations based on such ensembles accurate.

2.3. Kinetic Equations

An ensemble of microscopic particle configurations can be described by a phase space distribution function which gives the probability for each complete configuration. In studying macroscopic phenomena, it is, however, convenient to consider reduced distribution functions, in which an average has been taken over most degrees of freedom in the system. Thus, for example, the one-particle distribution function $f_a(\mathbf{x}, t)$ gives the probability of finding a particle with velocity $\mathbf{e}_a$ at position $\mathbf{x}$ and time t, averaged over all other features of the configuration (e.g., Ref. 23).

Two processes lead to changes in f_a with time: motion of particles from one cell to another, and interactions between particles in a given cell. A master equation can be constructed to describe these processes.

In the absence of collisions, the cellular automaton rules imply that all particles in a cell at position $\mathbf{X}$ with velocity $\mathbf{e}_a$ move at the next time step to the adjacent cell at position $\mathbf{X} + \mathbf{e}_a$. As a result, the distribution function evolves according to

$$f_a(\mathbf{X} + \mathbf{e}_a, T + 1) = f_a(\mathbf{X}, T) \tag{2.3.1}$$

For large lattices and long time intervals, position and time may be approximated by continuous variables. One may define, for example, scaled variables $\mathbf{x} = \delta_x \mathbf{X}$ and $t = \delta_t T$, where $\delta_x, \delta_t \ll 1$. In terms of these scaled variables, the difference equation (2.3.1) becomes

$$f_a(\mathbf{x} + \mathbf{e}_a \delta_x, t + \delta_t) - f_a(\mathbf{x}, t) = 0 \tag{2.3.2}$$

In deriving macroscopic transport equations, this must be converted to a differential equation. Carrying out a Taylor expansion, one obtains[24]

$$\delta_t \partial_t f_a + \delta_x \mathbf{e}_a \cdot \nabla f_a + \frac{1}{2}\delta_t^2 \partial_{tt} f_a + \delta_x \delta_t (\mathbf{e}_a \cdot \nabla)\partial_t f_a + \frac{1}{2}\delta_x^2 (\mathbf{e}_a \cdot \nabla)^2 f_a + O(\delta^3) = 0 \tag{2.3.3}$$

---

[2] The kinetic theory approach used in this paper concentrates on average particle distribution functions. An alternative but essentially equivalent approach concentrates on microscopic correlation functions (e.g., Refs. 21, 22).

If all variations in the f_a are assumed small, and certainly less than $O(1/\delta_x, 1/\delta_t)$, it suffices to keep only first-order terms in δ_x, δ_t. In this way one obtains the basic transport equation

$$\partial_t f_a(\mathbf{x}, t) + \mathbf{e}_a \cdot \nabla f_a(\mathbf{x}, t) = 0 \qquad (2.3.4)$$

This has the form of a collisionless Boltzmann transport equation for f_a (e.g., Ref. 25). It implies, as expected, that f_a is unaffected by particle motion in a spatially uniform system.

Collisions can, however, change f_a even in a uniform system, and their effect can be complicated. Consider, for example, collisions that cause particles in directions $\mathbf{e}_1$ and $\mathbf{e}_4$ to scatter in directions $\mathbf{e}_2$ and $\mathbf{e}_5$. The rate for such collisions is determined by the probability that particles in directions $\mathbf{e}_1$ and $\mathbf{e}_4$ occur together in a particular cell. This probability is defined as the joint two-particle distribution function $\tilde{F}_{14}^{(2)}$. The collisions deplete the population of particles in direction $\mathbf{e}_1$ at a rate $\tilde{F}_{14}^{(2)}$. Microscopic reversibility guarantees the existence of an inverse process, which increases the population of particles in direction $\mathbf{e}_1$ at a rate given in this case by $\tilde{F}_{25}^{(2)}$. Notice that in a model where there can be at most one particle on each link, the scattering to directions $\mathbf{e}_2$ and $\mathbf{e}_5$ in a particular cell can occur only if no particles are already present on these links. The distribution function $\tilde{F}$ is constructed to include this effect, which is mathematically analogous to the Pauli exclusion principle for fermions.

The details of collisions are, however, irrelevant to the derivation of macroscopic equations given in this section. As a result, the complete change due to collisions in a one-particle distribution function f_a will for now be summarized by a simple "collision term" Ω_a, which in general depends on two-particle and higher order distribution functions. (In the models considered here, Ω_a is always entirely local, and cannot depend directly on, for example, derivatives of distribution functions.) In terms of Ω_a, the kinetic equation (2.3.3) extended to include collisions becomes

$$\partial_t f_a + \mathbf{e}_a \cdot \nabla f_a = \Omega_a \qquad (2.3.5)$$

With the appropriate form for Ω_a, this is an exact statistical equation for f_a (at least to first order in δ).

But the equation is not in general sufficient to determine f_a. It gives the time evolution of f_a in terms of the two-particle and higher order distribution functions that appear in Ω_a. The two-particle distribution function then in turn satisfies an equation involving three-particle and higher order distribution functions, and so on. The result is the exact BBGKY hierarchy of equations,[23] of which Eq. (2.3.5) is the first level.

The Boltzmann transport equation approximates (2.3.5) by assuming that Ω_a depends only on one-particle distribution functions. In particular, one may make a "molecular chaos" assumption that all sets of particles are statistically uncorrelated before each collision, so that multiple-particle distribution functions can be written as products of one-particle ones. The distribution function $\tilde{F}_{14}^{(2)}$ is thus approximated as $f_1 f_4 (1 - f_2)(1 - f_3)(1 - f_5)(1 - f_6)$. The resulting Boltzmann equations will be used in Section 4. In this section, only the general form (2.3.5) is needed.

The derivation of Eq. (2.3.5) has been discussed here in the context of a cellular automaton model in which particles are constrained to lie on the links of a fixed array. In this case, the maintenance of terms in (2.3.3) only to first order in δ_x, δ_t is an approximation, and corrections can arise, as discussed in Section 2.5.[24] Equation (2.3.5) is, however, exact for a slightly different class of models, in which particles have a discrete set of possible velocities, but follow continuous trajectories with arbitrary spatial positions. Such "discrete velocity gases" have often been considered, particularly in studies of highly rarefied fluids, in which the mean distance between collisions is comparable to the overall system size.[11,14]

2.4. Conservation Laws

The one-particle distribution functions typically determine macroscopic average quantities. In particular, the total particle density n is given by

$$\sum_a f_a = n \tag{2.4.1}$$

while the momentum density $n\mathbf{u}$, where $\mathbf{u}$ is the average fluid velocity, is given by

$$\sum_a \mathbf{e}_a f_a = n\mathbf{u} \tag{2.4.2}$$

The conservation of these quantities places important constraints on the behavior of the f_a.

In a uniform system $\nabla f_a = 0$, so that Eq. (2.3.5) becomes

$$\partial_t f_a = \Omega_a \tag{2.4.3}$$

and Eqs. (2.4.1) and (2.4.2) imply

$$\sum_a \Omega_a = 0 \tag{2.4.4}$$

$$\sum_a \mathbf{e}_a \Omega_a = 0 \tag{2.4.5}$$

Using the kinetic equation (2.3.5), Eq. (2.4.4) implies

$$\partial_t \sum_a f_a + \sum_a \mathbf{e}_a \cdot \nabla f_a = 0 \tag{2.4.6}$$

With the second term in the form $\nabla \cdot (\sum_a \mathbf{e}_a f_a)$, Eq. (2.4.6) can be written exactly in terms of macroscopic quantities as

$$\partial_t n + \nabla \cdot (n\mathbf{u}) = 0 \tag{2.4.7}$$

This is the usual continuity equation, which expresses the conservation of fluid. It is a first example of a macroscopic equation for the average behavior of a cellular automaton fluid.

Momentum conservation yields the slightly more complicated equation

$$\partial_t \sum_a \mathbf{e}_a f_a + \sum_a \mathbf{e}_a(\mathbf{e}_a \cdot \nabla f_a) = 0 \tag{2.4.8}$$

Defining the momentum flux density tensor

$$\Pi_{ij} = \sum_a (\mathbf{e}_a)_i (\mathbf{e}_a)_j f_a \tag{2.4.9}$$

Eq. (2.4.8) becomes

$$\partial_t (nu_i) + \partial_j \Pi_{ij} = 0 \tag{2.4.10}$$

No simple macroscopic result for Π_{ij} can, however, be obtained directly from the definitions (2.4.1) and (2.4.2).

Equations (2.4.7) and (2.4.10) have been derived here from the basic transport equation (2.3.5). However, as discussed in Section 2.3, this transport equation is only an approximation, valid to first order in the lattice scale parameters δ_x, δ_t.[24] Higher order versions of (2.4.7) and (2.4.10) may be derived from the original Taylor expansion (2.3.3), and in some cases, correction terms are obtained.[24]

Assuming $\delta_x = \delta_t = \delta$, Eq. (2.4.6) to second order becomes

$$\sum_a \left[(\partial_t + \mathbf{e}_a \cdot \nabla) + \frac{1}{2}\delta(\partial_t + \mathbf{e}_a \cdot \nabla)^2 \right] = 0 \tag{2.4.11}$$

Writing the $O(\delta)$ term in the form

$$\partial_t \sum_a (\partial_t + \mathbf{e}_a \cdot \nabla) f_a + \nabla \cdot \sum_a (\partial_t + \mathbf{e}_a \cdot \nabla)\mathbf{e}_a f_a \tag{2.4.12}$$

this term is seen to vanish for any f_a which satisfy the first-order equations (2.4.7) and (2.4.10). Lattice discretization effects thus do not affect the continuity equation (2.4.7), at least to second order.

Corrections do, however, appear at this order in the momentum equation (2.4.10). To second order, Eq. (2.4.8) can be written as

$$\sum_a (\partial_t + \mathbf{e}_a \cdot \nabla)\mathbf{e}_a f_a + \frac{1}{2}\delta\partial_t \sum_a (\partial_t + \mathbf{e}_a \cdot \nabla)\mathbf{e}_a f_a$$
$$+ \frac{1}{2}\delta \sum_a [\mathbf{e}_a \cdot \nabla\partial_t + (\mathbf{e}_a \cdot \nabla)^2]\mathbf{e}_a f_a = 0 \tag{2.4.13}$$

The second term vanishes if f_a satisfies the first-order equation (2.4.8). The third term, however, contains a piece trilinear in the $\mathbf{e}_a$, which gives a correction to the momentum equation (2.4.10).[24]

2.5. Chapman-Enskog Expansion

If there is local equilibrium, as discussed in Section 2.2, then the microscopic distribution functions $f_a(\mathbf{x}, t)$ should depend, on average, only on the macroscopic parameters $\mathbf{u}(\mathbf{x}, t)$ and $n(\mathbf{x}, t)$ and their derivatives. In general, this dependence may be very complicated. But in hydrodynamic processes, $\mathbf{u}$ and n vary only slowly with position and time. In addition, in the subsonic limit, $|\mathbf{u}| \ll 1$.

With these assumptions, one may approximate the f_a by a series or Chapman-Enskog expansion in the macroscopic variables. To the order required for standard hydrodynamic phenomena, the possible terms are

$$f_a = f \left\{ 1 + c^{(1)} \mathbf{e}_a \cdot \mathbf{u} + c^{(2)} \left[(\mathbf{e}_a \cdot \mathbf{u})^2 - \frac{1}{2} |\mathbf{u}|^2 \right] \right.$$
$$\left. + c_\nabla^{(2)} \left[(\mathbf{e}_a \cdot \nabla)(\mathbf{e}_a \cdot \mathbf{u}) - \frac{1}{2} \nabla \cdot \mathbf{u} \right] + \cdots \right\} \qquad (2.5.1)$$

where the $c^{(i)}$ are undetermined coefficients. The first three terms here represent the change in microscopic particle densities as a consequence of changes in macroscopic fluid velocity; the fourth term accounts for first-order dependence of the particle densities on macroscopic spatial variations in the fluid velocity. The structures of these terms can be deduced merely from the need to form scalar quantities f_a from the vectors $\mathbf{e}_a$, $\mathbf{u}$, and ∇.

The relation

$$\sum_a (\mathbf{e}_a)_i (\mathbf{e}_a)_j = \frac{M}{d} \delta_{ij} \qquad (2.5.2)$$

where here $M = 6$ and $d = 2$, and i and j are space indices, has been used in Eq. (2.5.1) to choose the forms of the $|\mathbf{u}|^2$ and $\nabla \mathbf{u}$ terms so as to satisfy the constraints (2.4.1) and (2.4.2), independent of the values of the coefficients $c^{(2)}$ and $c_\nabla^{(2)}$. In terms of (2.5.1), Eq. (2.4.1) yields immediately

$$f = n/6 \qquad (2.5.3)$$

while (2.4.2) gives

$$c^{(1)} = 2 \qquad (2.5.4)$$

The specific values of $c^{(2)}$ and $c_\nabla^{(2)}$ can be determined only by explicit solution of the kinetic equation (2.3.5) including collision terms. (Some approximate results for these coefficients based on the Boltzmann transport equation will be given in Section 4.) Nevertheless, the structure of macroscopic equations can be derived from (2.5.1) without knowledge of the exact values of these parameters.

For a uniform equilibrium system with $\mathbf{u} = 0$, all the f_a are given by

$$f_a = f = n/6 \qquad (2.5.5)$$

In this case, the momentum flux tensor (2.4.9) is equal to the pressure tensor, given, as in the standard kinetic theory of gases, by

$$P_{ij} = \sum_a (\mathbf{e}_a)_i (\mathbf{e}_a)_j f = \frac{1}{2} n \delta_{ij} \tag{2.5.6}$$

where the second equality follows from Eq. (2.5.2). Note that this form is spatially isotropic, despite the underlying anisotropy of the cellular automaton lattice. This result can be deduced from general symmetry considerations, as discussed in Section 3. Equation (2.5.6) gives the equation of state relating the scalar pressure to the number density of the cellular automaton fluid:

$$p = n/2 \tag{2.5.7}$$

When $\mathbf{u} \neq 0$, Π_{ij} can be evaluated in the approximation (2.5.1) using the relations

$$\sum_a (\mathbf{e}_a)_i (\mathbf{e}_a)_j (\mathbf{e}_a)_k = 0 \tag{2.5.8}$$

and

$$\sum_a (\mathbf{e}_a)_i (\mathbf{e}_a)_j (\mathbf{e}_a)_k (\mathbf{e}_a)_l = \frac{M}{d(d+2)} (\delta_{ij}\delta_{kl} + \delta_{ik}\delta_{jl} + \delta_{il}\delta_{jk}) \tag{2.5.9}$$

The result is

$$\Pi_{ij} = \frac{n}{2}\delta_{ij} + \frac{n}{4}c^{(2)}\left[u_i u_j - \frac{1}{2}|\mathbf{u}|^2 \delta_{ij} \right] + \frac{n}{4}c_\nabla^{(2)}\left[\partial_i u_j - \frac{1}{2}\nabla \cdot \mathbf{u} \right] \tag{2.5.10}$$

Substituting the result into Eq. (2.4.10), one obtains the final macroscopic equation

$$\partial_t(n\mathbf{u}) + \frac{1}{4}nc^{(2)}\left\{ (\mathbf{u}\cdot\nabla)\mathbf{u} + \left[\mathbf{u}(\nabla\cdot\mathbf{u}) - \frac{1}{2}\nabla|\mathbf{u}|^2 \right] \right\}$$
$$= -\frac{1}{2}\nabla n - \frac{1}{8}nc_\nabla^{(2)}\nabla^2\mathbf{u} - \frac{1}{4}\Xi \tag{2.5.11}$$

where

$$\Xi = \mathbf{u}(\mathbf{u}\cdot\nabla)(nc^{(2)}) - \frac{1}{2}|\mathbf{u}|^2\nabla(nc^{(2)}) + (\mathbf{u}\cdot\nabla)(nc_\nabla^{(2)}) - \frac{1}{2}(\nabla\cdot\mathbf{u})\nabla(nc_\nabla^{(2)}) \tag{2.5.12}$$

The form (2.5.10) for Π_{ij} follows exactly from the Chapman-Enskog expansion (2.5.1). But to obtain Eq. (2.5.11), one must use the momentum equation (2.4.10). Equation (2.4.13) gives corrections to this equation that arise at second order in the lattice size parameter δ. These corrections must be compared with other effects included in Eq. (2.5.11). The rescaling $\mathbf{x} = \delta_x \mathbf{X}$ implies that spatial gradient terms in the Chapman-Enskog expansion can be of the same order as the $O(\delta_x)$ correction terms in Eq. (2.4.13). When the $\mathbf{e}_a \cdot \mathbf{u}$ term in the Chapman-Enskog expansion (2.5.1) for the f_a is substituted into the last term of Eq. (2.4.13), it gives a contribution[24]

$$\Psi = -\frac{1}{16}nc^{(1)}\nabla^2\mathbf{u} = -\frac{1}{8}n\nabla^2\mathbf{u} \tag{2.5.13}$$

to the right-hand side of Eq. (2.5.11). Note that Ψ depends solely on the choice of $\mathbf{e}_a$, and must, for example, vary purely linearly with the particle density f.

2.6. Navier-Stokes Equation

The standard Navier-Stokes equation for a continuum fluid in d dimensions can be written in the form

$$\partial_t(n\mathbf{u}) + \mu n(\mathbf{u} \cdot \nabla)\mathbf{u} = -\nabla p + \eta\nabla^2\mathbf{u} + \left(\zeta + \frac{1}{d}\eta\right)\nabla(\nabla \cdot \mathbf{u}) \tag{2.6.1}$$

where p is pressure, and η and ζ are, respectively, shear and bulk viscosities (e.g., Ref. 27). The coefficient μ of the convective term is usually constrained to have value 1 by Galilean invariance. Note that the coefficient of the last term in Eq. (2.6.1) is determined by the requirement that the term in Π_{ij} proportional to η be traceless.[27,57]

The macroscopic equation (2.5.11) for the cellular automaton fluid is close to the Navier-Stokes form (2.6.1). The convective and viscous terms are present, and have the usual structure. The pressure term appears according to the equation of state (2.5.7). There are, however, a few additional terms.

Terms proportional to $\mathbf{u}\nabla n$ must be discounted, since they depend on features of the microscopic distribution functions beyond those included in the Chapman-Enskog expansion (2.5.1). The continuity equation (2.4.7) shows that terms proportional to $\mathbf{u}(\nabla \cdot \mathbf{u})$ must also be neglected.

The term proportional to $\nabla|\mathbf{u}|^2$ remains, but can be combined with the ∇n term to yield an effective pressure term which includes fluid kinetic energy contributions.

The form of the viscous terms in (2.5.11) implies that for a cellular automaton fluid, considered here, bulk viscosity is given by

$$\zeta = 0 \tag{2.6.2}$$

The value of η is determined by the coefficient $c_\nabla^{(2)}$ that appears in the microscopic distribution function (2.5.1), according to

$$\eta = n\nu = -\frac{1}{8}nc_\nabla^{(2)} \tag{2.6.3}$$

where ν is the kinematic viscosity. An approximate method of evaluating $c_\nabla^{(2)}$ is discussed in Section 4.6.

The convective term in Eq. (2.5.11) has the same structure as in the Navier-Stokes equation (2.6.1), but includes a coefficient

$$\mu = \frac{1}{4}c^{(2)} \tag{2.6.4}$$

which is not in general equal to 1. In continuum fluids, the covariant derivative usually has the form $D_t = \partial_t + \mathbf{u} \cdot \nabla$ implied by Galilean invariance. The cellular automaton fluid acts, however, as a mixture of components, each with velocities $\mathbf{e}_a$, and these components can contribute with different weights to the covariant derivatives of different quantities, leading to convective terms with different coefficients.

The usual coefficient of the convective term can be recovered in Eq. (2.6.1) and thus Eq. (2.5.11) by a simple rescaling in velocity: setting

$$\tilde{\mathbf{u}} = \mu \mathbf{u} \tag{2.6.5}$$

the equation for $\tilde{\mathbf{u}}$ has coefficient 1 for the $(\tilde{\mathbf{u}} \cdot \nabla)\tilde{\mathbf{u}}$ term.

Small perturbations from a uniform state may be represented by a linearized approximation to Eqs. (2.4.7) and (2.5.11), which has the standard sound wave equation form, with a sound speed obtained from the equation of state (2.5.7) as

$$c = 1/\sqrt{2} \tag{2.6.6}$$

The form of the Navier-Stokes equation (2.6.1) is usually obtained by simple physical arguments. Detailed derivations suggest, however, that more elaborate equations may be necessary, particularly in two dimensions (e.g., Ref. 28). The Boltzmann approximation used in Section 4 yields definite values for $c^{(2)}$ and $c_\nu^{(2)}$. Correlation function methods indicate, however, that additional effects yield logarithmically divergent contributions to $c_\nu^{(2)}$ in two dimensions (e.g., Ref. 29). The full viscous term in this case may in fact be of the rough form $\nabla^2 \log(\nabla^2)\mathbf{u}$.

2.7. Higher Order Corrections

The derivation of the Navier-Stokes form (2.5.11) neglects all terms in the Chapman-Enskog expansion beyond those given explicitly in Eq. (2.5.1). This approximation is expected to be adequate only when $|\mathbf{u}| \ll c$. Higher order corrections may be particularly significant for supersonic flows involving shocks (e.g., Ref. 30).

Since the dynamics of shocks are largely determined just by conservation laws (e.g., Ref. 27), they are expected to be closely analogous in cellular automaton fluids and in standard continuum fluids. For $|\mathbf{u}|/c \gtrsim 2$, however, shocks become so strong and thin that continuum descriptions of physical fluids can no longer be applied in detail (e.g., Ref. 14). The structure of shocks in such cases can apparently be found only through consideration of explicit particle dynamics.[11,14]

In the transonic flow regime $|\mathbf{u}| \approx c$, however, continuum equations may be used, but corrections to the Navier-Stokes form may be significant. A class of such corrections can potentially be found by maintaining terms $O(u^3)$ and higher in the Chapman-Enskog expansion (2.5.1).

In the homogeneous fluid approximation $\nabla \mathbf{u} = 0$, one may take

$$\begin{aligned}
f_a = f\{1 &+ c^{(1)}\mathbf{e}_a \cdot \mathbf{u} + c^{(2)}[(\mathbf{e}_a \cdot \mathbf{u})^2 + \sigma_2|\mathbf{u}|^2] \\
&+ c^{(3)}[(\mathbf{e}_a \cdot \mathbf{u})^3 + \sigma_3|\mathbf{u}|^2(\mathbf{e}_a \cdot \mathbf{u})] \\
&+ c^{(4)}[(\mathbf{e}_a \cdot \mathbf{u})^4 + \sigma_{4,1}|\mathbf{u}|^2(\mathbf{e}_a \cdot \mathbf{u})^2 + \sigma_{4,2}|\mathbf{u}|^4] + \cdots\}
\end{aligned} \tag{2.7.1}$$

The constraints (2.4.1) and (2.4.2) imply

$$c^{(1)} = d \tag{2.7.2}$$

$$\sigma_2 = -\frac{1}{d} \tag{2.7.3}$$

$$\sigma_3 = -\frac{3}{d+2} \tag{2.7.4}$$

$$\frac{3}{d(d+2)} + \frac{1}{d}\sigma_{4,1} + \sigma_{4,2} = 0 \tag{2.7.5}$$

where d is the space dimension, equal to two for the model of this section.

Corrections to (2.5.11) can be found by substituting (2.7.1) in the kinetic equation (2.4.8). For the hexagonal lattice model, one obtains, for example,

$$\partial_t(nu_x) + \frac{1}{4}nc^{(2)}(u_x\partial_x u_x + u_x\partial_y u_y + u_y\partial_y u_x - u_y\partial_x u_y)$$

$$+ \frac{1}{8}nc^{(4)}\{[(5+4\sigma_{4,1})u_x^3 - 3u_x u_y^2]\partial_x u_x$$

$$+ [(3+2\sigma_{4,1})u_y^3 + (3+6\sigma_{4,1})u_x^2 u_y]\partial_y u_x$$

$$- [(3+4\sigma_{4,1})u_y^3 + 3u_x^2 u_y]\partial_x u_y$$

$$+ [(1+2\sigma_{4,1})u_x^3 + (9+6\sigma_{4,1})u_x u_y^2]\partial_y u_y\} = 0 \tag{2.7.6}$$

The $O(u^2)$ term in Eq. (2.7.6) has the isotropic form given in Eq. (2.5.11). The $O(u^4)$ term is, however, anisotropic.

To obtain an isotropic $O(u^4)$ term, one must generalize the model, as discussed in Section 3. One possibility is to allow vectors $\mathbf{e}_a$ corresponding to corners of an M-sided polygon with $M > 6$. In this case, the continuum equation deduced from the Chapman-Enskog expansion (2.7.1) becomes

$$\partial_t(n\mathbf{u}) + \frac{1}{4}nc^{(2)}\left[(\mathbf{u}\cdot\nabla)\mathbf{u} + \mathbf{u}(\nabla\cdot\mathbf{u}) - \frac{1}{2}\nabla|\mathbf{u}|^2\right]$$

$$+ \frac{1}{4}nc^{(4)}(1+\sigma_{4,1})\{|\mathbf{u}|^2[(\mathbf{u}\cdot\nabla)\mathbf{u} + \mathbf{u}(\nabla\cdot\mathbf{u}) - \nabla|\mathbf{u}|^2] \tag{2.7.7}$$

$$+ \mathbf{u}(\mathbf{u}\cdot\nabla)|\mathbf{u}|^2\} = 0$$

This gives a definite form for the next-order corrections to the convective part of the Navier-Stokes equation.

Corrections to the viscous part can be found by including terms proportional to $\nabla\mathbf{u}$ in the Chapman-Enskog expansion (2.7.1). The possible fourth-order terms are given by contractions of $u_i u_j \partial_k u_l$ with products of $(\mathbf{e}_a)_m$ or δ_{mn}. They yield a piece in the Chapman-Enskog expansion of the form

$$c_\nabla^{(4)}[\tau_1(\mathbf{e}_a\cdot\mathbf{u})^2(\mathbf{e}_a\cdot\nabla)(\mathbf{e}_a\cdot\mathbf{u}) + \tau_2|\mathbf{u}|^2(\mathbf{e}_a\cdot\nabla)(\mathbf{e}_a\cdot\mathbf{u})$$

$$+ \tau_3(\mathbf{e}_a\cdot\mathbf{u})(\mathbf{u}\cdot\nabla)(\mathbf{e}_a\cdot\mathbf{u}) + \tau_4(\mathbf{e}_a\cdot\mathbf{u})^2(\nabla\cdot\mathbf{u}) + \tau_5|\mathbf{u}|^2(\nabla\cdot\mathbf{u})] \tag{2.7.8}$$

where Eq. (2.4.1) implies the constraints (for $d = 2$)

$$\tau_1 + 2\tau_3 = 0 \tag{2.7.9}$$

$$\tau_1 + 4\tau_2 + 4\tau_4 + 8\tau_5 = 0 \tag{2.7.10}$$

The resulting continuum equations may be written in terms of vectors formed by contractions of $u_i u_j \partial_k \partial_l u_m$ and $u_i \partial_j u_k \partial_l u_m$. The complete result is

$$
\begin{aligned}
\partial_t(n\mathbf{u}) &+ \frac{1}{4}nc^{(2)}\left[(\mathbf{u}\cdot\nabla)\mathbf{u} + \mathbf{u}(\nabla\cdot\mathbf{u}) - \frac{1}{2}\nabla|\mathbf{u}|^2\right] \\
&+ \frac{1}{4}nc^{(4)}(1+\sigma_{4,1})\{|\mathbf{u}|^2[(\mathbf{u}\cdot\nabla)\mathbf{u} + \mathbf{u}(\nabla\cdot\mathbf{u}) - \nabla|\mathbf{u}|^2] + \mathbf{u}(\mathbf{u}\cdot\nabla)|\mathbf{u}|^2\} \\
&= -\frac{1}{8}nc_\nabla^{(2)}\nabla^2\mathbf{u} \\
&\quad -\frac{1}{32}nc_\nabla^{(4)}\left[\left((\tau_1 - 4\tau_2 + 12\tau_4)\mathbf{u}(\nabla\cdot\mathbf{u})^2 - (\tau_1 - 4\tau_2 + 4\tau_4)\right.\right. \\
&\quad \times \mathbf{u}\{\nabla[(\mathbf{u}\cdot\nabla)\mathbf{u}] - (\mathbf{u}\cdot\nabla)(\nabla\cdot\mathbf{u})\} + 8\tau_4\{[(\mathbf{u}\cdot\nabla)\mathbf{u}]\cdot\nabla\}\mathbf{u} \\
&\quad + \frac{1}{2}(\tau_1 + 4\tau_2)[(\nabla|\mathbf{u}|^2)\cdot\nabla]\mathbf{u} \\
&\quad + 2\tau_1\mathbf{u}\left[\frac{1}{2}\nabla\cdot(\nabla|\mathbf{u}|^2) - \mathbf{u}\cdot(\nabla^2\mathbf{u})\right] - 4\tau_4(\nabla\cdot\mathbf{u})\nabla|\mathbf{u}|^2\bigg) \\
&\quad + \left\{8\tau_4\left[\mathbf{u}(\mathbf{u}\cdot\nabla)(\nabla\cdot\mathbf{u}) - \frac{1}{2}|\mathbf{u}|^2\nabla(\nabla\cdot\mathbf{u})\right]\right. \\
&\quad \left.\left. + 2\tau_1\mathbf{u}[\mathbf{u}\cdot(\nabla^2\mathbf{u})] + 4\tau_2|\mathbf{u}|^2\nabla^2\mathbf{u}\right\}\right]
\end{aligned} \tag{2.7.11}
$$

where, on the right-hand side, the first group of terms are all $O((\nabla\mathbf{u})^2)$, while the second group are $O(\nabla\nabla\mathbf{u})$. Further corrections involve higher derivative terms, such as $u_i\partial_j\partial_k\partial_l u_m$.

For a channel flow with $u_x = ax^2$, $u_y = 0$, the time-independent terms in Eq. (2.7.11) have an x component

$$\frac{1}{4}ac_\nabla^{(2)} + \frac{5}{8}a^3x^4c_\nabla^{(4)}(\tau_1 + 2\tau_2 + 2\tau_4) + \frac{1}{2}a_2x^3c^{(2)} + a^4x^7c^{(4)}(1+\sigma_{4,1}) \tag{2.7.12}$$

and zero y component.

3. Symmetry Considerations

3.1. Tensor Structure

The form of the macroscopic equations (2.4.7) and (2.5.11) depends on few specific properties of the hexagonal lattice cellular automaton model. The most important properties relate to the symmetries of the tensors

$$\mathbf{E}^{(n)}_{i_1 i_2 \cdots i_n} = \sum_a (\mathbf{e}_a)_{i_1} \cdots (\mathbf{e}_a)_{i_n} \tag{3.1.1}$$

These tensors are determined in any cellular automaton fluid model simply from the choice of the basic particle directions $\mathbf{e}_a$. The momentum flux tensor (2.4.9) is given in terms of them by

$$\Pi_{ij} = f(\mathbf{E}^{(2)}_{ij} + c^{(1)}\mathbf{E}^{(3)}_{ijk} u_k + c^{(2)}[\mathbf{E}^{(4)}_{ijkl} u_k u_l + \sigma \mathbf{E}^{(2)}_{ij} u_k u_k]$$
$$+ c^{(2)}_{\nabla}[\mathbf{E}^{(4)}_{ijkl} \partial_k u_l + \sigma \mathbf{E}^{(2)}_{ij} \partial_k u_k]) \tag{3.1.2}$$

where repeated indices are summed, and to satisfy the conditions (2.4.1) and (2.4.2)

$$\sigma = -\mathbf{E}^{(4)}_{ijkk}/\mathbf{E}^{(2)}_{ij} \tag{3.1.3}$$

The basic condition for standard hydrodynamic behavior is that the tensors $\mathbf{E}^{(n)}$ for $n \le 4$ which appear in (3.1.2) should be isotropic. From the definition (3.1.1), the tensors must always be invariant under the discrete symmetry group of the underlying cellular automaton array. What is needed is that they should in addition be invariant under the full continuous rotation group.

The definition (3.1.1) implies that the $\mathbf{E}^{(n)}$ must be totally symmetric in their space indices. With no further conditions, the $\mathbf{E}^{(n)}$ could have $\binom{n+d-1}{n}$ independent components in d space dimensions. Symmetries in the underlying cellular automaton array provide constraints which can reduce the number of independent components.

Tensors that are invariant under all rotations and reflections (or inversions) can have only one independent component. Such invariance is obtained with a continuous set of vectors $\mathbf{e}_a$ uniformly distributed on the unit sphere. Invariance up to finite n can also be obtained with certain finite sets of vectors $\mathbf{e}_a$.

Isotropic tensors $\mathbf{E}^{(n)}$ obtained with sets of M vectors $\mathbf{e}_a$ in d space dimensions must take the form

$$\mathbf{E}^{(2n+1)} = 0 \tag{3.1.4}$$

$$\mathbf{E}^{(2n)} = \frac{M}{d(d+2)\cdots(d+2n-2)}\Delta^{(2n)} \tag{3.1.5}$$

where

$$\Delta^{(2)}_{ij} = \delta_{ij} \tag{3.1.6}$$

$$\Delta^{(4)}_{ijkl} = \delta_{ij}\delta_{kl} + \delta_{ik}\delta_{jl} + \delta_{il}\delta_{jk} \tag{3.1.7}$$

and in general $\Delta^{(2n)}$ consists of a sum of all the $(2n-1)!!$ possible products of Kronecker delta symbols of pairs of indices, given by the recursion relation

$$\Delta^{(2n)}_{i_1 i_2 \cdots i_{2n}} = \sum_{j=2}^{2n} \delta_{i_1 i_j} \Delta^{(2n-2)}_{i_2 \cdots i_{j-1} i_{j+1} \cdots i_{2n}} \tag{3.1.8}$$

The form of the $\Delta^{(2n)}$ can also be specified by giving their upper simplicial components (whose indices form a nonincreasing sequence). Thus, in two dimensions,

$$\Delta^{(4)} = [3, 0, 1, 0, 3] \tag{3.1.9}$$

where the 1111, 2111, 2211, 2221, and 2222 components are given. In three dimensions,

$$\Delta^{(4)} = [3, 0, 1, 0, 3, 0, 0, 0, 0, 1, 0, 1, 0, 0, 3] \tag{3.1.10}$$

Similarly,

$$\Delta^{(6)} = [5, 0, 1, 0, 1, 0, 5] \tag{3.1.11}$$

and

$$\Delta^{(6)} = [15, 0, 3, 0, 3, 0, 15, 0, 0, 0, 0, 0, 0, 3,$$
$$0, 1, 0, 3, 0, 0, 0, 0, 3, 0, 3, 0, 0, 15] \tag{3.1.12}$$

in two and three dimensions, respectively.

For isotropic sets of vectors $\mathbf{e}_a$, one finds from (3.1.5)

$$\frac{1}{M}\sum_a (\mathbf{e}_a \cdot \mathbf{v})^{2n} = Q_{2n}|\mathbf{v}|^{2n} = \frac{(2n-1)!!}{d(d+2)\cdots(d+2n-2)}|\mathbf{v}|^{2n} \tag{3.1.13}$$

so that for $d = 2$

$$Q_2 = \frac{1}{2}, \quad Q_4 = \frac{3}{8}, \quad Q_6 = \frac{5}{16}, \quad Q_8 = \frac{35}{128} \tag{3.1.14}$$

while for $d = 3$

$$Q_{2n} = \frac{1}{2n+1} \tag{3.1.15}$$

Similarly,

$$\frac{1}{M}\sum_a (\mathbf{e}_a \cdot \mathbf{v})^{2n}\mathbf{e}_a \cdot \mathbf{v} = Q_{2n}|\mathbf{v}|^{2n}\mathbf{v} \tag{3.1.16}$$

In the model of Section 2, all the particle velocities $\mathbf{e}_a$ are fundamentally equivalent, and so are added with equal weight in the tensor (3.1.1). In some cellular automaton fluid models, however, one may, for example, allow particle velocities $\mathbf{e}_a$ with unequal magnitudes (e.g., Ref. 31). The relevant tensors in such cases are

$$\mathbf{E}^{(n)}_{i_1 i_2 \cdots i_n} = \sum_a w(|\mathbf{e}_a|^2)(\mathbf{e}_a)_{i_1}\cdots(\mathbf{e}_a)_{i_n} \tag{3.1.17}$$

where the weights $w(|\mathbf{e}_a|^2)$ are typically determined from coefficients in the Chapman-Enskog expansion.

3.2. Polygons

As a first example, consider a set of unit vectors $\mathbf{e}_a$ corresponding to the vertices of a regular M-sided polygon:

$$\mathbf{e}_a = \left(\cos \frac{2\pi a}{M}, \sin \frac{2\pi a}{M} \right) \tag{3.2.1}$$

For sufficiently large M, any tensor $\mathbf{E}^{(n)}$ constructed from these $\mathbf{e}_a$ must be isotropic. Table 1 gives the conditions on M necessary to obtain isotropic $\mathbf{E}^{(n)}$. In general, it can be shown that $\mathbf{E}^{(n)}$ is isotropic if and only if M does not divide any of integers $n, n-2, n-4, \ldots .^{(32)}$ Thus, for example, $\mathbf{E}^{(n)}$ must be isotropic whenever $n > M$.

$\mathbf{E}^{(2)}$	$M > 2$
$\mathbf{E}^{(3)}$	$M \geq 2, M \neq 3$
$\mathbf{E}^{(4)}$	$M > 2, M \neq 4$
$\mathbf{E}^{(5)}$	$M \geq 2, M \neq 3, 5$
$\mathbf{E}^{(6)}$	$M > 4, M \neq 6$
$\mathbf{E}^{(7)}$	$M \geq 2, M \neq 3, 5, 7$

Table 1. Conditions for the tensors $\mathbf{E}^{(n)}$ of Eq. (3.1.1) to be isotropic with the lattice vectors $\mathbf{e}_a$ chosen to correspond to the vertices of regular M-sided polygons.

In the case $M = 6$, corresponding to the hexagonal lattice considered in Section 2, the $\mathbf{E}^{(n)}$ are isotropic up to $n = 5$. The macroscopic equations obtained in this case thus have the usual hydrodynamic form. However, a square lattice, with $M = 4$, yields an anisotropic $\mathbf{E}^{(4)}$, given by

$$\mathbf{E}^{(4)}|_{M=4} = 2\delta^{(4)} \tag{3.2.2}$$

where $\delta^{(n)}$ is the Kronecker delta symbol with n indices. The macroscopic equation obtained in this case is

$$\partial_t(nu_x) + \frac{1}{2}nc^{(2)}(u_x\partial_x u_x - u_y\partial_x u_y)$$
$$= -\frac{1}{2}\partial_x n - \frac{1}{8}nc_\nabla^{(2)}(\partial_{xx}u_x - \partial_{xy}u_y) - \frac{1}{4}(u_x^2 - u_y^2)\partial_x(nc^{(2)})$$
$$- \frac{1}{8}(\partial_x u_x - \partial_y u_y)\partial_x(nc_\nabla^{(2)}) \tag{3.2.3}$$

which does not have the standard Navier-Stokes form.[6],3

On a hexagonal lattice, $\mathbf{E}^{(4)}$ is isotropic, but $\mathbf{E}^{(6)}$ has the component form

$$\mathbf{E}^{(6)}|_{M=6} = \frac{1}{16}[33, 0, 3, 0, 9, 0, 27] \tag{3.2.4}$$

which differs from the isotropic result (3.1.11). The corrections (2.7.6) to the Navier-Stokes equation are therefore anisotropic in this case.

---

3 Note that even the linearized equation for sound waves is anisotropic on a square lattice. The waves propagate isotropically, but are damped with an effective viscosity that varies with direction, and can be negative.[33]

3.3. Polyhedra

As three-dimensional examples, one can consider vectors $\mathbf{e}_a$ corresponding to the vertices of regular polyhedra. Only for the five Platonic solids are all the $|\mathbf{e}_a|^2$ equal. Table 2 gives results for the isotropy of the $\mathbf{E}^{(n)}$ in these cases. Only for the icosahedron and dodecahedron is $\mathbf{E}^{(4)}$ found to be isotropic, so that the usual hydrodynamic equations are obtained. As in two dimensions, the $\mathbf{E}^{(2n)}$ for the cube are all proportional to a single Kronecker delta symbol over all indices.

In five and higher dimensions, the only regular polytopes are the simplex, and the hypercube and its dual.[34] These give isotropic $\mathbf{E}^{(n)}$ only for $n < 3$, and for $n < 4$ and $n < 4$, respectively.

In four dimensions, there are three additional regular polytopes,[34] specified by Schläfi symbols $\{3, 4, 3\}$, $\{3, 3, 5\}$, and $\{5, 3, 3\}$. (The elements of these lists give the number of edges around each vertex, face, and 3-cell, respectively.) The $\{3, 4, 3\}$ polytope has 24 vertices with coordinates corresponding to permutations of $(\pm 1, \pm 1, 0, 0)$. It yields $\mathbf{E}^{(n)}$ that are isotropic up to $n = 4$. The $\{3, 3, 5\}$ polytope has 120 vertices corresponding to $(\pm 1, \pm 1, \pm 1, \pm 1)$, all permutations of $(\pm 2, 0, 0, 0)$, and even-signature permutations of $(\pm \phi, \pm 1, \phi^{-1}, 0)$, where $\phi = (1 + \sqrt{5})/2$. The $\{5, 3, 3\}$ polytope is the dual of $\{3, 3, 5\}$. Both yield $\mathbf{E}^{(n)}$ that are isotropic up to $n = 8$.

3.4. Group Theory

The structure of the $\mathbf{E}^{(n)}$ was found above by explicit calculations based on particular choices for the $\mathbf{e}_a$. The general form of the results is, however, determined solely by the symmetries of the set of $\mathbf{e}_a$. A finite group $\mathbf{G}$ of transformations leaves the $\mathbf{e}_a$ invariant. (For the hexagonal lattice model of Section 2, it is the hexagonal group S_6.) In general $\mathbf{G}$ is a finite subgroup of the d-dimensional rotation group $O(d)$.

The $\mathbf{e}_a$ form the basis for a representation of $\mathbf{G}$, as do their products $\mathbf{E}^{(n)}$. If the representation $\mathbf{R}^{(n)}$ carried by the $\mathbf{E}^{(n)}$ is irreducible, then the $\mathbf{E}^{(n)}$ can have only one

	$\mathbf{e}_a$	M	$\mathbf{E}^{(2)}$	$\mathbf{E}^{(3)}$	$\mathbf{E}^{(4)}$	$\mathbf{E}^{(5)}$	$\mathbf{E}^{(6)}$
Tetrahedron	$(1, 1, 1)$, cyc: $(1, -1, -1)$	4	Y	N	N	N	N
Cube	$(\pm 1, \pm 1, \pm 1)$	8	Y	Y	N	Y	N
Octahedron	cyc: $(\pm 1, 0, 0)$	6	Y	Y	N	Y	N
Dodecahedron	$(\pm 1, \pm 1, \pm 1)$, cyc: $(0, \pm \phi^{-1}, \pm \phi)$	20	Y	Y	Y	Y	N
Icosahedron	cyc: $(0, \pm \phi, \pm 1)$	12	Y	Y	Y	Y	N

Table 2. Isotropy of the tensors $\mathbf{E}^{(n)}$ with $\mathbf{e}_a$ chosen as the M vertices of regular polyhedra. In the forms for $\mathbf{e}_a$ (which are given without normalization), the notation "cyc:" indicates all cyclic permutations. (All possible combinations of signs are chosen in all cases.) ϕ is the golden ratio $(1 + \sqrt{5})/2 \approx 1.618$.

independent component, and must be rotationally invariant. But $\mathbf{R}^{(n)}$ is in general reducible. The number of irreducible representations that it contains gives the number of independent components of $\mathbf{E}^{(n)}$ allowed by invariance under $\mathbf{G}$.

This number can be found using the method of characters (e.g., Refs. 35 and 36). Each class of elements of $\mathbf{G}$ in a particular representation $\mathbf{R}$ has a character that receives a fixed contribution from each irreducible component of $\mathbf{R}$. Characters for the representation $\mathbf{R}^{(n)}$ of $\mathbf{G}$ can be found by first evaluating them for arbitrary rotations, and then specializing to the particular sets of rotations (typically through angles of the form $2\pi/k$) that appear in $\mathbf{G}$. To find characters for arbitrary rotations, one writes the $\mathbf{E}^{(n)}$ as sums of completely traceless tensors $\mathbf{U}^{(n)}$ which form irreducible representations of $O(d)$ (e.g., Ref. 37):

$$\mathbf{E}^{(n)} = \mathbf{U}^{(n)} + \mathbf{U}^{(n-2)} + \cdots + \mathbf{U}^{(0)} \tag{3.4.1}$$

The characters of the $\mathbf{E}^{(n)}$ are then sums of the characters $\chi^{(m)}$ for the irreducible tensors $\mathbf{U}^{(m)}$. For proper rotations through an angle ϕ, the $\chi^{(m)}$ are given by (e.g., Ref. 37)

$$\chi^{(m)}(\phi) = e^{2\pi i m \phi} \qquad (d = 2) \tag{3.4.2}$$

$$\chi^{(m)}(\phi) = \frac{\sin[(2m+1)\phi/2]}{\sin\phi/2} \qquad (d = 3)$$

The resulting characters for the representations $\mathbf{R}^{(n)}$ formed by the $\mathbf{E}^{(n)}$ are given in Table 3.

The number of irreducible representations in $\mathbf{R}^{(n)}$ can be found as usual by evaluating the characters for each class in $\mathbf{R}^{(n)}$ (e.g., Ref. 35). Consider as an example the case of $\mathbf{R}^{(4)}$ with $\mathbf{G}$ the octahedral group $\mathbf{O}$. This group has classes $E, 8C_3, 9C_2, 6C_4$, where E represents the identity, and C_k represents a proper rotation by $\phi = 2\pi/k$ about a k-fold symmetry axis. The characters for these classes in the representation $\mathbf{R}^{(4)}$ can be found from Table 3. Adding the results, and dividing by the total number of classes in $\mathbf{G}$, one finds that $\mathbf{R}^{(4)}$ contains exactly two irreducible representations of $\mathbf{O}$. Rank 4 symmetric tensors can thus have up to two independent components while still being invariant under the octahedral group.[38]

Dimension	Rank	Character
2	2	$4c^2 - 1$
2	4	$(4c^2 + 2c - 1)(4c^2 - 2c - 1)$
3	2	$4c^2 + 2c$
3	4	$(2c + 1)(2c - 1)(4c^2 + 2c - 1)$

Table 3. Characters of transformations of totally symmetric rank n tensors $\mathbf{E}^{(n)}$ in d dimensions. $c = \cos(\phi)$, where ϕ is the rotation angle. For improper rotations in three dimensions, $\pi - \phi$ must be used.

In general, one may consider sets of vectors $\mathbf{e}_a$ that are invariant under any point symmetry group. Typically, the larger the group is, the smaller the number of independent components in the $\mathbf{E}^{(n)}$ can be. In two dimensions, there are an infinite number of point groups, corresponding to transformations of regular polygons. There are only a finite of nontrivial additional point groups in three dimensions. The largest is the group $\mathbf{Y}$ of symmetries of the icosahedron (or dodecahedron). Second largest is the cubic group $\mathbf{E}$. As seen in Table 2, only $\mathbf{Y}$ guarantees isotropy of all tensors $\mathbf{E}^{(n)}$ up to $n = 4$ (compare Ref. 39).

It should be noted, however, that such group-theoretic considerations can only give upper bounds on the number of independent components in the $\mathbf{E}^{(n)}$. The actual number of independent components depends on the particular choice of the $\mathbf{e}_a$, and potentially on the values of weights such as those in Eq. (3.1.16).

3.5. Regular Lattices

If the vectors $\mathbf{e}_a$ correspond to particle velocities, then the possible displacements of particles at each time step must be of the form $\sum_a k_a \mathbf{e}_a$. In discrete velocity gases, particle positions are not constrained. But in a cellular automaton model, they are usually taken to correspond to the sites of a regular lattice.

Only a finite number of such "crystallographic" lattices can be constructed in any space dimension (e.g., Refs. 40 and 41). As a result, the point symmetry groups that can occur are highly constrained. In two dimensions, the most symmetrical lattices are square and hexagonal ones. In three dimensions, the most symmetrical are hexagonal and cubic. The group-theoretic arguments of Section 3.4 suffice to show that in two dimensions, hexagonal lattices must give tensors $\mathbf{E}^{(n)}$ that are isotropic up to $n = 4$, and so yield standard hydrodynamic equations (2.5.11). In three dimensions, group-theoretic arguments alone fail to establish the isotropy of $\mathbf{E}^{(4)}$ for hexagonal and cubic lattices. A system with icosahedral point symmetry would be guaranteed to yield an isotropic $\mathbf{E}^{(4)}$, but since it is not possible to tesselate three-dimensional space with regular icosahedra, no regular lattice with such a large point symmetry group can exist.

Crystallographic lattices are classified not only by point symmetries, but also by the spatial arrangement of their sites. The lattices consist of "unit cells" containing a definite arrangement of sites, which can be repeated to form a regular tesselation. In two dimensions, five distinct such Bravais lattice structures exist; in three dimensions, there are 14 (e.g., Refs. 40 and 41).

Sites in these lattices can correspond directly to the sites in a cellular automaton. The links which carry particles in cellular automaton fluid models are obtained by joining pairs of sites, usually in a regular arrangement. The link vectors give the velocities $\mathbf{e}_a$ of the particles.

In the simplest cases, the links join each site to its nearest neighbors. The regularity of the lattice implies that in such cases, all the $\mathbf{e}_a$ are of equal length, so that all particles have the same speed.

For two-dimensional square and hexagonal lattices, the $\mathbf{e}_a$ with this nearest neighbor arrangement have the form (3.2.1). The results of Section 3.2 then show that with hexagonal lattices, such $\mathbf{e}_a$ give $\mathbf{E}^{(n)}$ that are isotropic up to $n = 4$, and so yield the standard hydrodynamic continuum equations (2.6.1).

Table 4 gives the forms of $\mathbf{E}^{(n)}$ for the most symmetrical three-dimensional lattices with nearest neighbor choices for the $\mathbf{e}_a$. None yield isotropic $\mathbf{E}^{(4)}$ (compare Ref. 38).

The hexagonal and face-centered cubic lattices, which have the largest point symmetry groups in two and three dimensions, respectively, are also the lattices that give the densest packings of circles and spheres (e.g., Ref. 42). One suspects that in more than three dimensions (compare Ref. 43) the lattices with the largest point symmetry continue to be those with the densest sphere packing. The spheres are placed on lattice sites; the positions of their nearest neighbors are defined by a Voronoi polyhedron or Wigner-Seitz cell. The densest sphere packing is obtained when this cell, and thus the nearest neighbor vectors $\mathbf{e}_a$, are closest to forming a sphere. In dimensions $d \leq 8$, it has been found that the optimal lattices for sphere packing are those based on the sets of root vectors for a sequence of simple Lie groups (e.g., Ref. 44). Results on the isotropy of the tensors $\mathbf{E}^{(n)}$ for these lattices are given in Table 5.

More isotropic sets of $\mathbf{e}_a$ can be obtained by allowing links to join sites on the lattice beyond nearest neighbors.[31] On a square lattice, one may, for example, include diagonal links, yielding a set of vectors

$$\mathbf{e}_a = (0, \pm 1), (\pm 1, 0), (\pm 1, \pm 1) \tag{3.5.1}$$

Including weights $w(|\mathbf{e}_a|^2)$ as in Eq. (3.1.16), this choice of $\mathbf{e}_a$ yields

$$\mathbf{E}^{(2)} = 2[w(1) + 2w(2)]\delta^{(2)} \tag{3.5.2}$$
$$\mathbf{E}^{(4)} = 4w(2)\Delta^{(4)} + 2[w(1) - 4w(2)]\delta^{(4)} \tag{3.5.3}$$

If the ratio of particles on diagonal and orthogonal links can be maintained so that

$$w(1) = 4w(2) \tag{3.5.4}$$

	$\mathbf{e}_a$	M	$\mathbf{E}^{(2)}$	$\mathbf{E}^{(4)}$	$\mathbf{E}^{(6)}$
Primitive cubic	cyc: $(\pm 1, 0, 0)$	6	$2\delta^{(2)}$	$2\delta^{(4)}$	$2\delta^{(6)}$
Body-centered cubic	$(\pm 1, \pm 1, \pm 1)$	8	$8\delta^{(2)}$	$8(\Delta^{(4)} - 2\delta^{(4)})$	$8(\Delta^{(6)} - 2\Delta^{(4,2)} + 16\delta^{(6)})$
Face-centered cubic	cyc: $(\pm 1, \pm 1, 0)$	12	$8\delta^{(2)}$	$4(\Delta^{(4)} - \delta^{(4)})$	$4(\Delta^{(4,2)} - 13\delta^{(6)})$

Table 4. Forms of the tensors $\mathbf{E}^{(n)}$ for the most symmetrical three-dimensional Bravais lattices. The basic vectors $\mathbf{e}_a$ (used here without normalization) are taken to join each site with its M nearest neighbors. $\delta^{(n)}$ represents the Kronecker delta symbol of n indices; $\Delta^{(n)}$ represents the rotationally invariant tensor defined in Eqs. (3.1.6)–(3.1.8). $\Delta^{(n,m)}$ is the sum of all possible products of pairs of Kronecker delta symbols with n and m indices, respectively.

d	Group		M	n_{max}
1	A_1	$SU(2)$	2	
2	A_2	$SU(3)$	6	4
3	A_3	$SU(4)$	12	2
4	D_4	$SO(8)$	24	4
5	D_5	$SO(10)$	40	2
6	E_6		72	0

Table 5. Sequence of simple Lie groups whose sets of root vectors yield optimal lattices for sphere packing in d dimensions. These lattices may also yield maximal isotropy for the tensors $\mathbf{E}^{(n)}$. Results are given for the maximum even n at which the $\mathbf{E}^{(n)}$ are found to be isotropic. The root vectors are given in Ref. 45.

then Eq. (3.5.3) shows that $\mathbf{E}^{(4)}$ will be isotropic. This choice effectively weights the individual vectors $(0, \pm 1)$ and $(\pm 1, 0)$ with a factor $\sqrt{2}$. As a result, the vectors (3.5.1) are effectively those for a regular octagon, given by Eq. (3.2.1) with $M = 8$.

Including all 24 $\mathbf{e}_a$ with components $|(\mathbf{e}_a)_i| \le 2$ on a square lattice, one obtains

$$\mathbf{E}^{(2)} = 2[w(1) + 2w(2) + 4w(4) + 10w(5) + 8w(8)]\delta^{(2)} \tag{3.5.5}$$

$$\mathbf{E}^{(4)} = 4[w(2) + 8w(5) + 16w(8)]\Delta^{(4)}$$
$$+ 2[w(1) - 4w(2) + 16w(4) - 14w(5) - 64w(8)]\delta^{(4)} \tag{3.5.6}$$

$$\mathbf{E}^{(6)} = \frac{4}{3}[w(2) + 20w(5) + 64w(8)]\Delta^{(6)}$$
$$+ 2[w(1) - 8w(2) + 64w(4) - 70w(5) - 512w(8)]\delta^{(6)} \tag{3.5.7}$$

With $w(5) = w(8) = 0$, $\mathbf{E}^{(4)}$ and $\mathbf{E}^{(6)}$ are isotropic if

$$\frac{w(2)}{w(1)} = \frac{3}{8}, \qquad \frac{w(4)}{w(1)} = \frac{1}{32} \tag{3.5.8}$$

They cannot both be isotropic if $w(4)$ also vanishes.

In three dimensions, one may consider a cubic lattice with sites at distances 1, $\sqrt{2}$, and $\sqrt{3}$ joined. The $\mathbf{e}_a$ in this case contain all those for primitive, face-centered, and body-centered cubic lattices, as given in Table 4. The $\mathbf{E}^{(n)}$ can then be deduced from the results of Table 4, and are given by

$$\mathbf{E}^{(2)} = 2[w(1) + 4w(2) + 4w(3)]\delta^{(2)} \tag{3.5.9}$$

$$\mathbf{E}^{(4)} = 4[w(2) + 2w(3)]\Delta^{(4)} + 2[w(1) - 2w(2) - 8w(3)]\delta^{(4)} \tag{3.5.10}$$

$$\mathbf{E}^{(6)} = 8w(2)\Delta^{(6)} + 4[w(2) - 4w(3)]\Delta^{(4,2)} + 2[w(1) - 26w(2)$$
$$+ 64w(3)]\delta^{(6)} \tag{3.5.11}$$

Isotropy of $\mathbf{E}^{(4)}$ is obtained when

$$w(1) = 2w(2) + 8w(3) \tag{3.5.12}$$

and of $\mathbf{E}^{(6)}$ when

$$w(1) = 10w(2) = 40w(3) \tag{3.5.13}$$

Notice that (3.5.12) and (3.5.13) cannot simultaneously be satisfied by any nonzero choice of weights. Nevertheless, so long as (3.5.12) holds, isotropic hydrodynamic behavior is obtained in this three-dimensional cellular automaton fluid. Isotropic $E^{(6)}$ can be obtained by including in addition vectors $\mathbf{e}_a$ of the form $(\pm 2, 0, 0)$ (and permutations), and choosing

$$w(2) = \frac{1}{2}w(1), \qquad w(3) = \frac{1}{8}w(1), \qquad w(4) = \frac{1}{16}w(1) \tag{3.5.14}$$

The weights in Eq. (3.1.17) give the probabilities for particles with different speeds to occur. These probabilities are determined by microscopic equilibrium conditions. They can potentially be controlled by using different collision rules on different time steps (as discussed in Section 4.9). Each set of collision rules can, for example, be arranged to yield each particle speed with a certain probability. Then the frequency with which different collision rules are used can determine the densities of particles with different speeds.

3.6. Irregular Lattices

The general structure of cellular automaton fluid models considered here requires that particles can occur only at definite positions and with definite discrete velocities. But the possible particle positions need not necessarily correspond with the sites of a regular lattice. The directions of particle velocities should be taken from the directions of links. But the particle speeds may consistently be taken independent of the lengths of links.

As a result, one may consider constructing cellular automaton fluids on quasilattices (e.g., Ref. 46), such as that illustrated in Fig. 2. Particle velocities are taken to follow the directions of the links, but to have unit magnitude, independent of the spatial lengths of the links. Almost all intersections involve just two links, and so

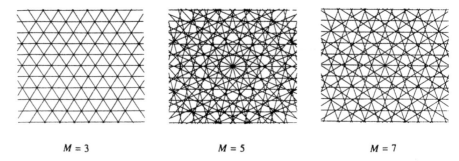

$M = 3$ $\qquad\qquad$ $M = 5$ $\qquad\qquad$ $M = 7$

Figure 2. Lattices and quasilattices constructed from grids oriented in the directions of the vertices of regular M-sided polygons. An appropriate dual of the $M = 5$ pattern is the Penrose aperiodic tiling.

can support only two-particle interactions. These intersections occur at a seemingly irregular set of points, perhaps providing a more realistic model of collisions in continuum fluids.

The possible $\mathbf{e}_a$ on regular lattices are highly constrained, as discussed in Section 3.5. But it is possible to construct quasilattices which yield any set of $\mathbf{e}_a$. Given a set of generator vectors $\mathbf{g}_a$, one constructs a grid of equally spaced lines orthogonal to each of them.[47] The directions of these lines correspond to the $\mathbf{e}_a$.

If the tangent of the angles between the $\mathbf{g}_a$ are rational, then these lines must eventually form a periodic pattern, corresponding to a regular lattice. But if, for example, the $\mathbf{g}_a$ correspond to the vertices of a pentagon, then the pattern never becomes exactly periodic, and only a quasilattice is obtained. A suitable dual of the quasilattice gives in fact the standard Penrose aperiodic tiling.[48]

In three dimensions, one may form grids of planes orthogonal to generator vectors $\mathbf{g}_a$. Possible particle positions and velocities are obtained from the lines in which these planes intersect.

Continuum equations may be derived for cellular automaton fluids on quasilattices by the same methods as were used for regular lattices above. But by appropriate choices of generator vectors, three-dimensional quasilattices with effective icosahedral point symmetry may be obtained, so that isotropic fluid behavior can be obtained even with a single particle speed.

Quasilattices yield an irregular array of particle positions, but allow only a limited number of possible particle velocities. An entirely random lattice would also allow arbitrary particle velocities. Momentum conservation cannot be obtained exactly with discrete collision rules on such a lattice, but may be arranged to hold on average.

4. Evaluation of Transport Coefficients

4.1. Introduction

Section 2 gave a derivation of the general form of the hydrodynamic equations for a sample cellular automaton fluid model. This section considers the evaluation of the specific transport coefficients that appear in these equations. While these coefficients may readily be found by explicit simulation, as discussed in the second paper in this series, no exact mathematical procedure is known for calculating them. This section considers primarily an approximation method based on the Boltzmann transport equation. The results obtained are expected to be accurate for certain transport coefficients at low particle densities.[4]

4.2. Basis for Boltzmann Transport Equation

The kinetic equation (2.3.5) gives an exact result for the evolution of the one-particle distribution function f_a. But the collision term Ω_a in this equation depends on two-particle distribution functions, which in turn depend on higher order distribution

---

[4] Some similar results have been obtained by a slightly different method in Ref. 49.

functions, forming the BBGKY hierarchy of kinetic equations. To obtain explicit results for the f_a one must close or truncate this hierarchy.

The simplest assumption is that there are no statistical correlations between the particles participating in any collision. In this case, the multiparticle distribution functions that appear in Ω_a can be replaced by products of one-particle distribution functions f_a, yielding an equation of the standard Boltzmann transport form, which can in principle be solved explicitly for the f_a.

Even if particles were uncorrelated before a collision, they must necessarily show correlations after the collision. As a result, the factorization of multiparticle distribution functions used to obtain the Boltzmann transport equation cannot formally remain consistent. At low densities, it may nevertheless in some cases provide an adequate approximation.

Correlations produced by a particular collision are typically important only if the particles involved collide again before losing their correlations. At low densities, particles usually travel large distances between collisions, so that most collisions involve different sets of particles. The particles involved in one collision will typically suffer many other collisions before meeting again, so that they are unlikely to maintain correlations. At high densities, however, the same particles often undergo many successive collisions, so that correlations can instead be amplified.

In the Boltzmann transport equation approximation, correlations and deviations from equilibrium decay exponentially with time. Microscopic perturbations may, however, lead to collective, hydrodynamic, effects, which decay only as a power of time.[29] Such effects may lead to transport coefficients that are nonanalytic functions of density and other parameters, as mentioned in Section 2.6.

4.3. Construction of Boltzmann Transport Equation

This subsection describes the formulation of the Boltzmann transport equation for the sample cellular automaton fluid model discussed in Section 2.

The possible classes of particle collisions in this model are illustrated in Fig. 3. The rules for different collisions within each class are related by lattice symmetries. But, as illustrated in Fig. 3, several choices of overall rules for each class are often allowed by conservation laws.

In the simplest case, the same rule is chosen for a particular class of collisions at every site. But it is often convenient to allow different choices of rules at different sites. Thus, for example, there could be a checkerboard arrangement of sites on which two-body collisions lead alternately to scattering to the left and to the right. In general, one may apply a set of rules denoted by k at some fraction γ_k of the sites in a cellular automaton. (A similar procedure was mentioned in Section 3.5 as a means for obtaining isotropic behavior on three-dimensional cubic lattices.) The randomness of microscopic particle configurations suggests that the γ_k should serve merely to change the overall probabilities for different types of collisions.

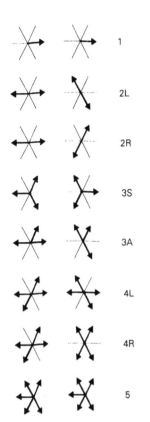

1

2L

2R

3S

3A

4L

4R

5

Figure 3. Possible types of initial and final states for collisions in the cellular automaton fluid model of Section 2.

The term Ω_a in the kinetic equation (2.3.5) for f_a is a sum of terms representing possible collisions involving particles of type a. Each term gives the change in the number of type a particles due to a particular type of collisions, multiplied by the probability for the arrangement of particles involved in the collision to occur. In the Boltzmann equation approximation, the probability for a particular particle arrangement is taken to be a simple product of the densities f_b for particles that should be present, multiplied by factors $(1 - f_c)$ for particles that should be absent.

The complete Boltzmann transport equation for the model of Section 2 thus becomes

$$\partial_t f_a + \mathbf{e}_a \cdot \nabla f_a = \Omega_a \qquad (4.3.1)$$

where

$$\begin{aligned}
\Omega = &[\gamma_{2L}\Lambda(1, 4) + (\gamma_2 - \gamma_{2L})\Lambda(2, 5)] - \gamma_2\Lambda(0, 3) \\
&+ \gamma_{3S}[\Lambda(1, 3, 5) - \Lambda(0, 2, 4)] \\
&+ \gamma_{3A}[\Lambda(2, 4, 5) + \Lambda(1, 2, 5) - \Lambda(0, 3, 5) - \Lambda(0, 2, 3) \\
&+ \Lambda(1, 4, 5) + \Lambda(1, 2, 4) - \Lambda(0, 3, 4) - \Lambda(0, 1, 3)] \\
&+ [\gamma_4\Lambda(1, 2, 4, 5) - \gamma_{4L}\Lambda(0, 2, 3, 5) - (\gamma_4 - \gamma_{4L})\Lambda(0, 1, 3, 4)] \qquad (4.3.2)
\end{aligned}$$

Here

$$\Lambda_a(i_1, i_2, \ldots, i_k) = \frac{f_{a+i_1}}{1 - f_{a+i_1}} \frac{f_{a+i_2}}{1 - f_{a+i_2}} \cdots \frac{f_{a+i_k}}{1 - f_{a+i_k}} \prod_{j=1}^{M} (1 - f_{a+j}) \qquad (4.3.3)$$

where all indices on the f_b are evaluated modulo M, and in this case $M = 6$. Note that in Eq. (4.3.2), the index a has been dropped on both Ω and Λ.

The Boltzmann transport equations for any cellular automaton fluid model have the overall form of Eqs. (4.3.1) and (4.3.2). In a more general case, the simple addition of constants i_j to the indices a in the definition of Λ can be replaced by transformations with appropriate lattice symmetry group operations.

Independent of the values of the γ_k, Ω_a is seen to satisfy the momentum and particle number constraints (2.4.4) and (2.4.5).

In the following calculations it is often convenient to maintain arbitrary values for the γ_k so as to trace the contributions of different classes of collisions. But to obtain a form for Ω_a that is invariant under the complete lattice symmetry group, one must take

$$\gamma_{2L} = \gamma_{2R} = \frac{1}{2}\gamma_2 \qquad (4.3.4)$$

$$\gamma_{4L} = \gamma_{4R} = \frac{1}{2}\gamma_4 \qquad (4.3.5)$$

4.4. Linear Approximation to Boltzmann Transport Equation

In studying macroscopic behavior, one assumes that the distribution functions f_a differ only slightly from their equilibrium values, as in the Chapman-Enskog expansion (2.5.1). The f_a may thus be approximated as

$$f_a = f(1 + \phi_a) \qquad (|\phi_a| \ll 1) \qquad (4.4.1)$$

With this approximation, the collision term Ω_a in the Boltzmann transport equation may be approximated by a power series expansion in the ϕ_a:

$$\Omega_a = \sum_b \omega_{ab}^{(1)} \phi_b + \sum_{b,c} \omega_{abc}^{(2)} \phi_b \phi_c + \cdots \qquad (4.4.2)$$

The matrix $\omega^{(1)}$ here is analogous to the usual linearized collision operator (e.g., Ref. 26). Notice that for a cellular automaton fluid model with collisions involving at most K particles, the expansion (4.4.2) terminates at $O(\phi^K)$.

Microscopic reversibility immediately implies that the tensors $\omega^{(n)}$ are all completely symmetric in their indices. The conservation laws (2.4.4) and (2.4.5) yield conditions on all the $\omega^{(n)}$ of the form

$$\sum_{abc..} \omega_{abc..}^{(n)} = 0 \qquad (4.4.3)$$

$$\sum_{abc..} e_a \omega_{abc..}^{(n)} = 0 \qquad (4.4.4)$$

In the particular case of $\omega^{(1)}$, the more stringent conditions

$$\sum_b \omega^{(1)}_{ab} = 0 \qquad (4.4.5)$$

and

$$\sum_b \mathbf{e}_a \omega^{(1)}_{ab} = 0 \qquad (4.4.6)$$

also apply.

In the model of Section 2, all particle types a are equivalent up to lattice symmetry transformations. As a result, $\omega^{(n)}_{(a+1)bc..}$ is always given simply by a cyclic shift of $\omega^{(n)}_{abc..}$, so that the complete form of $\omega^{(n)}$ can be determined from the first row $\omega^{(n)}_{1bc..}$. The $\omega^{(n)}$ are thus circulant tensors (e.g., Ref. 50), and the values of their components depend only on numerical differences between their indices, evaluated modulo M.

Expansion of (4.3.2) now yields

$$\begin{aligned}
\omega^{(1)}_{ab} = f^2(1-f)\,\mathrm{circ}\{&-[\gamma_2 \bar{f}^2 + (\gamma_{3S}+4\gamma_{3A})\bar{f}f + \gamma_4 f^2],\\
&\gamma_{2L}\bar{f}^2 + (\gamma_{3S}+2\gamma_{3A})\bar{f}f + \gamma_{4L}f^2,\\
&(1-\gamma_{2L})\bar{f}^2 + (-\gamma_{3S}+2\gamma_{3A})\bar{f}f + (\gamma_4-\gamma_{4L})f^2,\\
&-[\gamma_2\bar{f}^2 + (-\gamma_{3S}+4\gamma_{3A})\bar{f}f + \gamma_4 f^2],\\
&\gamma_{2L}\bar{f}^2 + (-\gamma_{3S}+2\gamma_{3A})\bar{f}f + \gamma_{4L}f^2,\\
&(1-\gamma_{2L})\bar{f}^2 + (\gamma_{3S}+2\gamma_{3A})\bar{f}f + (\gamma_4-\gamma_{4L})f^2\}
\end{aligned} \qquad (4.4.7)$$

where $\bar{f} = (1-f)$. Taking for simplicity $\gamma_2 = 1$, $\gamma_{2L} = \frac{1}{2}$, $\gamma_{3S} = 1$, $\gamma_{3A} = \gamma_{4i} = 0$, one finds

$$\omega^{(1)}_{ab} = f^2(1-f)^2\,\mathrm{circ}\left[-1,\ \frac{1}{2}(1+f),\ \frac{1}{2}(1-3f),\ 2f-1,\ \frac{1}{2}(1-3f),\ \frac{1}{2}(1+f)\right] \qquad (4.4.8)$$

$$\omega^{(2)}_{abc} = \frac{1}{2}f^2(1-f)\,\mathrm{circ}
\begin{bmatrix}
0 & -f(f-1) & f(3f-1) \\
-f(f-1) & 0 & 2f(f-1) \\
f(3f-1) & 2f(f-1) & 0 \\
-2(f-1)(2f-1) & -f(5f-3) & -f(f-1) \\
f(3f-1) & (f-1)(2f-1) & 2f(3f-2) \\
-f(f-1) & -2f^2 & (f-1)(2f-1)
\end{bmatrix}$$

$$\begin{bmatrix}
-2(f-1)(2f-1) & f(3f-1) & -f(f-1) \\
-f(5f-3) & (f-1)(2f-1) & -2f^2 \\
-f(f-1) & 2f(3f-2) & (f-1)(2f-1) \\
0 & -f(f-1) & -f(5f-3) \\
-f(f-1) & 0 & 2f(f-1) \\
-f(5f-3) & 2f(f-1) & 0
\end{bmatrix} \qquad (4.4.9)$$

4.5. Approach to Equilibrium

In a spatially uniform system close to equilibrium, one may use a linear approximation to the Boltzmann equation (4.3.1):

$$\partial_t(f\phi_a) = \sum_b \omega_{ab}^{(1)}\phi_b \tag{4.5.1}$$

This equation can be solved in terms of the eigenvalues and eigenvectors of the matrix $\omega_{ab}^{(1)}$. The circulant property of $\omega_{ab}^{(1)}$ considerably simplifies the computations required.

An $M \times M$ circulant matrix U_{ab} can in general be written in the form[50]

$$U_{ab} = \mathbf{u}[(a-b+1) \bmod M] = U_{11}I + U_{12}\Pi + \cdots + U_{1M}\Pi^{(M-1)} \tag{4.5.2}$$

where

$$\Pi = \text{circ}[0, 1, 0, 0, \ldots, 0] \tag{4.5.3}$$

is an $M \times M$ cyclic permutation matrix, and I is the $M \times M$ identity matrix. From this representation, it follows that all $M \times M$ circulants have the same set of right eigenvectors $\mathbf{v}_c$, with components given by

$$(\mathbf{v}_c)_a = \frac{1}{\sqrt{M}}\exp\frac{2\pi i(c-1)(a-1)}{M} \tag{4.5.4}$$

Writing

$$\Gamma(z) = \sum_{a=1}^{M} U_{1a}z^{a-1} \tag{4.5.5}$$

the corresponding eigenvalues are found to be

$$\lambda_c = \Gamma\left(\exp\left[\frac{2\pi i(c-1)}{M}\right]\right) \tag{4.5.6}$$

Using these results, the eigenvectors of $\omega_{ab}^{(1)}$ for the model of Section 2 are found to be

$$\mathbf{v}_1 = \frac{1}{\sqrt{6}}(1, 1, 1, 1, 1, 1)$$

$$\mathbf{v}_2 = \frac{1}{\sqrt{6}}(1, \sigma, -\sigma^*, -1, -\sigma, \sigma^*) = (\mathbf{v}_6)^*$$

$$\mathbf{v}_3 = \frac{1}{\sqrt{6}}(1, -\sigma^*, -\sigma, 1, -\sigma^*, -\sigma) = (\mathbf{v}_5)^* \tag{4.5.7}$$

$$\mathbf{v}_4 = \frac{1}{\sqrt{6}}(1, -1, 1, -1, 1, -1)$$

$$\frac{1}{2}(\mathbf{v}_2 + \mathbf{v}_6) = \frac{1}{2\sqrt{6}}(2, 1, -1, -2, -1, 1)$$

$$\frac{1}{2i}(\mathbf{v}_2 - \mathbf{v}_6) = \frac{\sqrt{3}}{2\sqrt{6}}(0, 1, 1, 0, -1, -1)$$

where

$$\sigma = \exp(i\pi/3) = \frac{1}{2}(1 + i\sqrt{3})$$

and the corresponding eigenvalues are

$$\lambda_1 = 0$$

$$\lambda_2 = 0$$

$$\lambda_3 = -3f^2(1-f)\left\{[\gamma_2(1-f)^2 + 4\gamma_{3A}f(1-f) + \gamma_4 f^2]\right.$$

$$\left. -\frac{4i}{\sqrt{3}}\left[(1-f)^2\left(\frac{\gamma_2}{2} - \gamma_{2L}\right) + f^2\left(\frac{\gamma_4}{2} - \gamma_{4L}\right)\right]\right\} \qquad (4.5.8)$$

$$\lambda_4 = -6\gamma_{3S}f^3(1-f)^2$$

$$\lambda_5 = (\lambda_3)^*$$

$$\lambda_6 = 0$$

Combinations of the ϕ_a corresponding to eigenvectors with zero eigenvalue are conserved with time according to Eq. (4.5.1). Three such combinations are associated with the conservation laws (2.4.1) and (2.4.2). $\mathbf{v}_1$ corresponds to $\sum_a \phi_a$, which is the total particle number density. $(\mathbf{v}_2 + \mathbf{v}_6)/2$ and $(\mathbf{v}_2 - \mathbf{v}_6)/2i$ correspond, respectively, to the x and y components of the momentum density $\sum_a \mathbf{e}_a \phi_a$.

The ϕ_a may always be written as sums of pieces proportional to each of the orthogonal eigenvectors $\mathbf{v}_c$ of Eq. (4.5.7):

$$\phi_a = \sum_c \psi_c(\mathbf{v}_c)_a \qquad (4.5.9)$$

The coefficients ψ_1, $(\psi_2 + \psi_6)/2$, and $(\psi_2 - \psi_6)/2i$ give the values of the conserved particle and momentum densities in this representation, and remain fixed with time.

The general solution of Eq. (4.5.1) is given in terms of Eq. (4.5.9) by

$$\psi_c(t) = \psi_c(0)e^{\lambda_c t} \qquad (4.5.10)$$

Equation (4.5.8) shows that for any positive choices of the γ_k, all nonzero λ_c have negative real parts. As a result, the associated ψ_c must decay exponentially with time. Only the combinations of ϕ_a associated with conserved quantities survive at large times.

This result supports the local equilibrium assumption used for the derivation of hydrodynamic equations in Section 2. It implies that regardless of the initial average densities ϕ_a, collisions bring the system to an equilibrium that depends only on the values of the macroscopic conserved quantities (2.4.1) and (2.4.2). One may thus expect to be able to describe the local state of the cellular automaton fluid on time scales large compared to $|\lambda_c|^{-1}$ ($\lambda_c \neq 0$) solely in terms of these macroscopic conserved quantities. [Section 4.2 nevertheless mentioned some effects not accounted for by the Boltzmann equation (4.3.1) that can slow the approach to equilibrium.]

One notable feature of the results (4.5.8) is that they imply that the final equilibrium values of the ϕ_a are not affected by the choice of the parameters γ_{2L} and γ_{4L}, which determine the mixtures of two- and four-particle collisions with different chiralities. When the rate for collisions with different chiralities are unequal, however, λ_3 and λ_5 acquire imaginary parts, which lead to damped oscillations in the ϕ_a as a function of time.

When all the types of collisions illustrated in Fig. 3 can occur, Eq. (4.5.8) implies that momentum and particle number are indeed the only conserved quantities. If, however, only two-particle collisions are allowed, then there are additional conserved quantities. In fact, whenever symmetric three-particle collisions are absent, so that $\gamma_{3S} = 0$, Eq. (4.5.8) implies that the quantities

$$Q_i = \sum_{a=i}^{M/2+i} f_a \tag{4.5.11}$$

where the index a is evaluated modulo $M = 6$, is conserved. Thus, independent of the value of γ_{2L}, the total momenta on the two sides of any line (not along a lattice direction) through the cellular automaton must independently be conserved.

If three-particle symmetric collisions are absent, the cellular automaton thus exhibits a spurious additional conservation law, which prevents the attainment of standard local equilibrium, and modifies the hydrodynamic behavior discussed in Section 2. Section 4.8 considers some general conditions which avoid such spurious conservation laws.

4.6. Equilibrium Conditions and Transport Coefficients

Section 4.5 discussed the solution of the Boltzmann transport equation for uniform cellular automaton fluids. This section considers nonuniform fluids, and gives some approximate results for transport coefficients.

The Chapman-Enskog expansion (2.5.1) gives the general form for approximations to the microscopic distribution functions f_a. The coefficients $c^{(2)}$ and $c_\nabla^{(2)}$ that appear in this expansion can be estimated using the Boltzmann transport equation (4.3.1) from the microscopic equilibrium condition

$$\partial_t f_a = 0 \tag{4.6.1}$$

In estimating $c^{(2)}$, one must maintain terms in Ω_a to the second order in ϕ_b, but one can neglect spatial variation in the ϕ_a. As a result, the Boltzmann equation (4.3.1) becomes

$$\sum_b \omega_{ab}^{(1)} \phi_b + \sum_{b,c} \omega_{abc}^{(2)} \phi_b \phi_c = 0 \tag{4.6.2}$$

Substituting forms for the ϕ_a from the Chapman-Enskog expansion (2.5.1), one obtains

$$c^{(2)} \sum_b \omega_{ab}^{(1)} (\mathbf{u} \cdot \mathbf{e}_b)^2 + [c^{(1)}]^2 \sum_{b,c} \omega_{abc}^{(2)} (\mathbf{u} \cdot \mathbf{e}_b)(\mathbf{u} \cdot \mathbf{e}_c) = 0 \qquad (4.6.3)$$

where $c^{(1)} = 2$ according to Eq. (2.5.4). Using the forms for $\omega^{(1)}$ and $\omega^{(2)}$ determined by the expansion of Eq. (4.3.2), one finds that the two terms in (4.6.3) show exactly the same dependence on the γ_k. The final result for $c^{(2)}$ is thus independent of the γ_k, and is given by

$$c^{(2)} = 2(1 - 2f)/(1 - f) \qquad (4.6.4)$$

In the Boltzmann equation approximation, this implies that the coefficient μ of the $n(\mathbf{u} \cdot \nabla)\mathbf{u}$ term in the hydrodynamic equation (2.6.1) is $(1 - 2f)/[2(1 - f)]$. Notice that, as discussed in Section 2.6, this coefficient is not in general equal to 1.

The value of the coefficient $c_\nabla^{(2)}$ can be found by a slightly simpler calculation, which depends only on the linear part $\omega_{ab}^{(1)}$ of the expansion of the collision term Ω_a. Keeping now first-order spatial derivatives of the ϕ_a, one can determine $c_\nabla^{(2)}$ from the equilibrium condition

$$\sum_b \omega_{ab}^{(1)} \phi_b = f \mathbf{e}_a \cdot \nabla \phi_b \qquad (4.6.5)$$

which yields

$$\sum_b c_\nabla^{(2)} \omega_{ab}^{(1)} (\mathbf{e}_b \cdot \nabla)(\mathbf{e}_b \cdot \mathbf{u}) = c^{(1)} f (\mathbf{e}_a \cdot \nabla)(\mathbf{e}_a \cdot \mathbf{u}) \qquad (4.6.6)$$

With the approximations used, Eq. (2.4.7) implies that $\nabla \cdot \mathbf{u} = 0$. Then Eq. (4.6.6) gives the result

$$c_\nabla^{(2)} = -2\{12f(1 - f)[\gamma_2(1 - f)^2 + 4\gamma_{3A} f(1 - f) + \gamma_4 f^2]\}^{-1} \qquad (4.6.7)$$

Using Eq. (2.6.3), this gives the kinematic viscosity of the cellular automaton fluid in the Boltzmann equation approximation as

$$\nu = \{12f(1 - f)[\gamma_2(1 - f)^2 + 4\gamma_{3A} f(1 - f) + \gamma_4 f^2]\}^{-1} \qquad (4.6.8)$$

Some particular values are

$$\begin{aligned} \nu &= [12f(1 - f)^3]^{-1} & (\gamma_2 = 1, \gamma_{3A} = \gamma_4 = 0) \\ \nu &= [12f(1 - f)(1 + 2f - 2f^2)]^{-1} & (\gamma_2 = \gamma_{3A} = \gamma_4 = 1) \end{aligned} \qquad (4.6.9)$$

For $f = 1/6$ one obtains in these cases $\nu \approx 0.86$ and $\nu \approx 0.47$, respectively, while for $f = 1/3$, $\nu \approx 0.84$ and $\nu \approx 0.26$.

4.7. A General Nonlinear Approximation

At least for homogeneous systems, Boltzmann's H theorem (e.g., Ref. 51) yields a general form for the equilibrium solution of the full nonlinear Boltzmann equation (4.3.1). The H function can be defined as

$$H = \sum_a \tilde{f}_a \log(\tilde{f}_a) \qquad (4.7.1)$$

where

$$\tilde{f}_a = \frac{f_a}{1 - f_a} \qquad (4.7.2)$$

This definition is analogous to that used for Fermi-Dirac particles (e.g., Refs. 51, 52): the factors $(1 - f_a)$ account for the exclusion of more than one particle on each link, as in Eq. (4.3.1). The microscopic reversibility of (4.3.1) implies that when the equilibrium condition $\partial_t H = 0$ holds, all products $\tilde{f}_{a_1} \tilde{f}_{a_2} \cdots$ must be equal for all initial and final sets of particles $\{a_1, a_2, \ldots\}$ that can participate in collisions. As a result, the $\log(\tilde{f}_a)$ must be simple linear combinations of the quantities conserved in the collisions. If only particle number and momentum are conserved, and there are no spurious conserved quantities such as (4.5.11), the $\tilde{f}_a$ can always be written in the form[7,49,54]

$$\tilde{f}_a = \exp(-\alpha - \beta \mathbf{u} \cdot \mathbf{e}_a) \qquad (4.7.3)$$

The one-particle distribution functions thus have the usual Fermi-Dirac form

$$f_a = [1 + \exp(\alpha + \beta \mathbf{u} \cdot \mathbf{e}_a)]^{-1} \qquad (4.7.4)$$

where α and β are in general functions of the conserved quantities n and $|\mathbf{u}|^2$.

For small $|\mathbf{u}|^2$, one may write

$$\alpha = \alpha_0 + \alpha_1 |\mathbf{u}|^2 + \cdots, \qquad \beta = \beta_0 + \beta_1 |\mathbf{u}|^2 + \cdots \qquad (4.7.5)$$

These expansions can be substituted into Eq. (4.7.4), and the results compared with the Chapman-Enskog expansion (2.7.1).

For $\mathbf{u} = 0$, one finds immediately the "fugacity relation"

$$\exp(-\alpha_0) = \frac{f}{1 - f} \qquad (4.7.6)$$

Then, from the expansion (related to that for generating Euler polynomials)

$$(1 + \xi e^x)^{-1} = \frac{1}{1 + \xi} - \frac{\xi}{(1 + \xi)^2} x - \frac{\xi(1 - \xi)}{2(1 + \xi)^3} x^2$$
$$- \frac{\xi(1 - 4\xi + \xi^2)}{6(1 + \xi)^4} x^3 - \frac{\xi(1 - \xi)(1 - 10\xi + \xi^2)}{24(1 + \xi)^5} x^4 + \cdots \qquad (4.7.7)$$

together with the constraints (2.7.3)–(2.7.5) one obtains (for $d = 2$)

$$\beta_0 = -\frac{2}{1-f}$$

$$\alpha_1 = \frac{1-2f}{(1-f)^2}$$

$$\beta_1 = \frac{1-2f+2f^2}{(1-f)^3} \tag{4.7.8}$$

$$\alpha_2 = -\frac{(1-2f)(3-4f+4f^2)}{16(1-f)^4}$$

where it has been assumed that the $\mathbf{e}_a$ form an isotropic set of unit vectors, satisfying Eq. (3.1.5). The complete Chapman-Enskog expansion (2.7.1) then becomes

$$
\begin{aligned}
f_a = f \Bigg\{ & 1 + d\mathbf{u}\cdot\mathbf{e}_a + \frac{d^2}{2}\frac{1-2f}{1-f}\left[(\mathbf{u}\cdot\mathbf{e}_a)^2 - \frac{1}{d}|\mathbf{u}|^2\right] \\
& + \frac{d^3}{6}\frac{1-6f+6f^2}{(1-f)^2}\left[(\mathbf{u}\cdot\mathbf{e}_a)^3 - \frac{3}{d+2}|\mathbf{u}|^2(\mathbf{u}\cdot\mathbf{e}_a)\right] \\
& + \frac{1}{48}\frac{1-2f}{(1-f)^3}[32(1-12f+12f^2)(\mathbf{u}\cdot\mathbf{e}_a)^4 \\
& + 384f(1-f)|\mathbf{u}|^2(\mathbf{u}\cdot\mathbf{e}_a)^2 + 3(11-36f+36f^2)|\mathbf{u}|^4] + \cdots \Bigg\}
\end{aligned}
\tag{4.7.9}
$$

where for the last term it has been assumed that $d = 2$.

The result (4.6.4) for $c^{(2)}$ follows immediately from this expansion. For cellular automaton fluid models with $\mathbf{E}^{(6)}$ isotropic, the continuum equation (2.7.7) holds. The results for the coefficients that appear in this equation can be obtained from the approximation (4.7.9), and have the simple forms

$$c^{(2)} = \frac{d^2(1-2f)}{2(1-f)} \tag{4.7.10}$$

$$c^{(4)}(1+\sigma_{4,1}) = \frac{2(1-2f)}{3(1-f)^3} \quad (d=2) \tag{4.7.11}$$

These results allow an estimate of the importance of the next-order corrections to the Navier-Stokes equations included in Eq. (2.7.7). They suggest that the corrections may be important whenever $|\mathbf{u}/(1-f)^2|^2$ is not small compared to 1. The corrections can thus potentially be important both at high average velocities and high particle densities.

The hexagonal lattice model of Section 2 yields a continuum equation of the form (2.7.6), with an anisotropic $O(\mathbf{u}^2\nabla\mathbf{u})$ term. Equation (4.7.9) gives in this case

$$\partial_t(6fu_x) + \frac{3f(1-2f)}{1-f}(u_x\partial_x u_x + u_x\partial_y u_y + u_y\partial_y u_x - u_y\partial_x u_y)$$

$$+ \frac{f(1-2f)}{4(1-f)^3}\{[(55 - 84f + 84f^2)u_x^2$$

$$+ 3(13 + 4f - 4f^2)u_y^2]u_x\partial_x u_x$$

$$+ 2[(1 + 12f - 12f^2)u_x^2 + 9(1 - 2f)^2 u_y^2]u_x\partial_y u_y$$

$$+ 6[(1 + 12f - 12f^2)u_x^2 + (1 - 2f)^2]u_y\partial_y u_x$$

$$+ 3[(13 + 4f - 4f^2)u_x^2 + (13 - 28f + 28f^2)u_y^2]u_y\partial_x u_y\} = 0 \qquad (4.7.12)$$

$$\partial_t(6fu_y) + \frac{3f(1-2f)}{1-f}(-u_x\partial_x u_x + u_x\partial_x u_y + u_y\partial_x u_x - u_y\partial_y u_y)$$

$$+ \frac{f(1-2f)}{4(1-f)^3}\{[(35 - 36f + 36f^2)u_x^2$$

$$+ 3(17 - 44f + 44f^2)u_y^2]u_x\partial_y u_x$$

$$+ 2[(1 + 12f - 12f^2)u_x^2 + 9(1 - 2f)^2 u_y^2]u_x\partial_x u_y$$

$$+ 6[(1 + 12f - 12f^2)u_x^2 + (1 - 2f)^2]u_y\partial_x u_x$$

$$+ 3[(17 - 44f + 44f^2)u_x^2 + (17 - 12f + 12f^2)u_y^2]u_y\partial_y u_y\} = 0 \qquad (4.7.13)$$

The $O(\mathbf{u}\nabla\mathbf{u})$ term is as given in Eq. (2.5.11). The $O(\mathbf{u}^3\nabla\mathbf{u})$ terms are anisotropic, and are not even invariant under exchange of x and y coordinates ($\pi/2$ rotation). For small densities f, Eqs. (4.7.12) and (4.7.13) become

$$\partial_t(u_x) + \frac{1}{2}(u_x\partial_x u_x + u_x\partial_y u_y + u_y\partial_y u_x - u_y\partial_x u_y)$$

$$+ \frac{1}{24}[(55u_x^2 + 39u_y^2)u_x\partial_x u_x + 2(u_x^2 + 9u_y^2)u_x\partial_y u_y$$

$$+ 6(u_x^2 + u_y^2)u_y\partial_y u_x + 39(u_x^2 + u_y^2)u_y\partial_x u_y] = 0 \qquad (4.7.14)$$

$$\partial_t(u_y) + \frac{1}{2}(-u_x\partial_x u_x + u_x\partial_x u_y + u_y\partial_x u_x - u_y\partial_y u_y)$$

$$+ \frac{1}{24}[(35u_x^2 + 51u_y^2)u_x\partial_y u_x + 2(u_x^2 + 9u_y^2)u_x\partial_x u_y$$

$$+ 6(u_x^2 + u_y^2)u_y\partial_x u_x + 51(u_x^2 + u_y^2)u_y\partial_y u_y] = 0 \qquad (4.7.15)$$

The results (4.7.10) and (4.7.11) follow from the Fermi-Dirac particle distribution (4.7.4). If instead an arbitrary number of particles were allowed at each site, the equilibrium particle distribution (4.7.4) would take on the Maxwell-Boltzmann form

$$f_a = \exp(-\alpha - \beta\mathbf{u}\cdot\mathbf{e}_a) \qquad (4.7.16)$$

With this simpler form, more complete results for f_a as a function of n and u can be found. Results which are isotropic to all orders in $\mathbf{u}$ can be obtained only for an infinite set of possible particle directions, parametrized, say, by a continuous angle θ. In this case, the number and momentum densities (2.4.1) and (2.4.2) may be obtained as integrals

$$\frac{1}{2\pi} \int_0^{2\pi} f(\theta) \, d\theta = f \tag{4.7.17}$$

$$\frac{1}{2\pi} \int_0^{2\pi} \mathbf{e}(\theta) f(\theta) \, d\theta = f \mathbf{u} \tag{4.7.18}$$

With the distribution (4.7.16), these integrals become

$$\frac{1}{2\pi} \int_0^{2\pi} \exp(-\alpha - \beta u \cos \theta) \, d\theta = e^{-\alpha} I_0(\beta u) = f \tag{4.7.19}$$

$$\frac{1}{2\pi} \int_0^{2\pi} \exp(-\alpha - \beta u \cos \theta) \cos \theta \, \mathbf{u}/u \, d\theta = -e^{-\alpha} I_1(\beta u) \mathbf{u}/u = f \mathbf{u} \tag{4.7.20}$$

where $u = |\mathbf{u}|$, and the $I_\nu(z)$ are modified Bessel functions (e.g., Ref. 53)

$$I_\nu(z) = \sum_{n=0}^{\infty} \frac{(z/2)^{\nu+2n}}{n!(\nu+n)!}, \quad I_0(0) = 1, \quad I_\nu(0) = 0 \quad (\nu > 0) \tag{4.7.21}$$

$$\int_{-1}^{1} (1-x^2)^{\nu-1/2} e^{-zx} \, dx = \pi z^{-\nu} (2\nu-1)!! I_\nu(z) \tag{4.7.22}$$

$$I_{\nu-1}(z) - I_{\nu+1}(z) = (2\nu/z) I_\nu(z) \tag{4.7.23}$$

The rapid convergence of the series (4.7.21) means that Eqs. (4.7.19) and (4.7.20) provide highly accurate approximations even for a small number of discrete directions $\mathbf{e}_a$. [For example, with $M = 6$, $\alpha = 0$, $\beta = 1$, and $\mathbf{u} = (1, 1)$, the error in Eq. (4.7.19) is less than 10^{-9}.]

For the simple distribution (4.7.16) the momentum flux density tensor (2.4.9) may be evaluated in direct analogy with Eqs. (4.7.19) and (4.7.20) as

$$\Pi_{ij} = e^{-\alpha} \left[\frac{1}{2} I_0(\beta u) \delta_{ij} + I_2(\beta u) \left(\frac{u_i u_j}{u^2} - \frac{1}{2} \delta_{ij} \right) \right] \tag{4.7.24}$$

Using the recurrence relation (4.7.23), and substituting the results (4.7.19) and (4.7.20), this may be rewritten in the form

$$\Pi_{ij} = f \left[\frac{1}{2} I_0(\beta u) \delta_{ij} + \left(1 + \frac{2}{\beta} \right) \left(\frac{u_i u_j}{u^2} - \frac{1}{2} \delta_{ij} \right) \right] \tag{4.7.25}$$

Combining Eqs. (4.7.19) and (4.7.20), one finds that the function $\beta(f, u)$ is independent of f, and can be determined from the implicit equation

$$\frac{I_1(\beta u)}{I_0(\beta u)} = \frac{I_0'(\beta u)}{I_0(\beta u)} = -u \tag{4.7.26}$$

Expanding in powers of u^2, as in Eq. (4.7.5), yields

$$\beta(u) = \sum_{n=0} \beta_n u^{2n} \tag{4.7.27}$$

$$\beta_0 = -2, \quad \beta_1 = -1, \quad \beta_2 = -\frac{5}{6}, \quad \beta_3 = -\frac{19}{24}, \quad \beta_4 = -\frac{143}{180}$$

Equation (4.7.19) then gives

$$\alpha(u, f) = \sum_{n=0} \alpha_n u^{2n} \tag{4.7.28}$$

$$\exp(-\alpha_0) = f, \quad \alpha_1 = 1, \quad \alpha_2 = \frac{3}{4}, \quad \alpha_3 = \frac{25}{36}, \quad \alpha_4 = \frac{133}{192}$$

In the limit $u \to 1$, $\beta \to -\infty$.

The above results immediately yield values for the transport coefficients $c^{(n)}$ in the Chapman-Enskog expansion:

$$c^{(2)} = 2 \tag{4.7.29}$$

$$c^{(4)}(1 + \sigma_{4,1}) = \frac{2}{3} \tag{4.7.30}$$

independent of density. Equation (4.7.29) imples that the coefficient μ of the convective term in the Navier-Stokes equation (2.6.1) is equal to $1/2$. The deviation from the Galilean invariant result 1 is associated with the constraint of fixed speed particles.

Figure 4 shows the exact result for $\beta(u)$ obtained from Eq. (4.7.26), compared with series expansions to various orders. Significant deviations from the $O(u^2)$ "Navier-Stokes" approximation are seen for $u \gtrsim 0.4$.

For Fermi-Dirac distributions of the form (4.7.4), the integrals (4.7.19) and (4.7.20) can only be expressed as infinite sums of Bessel functions.

4.8. Other Models

The results obtained so far can be generalized directly to a large class of cellular automaton fluid models.

In the main case considered in Section 3, particles have velocities corresponding to a set of M unit vectors e_a. If this set is invariant under inversion, then both e_a and $-e_a$ always occur. As a result, two particles colliding head on with velocities e_a and $-e_a$ can always scatter in any directions e_b and $-e_b$ with $b \neq a$. One simple possibility is to choose the rules at different sites so that each scattering direction occurs with equal probability. If only such two-particle collisions are possible (as in a low-density approximation), and only one particle is allowed on each link, then the

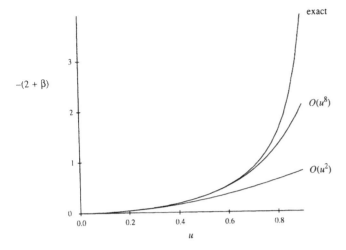

Figure 4. Dependence of $\beta(u)$ from Eq. (4.7.16) on the magnitude u of the macroscopic velocity. The results are for Maxwell-Boltzmann particles with unit speeds and arbitrary directions in two dimensions. The function $\beta(u)$ appears both in the microscopic distribution function (4.7.16) and in the macroscopic momentum flux tensor (4.7.25). The result for $\beta(u)$ from an exact solution of the implicit equation (4.7.26) is given, together with results from the series expansion (4.7.27). The $O(u^2)$ result corresponds to the Navier-Stokes approximation. Deviation from the exact result is seen for $u \gtrsim 0.4$.

Boltzmann transport equation becomes

$$\partial_t f_a + \mathbf{e}_a \cdot \nabla f_a = -f_a f_{\bar{a}} \prod_{c \neq a, \bar{a}} (1 - f_c) + \frac{1}{M-2} \sum_{b \neq a, \bar{a}} f_b f_{\bar{b}} \prod_{c \neq b, \bar{b}} (1 - f_c) \tag{4.8.1}$$

where $f_{\bar{a}}$ is the distribution function for particles with direction $-\mathbf{e}_a$. To second order in the expansion (4.4.2) this gives

$$\partial_t (f\phi_a) + \mathbf{e}_a \cdot \nabla (f\phi_a) = f^2 (1-f)^{M-3} \left[-(\phi_a + \phi_{\bar{a}}) + \frac{1}{M-2} \sum_{b \neq a, \bar{a}} (\phi_b + \phi_{\bar{b}}) \right]$$

$$+ f^2 (1-f)^{M-4} (1-2f) \left[-\phi_a \phi_{\bar{a}} + \frac{1}{M-2} {\sum_{b, \bar{b} \neq a, \bar{a}}}' \phi_b \phi_{\bar{b}} \right]$$

$$+ f^3 (1-f)^{M-4} \left[- {\sum_{b, c \neq a, \bar{a}; c \neq \bar{b}}}' \phi_b \phi_c + \frac{1}{M-2} \sum_{b \neq a} (\phi_a \phi_b + \phi_{\bar{a}} \phi_{\bar{b}}) \right] \tag{4.8.2}$$

where $\sum'$ denotes summation over the triangular region in which the indices form a strictly increasing sequence.

The form of the ϕ_a for a homogeneous system can be obtained from the general equilibrium conditions of Section 4.7. The coefficients $c^{(n)}$ in the Chapman-Enskog expansion are then given by Equations (4.7.10) and (4.7.11). The convective transport coefficient μ in the Navier-Stokes equation (2.6.1) is thus given by

$$\mu = \frac{d^2(1-2f)}{8(1-f)} \tag{4.8.3}$$

The $c_\nabla^{(n)}$ cannot be obtained by the methods of Section 4.7. But from Eq. (4.8.2) one may deduce immediately the linearized collision term

$$\omega_{ab}^{(1)} = f^2(1-f)^{M-3}\text{circ}[\chi_a] \tag{4.8.4}$$

$$\chi_1 = \chi_{1+M/2} = -1, \qquad \chi_a = \frac{2}{M-2}$$

Then, in analogy with Eq. (4.6.8), the kinematic viscosity for the cellular automaton fluid is found to be

$$\nu = \frac{M-2}{2d(d+2)Mf^2(1-f)^{M-3}} \tag{4.8.5}$$

For an icosahedral set of e_a, with $d=3$ and $M=12$, this yields

$$\nu = [36f^2(1-f)^9]^{-1} \tag{4.8.6}$$

Several generalizations may now be considered. First, one may allow not just one, but, say, up to κ particles on each link of the cellular automaton array. In the limit $\kappa \to \infty$ an arbitrary density of particles is thus allowed in each cell. The Boltzmann equation for this case is the same as (4.8.1), but with all $(1-f_c)$ factors omitted. The resulting transport coefficients are

$$\mu = \frac{1}{2} \tag{4.8.7}$$

$$\nu = \frac{M-2}{2d(d+2)Mf^2} \tag{4.8.8}$$

Another generalization is to allow collisions that involve more than two particles. The simplest such collisions are "composite" ones, formed by superposing collisions involving two or less particles. The presence of such collisions changes the values of transport coefficients, but cannot affect the basic properties of the model. The four-particle and asymmetric three-particle collisions in the hexagonal lattice model of Section 2 are examples of composite collisions. They increase the total collision rate, and thus, for example, decrease the viscosity, but do not change the overall macroscopic behavior of the model.

In general, collisions involving k particles can occur if the possible e_a are such that

$$\sum_{i=1}^{k} e_{a_i} = \sum_{i=1}^{k} e_{b_i} \tag{4.8.9}$$

for some sets of incoming and outgoing particles a_i and b_i. Cases in which all the

a_i and b_i are distinct may be considered "elementary" collisions. In the hexagonal lattice model of Section 2, only two-particle and symmetric three-particle collisions are elementary.

No elementary three-particle collisions are possible on primitive and body-centered cubic three-dimensional lattices, or with $\mathbf{e}_a$ corresponding to the vertices of icosahedra or dodecahedra. For a face-centered cubic lattice, however, eight distinct triples of $\mathbf{e}_a$ sum to zero [an example is $(1, -1, 0)+(0, 1, -1)+(-1, 0, 1)$], so that elementary three-particle collisions are possible.

One feature of the hexagonal lattice model discussed in Section 2 is the existence of the conservation law (4.5.11) when elementary three-body symmetric collisions are absent. Such spurious conservation laws exist in any cellular automaton fluid model in which all particles have the same speed, and only two-particle collisions can occur. Elementary three-particle collisions provide one mechanism for avoiding these conservation laws and allowing the equilibrium of Section 4.7 to be attained.

4.9. Multiple Speed Models

A further generalization is to allow particles with velocities $\mathbf{e}_a$ of different magnitudes. This generalization is significant not only in allowing two-particle collisions alone to avoid the spurious conservation laws of Section 4.5, but also in making it possible to obtain isotropic hydrodynamic behavior on cubic lattices, as discussed in Section 3.5.

One may define a kinetic energy $1/2|\mathbf{e}_a|^2$ that differs for particles of different speeds. In studies of processes such as heat conduction, one must account for the conservation of total kinetic energy. In many cases, however, one considers systems in contact with a heat bath, so that energy need not be conserved in individual collisions.

In a typical case, one may then take pairs of particles with speed s_i colliding head on to give pairs of particles with some other speed s_j. In general, different collision rules may be used on different sites, typically following some regular pattern, as discussed in Section 4.3. Thus, for example, collisions between speed s_i particles may yield speed s_j particles at a fraction $\gamma_{i \to j}$ of the sites.

The number m_i of possible particles with speed $s_i = (|\mathbf{e}_a|^2)^{1/2}$ that can occur at each site is determined by the structure of the lattice. The collision rules at different sites may be arranged, as in Section 4.8, to yield particles of a particular speed s_i with equal probabilities in each of the m_i possible directions.

In a homogenous system, the probability f_i for a link with speed s_i to be populated should satisfy the master equation

$$\partial_t \tilde{f}_i = \sum_j \Gamma_{ij} \tilde{f}_j^2 = \sum_{j \neq i} (-\gamma_{i \to j} m_i \tilde{f}_i^2 + \gamma_{j \to i} m_j \tilde{f}_j^2) \qquad (4.9.1)$$

where it assumed that

$$\sum_j \gamma_{i \to j} = \sum_j \gamma_{j \to i} = 1 \qquad (4.9.2)$$

399

and $\tilde{f}$ is the reduced particle density given by Eq. (4.7.2). With two speeds, Γ_{ij} becomes

$$\Gamma_{ij} = \begin{pmatrix} -\gamma_{1\to2}m_1 & \gamma_{2\to1}m_2 \\ \gamma_{1\to2}m_1 & -\gamma_{2\to1}m_2 \end{pmatrix} \tag{4.9.3}$$

The solutions of Eq. (4.9.1) can be found in terms of the eigenvalues and corresponding eigenvectors of this matrix:

$$\lambda = 0: \qquad\qquad (\gamma_{2\to1}m_2, \gamma_{1\to2}m_1) \tag{4.9.4}$$

$$\lambda = -(\gamma_{1\to2}m_1 + \gamma_{2\to1}m_2): \qquad (-1, 1) \tag{4.9.5}$$

In the large-time limit, only the equilibrium eigenvector (4.9.4) should survive, giving a ratio of reduced particle densities

$$\frac{\tilde{f}_2^2}{\tilde{f}_1^2} = \frac{\gamma_{1\to2}m_1}{\gamma_{2\to1}m_2} \tag{4.9.6}$$

For three particle speeds, one finds the equilibrium conditions

$$\frac{\tilde{f}_2^2}{\tilde{f}_1^2} = \frac{m_1}{m_2} \frac{\gamma_{1\to2}\gamma_{3\to1} + \gamma_{1\to2}\gamma_{3\to2} + \gamma_{1\to3}\gamma_{3\to2}}{\gamma_{2\to1}\gamma_{3\to1} + \gamma_{2\to1}\gamma_{3\to2} + \gamma_{2\to3}\gamma_{3\to1}}$$

$$\frac{\tilde{f}_3^2}{\tilde{f}_1^2} = \frac{m_1}{m_3} \frac{\gamma_{1\to2}\gamma_{2\to3} + \gamma_{1\to3}\gamma_{2\to1} + \gamma_{1\to3}\gamma_{2\to3}}{\gamma_{2\to1}\gamma_{3\to1} + \gamma_{2\to1}\gamma_{3\to2} + \gamma_{2\to3}\gamma_{3\to1}} \tag{4.9.7}$$

Different choices for the γ_k yield different equilibrium speed distributions. The probabilities f_i give the weights $w(s_i^2)$ that appear in Eq. (3.1.16). Equation (4.9.6) shows that by choosing

$$\gamma_{2\to1} \approx 2\gamma_{1\to2} \tag{4.9.8}$$

one obtains a ratio of weights for the model of Eq. (3.5.1) that satisfy the condition (3.5.5) for the isotropy of $\mathbf{E}^{(4)}$. [There is a small correction to equality in Eq. (4.9.8) associated with the difference between f and $\tilde{f}$.]

On a cubic lattice, one may similarly satisfy the condition (3.5.12) for the isotropy of $\mathbf{E}^{(4)}$ simply by taking $f_3 = 0$, and

$$\gamma_{2\to1} \approx 2\gamma_{1\to2} \tag{4.9.9}$$

In this way, one may obtain approximate isotropic hydrodynamic behavior on a three-dimensional cubic lattice.

4.10. Tagged Particle Dynamics

In the discussion above, all the particles in the cellular automaton fluid were assumed indistinguishable. This section considers the behavior of a small concentration of special "tagged" particles.

The density g_a of tagged particles with direction $\mathbf{e}_a$ satisfies an equation of the Fokker-Planck type (e.g., Ref. 26):

$$\partial_t g_a + (\mathbf{e}_a \cdot \nabla)g_a = \Theta_a \tag{4.10.1}$$

Assuming as in the Boltzmann equation approximation that there are no correlations between particles at different sites, the collision term of Eq. (4.10.1) may be written in the form

$$\Theta_a = \sum_b \theta_{ab} g_b \tag{4.10.2}$$

where θ_{ab} gives the probability that a particle that arrives at a particular site from direction $\mathbf{e}_b$ leaves in direction $\mathbf{e}_a$ with $a \neq b$. The probability is averaged over different arrangements of ordinary particles. Various deterministic rules may be chosen for collisions between ordinary and tagged particles. The simplest assumption is that on average the tagged particles take the place of any of the outgoing particles with equal probability.

Conservation of the total number of tagged particles implies

$$\sum_a g_a = g M \tag{4.10.3}$$

The total momentum of tagged particles is not conserved; the background of ordinary particles acts like a "heat bath" which can exchange momentum with the tagged particles through the noise term Θ_a. Assuming a uniform background fluid, one may make an expansion for the g_a of the form

$$g_a = (g + d^{(1)} \mathbf{e}_a \cdot \nabla g + \cdots) \tag{4.10.4}$$

The total number of tagged particles then satisfies the equation

$$\partial_t g + d^{(1)} \frac{1}{M} \mathbf{E}_{ij}^{(2)} \partial_i \partial_j g = 0 \tag{4.10.5}$$

where the collision term disappears as a result of Eq. (4.10.3). With the $\mathbf{e}_a$ chosen so that $\mathbf{E}_{ij}^{(2)}$ is isotropic, Eq. (4.10.5) becomes the standard equation for self-diffusion,

$$\partial_t g = D \nabla^2 g \tag{4.10.6}$$

with the diffusion coefficient D given by

$$D = -\frac{1}{d} d^{(1)} \tag{4.10.7}$$

The value of $d^{(1)}$ must be found by solving Eq. (4.10.1) for g_a using the approximation (4.10.4). The equilibrium condition for Eq. (4.10.1) in this case becomes

$$(\mathbf{e}_a \cdot \nabla) g = d^{(1)} \sum_b \theta_{ab} \mathbf{e}_b \cdot \nabla g \tag{4.10.8}$$

Thus $-d^{(1)}$ is given in this approximation by the mean free path λ for particle scattering, so that the diffusion coefficient is given by the standard kinetic theory formula

$$D = \frac{1}{d} \lambda \tag{4.10.9}$$

For the hexagonal lattice model of Section 2,

$$D = \{2f^2(1-f)^2[(1-f)^2 + (\gamma_{3S} + 4\gamma_{3A})f(1-f) + \gamma_4 f^2]\}^{-1} \qquad (4.10.10)$$

5. Some Extensions

The simple physical basis for cellular automaton fluid models makes it comparatively straightforward for them to include many of the physical effects that occur in actual fluid experiments.

Boundaries can be represented by special sites in the cellular automaton array. Collisions with boundaries conserve particle number, but not particle momentum. One possibility is to choose boundary collision rules that exactly reverse the velocities of all particles, so that particles in a layer close to the boundary have zero average momentum. This choice yields macroscopic "no slip" boundary conditions, appropriate for many solid surfaces (e.g., Ref. 27). For boundaries that consist of flat segments aligned along lattice directions, an alternative is to take particles to undergo "specular" reflection, yielding a zero average only for the transverse component of particle momentum, and giving "free slip" macroscopic boundary conditions. The roughness of surfaces may be modeled explicitly by including various combinations of these microscopic boundary conditions (corresponding, say, to different coefficients of accommodation).

Arbitrarily complex solid boundaries may be modeled by appropriate arrangements of boundary cells. To model, for example, a porous medium one can, for example, use a random array of "boundary" cells with appropriate statistical properties.

A net flux of fluid can be maintained by continually inserting particles on one edge with an appropriate average momentum and extracting particles on an opposite edge. The precise arrangement of the inserted particles should not affect the macroscopic properties of the system, since microscopic processes should rapidly establish a microscopically random state of local equilibrium. Large-scale inhomogeneities, perhaps representing "free stream turbulence" (e.g., Ref. 4), can be included explicitly.

External pressure and density constraints, whether static or time-dependent, can be modeled by randomly inserting or extracting particles so that local average particle densities correspond to the macroscopic distribution required.

External forces can be modeled by randomly changing velocities of individual particles so as to impart momentum to the fluid at the required average rate. Moving boundaries can then be modeled by explicit motion of the special boundary cells, together with the inclusion of an appropriate average momentum change for particles striking the boundary. Gravitational and other force fields can also be represented in a "quantized approximation" by explicit local changes in particle velocities.

Many other physical effects depend on the existence of surfaces that separate different phases of a fluid or distinct immiscible fluids. The existence of such surfaces requires collective ordering effects within the system. For some choices of parameters, no such ordering can typically occur. But as the parameters change, phase transitions may occur, allowing large correlated regions to form. Such phenomena will be studied elsewhere. (Surface tension effects have been observed in other two-dimensional cellular automata.[3])

6. Discussion

Partial differential equations have conventionally formed the basis for mathematical models of continuum systems such as fluids. But only in rather simple circumstances can exact mathematical solutions to such equations be found. Most actual studies of fluid dynamics must thus be based on digital computer simulations, which use discrete approximations to the original partial differential equations (e.g., Ref. 55).

Cellular automata provide an alternative approach to modeling fluids and other continuum systems. Their basic constituent cells are discrete, and ideally suited to simulation by digital computers. Yet collections of large numbers of these cells can show overall continuum behavior. This paper has given theoretical arguments that with appropriate rules for the individual cells, the overall behavior obtained should follow that described by partial differential equations for fluids.

The cellular automata considered give simple idealized models for the motion and collision of microscopic particles in a fluid. As expected from the second law of thermodynamics, precise particle configurations are rapidly randomized, and may be considered to come to some form of equilibrium. In this equilibrium, it should be adequate to describe configurations merely in terms of probabilities that depend on a few macroscopic quantities, such as momentum and particle number, that are conserved in the microscopic particle interactions. Such averaged macroscopic quantities change only slowly relative to the rate of particle interactions. Partial differential equations for their behavior can be found from the transport equations for the average microscopic particle dynamics.

So long as the underlying lattice is sufficiently isotropic, many cellular automata yield in the appropriate approximation the standard Navier–Stokes equations for continuum fluids. The essential features necessary for the derivation of these equations are the conservation of a few macroscopic quantities, and the randomization of all other quantities, by microscopic particle interactions. The Navier–Stokes equations follow with approximations of low fluid velocities and velocity gradients. The simplicity of the cellular automaton model in fact makes it possible to derive in addition next order corrections to these equations.

The derivation of hydrodynamic behavior from microscopic dynamics has never been entirely rigorous. Cellular automata can be considered as providing a simple example in which the necessary assumptions and approximations can be studied in detail. But strong support for the conclusions comes from explicit simulations of cellular automaton fluid models and the comparison of results with those from actual experiments. The next paper in this series will present many such simulations.

The cellular automaton method of this paper can potentially be applied to a wide variety of processes conventionally described by partial differential equations.

One example is diffusion. At a microscopic level, diffusion arises from random particle motions. The cellular automata used above can potentially reproduce diffusion phenomena, as discussed in Section 4.10. But much simpler cellular automaton rules should suffice. The derivation of the diffusion equation requires that the number of particles be conserved. But it is not necessary for total particle momentum to be conserved. Instead, particle directions should be randomized at each site. Such randomization can potentially be achieved by very simple cellular automaton rules, such as that of Ref. 20. Thus, one may devise cellular automaton methods for the solution of the diffusion equation,[56] which in turn gives a relaxation method for solving Laplace, Poisson, and related equations.

Whenever the physical basis for partial differential equations involves large numbers of particles or other components with local interactions, one can expect to derive an effective cellular automaton model. For systems such as electromagnetic or gravitational fields, such models can perhaps be obtained as analogues of lattice gauge theories.

Appendix: SMP Programs

This appendix contains a sample SMP[17] computation of the macroscopic equations for the hexagonal lattice cellular automaton fluid model of Section 2.

The SMP definitions are as follows:

```
/* two-dimensional case */
d:2

/* define position and velocity vectors */
r:{x,y}
u:{ux,uy}

/* generate polygonal set of lattice vectors */
<XTrig
polygon[$n] :: (e:Ar[$n,{Cos[2Pi $/$n],Sin[2Pi $/$n]}])

/* calculate terms in number density, momentum vector and stress tensor */
suma[$x] :: Ex[Sum[$x,{a,1,Len[e]}]]
nterm[$f] :: suma[$f[a]]
uterm[$f] :: suma[e[a] $f[a]]
piterm[$f] :: suma[e[a]**e[a] $f[a]]

/* define vector analysis operators */
egrad[$x,$a] :: Sum[e[$a][i] Dt[$x,r[i]],{i,1,d}]
div[$x] :: Sum[Dt[$x[i],r[i]],{i,1,d}]

/* terms in Chapman-Enskog expansion */
n : f Len[e]
ce0[$a] : f
ce1[$a] : f e[$a].u
```

```
ce2[$a] : f ((e[$a].u)^2 - u.u/2)
ce2d[$a] : f (egrad[e[$a].u,$a] - div[u]/2)
celist : {ce0,ce1,ce2,ce2d}

/* specify commutativity of second derivatives */
Dt[$f,$1,{$2_=(Ord[$2,$1]>0),1}] :: Dt[$f,$2,$1]

/* define printing of derivatives */
_Dt[Pr][[$1,$2]]::Fmt[{{0,0},{1,-1},{2,0}},D,$2,$1]
_Dt[Pr][[$1,$2{$3,1}]]::Fmt[{{0,0},{1,-1},{2,-1},{3,0}},D,$2,$3,$1]
```

The following is a transcript of an interactive SMP session:

```
#I[1]:: <"cafluid.smp"     /* load definitions */

#I[2]:: polygon[6]         /* set up for hexagonal lattice */

              1/2          1/2             1/2        1/2
           3            3              - 3         -3
#O[2]:  {{1/2,----},{-1/2,----},{-1,0},{-1/2,------},{1/2,------},{1,0}}
           2            2              2          2


#I[3]:: Map[nterm,celist]      /* find contributions to number density from
                                  terms in Chapman-Enskog expansion */

#O[3]:  {6f,0,0,0}

#I[4]:: Map[uterm,celist]   /* find contributions to momentum vector */

#O[4]:  {{0,0},{3f ux,3f uy},{0,0},{0,0}}

#I[5]:: Map[piterm,celist]  /* stress tensor */

#O[5]:* {{{3f,0},{0,3f}},{{0,0},{0,0}},
```

$$\{\{\frac{3f\ ux^2}{4} - \frac{3f\ uy^2}{4}, \frac{3f\ ux\ uy}{2}\}, \{\frac{3f\ ux\ uy}{2}, \frac{-3f\ ux^2}{4} + \frac{3f\ uy^2}{4}\}\},$$

$$\{\{\frac{3f\ D_x\ ux}{4} - \frac{3f\ D_y\ uy}{4}, \frac{3f\ D_y\ ux}{4} + \frac{3f\ D_x\ uy}{4}\},$$

$$\{\frac{3f\ D_y\ ux}{4} + \frac{3f\ D_x\ uy}{4}, \frac{-3f\ D_x\ ux}{4} + \frac{3f\ D_y\ uy}{4}\}\}\}$$

```
#I[6]:: Dt[f,$$]:0 ;  /* make incompressibility approximation */

#I[7]:: Fac[Map[div,@5]]  /* contributions to momentum equation */
```

$$\#0[7]:* \{\{0,0\},\{0,0\},\{\frac{3f \ (ux \ D_x \ ux + ux \ D_y \ uy + uy \ D_y \ ux - uy \ D_x \ uy)}{2},$$

$$\frac{-3f \ (ux \ D_y \ ux - ux \ D_x \ uy - uy \ D_x \ ux - uy \ D_y \ uy)}{2}\},$$

$$\{\frac{3f \ (D_{xx} \ ux + D_{yy} \ ux)}{4}, \frac{3f \ (D_{xx} \ uy + D_{yy} \ uy)}{4}\}\}$$

Acknowledgments

Many people have contributed in various ways to the material presented here. For general discussions I thank: Uriel Frisch, Brosl Hasslacher, David Levermore, Steve Orszag, Yves Pomeau, and Victor Yakhot. For specific suggestions I thank Roger Dashen, Dominique d'Humieres, Leo Kadanoff, Paul Martin, John Milnor, Steve Omohundro, Paul Steinhardt, and Larry Yaffe. Most of the calculations described here were made possible by using the SMP general-purpose computer mathematics system.[17] I thank Thinking Machines Corporation for much encouragement and partial support of this work.

References

1. S. Wolfram ed., *Theory and Applications of Cellular Automata* (World Scientific, 1986).
2. S. Wolfram, Cellular automata as models of complexity, *Nature* **311**:419 (1984).
3. N. Packard and S. Wolfram, Two-dimensional cellular automata, *J. Stat. Phys.* **38**:901 (1985).
4. D. J. Tritton, *Physical Fluid Dynamics* (Van Nostrand, 1977).
5. W. W. Wood, Computer studies on fluid systems of hard-core particles, in *Fundamental Problems in Statistical Mechanics 3*, E. D. G. Cohen, ed. (North-Holland, 1975).
6. J. Hardy, Y. Pomeau, and O. de Pazzis, Time evolution of a two-dimensional model system. I. Invariant states and time correlation functions, *J. Math. Phys.* **14**:1746 (1973); J. Hardy, O. de Pazzis, and Y. Pomeau, Molecular dynamics of a classical lattice gas: Transport properties and time correlation functions, *Phys. Rev. A* **13**:1949 (1976).
7. U. Frisch, B. Hasslacher, and Y. Pomeau, Lattice gas automata for the Navier–Stokes equation, *Phys. Rev. Lett.* **56**:1505 (1986).
8. J. Salem and S. Wolfram, Thermodynamics and hydrodynamics with cellular automata, *Theory and Applications of Cellular Automata*, S. Wolfram, ed. (World Scientific, 1986).
9. D. d'Humieres, P. Lallemand, and T. Shimomura, An experimental study of lattice gas hydrodynamics, Los Alamos preprint LA-UR-85-4051; D. d'Humieres, Y. Pomeau, and P. Lallemand, Simulation d'allees de Von Karman bidimensionnelles a l'aide d'un gaz sur reseau, *C. R. Acad. Sci. Paris II* **301**:1391 (1985).
10. J. Broadwell, Shock structure in a simple discrete velocity gas, *Phys. Fluids* **7**:1243 (1964).
11. H. Cabannes, The discrete Boltzmann equation, Lecture Notes, Berkeley (1980).
12. R. Gatignol, *Theorie cinetique des gaz a repartition discrete de vitesse* (Springer, 1975).
13. J. Hardy and Y. Pomeau, Thermodynamics and hydrodynamics for a modeled fluid, *J. Math. Phys.* **13**:1042 (1972).

14. S. Harris, *The Boltzmann Equation* (Holt, Rinehart and Winston, 1971).

15. R. Caflisch and G. Papanicolaou, The fluid-dynamical limit of a nonlinear model Boltzmann equation, *Commun. Pure Appl. Math.* **32**:589 (1979).

16. B. Nemnich and S. Wolfram, Cellular automaton fluids 2: Basic phenomenology, in preparation.

17. S. Wolfram, *SMP Reference Manual* (Inference Corporation, Los Angeles, 1983); S. Wolfram, Symbolic mathematical computation, *Commun. ACM* **28**:390 (1985).

18. A. Sommerfeld, *Thermodynamics and Statistical Mechanics* (Academic Press, 1955).

19. S. Wolfram, Origins of randomness in physical systems, *Phys. Rev. Lett.* **55**:449 (1985).

20. S. Wolfram, Random sequence generation by cellular automata, *Adv. Appl. Math.* **7**:123 (1986).

21. J. P. Boon and S. Yip, *Molecular Hydrodynamics* (McGraw-Hill, 1980).

22. E. M. Lifshitz and L. P. Pitaevskii, *Statistical Mechanics, Part 2* (Pergamon, 1980), Chapter 9.

23. R. Liboff, *The Theory of Kinetic Equations* (Wiley, 1969).

24. D. Levermore, Discretization effects in the macroscopic properties of cellular automaton fluids, in preparation.

25. E. M. Lifshitz and L. P. Pitaevskii, *Physical Kinetics* (Pergamon, 1981).

26. P. Resibois and M. De Leener, *Classical Kinetic Theory of Fluids* (Wiley, 1977).

27. L. D. Landau and E. M. Lifshitz, *Fluid Mechanics* (Pergamon, 1959).

28. M. H. Ernst, B. Cichocki, J. R. Dorfman, J. Sharma, and H. van Beijeren, Kinetic theory of nonlinear viscous flow in two and three dimensions, *J. Stat. Phys.* **18**:237 (1978).

29. J. R. Dorfman, Kinetic and hydrodynamic theory of time correlation functions, in *Fundamental Problems in Statistical Mechanics 3*, E. D. G. Cohen, ed. (North-Holland, 1975).

30. R. Courant and K. O. Friedrichs, *Supersonic Flows and Shock Waves* (Interscience, 1948).

31. D. Levermore, private communication.

32. J. Milnor, private communication.

33. V. Yakhot, B. Bayley, and S. Orszag, Analogy between hyperscale transport and cellular automaton fluid dynamics, Princeton University preprint (February 1986).

34. H. S. M. Coxeter, *Regular Polytopes* (Macmillan, 1963).

35. M. Hammermesh, *Group Theory* (Addison-Wesley, 1962), Chapter 9.

36. L. D. Landau and E. M. Lifshitz, *Quantum Mechanics* (Pergamon, 1977), Chapter 12.

37. H. Boerner, *Representations of Groups* (North-Holland, 1970), Chapter 7.

38. L. D. Landau and E. M. Lifshitz, *Theory of Elasticity* (Pergamon, 1975), Section 10.

39. D. Levine *et al.*, Elasticity and dislocations in pentagonal and icosahedral quasicrystals, *Phys. Rev. Lett.* **14**:1520 (1985).

40. L. D. Landau and E. M. Lifshitz, *Statistical Physics* (Pergamon, 1978), Chapter 13.

41. B. K. Vainshtein, *Modern Crystallography*, (Springer, 1981), Chapter 2.

42. J. H. Conway and N. J. A. Sloane, to be published.

43. R. L. E. Schwarzenberger, *N-Dimensional Crystallography* (Pitman, 1980).

44. J. Milnor, Hilbert's problem 18: On crystallographic groups, fundamental domains, and on sphere packing, *Proc. Symp. Pure Math.* **28**:491 (1976).

45. B. G. Wybourne, *Classical Groups for Physicists* (Wiley, 1974), p. 78; R. Slansky, Group theory for unified model building, *Phys. Rep.* **79**:1 (1981).

46. B. Grunbaum and G. C. Shephard, *Tilings and Patterns* (Freeman, in press); D. Levine and P. Steinhardt, Quasicrystals I: Definition and structure, Univ. of Pennsylvania preprint.

47. N. G. de Bruijn, Algebraic theory of Penrose's non-periodic tilings of the plane, *Nedl. Akad. Wetensch. Indag. Math.* **43**:39 (1981); J. Socolar, P. Steinhardt, and D. Levine, Quasicrystals with arbitrary orientational symmetry, *Phys. Rev. B* **32**:5547 (1985).

48. R. Penrose, Pentaplexity: A class of nonperiodic tilings of the plane, *Math. Intelligencer* **2**:32 (1979).

49. J. P. Rivet and U. Frisch, Automates sur gaz de reseau dans l'approximation de Boltzmann, *C. R. Acad. Sci. Paris II* **302**:267 (1986).

50. P. J. Davis, *Circulant Matrices* (Wiley, 1979).

51. L. D. Landau and E. M. Lifshitz, *Statistical Physics* (Pergamon, 1978), Chapter 5.

52. E. Kolb and S. Wolfram, Baryon number generation in the early universe, *Nucl. Phys. B* **172**:224 (1980), Appendix A.

53. I. S. Gradestyn and I. M. Ryzhik, *Table of Integrals, Series and Products* (Academic Press, 1965).

54. U. Frisch, private communication.

55. P. Roache, *Computational Fluid Mechanics* (Hermosa, Albuquerque, 1976).

56. S. Omohundro and S. Wolfram, unpublished (July 1985).

57. D. d'Humieres, private communication.

Additional and Survey Papers

Cellular Automata

1983

Introduction

It appears that the basic laws of physics relevant to everyday phenomena are now known. Yet there are many everyday natural systems whose complex structure and behavior have so far defied even qualitative analysis. For example, the laws that govern the freezing of water and the conduction of heat have long been known, but analyzing their consequences for the intricate patterns of snowflake growth has not yet been possible. While many complex systems may be broken down into identical components, each obeying simple laws, the huge number of components that make up the whole system act together to yield very complex behavior.

In some cases this complex behavior may be simulated numerically with just a few components. But in most cases the simulation requires too many components, and this direct approach fails. One must instead attempt to distill the mathematical essence of the process by which complex behavior is generated. The hope in such an approach is to identify fundamental mathematical mechanisms that are common to many different natural systems. Such commonality would correspond to universal features in the behavior of very different complex natural systems.

To discover and analyze the mathematical basis for the generation of complexity, one must identify simple mathematical systems that capture the essence of the process. Cellular automata are a candidate class of such systems. This article surveys their nature and properties, concentrating on fundamental mathematical features. Cellular automata promise to provide mathematical models for a wide variety of complex phenomena, from turbulence in fluids to patterns in biological growth. The general features of their behavior discussed here should form a basis for future detailed studies of such specific systems.

Originally published in *Los Alamos Science*, volume 9, pages 2–21 (Fall 1983).

The Nature of Cellular Automata and a Simple Example

Cellular automata are simple mathematical idealizations of natural systems. They consist of a lattice of discrete identical sites, each site taking on a finite set of, say, integer values. The values of the sites evolve in discrete time steps according to deterministic rules that specify the value of each site in terms of the values of neighboring sites. Cellular automata may thus be considered as discrete idealizations of the partial differential equations often used to describe natural systems. Their discrete nature also allows an important analogy with digital computers: cellular automata may be viewed as parallel-processing computers of simple construction.

As a first example of a cellular automaton, consider a line of sites, each with value 0 or 1 (Fig. 1). Take the value of a site at position i on time step t to be $a_i^{(t)}$. One very simple rule for the time evolution of these site values is

$$a_i^{(t+1)} = a_{i-1}^{(t)} + a_{i+1}^{(t)} \quad \mod 2, \tag{1}$$

where mod 2 indicates that the 0 or 1 remainder after division by 2 is taken. According to this rule, the value of a particular site is given by the sum modulo 2 (or, equivalently, the Boolean algebra "exclusive or") of the values of its left- and right-hand nearest neighbor sites on the previous time step. The rule is implemented simultaneously at each site.* Even with this very simple rule quite complicated behavior is nevertheless found.

Fractal Patterns Grown from Cellular Automata

First of all, consider evolution according to Eq. 1 from a "seed" consisting of a single site with value 1, all other sites having value 0. The pattern generated by evolution for a few time steps already exhibits some structure (Fig. 2). Figure 3 shows the pattern generated after 500 time steps. Generation of this pattern required application of Eq. 1 to a quarter of a million site values. The pattern of Figs. 2 and 3 is an intricate one but exhibits some striking regularities. One of these is "self-similarity." As illustrated in Fig. 3, portions of the pattern, when magnified, are indistinguishable from the whole. (Differences on small scales between the original pattern and the magnified portion disappear when one considers the limiting pattern obtained after an infinite number of time steps.) The pattern is therefore invariant under rescaling of lengths. Such a self-similar pattern is often called a fractal and may be characterized by a fractal dimension. The fractal dimension of the pattern in Fig. 3, for example, is

1 0 1 1 0 1 0 0 0 1 1 0 1 0 1 1 0 1 0 0

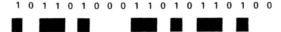

Figure 1. A typical configuration in the simple cellular automaton described by Eq. 1, consisting of a sequence of sites with values 0 or 1. Sites with value 1 are represented by squares; those with value 0 are blank.

---

* In the very simplest computer implementation a separate array of updated site values must be maintained and copied back to the original site value array when the updating process is complete.

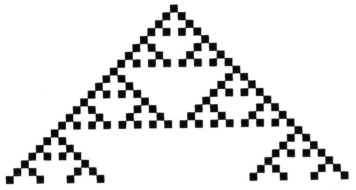

Figure 2. A few time steps in the evolution of the simple cellular automaton defined by Eq. 1, starting from a "seed" containing a single nonzero site. Successive lines are obtained by successive applications of Eq. 1 at each site. According to this rule, the value of each site is the sum modulo 2 of the values of its two nearest neighbors on the previous time step. The pattern obtained with this simple seed is Pascal's triangle of binomial coefficients, reduced modulo 2.

$\log_2 3 = \log 3 / \log 2 \simeq 1.59$. Many natural systems, including snowflakes, appear to exhibit fractal patterns. (See Benoit B. Mandelbrot, *The Fractal Geometry of Nature*, W. H. Freeman and Company, 1982.) It is very possible that in many cases these fractal patterns are generated through evolution of cellular automata or analogous processes.

Self-Organization in Cellular Automata

Figure 4 shows evolution according to Eq. 1 from a "disordered" initial state. The values of sites in this initial state are randomly chosen: each site takes on the value 0 or 1 with equal probability, independently of the values of other sites. Even though the initial state has no structure, evolution of the cellular automaton does manifest some structure in the form of many triangular "clearings." The spontaneous appearance of these clearings is a simple example of "self-organization."

The pattern of Fig. 4 is strongly reminiscent of the pattern of pigmentation found on the shells of certain mollusks (Fig. 5). It is quite possible that the growth of these pigmentation patterns follows cellular automaton rules.

In systems that follow conventional thermodynamics, the second law of thermodynamics implies a progressive degradation of any initial structure and a universal tendency to evolve with time to states of maximum entropy and maximum disorder. While many natural systems do tend toward disorder, a large class of systems, biological ones being prime examples, show a reverse trend: they spontaneously generate structure with time, even when starting from disordered or structureless initial states. The cellular automaton in Fig. 4 is a simple example of such a self-organizing system. The mathematical basis of this behavior is revealed by considering global properties of the cellular automaton. Instead of following evolution from a particular initial state, as in Fig. 4, one follows the overall evolution of an ensemble of many different initial states.

Figure 3. Many time steps in the evolution of the cellular automaton of Fig. 2, generated by applying the rule of Eq. 1 to about a quarter of a million site values. The pattern obtained is "self similar": a part of the pattern, when magnified, is indistinguishable from the whole. The pattern has a fractal dimension of $\log_2 3 \simeq 1.59$.

It is convenient when investigating global properties to consider finite cellular automata that contain a finite number N of sites whose values are subject to periodic boundary conditions. Such a finite cellular automaton may be represented as sites

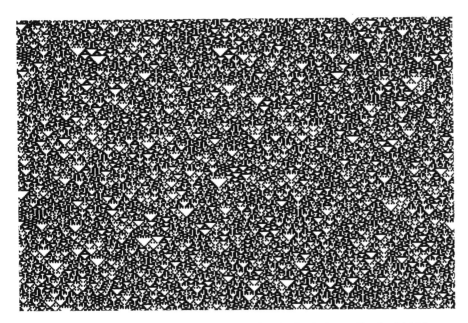

Figure 4. Evolution of the simple cellular automaton defined by Eq. 1, from a disordered initial state in which each site is taken to have value 0 or 1 with equal, independent probabilities. Evolution of the cellular automaton even from such a random initial state yields some simple structure.

arranged, for example, around a circle. If each site has two possible values, as it does for the rule of Eq. 1, there are a total of 2^N possible states, or configurations, for the complete finite cellular automaton. The global evolution of the cellular automaton may then be represented by a finite state transition graph plotted in the "state space" of the cellular automaton. Each of the 2^N possible states of the complete cellular automaton (such as the state 110101101010 for a cellular automaton with twelve sites) is represented by a node, or point, in the graph, and a directed line connects each node to the node generated by a single application of the cellular automaton rule. The trajectory traced out in state space by the directed lines connecting one particular node to its successors thus corresponds to the time evolution of the cellular automaton from the initial state represented by that particular node. The state transition graph of Fig. 6 shows all possible trajectories in state space for a cellular automaton with twelve sites evolving according to the simple rule of Eq. 1.

A notable feature of Fig. 6 is the presence of trajectories that merge with time. While each state has a unique successor in time, it may have several predecessors or no predecessors at all (as for states on the periphery of the state transition graph). The merging of trajectories implies that information is lost in the evolution of the cellular automaton: knowledge of the state attained by the system at a particular time is not sufficient to determine its history uniquely, so that the evolution is irreversible. Starting with an initial ensemble in which all configurations

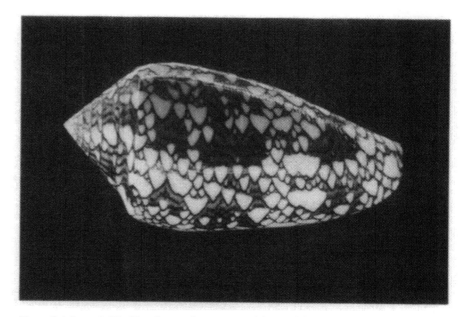

Figure 5. A "cone shell" with a pigmentation pattern reminiscent of the pattern generated by the cellular automaton of Fig. 4. (Shell courtesy of P. Hut.)

occur with any distribution of probabilities, the irreversible evolution decreases the probabilities for some configurations and increases those for others. For example, after just one time step the probabilities for states on the periphery of the state transition graph in Fig. 6 are reduced to zero; such states may be given as initial conditions, but may never be generated through evolution of the cellular automaton. After many time steps only a small number of all the possible configurations actually occur. Those that do occur may be considered to lie on "attractors" of the cellular automaton evolution. Moreover, if the attractor states have special "organized" features, these features will appear spontaneously in the evolution of the cellular automaton. The possibility of such self-organization is therefore a consequence of the irreversibility of the cellular automaton evolution, and the structures obtained through self-organization are determined by the characteristics of the attractors.

The irreversibility of cellular automaton evolution revealed by Fig. 6 is to be contrasted with the intrinsic reversibility of systems described by conventional thermodynamics. At a microscopic level, the trajectories representing the evolution of states in such systems never merge: each state has a unique predecessor, and no information is lost with time. Hence a completely disordered ensemble, in which all possible states occur with equal probabilities, remains disordered forever. Moreover, if nearby states are grouped (or "coarse-grained") together, as by imprecise measurements, then with time the probabilities for different groups of states will tend to equality, regardless of their initial values. In this way such systems tend with time to complete disorder and maximum entropy, as prescribed by the second law of

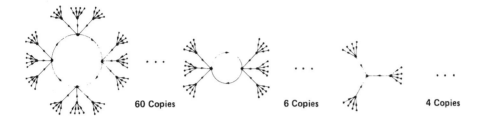

60 Copies   6 Copies   4 Copies

Figure 6. The global state transition graph for a finite cellular automaton consisting of twelve sites arranged around a circle and evolving according to the simple rule of Eq. 1. Each node in the graph represents one of the 4096 possible states, or sequences of the twelve site values, of the cellular automaton. Each node is joined by a directed line to a successor node that corresponds to the state obtained by one time step of cellular automaton evolution. The state transition graph consists of many disconnected pieces, many of identical structure. Only one copy of each structurally identical piece is shown explicitly. Possible paths through the state transition graph represent possible trajectories in the state space of the cellular automaton. The merging of these trajectories reflects the irreversibility of the cellular automaton evolution. Any initial state of this cellular automaton ultimately evolves to an "attractor" represented in the graph by a cycle. For this particular cellular automaton all configurations evolve to attractors in at most three time steps. (From O. Martin, A. Odlyzko, and S. Wolfram, "Algebraic Properties of Cellular Automata," Bell Laboratories report (January 1983) and to be published in *Communications in Mathematical Physics.*)

thermodynamics. Tendency to disorder and increasing entropy are universal features of intrinsically reversible systems in statistical mechanics. Irreversible systems, such as the cellular automaton of Figs. 2, 3, and 4, counter this trend, but universal laws have yet to be found for their behavior and for the structures they may generate. One hopes that such general laws may ultimately be abstracted from an investigation of the comparatively simple examples provided by cellular automata.

While there is every evidence that the fundamental microscopic laws of physics are intrinsically reversible (information-preserving, though not precisely time-reversal invariant), many systems behave irreversibly on a macroscopic scale and are appropriately described by irreversible laws. For example, while the microscopic molecular interactions in a fluid are entirely reversible, macroscopic descriptions of the average velocity field in the fluid, using, say, the Navier-Stokes equations, are irreversible and contain dissipative terms. Cellular automata provide mathematical models at this macroscopic level.

Mathematical Analysis of a Simple Cellular Automaton

The cellular automaton rule of Eq. 1 is particularly simple and admits a rather complete mathematical analysis.

The fractal patterns of Figs. 2 and 3 may be characterized in a simple algebraic manner. If no reduction modulo 2 were performed, then the values of sites generated from a single nonzero initial site would simply be the integers appearing in

Pascal's triangle of binomial coefficients. The pattern of nonzero sites in Figs. 2 and 3 is therefore the pattern of odd binomial coefficients in Pascal's triangle. (See Stephen Wolfram, "Geometry of Binomial Coefficients," to be published in *American Mathematical Monthly.*)

This algebraic approach may be extended to determine the structure of the state transition diagram of Fig. 6. (See O. Martin, A. Odlyzko, and S. Wolfram, "Algebraic Properties of Cellular Automata," Bell Laboratories report (January 1983) and to be published in *Communications in Mathematical Physics.*) The analysis proceeds by writing for each configuration a characteristic polynomial

$$A(x) = \sum_{i=0}^{N-1} a_i x^i,$$

where x is a dummy variable, and the coefficient of x^i is the value of the site at position i. In terms of characteristic polynomials, the cellular automaton rule of Eq. 1 takes on the particularly simple form

$$A^{(t+1)}(x) = T(x)A^{(t)}(x) \mod (x^N - 1),$$

where

$$T(x) = (x + x^{-1})$$

and all arithmetic on the polynomial coefficients is performed modulo 2. The reduction modulo $x^N - 1$ implements periodic boundary conditions. The structure of the state transition diagram may then be deduced from algebraic properties of the polynomial $T(x)$. For even N one finds, for example, that the fraction of states on attractors is $2^{-D_2(N)}$, where $D_2(N)$ is defined as the largest integral power of 2 that divides N (for example, $D_2(12) = 4$).

Since a finite cellular automaton evolves deterministically with a finite total number of possible states, it must ultimately enter a cycle in which it visits a sequence of states repeatedly. Such cycles are manifest as closed loops in the state transition graph. The algebraic analysis of Martin et al. shows that for the cellular automaton of Eq. 1 the maximal cycle length Π (of which all other cycle lengths are divisors) is given for even N by

$$\Pi_{N=2^j} = 1$$

or

$$\Pi_{N=2(2k+1)} = 2\Pi_{N=2k+1}.$$

For odd N, Π may be shown to divide

$$2^{\text{sord}_N(2)} - 1$$

and in fact is almost always equal to this value (the first exception occurs for $N = 37$). Here $\text{sord}_N(2)$ is a number theoretical function defined to be the minimum positive integer j for which $2^j = \pm 1$ modulo N. The maximum value of $\mathrm{sord}_N(2)$, typically achieved when N is prime, is $(N-1)/2$. The maximal cycle length is thus of order $2^{N/2}$, approximately the square root of the total number of possible states 2^N.

An unusual feature of this analysis is the appearance of number theoretical concepts. Number theory is inundated with complex results based on very simple premises. It may be part of the mathematical mechanism by which natural systems of simple construction yield complex behavior.

More General Cellular Automata

The discussion so far has concentrated on the particular cellular automaton rule given by Eq. 1. This rule may be generalized in several ways. One family of rules is obtained by allowing the value of a site to be an arbitrary function of the values of the site itself and of its two nearest neighbors on the previous time step:

$$a_i^{(t+1)} = F(a_{i-1}^{(t)}, a_i^{(t)}, a_{i+1}^{(t)}).$$

A convenient notation illustrated in Fig. 7 assigns a "rule number" to each of the 256 rules of this type. The rule number of Eq. 1 is 90 in this notation.

Further generalizations allow each site in a cellular automaton to take on an arbitrary number k of values and allow the value of a site to depend on the values of sites at a distance up to r on both sides, so that

$$a_i^{(t+1)} = F(a_{i-r}^{(t)}, \ldots, a_{i+r}^{(t)}).$$

The number of different rules with given k and r grows as $k^{k^{2r+1}}$ and therefore becomes immense even for rather small k and r.

Figure 8 shows examples of evolution according to some typical rules with various k and r values. Each rule leads to patterns that differ in detail. However, the examples suggest a very remarkable result: all patterns appear to fall into only four

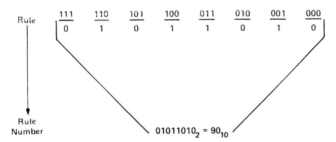

Figure 7. Assignment of rule numbers to cellular automata for which $k = 2$ and $r = 1$. The values of sites obtained from each of the eight possible three-site neighborhoods are combined to form a binary number that is quoted as a decimal integer. The example shown is for the rule given by Eq. 1.

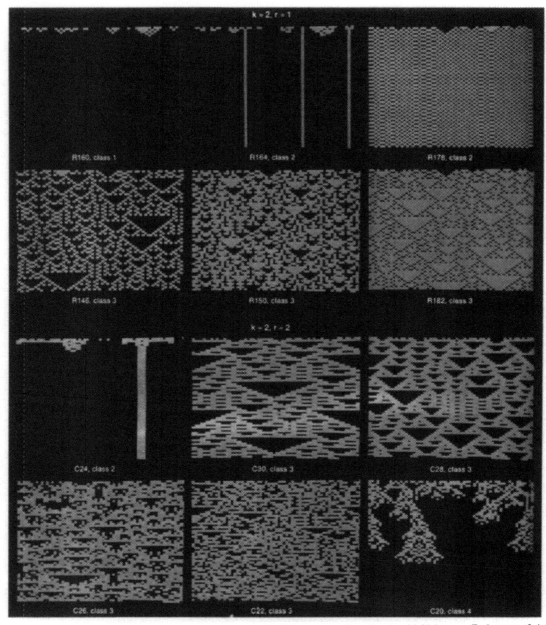

Figure 8. Evolution of some typical cellular automata from disordered initial states. Each group of six patterns shows the evolution of various rules with particular values of k and r. Sites take on k possible values, and the value of a site depends on the values of sites up to r sites distant on both sides. Different colors represent different site values: black corresponds to a value of 0, red to 1, green to 2, blue to 3, and yellow to 4. The fact that these and other examples exhibit only four qualitative classes of behavior (see text) suggests considerable universality in the behavior of cellular automata. The examples on page 420 for which $r = 1$ are labeled by rule number (in the notation of Fig. 7) and behavior class. The examples

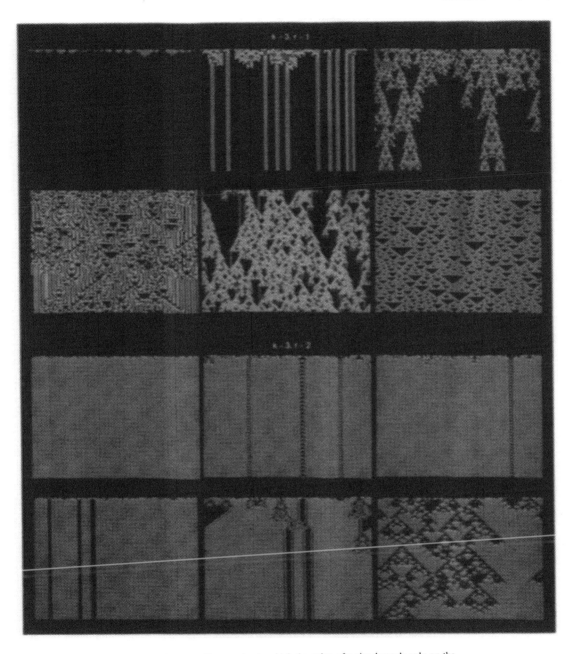

on page 420 for which $r = 2$ evolve according to rules in which the value of a site depends only on the sum of the values of the $2r + 1$ sites in its neighborhood on the previous time step. Such rules may be specified by numerical codes C such that the coefficient of 2^j in the binary decomposition of C gives the value attained by a site if its neighborhood had total value j on the previous time step. These examples are labeled by code number and behavior class. (I am grateful to R. Pike and J. Condon of Bell Laboratories for their help in preparing these and other color pictures of cellular automata.)

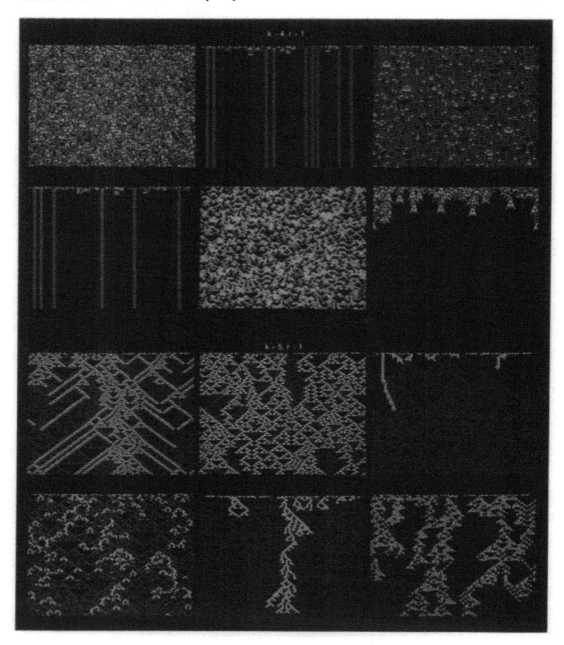

Figure 8 (continued).

qualitative classes. These basic classes of behavior may be characterized empirically as follows:

- Class 1—evolution leads to a homogeneous state in which, for example, all sites have value 0;
- Class 2—evolution leads to a set of stable or periodic structures that are separated and simple;
- Class 3—evolution leads to a chaotic pattern;
- Class 4—evolution leads to complex structures, sometimes long-lived.

Examples of these classes are indicated in Fig. 8.

The existence of only four qualitative classes implies considerable universality in the behavior of cellular automata; many features of cellular automata depend only on the class in which they lie and not on the precise details of their evolution. Such universality is analogous, though probably not mathematically related, to the universality found in the equilibrium statistical mechanics of critical phenomena. In that case many systems with quite different detailed construction are found to lie in classes with critical exponents that depend only on general, primarily geometrical features of the systems and not on their detailed construction.

Universality Classes in Cellular Automata

To proceed in analyzing universality in cellular automata, one must first give more quantitative definitions of the classes identified above. One approach to such definitions is to consider the degree of predictability of the outcome of cellular automaton evolution, given knowledge of the initial state. For class 1 cellular automata complete prediction is trivial: regardless of the initial state, the system always evolves to a unique homogeneous state. Class 2 cellular automata have the feature that the effects of particular site values propagate only a finite distance, that is, only to a finite number of neighboring sites. Thus a change in the value of a single initial site affects only a finite region of sites around it, even after an infinite number of time steps. This behavior, illustrated in Fig. 9, implies that prediction of a particular final site value requires knowledge of only a finite set of initial site values. In contrast, changes of initial site values in class 3 cellular automata, again as illustrated in Fig. 9, almost always propagate at a finite speed forever and therefore affect more and more distant sites as time goes on. The value of a particular site after many time steps thus depends on an ever-increasing number of initial site values. If the initial state is disordered, this dependence may lead to an apparently chaotic succession of values for a particular site. In class 3 cellular automata, therefore, prediction of the value of a site at infinite time would require knowledge of an infinite number of initial site values. Class 4 cellular automata are distinguished by an even greater degree of unpredictability, as discussed below.

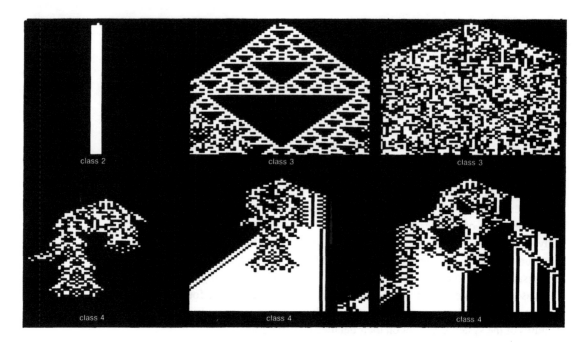

Figure 9. Difference patterns showing the differences between configurations generated by evolution, according to various cellular automaton rules, from initial states that differ in the value of a single site. Each difference pattern is labeled by the behavior class of the cellular automaton rule. The effects of changes in a single site value depend on the behavior class of the rule: for class 2 rules the effects have finite range; for class 3 rules the effects propagate to neighboring sites indefinitely at a fixed speed; and for class 4 rules the effects also propagate to neighboring sites indefinitely but at various speeds. The difference patterns shown here are analogues of Green's functions for cellular automata.

Class 2 cellular automata may be considered as "filters" that select particular features of the initial state. For example, a class 2 cellular automata may be constructed in which initial sequences 111 survive, but sites not in such sequences eventually attain value 0. Such cellular automata are of practical importance for digital image processing: they may be used to select and enhance particular patterns of pixels. After a sufficiently long time any class 2 cellular automaton evolves to a state consisting of blocks containing nonzero sites separated by regions of zero sites. The blocks have a simple form, typically consisting of repetitions of particular site values or sequences of site values (such as 101010...). The blocks either do not change with time (yielding vertical stripes in the patterns of Fig. 8) or cycle between a few states (yielding "railroad track" patterns).

While class 2 cellular automata evolve to give persistent structures with small periods, class 3 cellular automata exhibit chaotic aperiodic behavior, as shown in Fig. 8. Although chaotic, the patterns generated by class 3 cellular automata are not completely random. In fact, as mentioned for the example of Eq. 1, they may exhibit important self-organizing behavior. In addition and again in contrast to class 2 cellular automata, the statistical properties of the states generated by many time steps of class 3 cellular automaton evolution are the same for almost all possible initial states. The large-time behavior of a class 3 cellular automaton is therefore determined by these common statistical properties.

The configurations of an infinite cellular automaton consist of an infinite sequence of site values. These site values could be considered as digits in a real number, so that each complete configuration would correspond to a single real number. The topology of the real numbers is, however, not exactly the same as the natural one for the configurations (the binary numbers 0.111111... and 1.00000... are identical, but the corresponding configurations are not). Instead, the configurations of an infinite cellular automaton form a Cantor set. Figure 10 illustrates two constructions for a Cantor set. In construction (a) of Fig. 10, one starts with the set of real numbers in the interval 0 to 1. First one excludes the middle third of the interval, then the middle third of each interval remaining, and so on. In the limit the set consists of an infinite number of disconnected points. If positions in the interval are represented by ternimals (base 3 fractions, analogous to base 10 decimals), then the construction is seen to retain only points whose positions are represented by ternimals containing no 1's (the point 0.2202022 is therefore included; 0.2201022 is excluded). An important feature of the limiting set is its self-similarity or fractal form: a piece of the set, when magnified, is indistinguishable from the whole. This self-similarity is mathematically analogous to that found for the limiting two-dimensional pattern of Fig. 3.

In construction (b) of Fig. 10, the Cantor set is formed from the "leaves" of an infinite binary tree. Each point in the set is reached by a unique path from the "root" (top as drawn) of the tree. This path is specified by an infinite sequence of binary digits, in which successive digits determine whether the left- or right-hand branch is taken at each successive level in the tree. Each point in the Cantor set corresponds uniquely to one infinite sequence of digits and thus to one configuration of an infinite cellular automaton. Evolution of the cellular automaton then corresponds to iterated mappings of the Cantor set to itself. (The locality of cellular automaton rules implies that the mappings are continuous.) This interpretation of cellular automata leads to analogies with the theory of iterated mappings of intervals of the real line. (See Mitchell J. Feigenbaum, "Universal Behavior in Nonlinear Systems," *Los Alamos Science*, Vol. 1, No. 1(1980): 4–27.)

Cantor sets are parameterized by their "dimensions." A convenient definition of dimension, based on construction (a) of Fig. 10, is as follows. Divide the interval from 0 to 1 into k^n bins, each of width k^{-n}. Then let $N(n)$ be the number of these bins that contain points in the set. For large n this number behaves according to

$$N(n) \sim k^{dn} \tag{2}$$

and d is defined as the "set dimension" of the Cantor set. If a set contained all points in the interval 0 to 1, then with this definition its dimension would simply

425

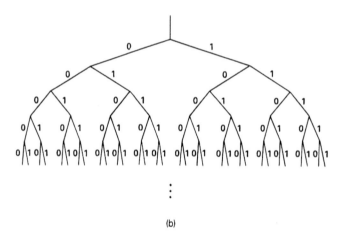

(a)

(b)

Figure 10. Steps in two constructions of a Cantor set. At each step in construction (a), the middle third of all intervals is excluded. The first step thus excludes all points whose positions, when expressed as base 3 fractions, have a 1 in the first "ternimal place" (by analogy with decimal place), the second step excludes all points whose positions have a 1 in the second ternimal place, and so on. The limiting set obtained after an infinite number of steps consists of an infinite number of disconnected points whose positions contain no 1's. The set may be assigned a dimension, according to Eq. 2, that equals $\log_3 2 \simeq 0.63$. Construction (b) reflects the topological structure of the Cantor set. Infinite sequences of digits, representing cellular automaton configurations, are seen to correspond uniquely with points in the Cantor set.

be 1. Similarly, any finite number of segments of the real line would form a set with dimension 1. However, the Cantor set of construction (a), which contains an infinite number of disconnected pieces, has a dimension according to Eq. 2 of $\log_3 2 \simeq 0.63$.

An alternative definition of dimension, agreeing with the previous one for present purposes, is based on self-similarity. Take the Cantor set of construction (a) in Fig. 10. Contract the set by a magnification factor k^{-m}. By virtue of its self-similarity, the whole set is identical to a number, say $M(m)$, of copies of this contracted copy. For large m, $M(m) \approx k^{dm}$, where again d is defined as the set dimension.

426

With these definitions the dimension of the Cantor set of all possible configurations for an infinite one-dimensional cellular automaton is 1. A disordered ensemble, in which each possible configuration occurs with equal probability, thus has dimension 1. Figure 11 shows the behavior of the probabilities for the configurations of a typical cellular automaton as a function of time, starting from such a disordered initial ensemble. As expected from the irreversibility of cellular automaton evolution, exemplified by the state transition graph of Fig. 6, different configurations attain different probabilities as evolution proceeds, and the probabilities for some configurations decrease to zero. This phenomenon is manifest in the "thinning" of configurations on successive time steps apparent in Fig. 11. The set of configurations that survive with nonzero probabilities after many time steps of cellular automaton evolution constitutes the "attractors" for the evolution. This set is again a Cantor set; for the example of Fig. 11 its dimension is $\log_2 \kappa \simeq 0.88$, where $\kappa \simeq 1.755$ is the real solution of the polynomial equation $z^3 - z^2 + 2z - 1 = 0$. (See D. A. Lind, "Applications of Ergodic Theory and Sofic Systems to Cellular Automata," University of Washington preprint (April 1983) and to be published in *Physica D*; see also Martin et al., op. cit.) The greater the irreversibility in the cellular automaton evolution, the smaller is the dimension of the Cantor set corresponding to the attractors for the evolution. If the set of attractors for a cellular automaton has dimension 1, then essentially all the configurations of the cellular automaton may occur at large times. If the attractor set has dimension less than 1, then a vanishingly small fraction of

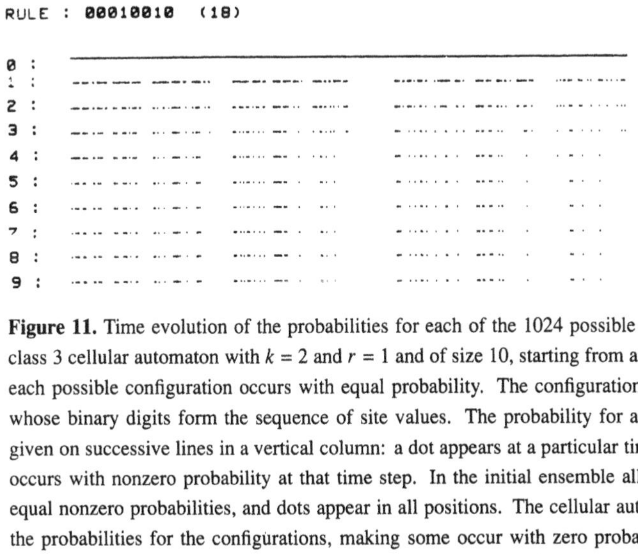

RULE : 00010010   (18)

Figure 11. Time evolution of the probabilities for each of the 1024 possible configurations of a typical class 3 cellular automaton with $k = 2$ and $r = 1$ and of size 10, starting from an initial ensemble in which each possible configuration occurs with equal probability. The configurations are specified by integers whose binary digits form the sequence of site values. The probability for a particular configuration is given on successive lines in a vertical column: a dot appears at a particular time step if the configuration occurs with nonzero probability at that time step. In the initial ensemble all configurations occur with equal nonzero probabilities, and dots appear in all positions. The cellular automaton evolution modifies the probabilities for the configurations, making some occur with zero probability and yielding gaps in which no dots appear. This "thinning" is a consequence of the irreversibility of the cellular automaton evolution and is reflected in a decrease of entropy with time. In the limit of cellular automata of infinite size, the configurations appearing at large times form a Cantor set. For the rule shown (rule 18 in the notation of Fig. 7) the limiting dimension of this Cantor set is found to be approximately 0.88.

all possible configurations are generated after many time steps of evolution. The attractor sets for most class 3 cellular automata have dimensions less than 1. For those class 3 cellular automata that generate regular patterns, the more regular the pattern, the smaller is the dimension of the attractor set; these cellular automata are more irreversible and are therefore capable of a higher degree of self-organization.

The dimension of a set of cellular automaton configurations is directly proportional to the limiting entropy (or information) per site of the sequence of site values that make up the configurations. (See Patrick Billingsley, *Ergodic Theory and Information*, John Wiley & Sons, 1965.) If the dimension of the set was 1, so that all possible sequences of site values could occur, then the entropy of these sequences would be maximal. Dimensions lower than 1 correspond to sets in which some sequences of site values are absent, so that the entropy is reduced. Thus the dimension of the attractor for a cellular automaton is directly related to the limiting entropy attained in its evolution, starting from a disordered ensemble of initial states.

Dimension gives only a very coarse measure of the structure of the set of configurations reached at large times in a cellular automaton. Formal language theory may provide a more complete characterization of the set. "Languages" consist of a set of words, typically infinite in number, formed from a sequence of letters according to certain grammatical rules. Cellular automaton configurations are analogous to words in a formal language whose letters are the k possible values of each cellular automaton site. A grammar then gives a succinct specification for a set of cellular automaton configurations.

Languages may be classified according to the complexity of the machines or computers necessary to generate them. A simple class of languages specified by "regular grammars" may be generated by finite state machines. A finite state machine is represented by a state transition graph (analogous to the state transition graph for a finite cellular automaton illustrated in Fig. 6). The possible words in a regular grammar are generated by traversing all possible paths in the state transition graph. These words may be specified by "regular expressions" consisting of finite length sequences and arbitrary repetitions of these. For example, the regular expression 1(00)*1 represents all sequences containing an even number of 0's (arbitrary repetition of the sequence 00) flanked by a pair of 1's. The set of configurations obtained at large times in class 2 cellular automata is found to form a regular language. It is likely that attractors for other classes of cellular automata correspond to more complicated languages.

Analogy with Dynamical Systems Theory

The three classes of cellular automaton behavior discussed so far are analogous to three classes of behavior found in the solutions to differential equations (continuous dynamical systems). For some differential equations the solutions obtained with any initial conditions approach a fixed point at large times. This behavior is analogous to class 1 cellular automaton behavior. In a second class of differential equations, the limiting solution at large times is a cycle in which the parameters vary periodically with time. These equations are analogous to class 2 cellular automata. Finally, some differential equations have been found to exhibit complicated, apparently chaotic behavior depending in detail on their initial conditions. With the initial conditions specified by decimals, the solutions to these differential equations depend on progressively higher and higher order digits in the initial conditions. This phenomenon is analogous to the dependence of a particular site value on progressively more distant initial site values in the evolution of a class 3 cellular automaton. The solutions to this final class of differential equations tend to "strange" or "chaotic" attractors (see Robert Shaw, "Strange Attractors, Chaotic Behavior, and Information Flow," *Zeitschrift für Naturforschung* 36A(1981):80), which form Cantor sets in direct analogy with those found in class 3 cellular automata. The correspondence between classes of behavior found in cellular automata and those found in continuous dynamical systems supports the generality of these classes. Moreover, the greater mathematical simplicity of cellular automata suggests that investigation of their behavior may elucidate the behavior of continuous dynamical systems.

A Universal Computation Class of Cellular Automata

Figure 12 shows patterns obtained by evolution from disordered initial states according to a class 4 cellular automaton rule. Complicated behavior is evident. In most cases all sites eventually "die" (attain value 0). In some cases, however, persistent structures that survive for an infinite time are generated, and a few of these persistent structures propagate with time. Figure 13 shows all the persistent structures generated from initial states with nonzero sites in a region of twenty or fewer sites. Unlike the periodic structures of class 2 cellular automata, these persistent structures have no simple patterns. In addition, the propagating structures allow site values at one position to affect arbitrarily distant sites after a sufficiently long time. No analogous behavior has yet been found in a continuous dynamical system.

The complexity apparent in the behavior of class 4 cellular automata suggests the conjecture that these systems may be capable of universal computation. A computer may be regarded as a system in which definite rules are used to transform an initial sequence of, say, 1's and 0's to a final sequence of 1's and 0's. The initial sequence may be considered as a program and data stored in computer memory, and part of the final sequence may be considered as the result of the computation. Cellular automata may be considered as computers; their initial configurations represent programs and initial data, and their configurations after a long time contain the results of computations.

A system is a universal computer if, given a suitable initial program, its time evolution can implement any finite algorithm. (See Frank S. Beckman, *Mathematical Foundations of Programming*, Addison-Wesley Publishing Co., 1980.) A universal computer need thus only be "reprogrammed," not "rebuilt," to perform each possi-

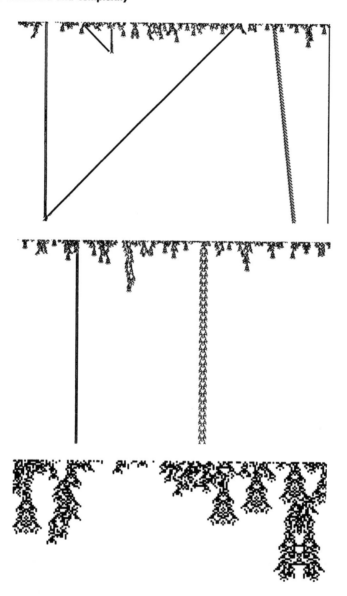

Figure 12. Evolution of a class 4 cellular automaton from several disordered initial states. The bottom example has been reproduced on a larger scale to show detail. In this cellular automaton, for which $k = 2$ and $r = 2$, the value of a site is 1 only if a total of two or four sites out of the five in its neighborhood have the value 1 on the previous time step. For some initial states persistent structures are formed, some of which propagate with time. This cellular automaton is believed to support universal computation, so that with suitable initial states it may implement any finite algorithm.

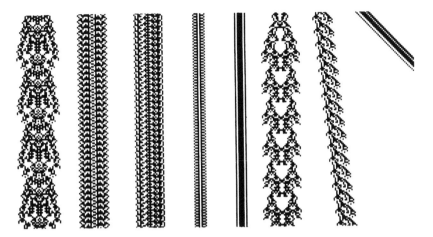

Figure 13. Persistent structures exhibited by the class 4 cellular automaton of Fig. 12 as it evolves from initial states with nonzero sites in a region of twenty or fewer sites. These structures are almost sufficient to demonstrate a universal computation capability for the cellular automaton.

ble calculation. (All modern general-purpose electronic digital computers are, for practical purposes, universal computers; mechanical adding machines were not.) If a cellular automaton is to be a universal computer, then, with a fixed rule for its time evolution, different initial configurations must encode all possible programs.

The only known method of proving that a system may act as universal computer is to show that its computational capabilities are equivalent to those of another system already classified as a universal computer. The Church-Turing thesis states that no system may have computational capabilities greater than those of universal computers. The thesis is supported by the proven equivalence of computational models such as Turing machines, string-manipulation systems, idealized neural networks, digital computers, and cellular automata. While mathematical systems with computational power beyond that of universal computers may be imagined, it seems likely that no such systems could be built with physical components. This conjecture could in principle be proved by showing that all physical systems could be simulated by a universal computer. The main obstruction to such a proof involves quantum mechanics.

A cellular automaton may be proved capable of universal computation by identifying structures that act as the essential components of digital computers, such as wires, NAND gates, memories, and clocks. The persistent structures illustrated in Fig. 13 provide many of the necessary components, strongly suggesting that the cellular automaton of Figs. 12 and 13 is a universal computer. One important missing component is a "clock" that generates an infinite sequence of "pulses"; starting from an initial configuration containing a finite number of nonzero sites, such a structure would give rise to an ever-increasing number of nonzero sites. If such a structure exists, it can undoubtedly be found by careful investigation, although it is probably too large to be found by any practical exhaustive search. If the cellular automaton of Figs. 12 and 13 is indeed capable of universal computation, then, despite its very simple construction, it is in some sense capable of arbitrarily complicated behavior.

Several complicated cellular automata have been proved capable of universal computation. A one-dimensional cellular automaton with eighteen possible values at each site (and nearest neighbor interactions) has been shown equivalent to the simplest known universal Turing machine. In two dimensions several cellular automata with just two states per site and interactions between nearest neighbor sites (including diagonally adjacent sites, giving a nine-site neighborhood) are known to be equivalent to universal digital computers. The best known of these cellular automata is the "Game of Life" invented by Conway in the early 1970s and simulated extensively ever since. (See Elwyn R. Berlekamp, John H. Conway, and Richard K. Guy, *Winning Ways*, Academic Press, 1982 and Martin Gardner, *Wheels, Life, and Other Mathematical Amusements*, W. H. Freeman and Company, October 1983. The Life rule takes a site to have value 1 if three and only three of its eight neighbors are 1 or if four are 1 and the site itself was 1 on the previous time step.) Structures analogous to those of Fig. 13 have been identified in the Game of Life. In addition, a clock structure, dubbed the glider gun, was found after a long search.

By definition, any universal computer may in principle be simulated by any other universal computer. The simulation proceeds by emulating the elementary operations in the first universal computer by sets of operations in the second universal computer, as in an "interpreter" program. The simulation is in general only faster or slower by a fixed finite factor, independent of the size or duration of a computation. Thus the behavior of a universal computer given particular input may be determined only in a time of the same order as the time required to run that universal computer explicitly. In general the behavior of a universal computer cannot be predicted and can be determined only by a procedure equivalent to observing the universal computer itself.

If class 4 cellular automata are indeed universal computers, then their behavior may be considered completely unpredictable. For class 3 cellular automata the values of particular sites after a long time depend on an ever-increasing number of initial sites. For class 4 cellular automata this dependence is by an algorithm of arbitrary complexity, and the values of the sites can essentially be found only by explicit observation of the cellular automaton evolution. The apparent unpredictability of class 4 cellular automata introduces a new level of uncertainty into the behavior of natural systems.

The unpredictability of universal computer behavior implies that propositions concerning the limiting behavior of universal computers at indefinitely large times are formally undecidable. For example, it is undecidable whether a particular universal computer, given particular input data, will reach a special "halt" state after a finite time or will continue its computation forever. Explicit simulations can be run only for finite times and thus cannot determine such infinite time behavior. Results may be obtained for some special input data, but no general (finite) algorithm or procedure may even in principle be given. If class 4 cellular automata are indeed universal computers, then it is undecidable (in general) whether a particular initial state will ultimately evolve to the null configuration (in which all sites have value 0) or will generate persistent structures. As is typical for such generally undecidable propositions, particular cases may be decided. In fact, the halting of the cellular automaton of Figs. 12 and 13 for all initial states with nonzero sites in a region of twenty sites has been determined by explicit simulation. In general, the halting probability, or fraction of initial configurations ultimately evolving to the null configuration, is a noncomputable number. However, the explicit results for small initial patterns suggest that for the cellular automaton of Figs. 12 and 13, this halting probability is approximately 0.93.

In an infinite disordered configuration all possible sequences of site values appear at some point, albeit perhaps with very small probability. Each of these sequences may be considered to represent a possible "program"; thus with an infinite disordered initial state, a class 4 automaton may be considered to execute (in parallel) all possible programs. Programs that generate structures of arbitrarily great complexity occur, at least with indefinitely small probabilities. Thus for example, somewhere on the infinite line a sequence that evolves to a self-reproducing structure should occur. After a sufficiently long time this configuration may reproduce many times, so that it ultimately dominates the behavior of the cellular automaton. Even though the *a priori* probability for the occurrence of a self-reproducing structure in the initial state is very small, its *a posteriori* probability after many time steps of cellular automaton evolution may be very large. The possibility that arbitrarily complex behavior seeded by features of the initial state can occur in class 4 cellular automata with indefinitely low probability prevents the taking of meaningful statistical averages over infinite volume (length). It also suggests that in some sense any class 4 cellular automaton with an infinite disordered initial state is a microcosm of the universe.

In extensive samples of cellular automaton rules, it is found that as k and r increase, class 3 behavior becomes progressively more dominant. Class 4 behavior occurs only for $k > 2$ or $r > 1$; it becomes more common for larger k and r but remains at the few percent level. The fact that class 4 cellular automata exist with only three values per site and nearest neighbor interactions implies that the threshold in complexity of construction necessary to allow arbitrarily complex behavior is very low. However, even among systems of more complex construction, only a small fraction appear capable of arbitrarily complex behavior. This suggests that some physical systems may be characterized by a capability for class 4 behavior and universal computation; it is the evolution of such systems that may be responsible for very complex structures found in nature.

The possibility for universal computation in cellular automata implies that arbitrary computations may in principle be performed by cellular automata. This suggests that cellular automata could be used as practical parallel-processing computers. The mechanisms for information processing found in most natural systems (with the exception of those, for example, in molecular genetics) appear closer to those of cellular automata than to those of Turing machines or conventional serial-processing digital computers. Thus one may suppose that many natural systems could be simulated more efficiently by cellular automata than by conventional computers. In practical terms, the homogeneity of cellular automata leads to simple implementation by integrated circuits. A simple one-dimensional universal cellular automaton with perhaps a million sites and a time step as short as a billionth of a second could perhaps be fabricated with current technology on a single silicon wafer (the one-dimensional homogeneous structure makes defects easy to map out). Conventional programming methodology is, of course, of little utility for such a system. The development of a new methodology is a difficult but important challenge. Perhaps tasks such as image processing, which are directly suitable for cellular automata, should be considered first.

A Basis for Universality?

The existence of four classes of cellular automata was presented above as a largely empirical result. Techniques from computation theory may provide a basis, and ultimately a proof, of this result.

The first crucial observation is that with special initial states one cellular automaton may behave just like another. In this way one cellular automaton may be considered to "simulate" another. A single site with a particular value in one cellular automaton may be simulated by a fixed block of sites in another; after a fixed number of time steps, the evolution of these blocks imitates the single time-step evolution of sites in the first cellular automaton. For example, sites with value 0 and 1 in the first cellular automaton may be simulated by blocks of sites 00 and 11, respectively, in the second cellular automaton, and two time steps of evolution in the second cellular automaton correspond to one time step in the first. Then, with a special initial state containing 11 and 00 but not 01 and 10 blocks, the second cellular automaton may simulate the first.

Figure 14 gives the network that represents the simulation capabilities of symmetric cellular automata with $k = 2$ and $r = 1$. (Only simulations involving blocks of length less than four sites were included in the construction of the network.) If a cellular automaton is computationally universal, then with a sufficiently long encoding it should be able to simulate any other cellular automaton, so that a path should exist from the node that represents its rule to nodes representing all other possible rules.

An example of the simulation of one cellular automaton by another is the simulation of the additive rule 90 (Eq. 1) by the class 3 rule 18. A rule 18 cellular automaton behaves exactly like a rule 90 cellular automaton if alternate sites in the initial configuration have value 0 (so that 0 and 1 in rule 90 are represented by 00 and 01 in rule 18) and alternate time steps are considered. Figure 15 shows evolution according to rule 18 from a disordered initial state. Two "phases" are clearly evident: one in which sites at even-numbered positions have value 0 and one in which sites at odd-numbered positions have value 0. The boundaries between these regions

434

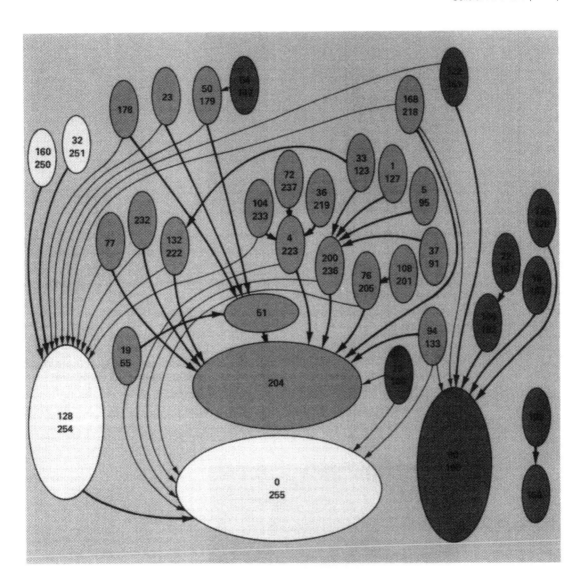

Figure 14. Simulation network for symmetric cellular automaton rules with $k = 2$ and $r = 1$. Each rule is specified by the number obtained as shown in Fig. 7, and its behavior class is indicated by shades of gray: light gray corresponds to class 1, medium gray to class 2, and dark gray to class 3. Rule A is considered to simulate rule B if there exist blocks of site values that evolve under rule A as single sites would evolve under rule B. Simulations are included in the network shown only when the necessary blocks are three or fewer sites long. Rules 90 and 150 are additive class 3 rules, rule 204 is the identity rule, and rules 170 and 240 are left- and right-shift rules, respectively. Attractive simulation paths are indicated by bold lines. (Network courtesy of J. Milnor.)

435

Figure 15. Evolution of the class 3 cellular automaton rule 18 from a disordered initial state with pairs of sites combined. The pair of site values 00 are shown as black, 01 as red, 10 as green, and 11 as blue. At large times two phases are clearly evident, separated by "defects" that execute approximately random walks and ultimately annihilate in pairs. In each phase alternate sites have value 0, and the other sites evolve according to the additive rule 90. Thus for almost all initial states rule 18 behaves like rule 90 at large times. Rule 18 therefore follows an attractive simulation path to rule 90.

execute approximately random walks and eventually annihilate in pairs, leaving a system consisting of blocks of sites that evolve according to the additive rule 90. (Cf. P. Grassberger, "Chaos and Diffusion in Deterministic Cellular Automata," to be published in *Physica D*.) Thus the simulation of rule 90 by rule 18 may be considered

Figure 16. Evolution of the class 2 cellular automaton rule 94 from an initial state in which the members of most pairs of sites have the same values, so that the digrams 00 and 11 predominate and the sequences 010 and 101 are nearly absent. (Color designations are the same as in Fig. 15.) Class 3 behavior occurs, but is unstable: class 2 behavior is "seeded" by 10 and 01 digrams and ultimately dominates. Rule 94 exhibits a repulsive simulation path to the class 3 additive rule 90 and an attractive path to the identity rule 204.

436

an "attractive" one: starting from almost all initial states, rule 18 evolves toward states in which it simulates rule 90. In general, one expects that some paths in the network of Fig. 14 are attractive, while the rest are repulsive. The consequences of a repulsive simulation path are illustrated in Fig. 16: with special initial states rule 94 behaves like rule 90, but any impurities in the initial states grow and eventually dominate the evolution of the system.

Class 1 cellular automata have an attractive simulation path to rule 0 (or its equivalents). Class 2 cellular automata have attractive simulation paths to the identity rule 204. A conjecture for which some evidence exists is that all class 3 rules exhibit attractive simulations to additive rules such as 90 or 150. Simulation by blocking of site values is analogous to a block spin or renormalization group transformation; additive rules have the special property that they are invariant under such transformations. As mentioned earlier, class 4 cellular automata are distinguished by the presence of simulation paths leading to every other cellular automaton rule. It is likely that no specific path is distinguished as attractive.

Cellular automata of different classes may thus be distinguished by their limiting behavior under simulation transformations. This approach suggests that classification of the qualitative behavior of cellular automata may be related to determinations of equivalence of systems and problem classes in computation theory. In general, one may hope for fundamental connections between computation theory and the theory of complex nonequilibrium statistical systems. Information theory forms a mathematical basis for equilibrium statistical mechanics. Computation theory, which addresses time-dependent processes, may be expected to play a fundamental role in nonequilibrium statistical mechanics.

Acknowledgments

I am grateful to many people for discussions and suggestions. I thank in particular my collaborators in various cellular automaton investigations: O. Martin, J. Milnor, and A. Odlyzko. The research described here was supported in part by the Office of Naval Research under contract number N00014-80-C-0657.

Further Reading

John von Neumann. Edited and completed by Arthur W. Burks. *Theory of Self-Reproducing Automata*. Urbana: University of Illinois Press, 1966.

Arthur W. Burks. editor, *Essays on Cellular Automata*. Urbana: University of Illinois Press, 1970.

Stephen Wolfram. "Statistical Mechanics of Cellular Automata." *Reviews of Modern Physics* 55(1983):601.

Stephen Wolfram. "Universality and Complexity in Cellular Automata." The Institute for Advanced Study preprint (May 1983) and to be published in *Physica D*.

Stephen Wolfram, J. Doyne Farmer, and Tommaso Toffoli, editors. "Cellular Automata: Proceedings of an Interdisciplinary Workshop (Los Alamos; March 7–11, 1983)." To be published in *Physica D* and to be available separately from North-Holland Publishing Company.

Computers in Science
and Mathematics

1984

Computation offers a new means of describing and investigating scientific and mathematical systems. Simulation by computer may be the only way to predict how certain complicated systems evolve.

Scientific laws give algorithms, or procedures, for determining how systems behave. The computer program is a medium in which the algorithms can be expressed and applied. Physical objects and mathematical structures can be represented as numbers and symbols in a computer, and a program can be written to manipulate them according to the algorithms. When the computer program is executed, it causes the numbers and symbols to be modified in the way specified by the scientific laws. It thereby allows the consequences of the laws to be deduced.

Executing a computer program is much like performing an experiment. Unlike the physical objects in a conventional experiment, however, the objects in a computer experiment are not bound by the laws of nature. Instead they follow the laws embodied in the computer program, which can be of any consistent form. Computation thus extends the realm of experimental science: it allows experiments to be performed in a hypothetical universe. Computation also extends theoretical science. Scientific laws have conventionally been constructed in terms of a particular set of mathematical functions and constructs, and they have often been developed as much for their mathematical simplicity as for their capacity to model the salient features of a phenomenon. A scientific law specified by an algorithm, however, can have any consistent form. The study of many complex systems, which have resisted analysis by traditional mathematical methods, is consequently being made possible through computer experiments and computer models. Computation is emerging as a major new approach to science, supplementing the long-standing methodologies of theory and experiment.

Originally published with illustrations under the title "Computer Software in Science and Mathematics" in *Scientific American*, volume 251, pages 188–203 (September 1984).

There are many scientific calculations, of course, that can be done by conventional mathematical means, without the aid of the computer. For example, given the equations that describe the motion of electrons in an arbitrary magnetic field, it is possible to derive a simple mathematical formula that gives the trajectory of an electron in a uniform magnetic field (one whose strength is the same at all positions). For more complicated magnetic fields, however, there is no such simple mathematical formula. The equations of motion still yield an algorithm from which the trajectory of an electron can be determined. In principle the trajectory could be worked out by hand, but in practice only a computer can go through the large number of steps necessary to obtain accurate results.

A computer program that embodies the laws of motion for an electron in a magnetic field can be used to perform computer experiments. Such experiments are more flexible than conventional laboratory experiments. For example, a laboratory experiment could readily be devised to study the trajectory of an electron moving under the influence of the magnetic field in a television tube. No laboratory experiment, however, could reproduce the conditions encountered by an electron moving in the magnetic field surrounding a neutron star. The computer program can be applied in both cases.

The magnetic field under investigation is specified by a set of numbers stored in a computer. The computer program applies an algorithm that simulates the motion of the electron by changing the numbers representing its position at successive times. Computers are now fast enough for the simulations to be carried out quickly, and so it is practical to explore a large number of cases. The investigator can interact directly with the computer, modifying various aspects of a phenomenon as new results are obtained. The usual cycle of the scientific method, in which hypotheses are formulated and then tested, can be followed much faster with the aid of the computer.

Computer experiments are not limited to processes that occur in nature. For example, a computer program can describe the motion of magnetic monopoles in magnetic fields, even though magnetic monopoles have not been detected in physical experiments. Moreover, the program can be modified to embody various alternative laws for the motion of magnetic monopoles. Once again, when the program is executed, the consequences of the hypothetical laws can be determined. The computer thus enables the investigator to experiment with a range of hypothetical natural laws.

The computer can also be used to study the properties of abstract mathematical systems. Mathematical experiments carried out by computer can often suggest conjectures that are subsequently established by conventional mathematical proof. Consider a mathematical system that can be introduced to model the path of a beam of electrons traveling through the magnetic fields in a circular particle accelerator. The transverse displacement of an electron as it passes a point on one of its revolutions around the accelerator ring is given by some fraction x between 0 and 1. The value of the fraction corresponding to the electron's displacement on the next revolution is then $ax(1 - x)$, where a is a number that can range between 0 and 4. The formula gives an algorithm from which the sequence of values for the electron's displacement can be worked out.

A few trials show how the properties of the sequence depend on the value of a. If a is equal to 2 and the initial value of x is equal to .8, the next value of x, which is given by $ax(1 - x)$, is equal to .32. If the formula is applied again, the value of x obtained is .4352. After several iterations the sequence of values for x converges to .5. Indeed, when a is small and x is any fraction between 0 and 1, the sequence quickly settles down to give the same value of x for each revolution of the electron.

As a increases, however, a phenomenon called period doubling can be observed. When a reaches 3, the sequence begins to alternate between two values of x. As a continues to increase, first four, then eight and finally, when it reaches about 3.57, an entire range of values for x appear. This behavior could not readily be guessed from the construction of the mathematical system, but it is immediately suggested by the computer experiment. The detailed properties of the system can then be established by a conventional proof.

The mathematical processes that can be described by a computer program are not limited to the operations and functions of conventional mathematics. For example, there is no conventional mathematical notation for the function that reverses the order of the digits in a number. Nevertheless, it is possible to define and apply the function in a computer program. The computer makes it practical to introduce scientific and mathematical laws that are intrinsically algorithmic in nature. Consider the chain of events set up when an electron accelerated to a high energy is fired into a block of lead. There is a certain probability that the electron emits a photon of a particular energy. If a photon is emitted, there is a certain probability that it gives rise to a second electron and a positron (the antiparticle of the electron). Each member of the pair can in turn emit more photons, so that a cascade of particles is eventually generated. There is no simple mathematical formula that can describe even the elements of the process. Nevertheless, an algorithm for the process can be incorporated into a computer program, and the outcome of the process can be deduced by executing the program. The algorithm serves as the basic law that describes the process.

The mathematical basis of most conventional models of natural phenomena is the differential equation. Such an equation gives relations between certain quantities and their rates of change. For example, a chemical reaction proceeds at a rate proportional to the concentrations of the reacting chemicals, and that relation can be expressed by a differential equation. A solution to the equation would give the concentration of each reactant as a function of time. In some simple cases it is possible to find a complete solution to the equation in terms of standard mathematical functions. In most cases, however, no such exact solution can be obtained, and one must resort to approximation.

The commonest approximations are numerical. Suppose one term of a differential equation gives the instantaneous rate of change of a quantity with time. The term can be approximated by the total change in the quantity over some small interval and then substituted into the differential equation. The resulting equation is in effect an algorithm that determines the approximate value of the quantity at the end of an interval, given its value at the beginning of the interval. By applying the algorithm repeatedly for successive intervals, the approximate variation of the quantity with time can be found. Smaller intervals yield more accurate results. The calculation required for each interval is quite simple, but in most cases it must be repeated many times to achieve an acceptable level of accuracy. Such an approach is practical only with a computer.

The numerical methods embodied in computer programs have been employed to find approximate solutions to differential equations in a wide variety of disciplines. In some cases the solutions have a simple form. In many cases, however, the solutions show complicated, almost random behavior, even though the differential equations from which they arise are quite simple. For such cases experimental mathematics must be used.

In practical applications one often finds not only that differential equations are complicated but also that there are many of them. For example, the theoretical models of nuclear explosions employed in the design of weapons and the study of supernovas involve hundreds of differential equations that describe the interactions of many isotopes. In practice such models are always used in the form of computer programs: only a computer can follow the interrelations among so many quantities.

The results of some numerical calculations, such as the abundance of helium in the universe, can be stated as single numbers. In most cases, however, one is concerned with the variation of certain quantities as the parameters of a calculation are changed. When the number of parameters is only one or two, the results can be displayed as a graph. When there are more than two parameters, however, the results often can be stated succinctly only as a mathematical formula. Exact formulas usually cannot be found, but it is often possible to derive approximate formulas. Such formulas are particularly convenient because, unlike graphs or tables of numbers, they can be inserted directly into other calculations.

A common form for an approximate formula is a series of terms. Each term includes a variable raised to some power; the power is larger in each successive term. When the value of the variable is small, the terms in the series become progressively smaller; thus for small values of x the sum of the first few terms in an infinite series such as $1 - x + x^2 - x^3 + \ldots$ gives an accurate approximation to the sum of the entire series, which is $1/(1+x)$. The first few terms in a series are usually easy to evaluate, but the complexity of the terms increases rapidly thereafter. In order to evaluate terms that include large powers of x the computer becomes essential.

In principle computer programs can operate with any well-defined mathematical construct. In practice, however, the kinds of construct that can be used in a particular program are largely determined by the computer language in which the program is written. Numerical methods require only a limited set of mathematical constructs, and the programs that embody such methods can be written in general-purpose computer languages such as C, FORTRAN or BASIC. The derivation and manipulation of formulas require operations on higher-level mathematical constructs such as algebraic expressions, for which new computer languages are needed. Among the languages of this kind now in use is the SMP language that I have developed.

SMP is a language for manipulating symbols. It operates not only with numbers but also with symbolic expressions that can represent mathematical formulas. For example, in SMP the algebraic expression $2x - 3y + 5x - y$ would be simplified to the form $7x - 4y$. This transformation is a general one, valid for any possible numerical values of x and y. The standard operations of algebra and mathematical analysis are among the fundamental instructions in SMP.

The SMP language also includes operations that allow higher-level mathematical constructs to be defined and manipulated, much as they are in ordinary mathematical work. Real numbers (which include all rational and irrational values) as well as complex numbers (which have both a real and an imaginary part) are fundamental in SMP. The mathematical constructs known as quaternions, which are generalizations of the complex numbers, are not fundamental. They can nonetheless be defined in SMP, and rules can be specified for their addition and multiplication. In this way the mathematical knowledge of SMP can be extended.

Some of the advantages of a language such as SMP can be compared to the advantages of using a calculator instead of a table of logarithms. By now the widespread availability of electronic calculators and computers has made such tables obsolete: it is far more convenient to call on an algorithm in a computer to obtain a logarithm than it is to look up the result in a table. Similarly, with a language such as SMP it has become possible to make the entire range of mathematical knowledge available in algorithmic form. For example, the calculation of integrals, conventionally done with the aid of a book of tables, can increasingly be left to a computer. The computer not only carries out the final calculations quickly and without error but also automates the process of finding the relevant formulas and methods.

In SMP an expanding collection of definitions is being assembled in order to provide for a wide variety of mathematical calculations. One can now find in SMP the definition of variance in statistics, and one can immediately apply the definition to calculate the variance in a particular case. Such definitions enable programs written in the SMP language to call on increasingly sophisticated mathematical knowledge.

Differential equations give adequate models for the overall properties of physical processes such as chemical reactions. They describe, for example, the changes in the total concentration of molecules; they do not, however, account for the motions of individual molecules. These motions can be modeled as random walks: the path of each molecule is like the path that might be taken by a person in a milling crowd. In the simplest version of the model the molecule is assumed to travel in a straight line

443

until it collides with another molecule; it then recoils in a random direction. All the straight-line steps are assumed to be of equal length. It turns out that if a large number of molecules are following random walks, the average change in the concentration of molecules with time can in fact be described by a differential equation called the diffusion equation.

There are many physical processes, however, for which no such average description seems possible. In such cases differential equations are not available and one must resort to direct simulation. The motions of many individual molecules or components must be followed explicitly; the overall behavior of the system is estimated by finding the average properties of the results. The only feasible way to carry out such simulations is by computer experiment: essentially no analysis of the systems for which analysis is necessary could be made without the computer.

The self-avoiding random walk is an example of a process that can apparently be studied only by direct simulation. It can be described by a simple algorithm that is similar to the ordinary random walk. It differs in that the successive steps in the self-avoiding random walk must not cross the path taken by any previous steps. The folding of long molecules such as DNA can be modeled as a self-avoiding random walk.

The introduction of the single constraint makes the self-avoiding random walk much more complicated than the ordinary random walk. Indeed, there is no simple average description, analogous to the diffusion equation, that is known for the self-avoiding random walk. In order to investigate its properties it seems one has no choice but to carry out a direct computer experiment. The procedure is to generate a large number of sample random walks, choosing a random direction at each step. The properties of all the walks are then averaged. Such a procedure is an example of the Monte Carlo method, so called because its application depends on the element of chance.

Several examples have been given of systems whose construction is quite simple but whose behavior is extremely complicated. The study of such systems is leading to a new field called complex-systems theory, in which the computational method plays a central role. The archetypal example is fluid turbulence, which develops, for example, when water flows rapidly around an obstruction. The set of differential equations satisfied by the fluid can easily be stated. Nevertheless, the patterns of fluid flow to which the equations give rise have largely defied mathematical analysis or description. In practice the patterns are found either through observation of the actual physical system or, as far as possible, through computer experiment.

It is suspected there is a set of mathematical mechanisms common to many systems that give rise to complicated behavior. The mechanisms can best be studied in systems whose construction is as simple as possible. Such studies have recently been done for a class of mathematical systems known as cellular automata. A cellular automaton is made up of many identical components; each component evolves according to a simple set of rules. Taken together, however, the components generate behavior of essentially arbitrary complexity.

The components of a cellular automaton are mathematical "cells," arranged in one dimension at a sequence of equally spaced points along a line or in two dimensions on a regular grid of squares or hexagons. Each cell carries a value chosen from a small set of possibilities, often just 0 and 1. The values of all the cells in the cellular automaton are simultaneously updated at each "tick" of a clock according to a definite rule. The rule specifies the new value of a cell, given its previous value and the previous values of its nearest neighbors or some other nearby set of cells.

Consider a one-dimensional cellular automaton in which each cell can have the value 0 or 1. Even in such a simple case the overall behavior of the cellular automaton can be quite complex; the most effective way to investigate the behavior is by computer experiment. Most of the properties of cellular automata have in fact been conjectured on the basis of patterns generated in computer experiments. In some cases they have later been established by conventional mathematical arguments.

Cellular automata can serve as explicit models for a wide variety of physical processes. Suppose ice is represented on a two-dimensional hexagonal grid by cells with the value 1 and water vapor is represented by cells with the value 0. A cellular-automaton rule can then be used to simulate the successive stages in the freezing of a snowflake. The rule states that once a cell is frozen it does not thaw. Cells exposed at the edge of the growing pattern freeze unless they have so many ice neighbors that they cannot dissipate enough heat to freeze. Snowflakes grown in a computer experiment from a single frozen cell according to this rule show intricate treelike patterns, which bear a close resemblance to real snowflakes. A set of differential equations can also describe the growth of snowflakes, but the much simpler model given by the cellular automaton seems to preserve the essence of the process by which complex patterns are created. Similar models appear to work for biological systems: intricate patterns of growth and pigmentation may be accounted for by the simple algorithms that generate cellular automata.

Simulation by computer is the only method now used for investigating many of the systems discussed so far. It is natural to ask whether simulation, as a matter of principle, is the most efficient possible procedure or whether there is a mathematical formula that could lead more directly to the results. In order to address the question the correspondence between physical and computational processes must be studied more closely.

It is presumably true that any physical process can be described by an algorithm, and so any physical process can be represented as a computational process. One must determine how complicated the latter process is. In cellular automata the correspondence between physical and computational processes is particularly clear. A cellular automaton can be regarded as a model of a physical system, but it can also be regarded as a computational system closely analogous to an ordinary digital computer. The sequence of initial cell values in a cellular automaton can be understood as abstract data or information, much like the sequence of binary digits in the memory of a digital computer. During the evolution of a cellular automaton the information is processed: the values of the cells are modified according to definite rules. Similarly, the digits stored in the memory of the digital computer are modified by rules built into the central processing unit of the computer.

The evolution of a cellular automaton from some initial configuration may thus be viewed as a computation that processes the information carried by the configuration. For cellular automata exhibiting simple behavior the computation is a simple one. For example, it may serve only to pick out sequences of three consecutive cells whose initial values are equal to 1. On the other hand, the evolution of cellular automata that show complicated behavior may correspond to a complicated computation.

It is always possible to determine the outcome of a given number of steps in the evolution of a cellular automaton by explicitly simulating each step. The problem is whether or not there can be a more efficient procedure. Can there be a short cut to step-by-step simulation, an algorithm that finds the outcome after many steps in the evolution of a cellular automaton without effectively tracing through each step? Such an algorithm could be executed by a computer, and it would predict the evolution of a cellular automaton without explicitly simulating it. The basis of its operation would be that the computer could carry out a more sophisticated computation than the cellular automaton could and so achieve the same result in fewer steps. It would be as if the cellular automaton were to calculate 7 times 18 by explicitly finding the sum of seven 18's, while the computer found the same product according to the standard method for multiplication. Such a short cut is available only if the computer is able to carry out a calculation that is intrinsically more sophisticated than the calculation embodied in the evolution of the cellular automaton.

One can define a certain class of problems called computable problems that can be solved in a finite time by following definite algorithms. A simple computer such as an adding machine can solve only a small subset of these problems. There exist universal, or general-purpose, computers, however, that can solve any computable problem. A real digital computer is essentially such a universal machine. The instructions that can be executed by the central processing unit of the computer are rich enough to serve as the elements of a computer program that can embody any algorithm. A number of systems in addition to the digital computer have been shown to be capable of universal computation. Several cellular automata are among them: for example, universal computation has been proved for a simple two-dimensional cellular automaton with a 0 or a 1 in each cell. It is strongly suspected that several one-dimensional cellular automata are also universal computers. The simplest candidates have three possible values at each cell and rules of evolution that take account only of the nearest-neighbor cells.

Cellular automata that are capable of universal computation can mimic the behavior of any possible computer; since any physical process can be represented as a computational process, they can mimic the action of any possible physical system as well. If there were an algorithm that could work out the behavior of these cellular automata faster than the automata themselves evolve, the algorithm would allow any computation to be speeded up. Because this conclusion would lead to a logical contradiction, it follows there can be no general short cut that predicts the evolution of an arbitrary cellular automaton. The calculation corresponding to the evolution is irreducible: its outcome can be found effectively only by simulating the evolution explicitly. Thus direct simulation is indeed the most efficient method for determining the behavior of some cellular automata. There is no way to predict their evolution; one must simply watch it happen.

It is not yet known how widespread the phenomenon of computational irreducibility is among cellular automata or among physical systems in general. Nevertheless, it is clear that the elements of a system need not be very complicated for the overall evolution of the system to be computationally irreducible. It may be that computational irreducibility is almost always present when the behavior of a system appears complicated or chaotic. General mathematical formulas that describe the overall behavior of such systems are not known, and it is possible no such formulas can ever be found. In that case, explicit simulation in a computer experiment is the only available method of investigation.

Much of physical science has traditionally focused on the study of computationally reducible phenomena, for which simple overall descriptions can be given. In real physical systems, however, computational reducibility may well be the exception rather than the rule. Fluid turbulence is probably one of many examples of computational irreducibility. In biological systems computational irreducibility may be even more widespread: it may turn out that the form of a biological organism can be determined from its genetic code essentially only by following each step in its development. When computational irreducibility is present, one must adopt a methodology that depends heavily on computation.

One of the consequences of computational irreducibility is that there are questions that can be asked about the ultimate behavior of a system but that cannot be answered in full generality by any finite mathematical or computational process. Such questions must therefore be considered undecidable. An example of such a question is whether a particular pattern ever dies out in the evolution of a cellular automaton. It is straightforward to answer the question for some definite number of steps, say 1,000: one need only simulate 1,000 steps in the evolution of the cellular automaton. In order to determine the answer for any number of steps, however, one must simulate the evolution of the cellular automaton for a potentially infinite number of steps. If the cellular automaton is computationally irreducible, there is no effective alternative to such direct simulation.

The upshot is that no calculation of any fixed length can be guaranteed to determine whether a pattern will ultimately die out. It may be possible to tell the fate of a particular pattern after tracing only a few steps in its evolution, but there is no general way to tell in advance how many steps will be required. The ultimate form of a pattern is the result of an infinite number of steps, corresponding to an infinite computation; unless the evolution of the pattern is computationally reducible, its consequences cannot be reproduced by any finite computational or mathematical process.

The possibility of undecidable questions in mathematical models for physical systems can be viewed as a manifestation of Gödel's theorem on undecidability in mathematics, which was proved by Kurt Gödel in 1931. The theorem states that in all but the simplest mathematical systems there may be propositions that cannot be proved or disproved by any finite mathematical or logical process. The proof of a given proposition may call for an indefinitely large number of logical steps. Even propositions that can be stated succinctly can require an arbitrarily long proof. In practice there are many simple mathematical theorems for which the only known proofs are very long. In addition the cases that must be examined to prove or refute conjectures are often quite complicated. In number theory, for example, there are many cases in which the smallest number having some special property is extremely large; the number can often be found only by testing each whole number in turn. Such phenomena are making the computer an essential tool in many mathematical investigations.

Computational irreducibility implies many fundamental limitations on the scope of theories for physical systems. It may be possible to model a system at many levels, from simulating the motions of individual molecules to solving differential equations for overall properties. Computational irreducibility implies there is a highest level at which abstract models can be made; above that level results can be found only by explicit simulation.

When the level of description becomes computationally irreducible, undecidable questions also begin to appear. Such questions must be avoided in the formulation of a theory, much as the simultaneous measurement of the position and velocity of an electron—impossible according to the uncertainty principle—is avoided in quantum mechanics. Even if such questions are eliminated, there is still the practical difficulty of answering questions that in principle can be answered. The degree of difficulty depends strongly on the nature of the objects involved in the simulation. If the only way to predict the weather were to simulate the motions of every molecule in the atmosphere, no practical calculations could be carried out. Nevertheless, the relevant features of the weather can probably be studied by considering the interactions of large volumes of the atmosphere, and so useful simulations should be possible.

The efficiency with which a computationally irreducible system can be simulated depends on the computational sophistication of each step in its evolution. The steps in the evolution of the system can be simulated by instructions in a computer program. The fewer the instructions needed to reproduce each step, the more efficient the simulation. Higher-level descriptions of physical systems typically call for more sophisticated steps, much as single instructions in higher-level computer languages correspond to many instructions in lower-level ones. One time step in the numerical approximation of a differential equation that describes a jet of gas requires a computation more sophisticated than the one needed to follow a collision between two molecules in the gas. On the other hand, each step in the higher-level description given by a differential equation accounts for an immense number of steps in the lower-level description of molecular collisions. The resulting gain in efficiency more than makes up for the fact that the individual steps are more sophisticated.

In general the efficiency of a simulation increases with higher levels of description, until the operations needed for the higher-level description are matched with the operations carried out directly by the computer doing the simulation. It is most efficient for the computer to be as close an analogue to the system being simulated as possible.

There is one major difference between most existing computers and physical systems or models of them: computers process information serially, whereas physical systems process information in parallel. In a physical system modeled by a cellular automaton the values of all the cells are updated together at each time step. In a standard computer program, however, the simulation of the cellular automaton is carried out by a loop that updates the value of each cell in turn. In such a case it is straightforward to write a computer program that performs a fundamentally parallel process with a serial algorithm. There is a well-established framework in which algorithms for the serial processing of information can be described. Many physical systems, however, seem to require descriptions that are essentially parallel in nature. A general framework for parallel processing does not yet exist, but when it is developed, more effective high-level descriptions of physical phenomena should become possible.

The introduction of the computer in science is comparatively recent. Already, however, computation is establishing a new approach to many problems. It is making possible the study of phenomena far more complex than the ones that could previously be considered, and it is changing the direction and emphasis of many fields of science. Perhaps most significant, it is introducing a new way of thinking in science. Scientific laws are now being viewed as algorithms. Many of them are studied in computer experiments. Physical systems are viewed as computational systems, processing information much the way computers do. New aspects of natural phenomena have been made accessible to investigation. A new paradigm has been born.

Geometry of Binomial Coefficients

1984

This note describes the geometrical pattern of zeroes and ones obtained by reducing modulo two each element of Pascal's triangle formed from binomial coefficients. When an infinite number of rows of Pascal's triangle are included, the limiting pattern is found to be "self-similar," and is characterized by a "fractal dimension" $\log_2 3$. Analysis of the pattern provides a simple derivation of the result that the number of even binomial coefficients in the nth row of Pascal's triangle is $2^{\#_1(n)}$, where $\#_1(n)$ is a function which gives the number of occurrences of the digit 1 in the binary representation of the integer n.

Pascal's triangle modulo two appears in the analysis of the structures generated by the evolution of a class of systems known as "cellular automata." (See [1], [2], [3] for further details and references.) These systems have been investigated as simple mathematical models for natural processes (such as snowflake growth) which exhibit the phenomenon of "self-organization." The self-similarity of the patterns discussed below leads to self-similarity in the natural structures generated.

Figure 1 shows the first few rows of Pascal's triangle, together with the figure obtained by reducing each element modulo two, and indicating ones by black squares and zeroes by white (blank) squares. Figure 2 gives sixty-four rows of Pascal's

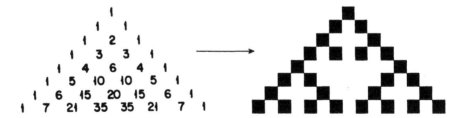

Figure 1. The first few lines of Pascal's triangle modulo two.

Originally published in the *American Mathematical Monthly*, volume 91, pages 566–571 (November 1984).

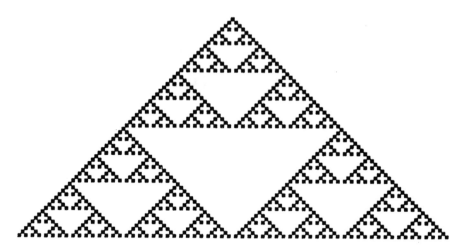

Figure 2. The first sixty-four lines of Pascal's triangle modulo two (black squares indicate ones, white squares indicate zeroes).

triangle reduced modulo two. A regular pattern of inverted triangles with various sizes differing by powers of two is clear. Large inverted triangles spanning the whole of Pascal's triangle begin at rows $n = 2^j$. Consider the pattern down to the beginning of one such large inverted triangle (say down to the sixty-third row). A striking feature of the pattern is that the largest upright triangle contains three smaller triangles whose contents are similar (except at the scale of very small triangles) to those of the largest triangle, but reduced in size by a factor of two. Inspection of each of these three smaller triangles reveals that each is built from three still smaller similar triangles. This "self-similarity" continues down to the smallest triangles. At each stage, one upright triangle from the pattern could be magnified by one or more factors of two to obtain essentially the complete pattern. The pattern obtained differs from the original complete pattern at the scale of very small triangles. If, however, Pascal's triangle were extended to an infinite number of rows, then for all finite triangles this effect would disappear, and the original and magnified patterns would be identical. In fact, triangles of any size could be reproduced by taking smaller triangles and then magnifying them. The limiting pattern obtained from Pascal's triangle modulo two is thus "self-similar" or "scale invariant," and may be considered to exhibit the same structure at all length scales. Many examples of other "self-similar" figures are given in [4], [5].

If the number of inverted triangles with base length i is denoted T_i, then Fig. 2 indicates that $T_{i/2} = 3T_i$. For large i, therefore

$$T_i \sim i^{\log_2 3}.$$

$$(1)$$

The exponent $\log_2 3 \simeq 1.59$ appearing here gives the "fractal dimensionality" [4], [5] of the self-similar pattern.

Consider a ("filled in") square. Reduce the square by a factor of two in each of its linear dimensions. Four copies of the resulting reduced square are then required to cover the original square. Alternatively, one may write that the number of squares S_i with side length i contained in the original square satisfies $S_{i/2} = 4S_i$, so that $s_i \sim i^{-2}$. The exponent two here gives the usual dimensionality of the square. One may then by analogy identify the exponent $\simeq 1.59$ in Equation (1) as the generalized or "fractal" dimension of the figure formed from Pascal's triangle modulo two.

Figure 2 suggests that the number $N(n)$ of ones in the nth row of Pascal's triangle modulo two (or, equivalently, the number of odd binomial coefficients of the form $\binom{n}{i}$ is a highly irregular function of n. However, when n is of the form 2^i, the simple result $N(2^i) = 2$ is obtained. This can be considered a consequence of the algebraic relation $\binom{p^j}{i} = 0 \bmod p$ for $0 < i < p^j$ and all primes p, which may be proved by considering the base p representations of factorials. Algebraic methods [6]–[12] have been used to obtained the general result

$$N(n) = 2^{\#_1(n)}. \qquad (2)$$

The function $\#_1(n)$ gives the number of occurrences of the digit 1 in the binary representation of the integer n. Hence, for example, $\#_1(1) = 1$, $\#_1(2) = \#_1(10_2) = 1$, $\#_1(3) = \#_1(11_2) = 2$, $\#_1(4) = 1$, and so on. A graph of $\#_1(n)$ for n up to 128 is given in Fig. 3. Note that although the function is defined only for integer n, values at successive integers have been joined by straight lines on the graph. For $n > 0$, $1 \le \#_1(n) \le [\log_2 n]$. The lower bound is reached when n is of the form 2^j; the upper one when $n = 2^j - 1$. Clearly $\#_1(2^j n) = \#_1(n)$ (since multiplication by 2^j simply appends zeroes, not affecting the number of 1 digits), and for $n < 2^j$,

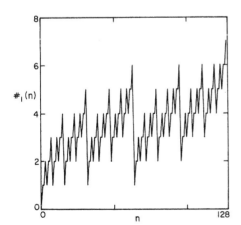

Figure 3. The number of ones in the binary representation of the integer n.

453

$\#_1(n+2^j) = \#_1(n)+1$ (since the addition of 2^j in this case prepends a single 1, without affecting the remaining digits).

The result (2) for $N(n)$ may be obtained by consideration of the geometrical pattern of Fig. 2, continued for $2^{\lceil \log_2 n \rceil}$ rows, so as to include the complete upright triangle containing the nth row. By construction, the nth row corresponds to a line which crosses the lower half of the largest upright triangle. Each successive digit in the binary decomposition of n determines whether the line crosses the upper (0) or lower (1) halves of successively smaller upright triangles. The upper halves always contain one upright triangle smaller by a factor two; the lower halves contain two such smaller triangles. The total number of triangles crossed by the line corresponding to the nth row is thus multiplied by a factor of two each time the lower half is chosen. The total number of ones in the nth row is therefore a product of the factors of two associated with each 1 digit in the binary representation of n, as given by Equation (2).

There are several possible extensions and generalizations of the results discussed above.

One may consider Pascal's triangle reduced modulo some arbitrary integer k. Figure 4 shows the resulting patterns for a few values of k. In all cases, a self-similar pattern is obtained when sufficiently many rows are included. For k prime, a very regular pattern is found, with fractal dimension

$$D_k = \log_k \sum_{i=1}^{k} i = 1 + \log_k \left(\frac{k+1}{2} \right),$$

so that $D_3 = 1 + \log_3 2 \simeq 1.631$. $D_5 \simeq 1.683$, and so on. In general, for large k, one finds that $D_k \sim 2 - 1/\log_2 k$; when $k \to \infty$, the elements of Pascal's triangle modulo k become ordinary integers, which are all nonzero by virtue of the nonzero values of binomial coefficients. By a simple generalization of Equation (2), the number of entries with value r in the nth row of Pascal's triangle modulo k is found to be $N^{(r)}(n) = 2^{\#_r^{[k]}(n)}$, where now $\#_r^{[k]}(n)$ gives the number of occurrences of the digit r in the base-k representation of the integer n.

One may also consider the generalization of Pascal's triangle to a three-dimensional pyramid of trinomial coefficients. Successive rows in the triangle are generalized to planes in the pyramid, with each plane carrying a square grid of integers. The apex of the pyramid is formed from a single 1. In each successive plane, the integer at each grid point is the sum of the integers at the four neighbouring grid points in the preceding plane. When the integers in the resulting three-dimensional array are reduced modulo k, a self-similar pattern is again obtained. With $k = 2$, the fractal dimension of the pattern is $\log_2 5 \simeq 2.32$. In general, the pattern obtained from the d-dimensional generalization of Pascal's triangle, reduced modulo two, has fractal dimension $\log_2(2d + 1)$.

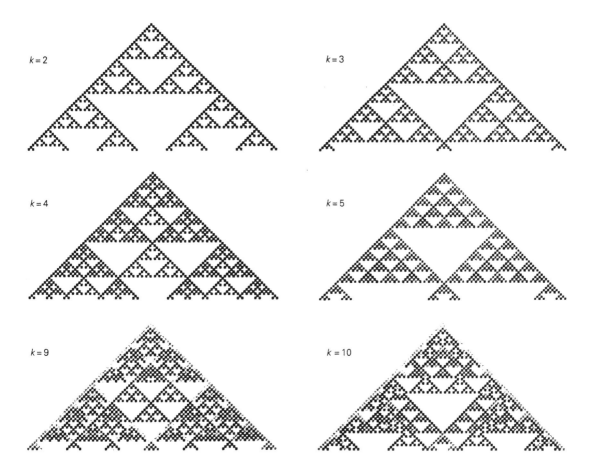

Figure 4. Patterns obtained by reducing Pascal's triangle modulo k for several values of k. White squares indicate zeroes; progressively blacker squares indicate increasing values, up to $k - 1$.

References

1. S. Wolfram, Statistical mechanics of cellular automata, Rev. Modern Phys., 55 (1983) 601.
2. O. Martin, A. Odlyzko and S. Wolfram, Algebraic properties of cellular automata, Comm. Math. Phys., 93 (1984) 219.
3. S. Wolfram, Universality and complexity in cellular automata, Physica 10D (1984) 1.
4. B. Mandelbrot, Fractals: Form, Chance and Dimension, Freeman, 1977.
5. B. Mandelbrot, The Fractal Geometry of Nature, Freeman, 1982.
6. J. W. L. Glaisher, On the residue of a binomial-theorem coefficient with respect to a prime modulus, Quart. J. Math., 30 (1899) 150.
7. N. J. Fine, Binomial coefficients modulo a prime, this Monthly, 54 (1947) 589.
8. S. H. Kimball et al., Odd binomial coefficients, this Monthly, 65 (1958) 368.
9. J. B. Roberts, On binomial coefficient residues, Canad. J. Math., 9 (1957) 363.
10. R. Honsberger, Mathematical Gems II, Dolciani Math. Expositions, Mathematical Association of America, 1976, p. 1.

11. K. B. Stolarsky, Power and exponential sums of digital sums related to binomial coefficient parity, SIAM J. Appl. Math., 32 (1977) 717.

12. M. D. McIlroy, the numbers of 1's in binary integers; bounds and extremal properties, SIAM J. Comput., 3 (1974) 255.

Twenty Problems in the Theory of Cellular Automata

1985

Cellular automata are simple mathematical systems that exhibit very complicated behaviour. They can be considered as discrete dynamical systems or as computational systems. Progress has recently been made in studying several aspects of them. Twenty central problems that remain unsolved are discussed.

Many of the complicated systems in nature have been found to have quite simple components. Their complex overall behaviour seems to arise from the cooperative effect of a very large number of parts that each follow rather simple rules. Cellular automata are a class of mathematical models that seem to capture the essential features of this phenomenon. From their study one may hope to abstract some general laws that could extend the laws of thermodynamics to encompass complex and self-organizing systems.

There has been recent progress in analysing some aspects of cellular automata. But many important problems remain. This paper discusses some of the ones that have so far been identified. The problems are intended to be broad in scope, and are probably not easy to solve. To solve any one of them completely will probably require a multitude of subsidiary questions to be asked and answered. But when they are solved, substantial progress towards a theory of cellular automata and perhaps of complex systems in general should have been made.

The emphasis of the paper is on what is not known: for expositions of what is already known about cellular automata, see [1–4]. The paper concentrates on theoretical aspects of cellular automata. There is little discussion of models for actual natural systems. But many of the theoretical issues discussed should have direct consequences for such models.

Originally published in *Physica Scripta*, volume T9 [proceedings of the fifty-ninth Nobel Symposium], pages 170–183 (1985).

Cellular automata consist of a homogeneous lattice of sites, with each site taking on one of k possible values. The sites are updated according to a definite rule that involves a neighbourhood of sites around each one. So in a one-dimensional cellular automaton the value $a_i^{(t)}$ of a site at position i evolves according to

$$a_i^{(t+1)} = \phi[a_{i-r}^{(t)}, a_{i-r+1}^{(t)}, \ldots, a_{i+r}^{(t)}].$$

The local rule ϕ has a range of r sites. Its form determines the behaviour of the cellular automaton. Some examples of patterns generated by cellular automata are shown in Figs. 1 and 2. Figure 1 shows examples of the four basic classes of behaviour seen in the evolution of cellular automata from disordered initial states. Figure 2 shows patterns generated by evolution from initial configurations containing a single nonzero site.

Cellular automata may be considered as discrete dynamical systems. Their global properties are studied by considering evolution from the set of all possible initial configurations (e.g., [5]). Since most cellular automata are irreversible, the set of configurations that is generated typically contracts with time. Its limiting form at large times determines the asymptotic behaviour of the cellular automaton, and is dominated by the attractors for the evolution. Some of the properties of cellular

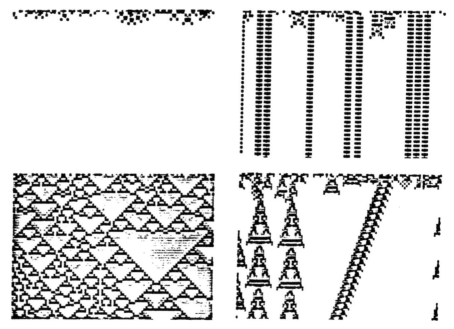

Figure 1. Examples of the four qualitative classes of behaviour seen in the evolution of one-dimensional cellular automata from disordered initial states. Successive time steps are shown on successive lines. Complex and varied behaviour is evident. The sites in the cellular automata illustrated have three possible values ($k = 3$); value 0 is shown blank, 1 is grey, and 2 is black. The value of each site at each time step is given by rules that depend on the sum of its own and its nearest neighbours' old values ($r = 1$ totalistic). The cases shown have rules specified by code numbers [5] 1302, 1005, 444 and 792, respectively.

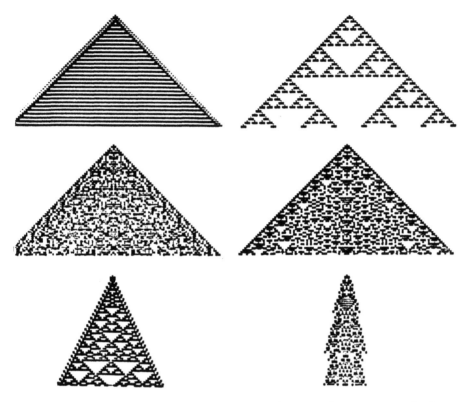

Figure 2. Examples of patterns generated by the evolution of various cellular automata starting from single site seeds. In the second case shown, a fractal pattern is generated. The subsequent cases shown illustrate the remarkable phenomenon that complicated and in some cases apparently random patterns can be generated by cellular automaton rules even from simple initial states. The cellular automata shown have $k = 3$, $r = 1$ totalistic rules with code numbers 1443, 312, 1554, 1617, 1410 and 600, respectively.

automata may be characterized in terms of quantities such as entropies and Lyapunov exponents that are used in studies of continuous dynamical systems (e.g., [6]).

An alternative view of cellular automata is as information-processing systems [7]. Cellular automaton evolution may be considered to carry out a computation on data represented by the initial sequence of site values. The nature of the evolution may then be characterized using methods from the theory of computation (e.g., [8]). So for example the sets of configurations generated in the evolution may be described as formal languages: a one-dimensional cellular automaton gives a regular formal language after any finite number of time steps [7]. One suspects that in many cases the computations corresponding to cellular automaton evolution are sufficiently complicated as to be irreducible (cf. [9]). In that case, there can be essentially no short-cut to determining the outcome of the cellular automaton evolution by explicit simulation or observation of each step. This implies that certain limiting properties of the cellular automaton are undecidable, since to find them would require an infinite computation.

The problems discussed here address both dynamical systems theory and computation theory aspects of cellular automata. But probably the most valuable insights will come from the interplay between these two aspects.

Problem 1

What overall classification of cellular automaton behaviour can be given?

Experimental mathematics provides a first approach to this problem. One performs explicit simulations of cellular automata, and tries to find empirical rules for their behaviour. These may then suggest results that can be investigated by more conventional mathematical methods.

An extensive experimental study [5] suggests that the patterns generated in the evolution of cellular automata from disordered initial states can be grouped into four general classes, illustrated in Fig. 1:

(1) Evolves to homogeneous state.

(2) Evolves to simple separated periodic structures.

(3) Yields chaotic aperiodic patterns.

(4) Yields complex pattern of localized structures.

The classification is at first qualitative. But there are several ways to make it more quantitative, and to formulate precise definitions for the four classes. For some cellular automaton rules, one expects that all definitions will agree. But there are likely to be borderline cases where definitions will disagree.

Continuous dynamical systems provide analogues for the classes of behaviour seen in cellular automata. Class 1 cellular automata show limit points, while class 2 cellular automata may be considered to evolve to limit cycles. Class 3 cellular automata exhibit chaotic behaviour analogous to that found with strange attractors. Class 4 cellular automata effectively have very long transients, and no direct analogue for them has been identified among continuous dynamical systems.

Dynamical systems theory gives a first approach to the quantitative characterization of cellular automaton behaviour. Various kinds of entropy may be defined for cellular automata. Each counts the number of possible sequences of site values corresponding to some spacetime region. For example, the spatial entropy gives the dimension of the set of configurations that can be generated at some time step in the evolution of the cellular automaton, starting from all possible initial states. There are in general $N(X) \leq k^X$ (k is the number of possible values for each site) possible sequences of values for a block of X sites in this set of configurations. The spatial topological entropy $d^{(x)}$ is given by $\lim_{X \to \infty} (1/X) \log_k N(X)$. One may also define a spatial measure entropy $d_\mu^{(x)}$ formed from the probabilities of possible sequences. Temporal entropies $d^{(t)}$ may then be defined to count the number of sequences that occur in the time series of values taken on by each site. Topological entropies reflect the possible configurations of a system; measure entropies reflect those that are probable, and are insensitive to phenomena that occur with zero probability. A tentative definition of the four classes of cellular automaton behaviour may be given in terms of measure entropies. Class 1 has zero spatial and temporal measure entropy. Class 2 has zero temporal measure entropy, since it almost always yields periodic structures,

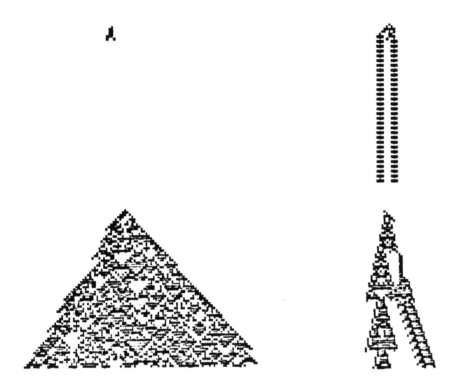

Figure 3. Patterns of differences generated by changing a single initial site value in the cellular automata of Fig. 1. In the first two cases, the difference (shown modulo three) is seen to remain localized. In the second two cases, it grows progressively with time.

but has positive spatial measure entropy. Class 3 has positive spatial and temporal measure entropies.

Another property of cellular automata is their stability under small perturbations in initial conditions. Figure 3 shows differences in patterns generated by cellular automata induced by changes in a single initial site value. Such differences almost always die out in class 1 cellular automata. In class 2 cellular automata, they may persist, but remain localized. In class 3 cellular automata, however, they typically expand at an asymptotically constant rate. The rate of this expansion gives the Lyapunov exponent for the evolution [5, 10], and measures the speed of propagation of information about the initial configuration in the cellular automaton. Class 4 cellular automata give rise to a pattern of differences that typically expands irregularly with time.

The four classes of cellular automaton behaviour identified here can be defined to be complete. But there are some cellular automata whose behaviour should probably be considered intermediate between the classes. In particular, there are many where there is a clear superposition of two classes of behaviour. So for example sites with values 0 and 1 can exhibit class 2 behaviour, while sites with values 0 and 2 show class 3 behaviour. The result is a sequence of chaotic regions separated by rigid "walls".

Even at a qualitative level, it is possible that definite subclasses of the four classes of cellular automaton behaviour may be identified. Some class 3 cellular automata in one dimension seem to give patterns with large triangular clearings and low but presumably nonzero entropies; others give highly irregular patterns with no long-range structure. No clear statistical difference between these kinds of class 3 cellular automata has yet been found. But it is possible that one exists. Among class 4 cellular automata there seem to be some definite subclasses in which persistent or almost persistent structures of rather particular kinds occur.

Problem 2

What are the exact relations between entropies and Lyapunov exponents for cellular automata?

Using the finite information density of cellular automaton configurations, and the finite rate of information propagation in cellular automata, a number of inequalities may be derived between entropies and Lyapunov exponents (λ). An example is $d^{(t)}/d^{(x)} \leq 2\lambda$ [5]. Preliminary numerical evidence suggests that for some cellular automata these inequalities may in fact be equalities. This would imply an important connection between the static properties of cellular automata, as embodied in entropies, and their dynamic properties, as measured by Lyapunov exponents. One is hampered in these studies by the lack of an efficient method for computing entropies. The best approach so far uses a conditional entropy method [11].

Lyapunov exponents can be considered to measure the rate of divergence of trajectories in the space of configurations. In continuous dynamical systems, a geometry is defined for this space, and one can identify Lyapunov exponents for various directions.

Problem 3

What is the analogue of geometry for the configuration space of a cellular automaton?

Several simple observations may be made. First, if the cellular automaton lattice is more than one-dimensional, one may consider Lyapunov exponents in different directions on this lattice. A remarkable empirical observation is that for cellular automata these exponents are approximately equal in all directions, even those not along the axes of the lattice, and even for cellular automata with asymmetric rules [12]. Second, in a one-dimensional cellular automaton one may consider Lyapunov exponents for subsets of configurations, or for particular components of configurations. For example, for a cellular automaton in which a class 1 component involving sites with values 0 and 2 is superimposed on class 3 behaviour involving sites with values 0 and 1, the Lyapunov exponent is positive in the "value 1" direction, and negative in the "value 2" direction. In general it seems that the cellular automaton evolution induces a form of geometry on the configuration space [13]. But the details are unclear; one does not know, for example, the analogue of the tangent space considered in continuous dynamical systems.

Problem 4

What statistical quantities characterize cellular automaton behaviour?

There are several direct statistical measurements that can be made on cellular automaton configurations. Very simple examples are densities of sites or blocks of sites with particular values. Such densities are closely related to block entropies; their limit for large block sizes is the spatial entropy of the cellular automaton configurations, equal to the dimension of the Cantor set formed by the configurations (e.g., [5]). Another direct statistical measurement that can be made is of correlation functions, which describe the interdependence of the values of separated sites [2]. For class 1 and 2 cellular automata, one expects that the correlation functions vanish beyond some critical distance. For class 3 cellular automata there are indications that the correlation functions typically fall off exponentially with distance. For class 4 cellular automata, the large distance part of the correlation function is dominated by propagating persistent structures, and may decrease slowly.

Power spectra or Fourier transforms provide other statistical measures of cellular automaton configurations. (Entirely discrete Walsh-Hadamard transforms [14] may be slightly more suitable.) Their form is not yet known. But many processes in cellular automata occur on a variety of spatial or temporal scales, so one expects definite structure in their transforms.

Beyond entropies and Lyapunov exponents, dynamical systems theory suggests that zeta functions may give a characterization of the global behaviour of cellular automata. Zeta functions measure the density of periodic sequences in cellular automaton configurations, and may possibly be related to Fourier transforms. The fact that the set of configurations generated from all possible initial states at a particular time step in the evolution of a cellular automaton forms a regular language (or "sofic system") implies that the corresponding zeta function is rational [15].

Problem 5

What invariants are there in cellular automaton evolution?

The existence of invariants or conservation laws in the evolution of a cellular automaton would imply a partitioning of its state space, much as energy provides a partitioning of the state space for Hamiltonian (energy-conserving) dynamical systems. For some class 1 and 2 cellular automata it is straightforward to identify invariants. In other cases, one can specifically construct cellular automaton rules that exhibit certain conservation laws [16–18]. For example, the cellular automata may evolve as if on several disjoint spatial lattices. Or they may support a set of persistent structures or "particles" that interact in simple ways. But in general, the identification of numerical invariants in cellular automata will probably be as difficult as it is in other non-linear dynamical systems.

It is nevertheless often possible to find partitionings of the state space for a cellular automaton that are left invariant by its evolution. The partitionings may be formed for example from sets of configurations corresponding to particular regular formal languages (cf. [7]). For example, the set of configurations with a particular period under a cellular automaton mapping is invariant, and in one dimension forms a finite-complement regular language (or "subshift of finite type"). Different elements in such partitionings may be considered to carry different values of what is often an infinite set of conserved quantities.

A particular cellular automaton rule usually evolves to give qualitatively similar behaviour from almost all initial states (each site is chosen to have each of the k possible values with equal probabilities). Often there are sets of initial states that occur with probability zero (for example, states in which all sites have the same value) that evolve differently from the rest. Such states may be distinguished by invariant or conserved quantities. But most initial states evolve to configurations with the same statistical properties. This suggests that even if the possible states could be partitioned according to the value of some invariant, they would be essentially equivalent. It remains conceivable, however, that there exist cellular automata in which two sets of initial states that occur with nonzero probabilities could lead to two qualitatively different forms of behaviour.

Problem 6

How does thermodynamics apply to cellular automata?

Thermodynamics is supposed to describe the average overall behaviour of physical systems with many components. The microscopic dynamics of these systems is assumed to be reversible, so that the mapping from one state to another with time is invertible. Most cellular automata are irreversible, so that a particular configuration may arise from several distinct predecessors. However, a small subset of cellular automaton rules are bijective or invertible. Complete tables of invertible rules exist for $k = 2, r \le 2$ [19, 20] and for $k = 3, r = 1$ [20], but in general no efficient procedure for finding such rules is known. Nevertheless, it is possible to construct particular classes of invertible rules [16, 21].

To apply thermodynamics one must also "coarse-grain" the system, grouping together many microscopically-different states to mimic the effect of imprecise measurements. Coarse-graining in cellular automata may be achieved by applying an irreversible transformation, perhaps a cellular automaton rule, to the cellular automaton configurations. A simple example would be to map the value of every other site to zero.

Coarse-grained entropy in reversible cellular automata should follow the second law of thermodynamics, and be on average non-decreasing with time. One may start from a set or ensemble of configurations with non-maximal coarse-grained entropy. The degrees of freedom that do not affect the coarse-grained entropy are undetermined, and are assumed to have maximal (fine-grained) entropy. In reversible class 2 cellular automata, the determined and undetermined degrees of freedom do not mix significantly with time, and the coarse-grained entropy remains essentially constant. But for class 3 and 4 cellular automata, the degrees of freedom mix, and the coarse-grained entropy increases towards its maximum possible value.

As in all applications of thermodynamics, the question arises of what coarse-graining prescriptions and ensembles of initial states are permissible. The initial states could for example be specially chosen so as to be the predecessors of a low coarse-grained entropy ensemble. The coarse-grained entropy would then decrease. Such examples do not seem physically reasonable. But it has never been clear exactly what mathematical criteria should be imposed to exclude them. One possibility is that one could require the coarse-graining procedure and the initial ensemble to be computationally simple (cf. [22]). If the cellular automaton evolution were computationally irreducible, then such a criterion could exclude ensembles obtained by reversing the evolution for many steps.

For the usual case of irreversible cellular automata, coarse-graining is usually of little consequence: the progressive contraction in the number of states generated by the cellular automaton evolution soon far outweighs the reduction associated with coarse-graining.

Problem 7

How is different behaviour distributed in the space of cellular automaton rules?

Random sampling yields some empirical indications of the frequencies of different classes of behaviour among cellular automaton rules of various kinds. For symmetric one-dimensional cellular automata, class 1 and 2 cellular automata appear to become progressively less common as k and r increase; class 3 becomes more common, and class 4 slowly becomes less common. In two-dimensional cellular automata, class 3 is overwhelmingly the most common; class 4 is very rare [12]. It seems that class

3 behaviour in any "direction" in the cellular automaton state space leads to overall class 3 behaviour. And as the number of degrees of freedom in the rules increases, the chance that this happens for one of the directions increases. For very large k and r a direct statistical treatment of the set of cellular automaton rules may well be possible.

There are many common features in the behaviour of cellular automata with apparently very different rules. It is not clear to what extent a direct equivalence exists between rules with qualitatively similar behaviour. In some cases, different rules may be related through invertible cellular automaton mappings. The nature of the equivalence classes of cellular automata generated in this way is presumably determined largely by the structure of the group of invertible cellular automaton mappings.

There are various ways to define distances in the space of cellular automaton rules. There are often cellular automata whose rules differ only slightly, but whose behaviour is very different. Nevertheless, it should be possible to find families of cellular automaton rules with closely related behaviour. For example, one may consider totalistic rules [5] in which the function that gives the new value of a site in terms of the sum of the old values in its neighbourhood is a discrete approximation to a function that involves a continuous parameter [23]. The behaviour of different cellular automaton rules obtained by changing this parameter may be compared with the behaviour found in iterated mappings of an interval of the real line (e.g., [24]) according to the same function. There are indications of a significant correspondence [23]. As the parameter is increased, regular periodic (class 2) cellular automaton behaviour can exhibit period doubling. Then as the parameter is further increased, chaotic (class 3) behaviour can occur. Class 4 seems to appear as an intermediate phenomenon.

Problem 8

What are the scaling properties of cellular automata?

Scaling transformations change the number of sites in a cellular automaton. Under such transformations, one cellular automaton rule may simulate another one. For example, if each site with value 0 is replaced by a pair of sites 00, and each 1 is replaced by 01, a new cellular automaton rule is obtained [2]. In some cases, this rule may have the same k and r as the original rule; in other cases it may not. The inverse transformation, in which 00 is replaced by 0, and 01 by 1, may be considered as a "blocking transformation" analogous to a block spin transformation (e.g., [25]), and yields a cellular automaton with fewer degrees of freedom. However, the transformation may be applied only to those special configurations in which just 00 and 01 site value pairs occur.

One may develop a network that shows the results of blocking transformations on rules of a particular kind, say with $k = 2$ and $r = 1$ [4, 26]. Some rules are found to be invariant under blocking transformations. Examples are the additive rules numbers 90 and 150 with $k = 2$ and $r = 1$. Patterns generated by these rules are thus scale invariant, so that they have the same form when viewed with different magnifications. If the initial configuration consists of a simple seed, say a single nonzero site, then regular scale-invariant patterns are obtained. These fractal patterns [27] have the property that pieces of them, when magnified, are indistinguishable from the whole pattern. (The fractal dimensions of the patterns are related to the parameters of the blocking transformations.) When the initial state is disordered, the patterns generated are instead statistically scale invariant, in the sense that their statistical properties are invariant under blocking transformations. So, for example, the pattern obtained by considering every site in the cellular automaton may have the same statistical properties as the pattern obtained by considering only every other site on every other time step.

Blocking transformations typically apply only to configurations that contain specific blocks in a given cellular automaton. So for example, different simple initial seeds in a cellular automaton may lead to rather different behaviour if they contain blocks that allow for different blocking transformations. Under certain blocking transformations, many of the $k = 2$, $r = 1$ cellular automata simulate the additive rules 90 or 150, which are invariant under blocking transformations. An initial state containing a single nonzero site is often one for which this simulation occurs, so that the pattern to which it leads is self-similar, just as for rule 90 or rule 150. With more complicated initial states, however, patterns with different forms may be obtained.

Starting from a disordered initial state, in which all possible sequences of site values occur with equal probabilities, the irreversible evolution of many cellular automata leads to states in which only particular sequences actually occur. If these sequences correspond to those for which some blocking transformation applies, then the overall behaviour of the cellular automaton will be given by the result of this blocking transformation. In a typical case, a cellular automaton rule supports a number of "phases". Each phase consists of sequences to which some blocking transformation applies, and under which the cellular automaton behaves just like one with a different rule. So for example [28], in the $k = 2$, $r = 1$ rule number 18, sequences containing only 00 and 01, or only 00 and 10, constitute two phases with behaviour just like the additive rule 90. An arbitrary disordered state consists of a series of small domains, each in one of these phases, separated by "domain walls", consisting of 11 blocks. These domain walls execute approximately random walks with time, and annihilate in pairs, leaving larger and larger domains in a pure phase [28]. In two and higher dimensional cellular automata, the domains may have complicated geometrical structures [12]. The domain walls often behave as if they have a surface tension. When the surface tension is positive, the domains tend to become spherical. When the surface tension is negative, the domains take on a highly-convoluted labyrinthine form.

It seems that one may in general define a quantity analogous to free energy, or essentially pressure, for each possible phase in a cellular automaton. Domains containing phases with higher pressures typically expand linearly with time through domains with lower pressures, sometimes following biased random walks. The

467

walls between domains with equal pressures typically execute unbiased random walks. After a long time, the phases with the highest pressure (or lowest free energy) dominate the behaviour of the cellular automaton, and thus determine the form of the limiting set of configurations. One may speculate that the phases that survive in this limit should be fixed points of the blocking transformation, and thus should exhibit some form of scale invariance. This is evident in some cases, where there are phases that behave say like rule 90. It is not clear how general the phenomenon is. If, however, it were widespread, then the overall large time behaviour of cellular automata would be dominated by fixed points of the blocking transformations, much as critical phenomena in spin systems are dominated by fixed points of the renormalization group or block spin transformation. Then there would be a universality in the properties of the many different cellular automata attracted to a particular fixed point rule. (So far the only fixed points of the blocking transformation that have been found are additive rules, but one suspects that not all fixed point rules must in fact be additive.) The spatial measure entropies for the different cellular automata would for example presumably then be related by simple rational factors.

One rule whose scaling properties remain unclear is the $k = 2$, $r = 1$ rule number 22. This rule simulates rule 90 under the blocking transformation $0000 \rightarrow 0$, $0001 \rightarrow 1$, and its rotated equivalents. But the simulation is not an attractive one: starting from a disordered initial state, domains of these phases do not grow. It may be possible to describe the configurations obtained as domains of phases corresponding to some other blocking transformation. A generalization of blocking transformations may be required. One may consider a blocking transformation as a translation from one formal language to another. In simple cases, such a translation may be achieved with a finite automaton that reads symbols sequentially from the "input" configuration, and writes symbols into the "output" configuration according to the internal state that it reaches. Blocking transformations that consist of simple substitutions correspond to very simple finite automata of this kind. More complicated finite automata may be necessary to describe phases in cellular automata such as rule number 22. In general, the irreversible nature of most cellular automata implies that only a subset of possible configurations are generated with time. As a consequence, only certain neighbourhoods of site values may appear, so that some of the elements of the cellular automaton rule are never used, and a different rule would give identical results.

The description of cellular automaton configurations in terms of domains of different phases is related to a description in terms of "elementary excitations". Just as for a spin system, one may consider decomposing a cellular automaton configuration into a "ground state" part, together with "phonons" or excitations. The excitations may for example correspond to domain walls. Or they could be persistent structures in class 4 cellular automata. But if their interactions are comparatively simple, then they can be used to provide an overall description of the cellular automaton behaviour, and perhaps allow for example a computation of entropies.

Problem 9

What is the correspondence between cellular automata and continuous systems?

Cellular automata are discrete in several respects. First, they consist of a discrete spatial lattice of sites. Second, they evolve in discrete time steps. And finally, each site has only a finite discrete set of possible values.

The first two forms of discreteness are addressed in the numerical analysis of approximate solutions to, say, differential equations. It is known that so long as a "stable" discretization is used, the exact continuum results are approximated more and more closely as the number of sites and the number of time steps is increased. It is possible to devise cellular automaton rules that provide approximations to partial differential equations in this way. In the simplest cases, however, the approximations are of the Jacobi, rather than the Gauss-Seidel kind, in that the algorithm for calculating new site values uses the old values of all the neighbours, rather than the new values of some of them. This can lead to slow convergence and instabilities in some cases.

The third form of discreteness in cellular automata is not so familiar from numerical analysis. It is an extreme form of round-off, in which each "number" can have only a few possible values (rather than the usual say 2^{16} or 2^{32}). It is not clear what aspects of, say, differential equations are preserved in such an approximation. However, preliminary studies in a few cases suggest that the overall structure of solutions to the equations are remarkably insensitive to such approximations. If the cellular automaton approximates for example a continuous field, then the value of the field at a particular point could correspond roughly to the density of say nonzero sites around that point: the values of individual field points would be represented in a distributed manner, just as they often are in actual physical systems. Explicit examples of cellular automaton approximations to partial differential equations of physical importance would be valuable.

There are some aspects of nonlinear differential equations that may well have rather direct analogues in cellular automata. For example, the persistent propagating structures found in class 4 cellular automata may well be related to solitons in nonlinear differential equations, at least in their solitary persistence, if not in their interactions. Similarly, the overall topological forms of some of the patterns generated by two and higher dimensional cellular automata [29] may correspond to those generated say by reaction-diffusion equations [30]. Moreover, many highly-nonlinear partial differential equations give solutions that exhibit discrete or cellular structure on some characteristic length scale (e.g., [31]). The interactions between components in the cellular structure cannot readily be described by a direct discretization of the original differential equation, but a cellular automaton model for them can be constructed.

Continuum descriptions may be given of many of the large-scale structures that occur in cellular automata. For example, the motion of domain walls between phases may be described by diffusion-like differential equations. A very direct continuum approximation to a cellular automaton is provided by a mean field theory, in which only the average density of sites, and not their individual values, is considered [2]. Presumably in the limit of large spatial dimensionality, this approximation should become accurate. But in one or two dimensions, it is usually quite inadequate, and gives largely misleading results. Large-scale phenomena in cellular automata occur as collective effects involving many individual sites, and the particular rules that relate the values of these sites are significant.

Problem 10

What is the correspondence between cellular automata and stochastic systems?

Cellular automata satisfy deterministic rules. But their initial states can have a random form. And the patterns they generate can have many of the properties of statistical randomness. As a consequence, the behaviour of cellular automata may have a close correspondence with the behaviour of systems usually described by basic rules that involve noise or probabilities. So for example domain walls in cellular automata execute essentially random walks, even though the evolution of the cellular automaton as a whole is entirely deterministic. Similarly, one can construct a cellular automaton that mimics say an Ising spin system with a fixed total energy (microcanonical ensemble) [32]. Apparently random behaviour occurs as a consequence of randomly-chosen initial conditions, just as in many systems governed by the deterministic laws of classical physics.

Even models that involve explicit randomness are in practice simulated in computer experiments using pseudorandom sequences generated by some definite algorithm. These sequences are not unlike the sequences of site values produced by many cellular automata. In fact, the linear feedback shift registers often used in practice to produce pseudorandom sequences are exactly equivalent to certain additive cellular automata (cf. [33]). Empirical evidence suggests that the properties of many supposedly stochastic models are quite insensitive to the detailed form of the randomness used in their simulation. It should be possible to find entirely deterministic forms for such models, based say on cellular automata. One expects in general that just as with algorithms say for primality testing the fundamental capabilities of stochastic and deterministic models should be equivalent.

Problem 11

How are cellular automata affected by noise and other imperfections?

Many mathematical approaches to the analysis of cellular automata make essential use of their simple deterministic structure. One must find out to what extent results for the overall behaviour of cellular automata are changed when imperfections are introduced into them. The imperfections can be of several kinds. First, the cellular automaton rules can have a probabilistic element (e.g., [17, 34, 35]). Then for example each site may be updated at each time step according to one rule with probability p, and according to another rule with probability $1 - p$. A second class of imperfections modifies the homogeneous cellular automaton lattice. One may for example take different sites to follow different rules. Or one may take the connections that specify the rules on the lattice to be different at different sites. In an ordinary cellular automaton, the values of all the sites are updated simultaneously, using the previous values of the sites in their neighbourhoods. One may consider the effect of deviations from this synchronization, allowing different sites to be updated at different times [36]. Finally, each site is usually taken to have a discrete set of possible values. One could instead allow the sites to have a continuum of values, but take the rules to be continuous functions with sharp thresholds.

Several classes of models can be considered as imperfect cellular automata. Directed percolation is directly analogous to certain cellular automata in the presence of noise [35]. The patterns generated with time by noisy cellular automata also correspond to the equilibrium configurations of spin systems at finite temperature [35]. And if inhomogeneities are introduced into the cellular automata, they give spin glass configurations. When nonlocal connections and asynchronous updates are introduced, models analogous to Boolean or neural networks are obtained (e.g., [37]).

Even an arbitrarily small imperfection in a cellular automaton can have a large effect at arbitrarily large times. However, small imperfections very often do not affect the overall behaviour of a cellular automaton. There is often a critical magnitude of imperfection at which essentially a phase transition occurs, and the behaviour of the cellular automaton changes suddenly. One can presumably find such transitions as a function of noise and other imperfections in many different cellular automata (cf. [34, 35]). Often the transitions should be associated with critical exponents; one expects that several universality classes may be identified. Note that even one-dimensional cellular automata can exhibit phase transitions at nonzero values of imperfection parameters if imperfections are introduced in such a way that for example certain initial states still evolve as they would without the imperfections.

Given a pattern generated by a cellular automaton with imperfections, as might be obtained in a physical experiment, one may consider how the basic cellular automaton rule could be deduced. One could lay down a definite grid, and then accumulate histograms of the new site values obtained with all neighbourhoods, and thereby deduce the cellular automaton rule (it will not necessarily be unique, since certain neighbourhoods may never appear) [13]. This procedure accounts for imperfections due to noise, but not for imperfections such as deformations of the lattice. It appears that an iterative optimization approach must be used to treat such imperfections.

Problem 12

Is regular language complexity generically non-decreasing with time in one-dimensional cellular automata?

The sets of configurations generated by cellular automaton evolution, starting say from all possible initial states, can be considered as formal languages. Each config-uration corresponds to a word in the language, formed from a sequence of symbols representing site values, according to a definite set of grammatical rules. For one-dimensional cellular automata, it can be shown that the set of configurations generated after any finite number of time steps forms a regular formal language [7]. Thus the configurations correspond to the possible paths through a finite directed graph, whose arcs are labelled by the values that occur at each site. There is an algorithm to find the graph with the minimal number of nodes that represents a particular regular language [8, 38], in such a way that each word in the language corresponds to a unique path through the graph (deterministic finite automaton). This minimal graph provides a complete canonical description of the set generated by the cellular automaton evolu-tion. From it properties such as topological entropy may be deduced. The entropy is in fact given by the logarithm of the largest eigenvalue of the adjacency matrix for the graph, which is an algebraic integer.

One characteristic of a regular language is the total size or number of nodes Ξ in its minimal graph. This quantity can be considered as a measure of the complexity of the regular language. The larger it is, the more complicated a subset of the space of possible symbol sequences the language corresponds to. Ξ gives in a sense the size of the shortest description of this subset, at least in terms of regular languages. The value of Ξ is in general bounded above by $2^{k^{2rt}} - 1$. The empirical studies done so far suggest that for class 1 and 2 cellular automata, Ξ in fact becomes constant after a few time steps, or increases at most as a polynomial with t. For most class 3 and 4 cellular automata, however, Ξ appears to increase rapidly with time, though it usually stays far below the upper bound. There are a few cases where Ξ decreases slightly at a particular time step, but in general it seems that Ξ is usually non-decreasing with time. If this is indeed a general result, it gives a quantitative form to the qualitative statement that complexity seems to increase with time. It could be a principle for self-organizing systems analogous in generality but complementary in content to the law of entropy increase in thermodynamic systems.

If the non-decrease of Ξ is indeed a general result, then it should have a simple proof that depends on few of the properties of the system considered. A crucial property of cellular automata may be irreversibility, which leads to a progressive contraction in the set of configurations generated. As a consequence of this con-traction, the set generated at each time step must correspond to a different regular language. But there are only a limited number of regular languages with complexi-ties less than any particular value, and so the complexity of the language generated must increase, albeit slowly, with time. To find a complete bound, one must study the structure of the space of possible regular languages. It is clear that the number of regular languages of complexity Ξ is less than the number of labelled directed graphs with Ξ nodes, $2^{k\Xi^2}$. The minimal graph for a regular language must have a trivial automorphism group; but the number of graphs with a given automorphism group does not appear to be known (e.g., [39]). Beyond the total number of regular languages, one may consider the network that represents the containment of regular languages, divided into zones of different Ξ. One suspects that this network is close to a tree, with a number of nodes increasing perhaps exponentially with depth Ξ.

Problem 13

What limit sets can cellular automata produce?

Not all possible sets of configurations can be produced as limit sets of cellular automata. For the number of distinct cellular automaton rules, while infinite, is countable. Yet the number of possible sets of configurations is uncountable.

At each step in the evolution of an irreversible cellular automaton, a new set of configurations is excluded. The limit set consists of those configurations that are never excluded. The set of all excluded configurations is recursively enumerable, since each of its elements is found by a finite computation. Thus the limit sets for cellular automata are always the complements of recursively enumerable (co-r.e.) sets, and are therefore countable in number. Nevertheless, not every co-r.e. set is the limit set for a cellular automaton: one additional condition is that they must be translationally invariant. Thus for example, cellular automaton limit sets must contain either one configuration, or an infinite number of distinct configurations, and cannot consist of some other finite number of configurations [40]. Not every possible real number value of dimension or entropy can be realized by cellular automata; but the set that is realized presumably includes some values that are non-computable.

After any finite number of time steps, the set of configurations generated by a one-dimensional cellular automaton forms a regular formal language. For some cellular automata (essentially those in classes 1 and 2), the limit set is also a regular language. But in other cases, the limit set probably corresponds to a more complicated formal language. Explicit examples are known in which context-free and context-sensitive languages are obtained as limit sets [40]. In addition, cellular automata that are capable of universal computation can generate limit sets that are not recursive [40]. The generic behaviour is however not known: some more examples would be valuable.

When the limit set forms a regular language, the simplest description of it, in terms of a regular grammar or graph, can be found by a finite algorithm. The size Ξ of this description can be used as a measure of the complexity of the set. However, for languages more complicated than regular ones, there is in general no finite algorithm to find the simplest grammar (e.g., [8]). The size of such a minimal grammar is thus formally non-computable. One may test a sequence of grammars, but the languages to which they lead cannot in general be enumerated by a computation of any bounded length.

Minimum grammar size is thus not a useful measure of complexity for complicated cellular automaton limit sets. Some other measure must be found. And in terms of this measure, one should be able to determine how the complexity of the behaviour of a cellular automaton, as revealed by the structure of its limit set, depends on the complexity of its local rule, or the values of k and r.

One may wonder what features of the local rule for a cellular automaton determine its global properties, and the structure of its limit set. Some simple observations may be made. For example, unless the local rule contains elements that give value 1 with neighbourhoods such as 001, no information can propagate in the cellular automaton, and class 1 or 2 behaviour must occur. But in general one expects that the problem is undecidable: the only way to determine many of the limiting properties of a cellular automaton is probably by explicit simulation of its evolution, for an infinite time.

As a practical matter, one may ask whether cellular automaton rules may be constructed to yield particular limit sets (cf. [41]), so that their evolution serves to filter out the components that appear in these limit sets. It is probably possible to construct cellular automata that yield any of some class of regular languages as limit sets. But one suspects that a construction for more complicated limit sets can be carried out only in very special cases.

Problem 14

What are the connections between the computational and statistical characteristics of cellular automata?

The rate of information transmission is one attribute of cellular automata that potentially affects both computational and statistical properties. On the statistical side, the rate of information transmission gives the Lyapunov exponent for the cellular automaton evolution. Class 1 and 2 cellular automata have zero Lyapunov exponents, so that information almost always remains localized, and the value of a particular site at any time can almost always be determined from the initial values of a bounded neighbourhood of initial sites. As a consequence, the limit sets for one-dimensional such cellular automata correspond to regular languages. The configurations can thus be generated by an essentially Markovian process, in which there are no long-range correlations between different parts.

Class 3 and 4 cellular automata have positive Lyapunov exponents, so that a small initial change expands with time. The value of a particular site after many time steps thus depends in general on an ever-increasing region in the initial state. The limit sets for such cellular automata can thus involve long-range correlations, and need not correspond to regular languages. If class 4 cellular automata are generically capable of universal computation, then their limit sets should be unrestricted, in general non-recursive, formal languages. Some arguments can be given that class 3 cellular automata should yield limit sets that correspond to context-sensitive languages. In general, one suspects that dynamical systems that exhibit chaotic behaviour characterized by positive Lyapunov exponents should yield limit sets that are more complicated than regular languages.

When the limit set for a cellular automaton is a regular language, its spatial entropy can be computed, and is given by the logarithm of an algebraic integer. If the limit set is a context-free language, then it seems that the entropy is always the logarithm of some algebraic number. But for context-sensitive and more complicated languages, the entropy is in general non-computable. It may thus be common to find class 3 and 4 cellular automata for which the entropy of their limit sets is non-computable.

The computational structure of sets generated in the evolution of two and higher dimensional cellular automata can be very complicated even after a finite number of time steps. In particular, while in one-dimensional cellular automata the set of configurations that can be generated at any finite time forms a regular formal language, this set can be non-recursive in two-dimensional cellular automata [12, 42]. The essential origin of this difference is that there is an iterative procedure to find the possible predecessors of arbitrarily long sequences in one-dimensional cellular automata, but no such procedure exists for two-dimensional cellular automata. In fact, even the problem of finding configurations that evolve periodically in time in a two-dimensional cellular automaton appears to be equivalent to the domino tiling problem, which is known to be formally undecidable [43]. Nevertheless, it seems likely that only two-dimensional cellular automata in which information transmission can occur throughout the plane, as revealed by positive Lyapunov exponents in all directions, exhibit such complications, and give non-recursive sets at finite times.

The grammar for a formal language specifies which sequences occur in the language, but not how often they occur. It does not for example distinguish sequences that occur with zero probability from those that occur with positive probability. However, it is the probable, rather than the possible, behaviour of cellular automata that is most significant in determining their statistical properties, such as Lyapunov exponents and measure entropies. There are class 1 and 2 cellular automata in which a set of states of measure zero yields class 3 behaviour: this is irrelevant in the Lyapunov exponent or the measure entropy, but affects the topological entropy, and the structure of the grammar for the limit set. One should construct formal languages that include probabilities for configurations. A suitable approach may be to consider stochastic automata, closely related to standard Markov chains.

Problem 15

How random are the sequences generated by cellular automata?

The spatial sequences obtained after a finite number of steps in the evolution of a one-dimensional cellular automaton starting from all possible initial states are known to form a regular formal language. But no such characterization is known for the temporal sequences generated by cellular automata. At least for cellular automata capable of universal computation, these sequences can be non-recursive. But the generic behaviour is not known, and no non-trivial examples have yet been given.

One question is to what extent the initial state of a cellular automaton can be reconstructed from a knowledge of the time series of values of a few sites. An essentially equivalent question is how wide a patch of sites need to be considered to compute the invariant entropy of the cellular automaton mapping. When the mapping is surjective and expansive (so that roughly information transmission occurs at a positive rate), only a finite width is required (e.g., [44]). Nevertheless, the transformation necessary to find the initial state from the temporal sequence may be very complicated. In particular, there may be effectively no better method than to try all exponentially many possible initial states. Temporal sequences in cellular automata are thus candidates for use in pseudorandom number generation and in cryptography [20].

The patterns generated by some cellular automata evolving from initial states consisting of simple seeds have a simple form. They may be asymptotically homogeneous, or may correspond to regular fractals. But many cellular automata yield complicated patterns even starting from an initial state as simple as a single nonzero site. Some examples are shown in Fig. 2. It is remarkable that such complicated and intricate patterns can be generated in such a simple system.

Often the temporal sequences that appear in these patterns have a seemingly random form, and satisfy many statistical tests for randomness. There is empirical evidence that in many cases the sequence of values taken on say by the centre site in the pattern contains all possible subsequences with equal frequencies, so that the whole sequence effectively has maximal measure entropy. A simple example of this phenomenon occurs in the $k = 2, r = 1$ rule number 30 ($a_i^{(t+1)} = a_{i-1}^{(t)} \oplus \max (a_i^{(t)}, a_{i+1}^{(t)})$).

Systems that exhibit chaotic behaviour usually start from initial conditions that contain an infinite amount of information, either in the form of an infinite sequence of cellular automaton site values, or the infinite sequence of digits in a real number. Their irregular behaviour with time can then be viewed as a progressive excavation of the initial conditions. The chaotic behaviour seen in Fig. 2 is however of another kind: it occurs as a consequence of the dynamics of the system, even though the initial conditions are simple. It may well be that this kind of chaos is central to physical phenomena such as fluid turbulence.

It is important to investigate the mathematical bases for such behaviour. The closest analogies seem to lie in number theory. The integers generated for example by repeated application of a linear congruence transformation form a pseudorandom sequence (e.g., [45]), often used in practical applications. The linearity of this system makes it amenable to a rather complete number theoretical analysis, which provides formulae for computing the nth integer in the sequence directly from the original seed, with working out all the intermediates. It seems likely that such analyses, and the resulting short cuts, are not possible in most nonlinear cellular automata. The randomness produced in these systems may be more like the randomness say of the digits of π. In some cases it is in fact possible to cast essentially number theoretical problems in terms of questions about patterns generated by cellular automata. One example concerns the sequence of leading binary digits in the fractional parts of successive powers of 3/2 [46]. There is empirical evidence that all possible blocks of digits occur in this sequence, so that in a sense it has maximal entropy. The sequence corresponds to the time series of values of the central site in the pattern generated by a particular cellular automaton from a simple initial state.

Problem 16

How common are computational universality and undecidability in cellular automata?

If a system is capable of universal computation, then with appropriate initial conditions, its evolution can carry out any finite computational process. A computationally universal system can thus mimic the behaviour of any other system, and so can in a sense exhibit the most complicated possible behaviour.

Several specific cellular automata are known to be capable of universal computation. The two-dimensional nearest-neighbour cellular automaton with two possible values at each site known as the "Game of Life" has been proved computation universal [47]. The proof was carried out by showing that the cellular automaton could support structures that correspond to all the components of an idealized digital electronic computer, and that these components could be connected so to implement any algorithm. Some one-dimensional nearest-neighbour cellular automata with $k = 18$ have been shown to be computationally equivalent to the simplest known universal Turing machines, and are thus capable of universal computation [48].

One speculates that cellular automata identified on statistical grounds as class 4 are in fact generically capable of universal computation. This would imply that there exist one-dimensional computationally universal cellular automata in cases as simple as $k = 2, r = 2$ or $k = 3, r = 1$. But it remains to prove the computational universality of any particular such rule. Several methods could be used for such a proof. One is to identify a set of persistent structures in the cellular automaton that could act as the components of a digital computer, or like combinations of symbols and internal states for a Turing machine. Structures that remain fixed, propagate, and interact in various ways have been found. A structure that can act as a "clock", producing an infinite sequence of "signals", has not yet been found in such cellular automata. Another method of proving universality would be a direct demonstration that this cellular automaton rule could simulate any other cellular automaton rule with an appropriate encoding of initial states. Blocking transformations may provide the necessary encodings: so one must find out whether a particular cellular automaton rule is connected to all others in the simulation networks constructed from blocking transformations.

If class 4 cellular automata are indeed capable of universal computation, then the capability for universal computation is quite common among one-dimensional cellular automata. Class 4 behaviour is however much rarer in two dimensional cellular automata—the "Game of Life" is almost the only known example (cf. [12]).

There may well be cellular automata whose behaviour is usually computationally simple, but which with very special initial states can perform arbitrary computations. It is certainly possible to construct cellular automata in which universal computation occurs only with initial states in which say every other site has value zero (cf. [49]), a condition that occurs in disordered states with probability zero. Such phenomena may be common in class 3 cellular automata.

Any predictions about the behaviour of a cellular automaton must be made by performing some computation. But if the cellular automaton is capable of universal computation, then this computation must in general reduce to a direct simulation of the cellular automaton evolution. So questions about the infinite time limiting behaviour of cellular automata may require infinite computations, and therefore be formally undecidable.

For example, one may consider the question of whether the patterns generated from particular finite initial seeds ever die out in the evolution of the cellular automaton. One may simulate the evolution explicitly to find out whether a pattern dies out after say a thousand time steps; but to determine its ultimate fate in general requires a computation of unbounded length. The question is therefore formally undecidable.

The set of finite configurations that evolve to the null configuration after a fixed finite time can be specified by a regular formal language (cf. [50]). But there is no such finite specification for the set of finite configurations that evolve after any time to the null configuration. Even the fraction of configurations in this set is in general a non-computable number.

A similar problem is to determine whether a particular finite sequence of site values occurs in any configurations in the limit set for a cellular automaton. Again this problem is in general undecidable [40]. An explicit finite calculation can show that a sequence is forbidden after say three time steps. But a particular sequence may only be forbidden after some arbitrarily large number of time steps. In a one-dimensional cellular automaton, the length $L^{(t)}$ of the shortest sequence newly excluded at a given time step in the evolution is bounded by $L^{(t)} \geq L^{(t-1)} - 2r$. In most actual examples $L^{(t)}$ seems to increase monotonically with time, so that the exclusion of a particular finite sequence must occur before some predictable finite time. But in some cases $L^{(t)}$ is not monotonic, and the occurrence of particular sequences may be undecidable.

The capability for universal computation can be used to establish the undecidability of questions about the behaviour of a system. But undecidability can occur even in systems not capable of full universal computation. For example, one may arrange to disable all computations that give results of a certain form. In this way, the system fails to be able to perform arbitrary computations. Nevertheless, there may be undecidable questions about the class of computations that it still can perform. These may well occur in cellular automata. Proofs of undecidability usually use a diagonal argument based essentially on universal computation. To establish undecidability in a system not itself capable of universal computation, one must usually find another system that is capable of universal computation, and show that a reduction of its capabilities does not affect undecidability.

Rice's theorem states that almost all questions about an arbitrary recursively-enumerable set are undecidable (e.g., [8]). However, it may be that natural or simple questions, which can be stated in say a few logical symbols, are usually decidable. So for example the halting of all simple initial seeds in a particular cellular automaton might be easy to determine, and it might only be very large and specially-chosen initial seeds whose halting was difficult to determine. There are certainly examples in which the halting problem appears to be difficult to answer even for simple seeds. One must establish in general not only whether there are any undecidable propositions about the behaviour of a particular cellular automaton, but whether simple propositions about it are in fact undecidable.

Problem 17

What is the nature of the infinite size limit for cellular automata?

Statistical averages in many systems converge to definite values when the infinite size or thermodynamic limit is taken. Several complications can however arise in cellular automata.

Different seeds can lead to very different behaviour in class 4 cellular automata. Some may die out; others may yield periodic patterns; still others may produce propagating structures. Propagating structures usually involve at least five or ten sites, and appear only with seeds of such a size. One expects that when larger seeds are used, new kinds of structures can begin to occur. For example, there may be structures that periodically generate propagating patterns, giving an asymptotically infinite number of nonzero sites. If the cellular automaton is capable of universal computation, then it should support structures with arbitrarily complicated behaviour. So for example there may be self-reproducing structures, which replicate even in the presence of a disordered background. Any such structure present in an initial state would yield offspring that could eventually dominate the behaviour of the system. In a given class 4 cellular automaton, the simplest self-reproducing structure may have a size of say 100 sites. The density at which the structure would occur in a disordered state is then k^{-100}. So in practical simulations, there is an overwhelming probability that no such structure would ever be seen. But if configurations of size much larger than k^{100} were considered, such a structure would occur in almost every case. And after a long time, the behaviour of the system would almost always be dominated by the self-reproducing structures. Statistical results obtained with smaller configurations would then be misleading. And as the idealized limit of infinite size is taken, more and more complicated phenomena may occur, and statistical quantities have no simple limits.

Since a finite description in terms of regular formal languages can be given for the set of configurations generated at any finite time in the evolution of a one-dimensional cellular automaton, definite infinite size limits for statistical quantities presumably exist in this case. With time the limits may however become more complicated, and be reached more slowly. One expects that most statistical quantities will continue to show simple behaviour for class 3 cellular automata. But for class 4 cellular automata, in which different structure appears to be manifest on every different scale, the limits may become progressively more complicated, and may not exist at infinite times.

Two-dimensional cellular automata exhibit complicated infinite size limits even after a finite number of time steps. The sets of configurations that they generate can be non-recursive in the infinite size limit [12, 42], and some statistical quantities may have no limits as a consequence.

It is in general undecidable how large the smallest structure with some property such as self-reproduction can be in a particular cellular automaton. In some cases, the cellular automaton rule may be specially constructed to allow such structures. But for simple rules, one is reduced to an essentially experimental search for the structures. In several class 4 one-dimensional cellular automata with $k = 2$, all configurations of less than 21 sites have been tested, and all those up to about 30 sites are probably accessible with special-purpose computer hardware [51]. In the Game of Life, a number of complex structures were found through extensive experimentation. Further examples, particularly in one-dimensional cellular automata, would be valuable. One may imagine that each capability such as self-reproduction has a logical description of some length. Then the size of the smallest configuration that has the capability may be related in some way to this length. Obviously particular cellular automata may have special properties with respect to particular capabilities, but the result may hold as some average over all possible capabilities. If so, the very large number of particles in the universe could be essential for very complex physical and biological phenomena to occur.

For direct simulation and other practical purposes one is often concerned with cellular automata of finite size. When an infinite size limit exists, the local properties deduced from studies of finite cellular automata are likely to correspond directly with the infinite size case. But for global properties the correspondence is less clear. For the rather special case of finite cellular automata with additive rules, algebraic methods provide a complete description of the state transition diagram [33]. There are typically about $k^{N/2}$ cycles, each of length about $k^{N/2}$ steps. The cycles are reached after transients of length less than N. In the limit $N \to \infty$, the system exhibits chaotic behaviour, but the mapping is surjective, so that all configurations are generated. Presumably in this limit there are an infinite number of infinite cycles, perhaps each characterized by a particular form of some invariant algebraic function. In general, some cellular automata that show chaotic behaviour in the infinite size limit exhibit exponentially long cycles at small finite sizes. Others exhibit exponentially long transients. Some show neither. The general connections between the structure of finite state transition diagrams, and the behaviour of cellular automata in the infinite size limit remain to be established.

Problem 18

How common is computational irreducibility in cellular automata?

One way to find out the behaviour of a cellular automaton is to simulate each step in its evolution explicitly. The question is how often there are better ways.

Cellular automaton evolution can be considered as a computation. A procedure can short cut this evolution only if it involves a more sophisticated computation. But there are cellular automata capable of universal computation that can perform arbitrarily sophisticated computations. So at least in these cases no short cut procedure can in general be found. The cellular automaton evolution corresponds to an irreducible computation, whose outcome can be found effectively only by carrying it out explicitly.

A number of complications arise in giving a precise definition of such computational irreducibility. In general one should compare the number of steps in the evolution of a system such as a cellular automaton with the number of steps required to reproduce the evolution using another computational system. However, by making the computational system more complicated, it is always possible to reduce the number of steps required by an arbitrary constant factor, or even an arbitrary function. For example, if a computer can apply the square of a cellular automaton mapping at each step, then it can always simulate T steps of cellular automaton evolution in $T/2$ steps.

Nevertheless, no amount of additional complication in the computer can allow it to find in a finite time the outcome of an infinite number of steps in the evolution of a cellular automata that is for example capable of universal computation. As a consequence, there are undecidable propositions about the ultimate behaviour of the cellular automaton. The occurrence of such undecidable propositions may be viewed as a consequence of computational irreducibility. But to give a complete definition of computational irreducibility for finite time processes, one must in some way exclude arbitrary complication in the computer used for predictions.

One approach is to consider finite cellular automata and to use methods from computational complexity theory. A cellular automaton with N sites can evolve for a time up to k^N before retracing its steps. The computation corresponding to this evolution is performed in a bounded space, and is therefore in the class *PSPACE* (e.g., [8]), but it can take a time exponential in N. However if the computation were reducible, then it could be possible to find the outcome of the evolution in a time polynomial in N, or in other words to reduce the problem to one in the class P. It is believed that $PSPACE \neq P$, so that there exist problems that can be solved in polynomial space that cannot be solved in polynomial time. Determining the outcome of the evolution of some cellular automata may be a problem of this kind (cf. [52]).

Conventional computational complexity theory concerns computations in finite systems. It may well be that the definition of computational irreducibility for cellular automata can be sharpened in the infinite size limit.

The evolution of class 1 and 2 cellular automata yielding periodic configurations is computationally reducible. But one suspects that the evolution of most class 3 and 4 cellular automata is computationally irreducible. In fact, it may well be in general that most systems that show apparently complex or chaotic behaviour are computationally irreducible.

Even if the detailed behaviour of a system can effectively be found only by direct simulation, it could be that many of its overall properties can be found by more efficient procedures. It is this possibility that makes investigations of cellular automata worthwhile even when computational irreducibility is present. But what should be done is to find a characterization of those properties whose behaviour can be found by efficient methods, and those for which computational irreducibility makes explicit simulation the only possible approach, and precludes a simple description.

Problem 19

How common are computationally intractable problems about cellular automata?

Questions concerning the finite time behaviour of finite cellular automata can always be answered by finite computations. But as the phenomenon of computational irreducibility suggests, there may be questions for which the computations are necessarily very long. One may consider for example the question of whether a particular sequence of X site values can occur after T time steps in the evolution of a one-dimensional cellular automaton, starting from any initial state. Then one may ask whether there exists any algorithm that can determine the answer in a time given by some polynomial in X and T. The question can certainly be answered by testing all k^{X+2rT} sequences of initial site values that determine the length X sequence, but this procedure requires a time that grows exponentially with X and T. Nevertheless, if an initial sequence could be guessed, then it could be tested in a time polynomial in X and T. As a consequence, the problem is in the class NP. Now if $P \neq NP$, then there may be no polynomial time algorithm for the problem, and the best method of solution may essentially be to try all the exponentially many possible cases explicitly, so that the problem rapidly becomes intractable. In the infinite time limit, the analogous problem is in general undecidable.

Just as undecidability in a system can be proved by establishing a capability for universal computation, so, assuming $P \neq NP$, computational intractability can be proved through NP-completeness. A problem is NP-complete if specific instances of it correspond to arbitrary problems in the class NP [8, 53]. This can be shown by establishing equivalence to a known NP-complete problem. Thus for example it has been possible to give a specific example of a cellular automaton in which the problem of determining whether particular sequences can occur after T time steps is equivalent to the *NP*-complete problem of finding a set of truth values for variables so that a particular logical expression is satisfied [54]. How widespread *NP*-completeness is in problems concerning cellular automata has yet to be established. But one suspects that it is common in many class 3 and 4 systems.

Problem 20

What higher-level descriptions of information processing in cellular automata can be given?

Cellular automaton evolution can in principle carry out arbitrary information processing. An important problem of theory and practice is to find a way of organizing this information processing. In specific cases one can devise cellular automaton rules that allow particular computations to be carried out (e.g., [55]). Or one can identify within a cellular automaton structures that can interact so as to mimic the components of conventional digital computers. But all these approaches are strongly based on analogues with conventional serial-processing computers. Information processing in cellular automata occurs however in a fundamentally distributed and parallel fashion, and one must invent a new framework to make use of it. Such a framework would likely be valuable in studying the many physical systems in which information processing is also distributed.

One approach is statistical in nature. It consists in devising and describing attractors for the global evolution of cellular automata. All initial configurations in a particular basin of attraction may be thought of as instances of some pattern, so that their evolution towards the same attractor may be considered as a recognition of the pattern. This approach is probably effective when the basins of attraction are local in space, as in image processing (e.g., [56]). But the construction of attractors for more general problems is likely to be very difficult. An attempt in this direction might be made by considering basins of attraction as sets of sequences corresponding to particular formal languages (cf. [50]).

Another approach is to use symbolic representations for various attributes or components of cellular automaton configurations. But the structures used in conventional computer languages are largely inappropriate. The definite organization of computer memory into named areas, stacks, and so on, is not suitable for cellular automata in which processing elements are not distinguished from memory elements. Rather perhaps data could be represented by an object like a graph, on which transformations can be performed in parallel. But the simple organizing principles that are required still remain to be found. It seems likely that a radically new approach is needed [57].

Acknowledgements

I have benefitted from discussions about cellular automata with many people, too numerous to list here. For recent discussions I am particularly grateful to: C. Bennett, M. Feigenbaum, E. Fredkin, D. Hillis, L. Hurd, L. Kadanoff, D. Lind, O. Martin, J. Milnor, N. Packard, D. Ruelle, R. Shaw, and K. Steiglitz.

References

1. Wolfram, S., Nature **311**, 419 (1984).
2. Wolfram, S., Rev. Mod. Phys. **55**, 601 (1983).
3. Farmer, D., Toffoli, T. and Wolfram, S. (editors), "Cellular automata: proceedings of an interdisciplinary workshop", Physica **10D** numbers 1 and 2 (1984), North-Holland Publishing Co. (1984).
4. Wolfram, S., "Cellular automata", Los Alamos Science (fall 1983).
5. Wolfram, S., Physica **10D**, 1 (1984).
6. Guckenheimer, J. and Holmes, P., Nonlinear Oscillations, Dynamical Systems, and Bifurcations of Vector Fields, Springer-Verlag (1983).
7. Wolfram, S., Comm. Math. Phys., **96**, 15 (1984).
8. Hopcroft, J. E. and Ullman, J. D., Introduction to Automata Theory, Languages, and Computation, Addison-Wesley (1979).
9. Wolfram, S., "Computer software in science and mathematics", Scientific American (September 1984).
10. Packard, N., "Complexity of growing patterns in cellular automata", Institute for Advanced Study preprint (October 1983).
11. Milnor, J., "Entropy of cellular automaton-maps", Institute for Advanced Study preprint (May 1984).
12. Packard, N. and Wolfram, S., "Two-dimensional cellular automata", J. Stat. Phys. **38**, 901 (1985).
13. Packard, N., Private communication.
14. Ahmed, N. and Rao, K. R., Orthogonal Transforms for Digital Signal Processing, Springer-Verlag (1975).
15. Franks, J. and Fried, D., Private communications.
16. Margolus, N., Physica **10D**, 81 (1984).
17. Vichniac, G., Physica **10D**, 96 (1984).
18. Pomeau, Y., J. Phys. **A17**, L415 (1984).
19. Hedlund, G., Private communication.
20. Milnor, J. and Wolfram, S., In preparation.
21. Milnor, J., "Notes on surjective cellular automaton-maps", Institute for Advanced Study preprint (June 1984).
22. Chaitin, G., "Towards a mathematical definition of life", in The Maximum Entropy Formalism", R. D. Levine and M. Tribus (eds.), MIT press (1979).
23. Crutchfield, J. and Packard, N., Private communication.
24. Collet, P. and Eckmann, J.-P., Iterated Maps on the Interval as Dynamical Systems, Birkhauser (1980).
25. Amit, D., Field Theory, the Renormalization Group, and Critical Phenomena, McGraw-Hill (1978).
26. Milnor, J., Unpublished.
27. Mandelbrot, B., The Fractal Geometry of Nature, Freeman (1982).

28. Grassberger, P., Physica **10D**, 52 (1984).

29. Greenberg, J. M., Hassard, B. D., and Hastings, S. P., Bull. Amer. Math. Soc. **84**, 1296 (1975); Madore, B. and Freedman, W., Science **222**, 615 (1983).

30. Winfree, A. and Winfree, E., "Organizing centers in a cellular excitable medium", to be published.

31. Packard, N., "Cellular automaton models for dendritic growth", Institute for Advanced Study preprint, in preparation.

32. Bennett, C., Neuberger, H., Pomeau, Y. and Vichniac, G., Private communications.

33. Martin, O., Odlyzko, A. and Wolfram, S., Comm. Math. Phys. **93**, 219 (1984).

34. Grassberger, P., Krause, F. and von der Twer, T., "A new type of kinetic critical phenomena", University of Wuppertal preprint WU B 83-22 (October 1983).

35. Domany, E. and Kinzel, W., Phys. Rev. Lett. **53**, 311 (1984).

36. Ingerson, T. E. and Buvel, R. L., Physica **10D**, 59 (1984).

37. Kauffman, S., Physica **10D**, 145 (1984).

38. Hopcroft, J., "An $n \log n$ algorithm for minimizing states in a finite automaton", in Proc. Int. Symp. on the Theory of Machines and Computations, Academic Press (1971).

39. Harary, F., Graph Theory, Chapter 15, Addison-Wesley (1972).

40. Hurd, L., "Formal language characterizations of cellular automaton limit sets", to be published.

41. Hasslacher, B., Private communication.

42. Yaku, T., J. Comput. System Sci. **7**, 481 (1973); Golze, U., "Differences between 1- and 2-dimensional cell spaces", in A. Lindenmayer and G. Rozenberg (eds.), Automata, Languages, Development, North-Holland (1976).

43. Berger, R., Mem. Amer. Math. Soc., no. 66 (1966); Robinson, R., Inventiones Math. **12**, 177 (1971).

44. Walters, P., An Introduction to Ergodic Theory, Springer (1982).

45. Knuth, D., Seminumerical Algorithms, 2nd. ed., Addison-Wesley (1981).

46. Furstenberg, H. and Lind, D., Private communication.

47. Berlekamp, E. R., Conway, J. H. and Guy, R. K., Winning Ways for Your Mathematical Plays, vol. 2, chap. 25, Academic Press (1982); Gardner, M., Wheels, Life and Other Mathematical Amusements, Freeman (1983).

48. Smith, A. R., J. ACM **18**, 339 (1971).

49. Banks, E. R., "Information processing and transmission in cellular automata", MIT project MAC report no. TR-81 (1971).

50. Smith, A. R., J. Comput. Sys. Sci. **6**, 233 (1972); Sommerhalder, R. and van Westrhenen, S. C., Acta Inform. **19**, 397 (1983).

51. Steiglitz, K., Private communication.

52. Bennett, C. H., "On the logical "depth" of sequences and their reducibilities to random sequences", Info. & Control, to be published.

53. Garey, M. R. and Johnson, D. S., Computers and Intractability: a Guide to the Theory of *NP*-completeness, Freeman (1979).

54. Sewelson, V., Private communication.

55. Cole, S. N., IEEE Trans. Comput. **C-18**, 349 (1969).

56. Preston, K. *et al.*, Proc. IEEE **67**, 826 (1979).

57. Hillis, D. and Wolfram, S., Work in progress.

Cryptography
with Cellular Automata

1986

This abstract discusses a stream cipher based on a simple one-dimensional cellular automaton. The cellular automaton consists of a circular register with N cells, each having a value a_i equal to 0 or 1. The values are updated synchronously in discrete time steps according to the rule

$$a_i' = a_{i-1} \text{ XOR } (a_i \text{ OR } a_{i+1}),\qquad\text{(1a)}$$

or, equivalently,

$$a_i' = (a_{i-1} + a_i + a_{i+1} + a_i a_{i+1}) \bmod 2.\qquad\text{(1b)}$$

The initial state of the register is used as a seed or key. The values $a^{(t)}$ attained by a particular cell through time can then serve as a random sequence. Ciphertext C can be obtained from binary plaintext P as usual according to $C_i = P_i \text{ XOR } a^{(i)}$; the plaintext can be recovered by repeating the same operation, but only if the sequence $a^{(i)}$ is known.

Cellular automata such as (1) have been investigated in studies of the origins of randomness in physical systems [2]. They are related to non-linear feedback shift registers, but have slightly different boundary conditions.

Figure 1 shows the pattern of cell values produced by (1) with a seed consisting of a single nonzero cell in a large register. The time sequence of values of the centre cell shows no statistical regularities under the tests of ref. [3] (for sequence lengths up to $2^{19} \simeq 5 \times 10^5$). Some definite spacetime patterns are nevertheless produced by the cellular automaton rule.

In the limit $N \to \infty$, the cellular automaton evolution is like an iterated continuous mapping of the Cantor set, and can be studied using dynamical systems theory [4]. One result is that the evolution is unstable with respect to small perturbations in

Originally published as an abstract in *Advances in Cryptology: Crypto '85 Proceedings*, Lecture Notes in Computer Science, volume 218, pages 429–432 (Springer-Verlag, 1986).

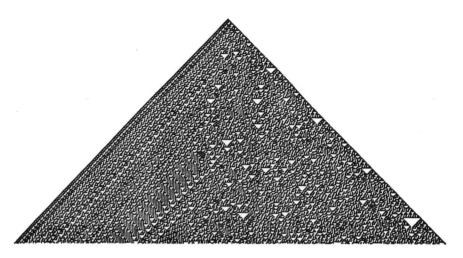

Figure 1. Pattern produced by evolution according to the cellular automaton of eqn. (1) from a simple seed containing a single nonzero bit. 250 successive states of an arbitrarily large register are shown; black squares represent nonzero cells. Columns of cell values, say in the centre, seem random for practical purposes.

the initial seed. A change produced by reversing a single cell value typically expands at a rate given by Lyapunov exponents, equal to 0.25 on the left, and 1 on the right. Length T time sequences of cell values are found however to be affected on average only by about $1.19T$ initial values.

Iterations of the cellular automaton rule (1) can be considered as Boolean functions of initial cell values. Disjunctive normal forms (minimized using [5]) for these functions are found to increase in size roughly as $4^{0.65t}$, giving some indication of the complexity of the cellular automaton evolution.

Figure 2 shows the complete state transition diagram for the cellular automaton (1) in a register of size $N = 11$. For large N, an overwhelming fraction of states lie on the longest cycle. But there are also shorter cycles, often corresponding to states with special symmetries. Figure 3 shows the length of the longest cycle as a function of N. The results (up to $N = 53$, which gives cycle length 40114679273) fit approximately $2^{0.61N}$. The mapping (1) is not a bijection, but is almost so; only a fraction $(\kappa/2)^N \simeq 0.85^N$ of states do not have unique predecessors [6] (κ is the real root of $4\kappa^3 - 2\kappa^2 - 1 = 0$).

The security of a cryptographic system based on (1) relies on the difficulty of finding the seed from a time sequence of cell values. This problem is in the class NP. No systematic algorithm for its solution is currently known that takes a time less than exponential in N. No statistical regularities have been found in sequences shorter than the cycle length.

One approach to the problem of finding the seed [6] uses the near linearity of the rule (1). Equation (1) can be written in the alternative form $a_{i-1} = a'_i \,\text{XOR}\, (a_i \,\text{OR}\, a_{i+1})$.

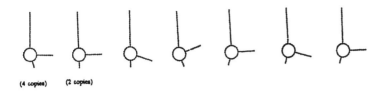

(4 copies)   (2 copies)

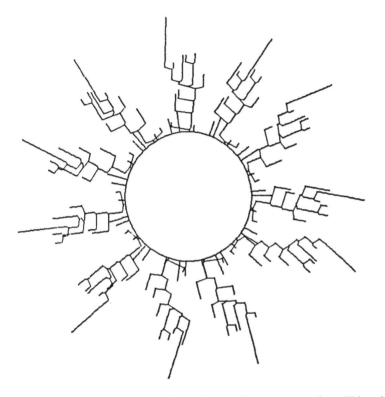

Figure 2. Complete state transition diagram for the cellular automaton of eqn. (1) in a circular register of size $N = 11$. There are 2^N states, each represented by dots. Evolution from any state leads eventually to one of the cycles shown.

Given the values of cells in two adjacent columns, this allows the values of all cells in a triangle to the left to be reconstructed. But the sequence provided gives only one column. Values in the other column can be guessed, and then determined from the consistency of Boolean equations for the seed. But in disjunctive normal form the number of terms in these equations increases linearly with N, presumably making their solution take a time more than polynomial in N.

The cellular automaton (1) can be implemented efficiently on an integrated circuit; it requires less than ten gate delay times to generate each output bit, and can thus potentially be used in a variety of high-bandwidth cryptographic applications.

489

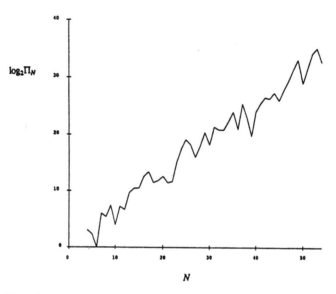

Figure 3. Length Π_N of the longest cycle as a function of register size N.

Much of the work summarized here was done while I was consulting at Thinking Machines Corporation (Cambridge, MA). I am grateful for discussions with many people, including Persi Diaconis, Carl Feynman, Richard Feynman, Shafi Goldwasser, Erica Jen and John Milnor.

References

1. S. Wolfram, "Random sequence generation by cellular automata", *Advances in Applied Mathematics*, **7**, 123 (1986).

2. S. Wolfram, "Origins of randomness in physical systems", Phys. Rev. Lett. **55**, 449 (1985); S. Wolfram, "Cellular automata as models of complexity", Nature **311**, 419 (1984).

3. D. Knuth, *Seminumerical Algorithms*, (Addison-Wesley, 1981).

4. S. Wolfram, "Universality and complexity in cellular automata", Physica **10D**, 1 (1984).

5. R. Rudell, *espresso* software program, Computer Science Dept., University of California, Berkeley (1985).

6. C. Feynman and R. Feynman, private communication.

Complex Systems Theory

1988

Some approaches to the study of complex systems are outlined. They are encompassed by an emerging field of science concerned with the general analysis of complexity.

Throughout the natural and artificial world one observes phenomena of great complexity. Yet research in physics and to some extent biology and other fields has shown that the basic components of many systems are quite simple. It is now a crucial problem for many areas of science to elucidate the mathematical mechanisms by which large numbers of such simple components, acting together, can produce behaviour of the great complexity observed. One hopes that it will be possible to formulate universal laws that describe such complexity.

The second law of thermodynamics is an example of a general principle that governs the overall behaviour of many systems. It implies that initial order is progressively degraded as a system evolves, so that in the end a state of maximal disorder and maximal entropy is reached. Many natural systems exhibit such behaviour. But there are also many systems that exhibit quite opposite behaviour, transforming initial simplicity or disorder into great complexity. Many physical phenomena, among them dendritic crystal growth and fluid turbulence are of this kind. Biology provides the most extreme examples of such self-organization.

The approach that I have taken over the last couple of years is to study mathematical models that are as simple as possible in formulation, yet which appear to capture the essential features of complexity generation. My hope is that laws found to govern these particular systems will be sufficiently general to be applicable to a wide range of actual natural systems.

The systems that I have studied are known as cellular automata. In the simplest case, a cellular automaton consists of a line of sites. Each site carries a value 0 or 1.

Based on a talk given Oct. 6, 1984 at the Founding Workshop of the Santa Fe Institute. Published in *Emerging Syntheses in Science: Proceedings of the Founding Workshops of the Santa Fe Institute*, pages 183–189 (Addison-Wesley, 1988).

The configurations of the system are thus sequences of zeroes and ones. They evolve in a series of time steps. At each step, the value of each site is updated according to a specific rule. The rule depends on the value of a site, and the values of say its two nearest neighbours. So for example, the rule might be that the new site value is given by the sum of the old value of the site and its nearest neighbours, reduced modulo two (i.e. the remainder after division of the sum by two).

Even though the construction of cellular automata is very simple, their behaviour can be very complicated. And as a consequence, their analysis can be correspondingly difficult. In fact, there are reasons of principle to expect that there are no general methods that can universally be applied.

The first step in studying cellular automata is to simulate them, and see explicitly how they behave. Figure 1 shows some examples of cellular automata evolving from simple seeds. In each picture, the cellular automaton starts on the top line from an initial state in which all the sites have value zero, except for one site in the middle, which has value one. Then successive lines down the page are calculated from the lines above by applying the cellular automaton rule at each site.

Figure 1(a) shows one kind of pattern that can be generated by this procedure. Even though the rule is very simple (it can be stated in just one sentence, or a simple formula), and the initial seed is likewise simple, the pattern produced is quite complicated. Nevertheless, it exhibits very definite regularities. In particular, it is self-similar or fractal, in the sense that parts of it, when magnified, are similar to the whole.

Figure 2 illustrates the application of a cellular automaton like the one in figure 1(a) to the study of a natural phenomenon: the growth of dendritic crystals, such as snowflakes (as investigated by Norman Packard). The cellular automaton of figure 1(a) is generalized to be on a planar hexagonal grid, rather than a line. Then a cellular automaton rule is devised to reproduce the microscopic properties of solidification. A set of partial differential equations provide a rather complete model for solidification. But to study the overall patterns of growth produced, one can use a model that includes only some specific features of the microscopic dynamics. The most significant feature is that a planar interface is unstable, and produces protrusions

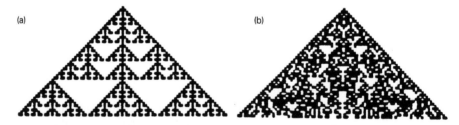

Figure 1. Patterns generated by evolution according to simple one-dimensional cellular automaton rules from simple initial conditions.

Figure 2. Snowflake growth simulation with a two-dimensional cellular automaton (courtesy of Norman H. Packard).

with some characteristic length scale. The sizes of the sites in the cellular automaton correspond to this length scale, and the rules that govern their evolution incorporate the instability. With this simple caricature of the microscopic laws, one obtains patterns apparently very similar to those seen in actual snowflakes. It remains to carry out an actual experiment to find out whether the model indeed reproduces all the details of snowflakes.

Figure 1(b) shows a further example of a pattern generated by cellular automaton evolution from simple initial seeds. It illustrates a remarkable phenomenon: even though the seed and the cellular automaton rules are very simple, the pattern produced is very complicated. The specification of the seed and cellular automaton rule requires little information. But the pattern produced shows few simplifying features, and looks as if it could only be described by giving a large amount of information, explicitly specifying its intricate structure.

Figure 1 is a rather concrete example of the fact that simple rules can lead to very complicated behaviour. This fact has consequences for models and methodologies in many areas of science. I suspect that the complexity observed in physical processes such as turbulent fluid flow is of much the same mathematical character as the complexity of the pattern in figure 1(b).

The phenomenon of figure 1 also has consequences for biology. It implies that complicated patterns of growth or pigmentation can arise from rather simple basic processes. In practice, however, more complicated processes may often be involved. In physics, it is a fair principle that the simplest model for any particular phenomenon is usually the right one. But in biology, accidents of history often invalidate this principle. It is only the improbability of very complicated arrangements that have been reached by biological evolution that makes a criterion of simplicity at all relevant. And in fact it may no more be possible to understand the construction of a biological organism than a computer program: each is arranged to work, but a multitude of arbitrary choices is made in its construction.

The method of investigation exemplified by figures 1 and 2 is what may be called "experimental mathematics". Mathematical rules are formulated, and then their consequences are observed. Such experiments have only recently become feasible, through the advent of interactive computing. They have made a new approach to science possible.

Through computers, many complex systems are for the first time becoming amenable to scientific investigation. The revolution associated with the introduction of computers in science may well be as fundamental as, say, the revolution in biology associated with the introduction of the telescope. But the revolution is just beginning. And most of the very easy questions have yet to be answered, or even asked. Like many other aspects of computing, the analysis of complex systems by computer is an area where so little is known that there is no formal training that is of much advantage. The field is in the exciting stage that anyone, whether a certified scientist or not, can potentially contribute.

Based on my observations from computer experiments such as those of figure 1, I have started to formulate a mathematical theory of cellular automata. I have had to use ideas and methods from many different fields. The two most fruitful so far are dynamical systems theory and the theory of computation.

Dynamical systems theory was developed to describe the global properties of solutions to differential equations. Cellular automata can be thought of as discrete idealizations of partial differential equations, and studied using dynamical systems theory. The basic method is to consider the evolution of cellular automata from all its possible initial states, not just say those consisting of a simple seed, as in figure 1. Figure 3 shows examples of patterns produced by the evolution of cellular automata with typical initial states, in which the value of each site is chosen at random. Even though the initial states are disordered, the systems organizing itself through its dynamical evolution, spontaneously generating complicated patterns. Four basic classes of behaviour are found, illustrated by the four parts of figure 3. The first three are analogous to the fixed points, limit cycles and strange attractors found in differential equations and other dynamical systems. They can be studied using quantities from dynamical systems theory such as entropy (which measures the information content of the patterns), and Lyapunov exponents (which measure the instability, or rate of information propagation).

Cellular automata can not only be simulated by computers: they can also be considered as computers in their own right, processing the information corresponding to their configurations. The initial state for a cellular automaton is a sequence of digits, say ones and zeroes. It is directly analogous to the sequence of digits that appears in the memory of a standard digital electronic computer. In both cases the

Figure 3. Four classes of behaviour found in evolution of one-dimensional cellular automata from disordered initial states.

sequences of digits are then processed according to some definite rules: in the first case the cellular automaton rules, and in the second case the instructions of the computer's central processing unit. Finally some new sequence of digits is produced that can be considered as the result or output of the computation.

Different cellular automata carry out computations with different levels of complexity. Some cellular automata, of which figure 3(d) is probably an example, are capable of computations as sophisticated as any standard digital computer. They can act as universal computers, capable of carrying out any finite computation, or of performing arbitrary information processing. The propagating structures in figure 3(d) are like signals, interacting according to particular logical rules.

If cellular automata such as the one in figure 3(d) can act as universal computers, then they are in a sense capable of the most complicated conceivable behaviour. Even though their basic structure is simple, their overall behaviour can be as complex as in any system.

This complexity implies limitations of principle on analyses which can be made of such systems. One way to find out how a system behaves in particular circumstances is always to simulate each step in its evolution explicitly. One may ask whether there can be a better way. Any procedure for predicting the behaviour of a system can be considered as an algorithm, to be carried out using a computer. For the prediction to be effective, it must short cut the evolution of the system itself. To do this it must perform a computation that is more sophisticated than the system itself is capable of. But if the system itself can act as a universal computer, then this is impossible. The behaviour of the system can thus be found effectively only by explicit simulation. No computational short cut is possible. The system must be considered "computationally irreducible".

Theoretical physics has conventionally been concerned with systems that are computationally reducible, and amenable for example to exact solution by analytical methods. But I suspect that many of the systems for which no exact solutions are now known are in fact computationally irreducible. As a consequence, at least some aspects of their behaviour, quite possibly including many of the interesting ones, can be worked out only through explicit simulation or observation. Many asymptotic questions about their infinite time behaviour thus cannot be answered by any finite computations, and are thus formally undecidable.

In biology, computational irreducibility is probably even more generic than in physics, and as a result, it may be even more difficult to apply conventional theoretical methods in biology than in physics. The development of an organism from its genetic code may well be a computational irreducible process. Effectively the only way to find out the overall characteristics of the organism may be to grow it explicitly. This would make large-scale computer-aided design of biological organisms, or "biological engineering", effectively impossible: only explicit search methods analogous to Darwinian evolution could be used.

Complex systems theory is a new and rapidly developing field. Much remains to be done. The ideas and principles that have already been proposed must be studied in a multitude of actual examples. And new principles must be sought.

Complex systems theory cuts across the boundaries between conventional scientific disciplines. It makes use of ideas, methods and examples from many disparate fields. And its results should be widely applicable to a great variety of scientific and engineering problems.

Complex systems theory is now gaining momentum, and is beginning to develop into a scientific discipline in its own right. I suspect that the sociology of this process is crucial to the future vitality and success of the field. Several previous initiatives in the direction of complex systems theory made in the past have failed to develop their potential for largely sociological reasons. One example is cybernetics, in which the detailed mathematical results of control theory came to dominate the field, obscuring the original more general goals. One of the disappointments in complex systems theory so far is that the approaches and content of most of the papers that appear reflect rather closely the training and background of their authors. Only time will ultimately tell the fate of complex systems theory. But as of now the future looks bright.

Bibliography

S. Wolfram, "Computer software in science and mathematics", *Scientific American* (September 1984).

S. Wolfram, "Cellular automata as models of complexity", *Nature* **311** (1984) 419–424.

S. Wolfram, "Undecidability and intractability in theoretical physics", *Physical Review Letters* **54** (1985) 735–738.

S. Wolfram, "Twenty problems in the theory of cellular automata", *Physica Scripta* **T9** (1985) 170–183.

S. Wolfram, "Origins of randomness in physical systems", *Physical Review Letters* **55** (1985) 449–452.

Cellular Automaton Supercomputing

1988

Many of the models now used in science and engineering are over a century old. Most of them can be implemented on modern digital computers only with considerable difficulty. This article discusses new basic models which are much more directly suitable for digital computer simulation.

The ultimate purpose of most scientific investigations is to determine how physical or other systems will behave in particular circumstances. Over the last few years, computer simulation has been emerging as the most effective method in many different cases. The basic approach is to use an algorithm which operates on data in the computer so as to emulate the behavior of the system studied (e.g., [1]). This algorithm can be considered to provide a "computational model" for the system.

Theoretical investigations of physical systems have conventionally been based on a few definite classes of mathematical models. By far the most common are partial differential equations (e.g., [2]). These equations were designed to describe systems such as fluids which can be considered as continuous media. Calculus was used as a tool to find mathematical formulae for the solutions to these equations. This allowed great progress to be made in the understanding of many phenomena, particularly those such as electro magnetism, which are described and/or approximated by linear partial differential equations and laminar (regular) fluid flows, which can be approximated by linear partial differential equations. But the standard methods of mathematical analysis made little headway on problems such as fluid turbulence, for which nonlinear partial differential equations are essential.

When digital computers became available, it was natural that they should be used to try to find solutions to such partial differential equations. But digital computers can represent such equations only approximately. While equations involve continuous variables, digital computers can treat only discrete digital quantities. The real

Originally published in *High-Speed Computing: Scientific Applications and Algorithm Design*, ed. Robert B. Wilhelmson (University of Illinois Press, 1988).

numbers that correspond to continuous variables in the equations must be represented on the computer by packets of bits, typically in the form of 32- or 64-bit numbers in floating point format. In addition, the derivates which appear in the equations must be approximated by finite differences on a discrete grid. Much effort has been spent in numerical analysis to show, for example, that with sufficiently fine grids, exact solutions to the continuum equations can be found. Unfortunately, such theorems have been proved almost exclusively only in cases where exact solutions to the continuum equations are known. For most important nonlinear equations, quite ad hoc methods must be used to gauge the accuracy of approximations.

Nevertheless, the thrust in scientific computation has been to develop computer hardware and algorithms which allow more and more extensive approximations to partial differential equations to be made. Thus, for example, the performance of computers is often measured in terms of the rate at which they can carry out the floating point operations needed. In many cases, there seem to be limitations that will prevent rapid increases in such performance.

Significant progress may perhaps more easily be made by somewhat shifting the emphasis. The kinds of operations which can efficiently be carried out by digital electronic circuits and thus digital computers, are quite clear. Large numbers of simple logical operations can be performed, potentially in parallel on many elements of a regular grid. Given the structure, one may then ask the question of whether accurate computational models based on this structure can be found for physical and other systems.

Cellular automata (e.g., [3,4]) provide one class of examples. A cellular automaton consists of a discrete lattice of sites. Each site carries a discrete value chosen from a small set of possibilities. The values are updated in a sequence of discrete timesteps according to logical rules which depend on the values of neighboring sites. Cellular automata are thus, by construction, almost ideal for simulation on digital electronic computers. They are particularly well suited for the coming generation of massively parallel machines, such as the Connection Machine computer [5], in which a very large number (currently 65,536) of separate processors, each simple, act in parallel.

One of the most remarkable results of recent studies on cellular automata is that even with very simple rules, it is possible to obtain behavior of considerable complexity [3,4]. Figure 1 shows a few examples. The rules consist of just a few simple logical operations, but when they are applied over and over again, their collective effect can yield very complex patterns of behavior. Often these show striking similarities to forms seen in many natural systems, and in other mathematical models for these systems. Chaotic behavior, corresponding to strange attractors, is common in cellular automata. Fractal patterns are also often produced, for example.

One thus expects that very simple computational models, based, for example, on cellular automata, should be sufficient to reproduce many different natural phenomena. The challenge is to abstract the essential mathematical features of the phenomena, so as to be able to capture them in as simple a model as possible.

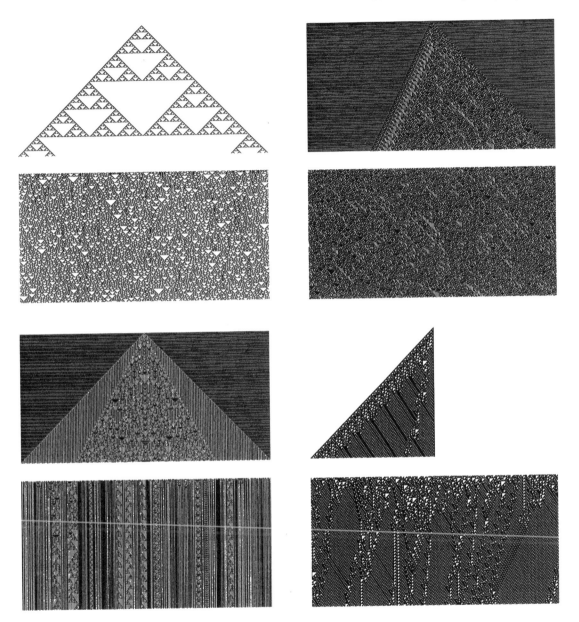

Figure 1. Examples of patterns generated by simple one-dimensional cellular automata. The cellular automaton consists of a row of about 600 sites whose values evolve with time down the page according to simple logical rules. The value 0 or 1 of each site (represented by white or black) is determined from its own value, and the values of its two nearest neighbors on the step before. Patterns generated by four different rules are shown. In each case, the pattern obtained with an initial state containing a single nonzero site is shown above, and a pattern generated with a random initial state is shown below. (In the notation of ref. [3], the rules are numbers 18, 45, 73, and 110.) Despite the simplicity of these cellular automata, the patterns generated show considerable complexity.

As one example, I shall discuss some recent models for fluid flow phenomena based on cellular automata (e.g., [6]).

Fluids are conventionally described by the Navier-Stokes partial differential equations (e.g., [7]). These equations can presumably describe in principle the important phenomenon of fluid turbulence. But digital computer simulations based on the Navier-Stokes equations are barely able to reach the regime needed to reproduce turbulence accurately. Of course, the Navier-Stakes equations are themselves an approximation. At a fundamental level, fluids consist of discrete particles, usually molecules. The Navier-Stokes equations give an approximate continuum description of the average behavior of large numbers of such discrete particles. When the Navier-Stokes are simulated on digital computers, however, discrete approximations must again be made. These approximations, perhaps in the form of finite differences, bear little resemblance to the original system of discrete particles. Yet in the limit of a large number of discrete elements, they too should correspond to the continuum Navier-Stokes equations.

A wide variety of systems, with very different microscopic dynamics, in fact appear to follow the Navier-Stokes equations in the large-scale limit. Thus, for example, air and water, despite their very different molecular constitution, can both be described by the Navier-Stokes equations, albeit with different values of parameters such as viscosity.

In an attempt to devise the most efficient computational models for fluids, one may try to find the simplest microscopic dynamics that reproduces the Navier-Stokes equations in the macroscopic limit. Such models may correspond to optimal algorithms for determining the behavior of a fluid using a digital computer.

One class of computational models is based on a simple discrete idealization of molecular dynamics [6]. Particles move in discrete steps along the links of a fixed lattice, with each link supporting, for instance, at most one particle. The particles collide and scatter according to simple logical rules. The rules are arranged so as to conserve the total number of particles, and the total momentum carried by these particles. Fluid behavior can potentially be obtained in this system by considering the values of bulk quantities such as particle density or momentum density, averaged over a large lattice region.

Figure 2 shows some results obtained in this way. Detailed studies have demonstrated that many of the phenomena seen in actual fluid experiments can accurately be reproduced by this simple cellular automaton model. Figure 2 shows calculations of two-dimensional flow past a cylinder. The standard transition from steady flow to a regular vortex street is observed. Then at higher Reynolds numbers (dimensionless fluid flow rates) the vortex street is seen to become aperiodic, corresponding to the onset of turbulent behavior.

The cellular automaton method used in figure 2 may well be practical for many fluid dynamics computations. Through its close correspondence with the underlying physics of fluids, it is straightforward to include many physical effects and con-

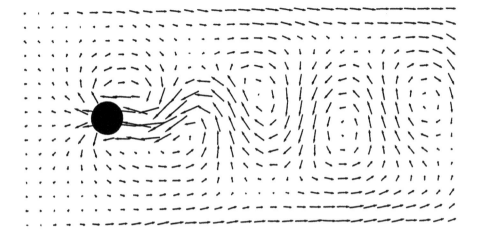

Figure 2. Fluid flow pattern obtained from a simple two-dimensional cellular automaton, simulated on a Connection Machine computer. The cellular automaton consists of 4096×2048 site hexagonal grid. Each site carries up to six discrete particles, which move and collide according to a simple discrete idealization of molecular dynamics. On a small scale, the particle motions appear random. But on a large scale, there is evidence that their average motion corresponds to that expected from a fluid which obeys the usual Navier-Stokes partial differential equations. In this figure, particles are injected on the left, leading to a net fluid motion from left to right. A circular obstacle is inserted in the fluid, and the resulting fluid velocities are computed by averaging individual particle velocities over 96×96 site regions. The velocities in the figure are shown transformed to the frame in which the obstacle is moving, and distant fluid is at rest. The simulation corresponds to a dimensionless Reynolds number around 100, and shows the formation of a "vortex street" behind the cylinder, as observed in physical experiments. The computations were performed with help from Bruce Nemnich and Jim Salem, on a Connection Machine computer with 65,536 Boolean processors. The results shown were obtained after 10^5 time steps.

straints. Thus, for example, solid objects with arbitrary shapes and, possibly, flexible boundaries can be treated easily. In our current implementation on a Connection Machine computer with 65,536 processors, lattices of size 4096×8192 can be updated at a rate of about 10^9 sites per second, allowing the fluid flow patterns around objects to be found interactively up to Reynolds numbers of several hundred. The readily scalable architecture of the Connection Machine computer makes much larger simulations with the same method quite feasible in the future.

At a theoretical level, cellular automaton fluid models can be analyzed by much the same methods of statistical mechanics as have been used in trying to derive the Navier-Stokes equations for physical fluids from the microscopic dynamics of real molecules. One approach is to use kinetic theory to derive transport equations for the average densities of particles with particular positions and directions (e.g., [8]). In the hydrodynamic limit, these microscopic average densities can be approximated through a Chapman-Enskog expansion in terms of macroscopic fluid densities and velocities. The resulting equations for these macroscopic quantities correspond closely with the usual Navier-Stokes equations. Just like a real fluid, however, the cellular automaton model contains definite higher-order corrections, not included in the Navier-Stokes equations. In addition, analytical methods provide only approximate values for parameters such as viscosity; accurate values must be obtained from explicit computer simulations.

A fundamental assumption of the kinetic theory method is that the microscopic configurations of particles can be specified purely in terms of probabilities, which are in turn determined by the values of averaged quantities. This is essentially equivalent to the assumption of thermodynamic equilibrium and is related to the fundamental principles of thermodynamics.

The Second Law of thermodynamics suggests that even if the initial configuration of particles is orderly, it will become progressively more disordered as a result of the motion and collisions of particles and will show, for example, an increasingly coarse-grained entropy. This phenomenon occurs if the evolution of the cellular automaton, even from "simple" initial conditions, yields behavior that is so complicated as to seem random for practical purposes.

Very simple examples of cellular automata are known in which such apparent randomness can be produced. Figure 3 shows a one-dimensional example [9,10]. Even starting from an initial state containing a single nonzero site, many features of the pattern produced, such as the sequences of values in the center vertical column, are sufficiently random that they pass standard statistical tests of randomness [9]. The cellular automaton evolution thus acts like a pseudorandom number generator: even though a simple seed is given, the algorithm yields sequences whose simple origins cannot be discerned. The evolution of the system thus effectively "encrypts" the initial data: given just the output sequence, it is very difficult to deduce the original seed. The cellular automaton of figure 3 can in fact be used as an efficient practical random sequence generator or stream encryption algorithm [11] (it is, for example, the primary pseudorandom generator used on the Connection Machine computer).

There are many mathematical systems which act in this way. It is, for example, easy to specify π, or to generate its digits. Yet once generated, the sequence of digits seems random for all practical purposes. Observations of this kind are related to the general conjecture of computational complexity theory (e.g., [12]) that $P \neq NP$. Computations that can be performed in polynomial time (P) seem to have inverses (which must be in the class NP) that require more than polynomial time, and probably often correspond to computations that are infeasible in practice.

Many mathematical models of physical processes probably show such behavior [10]. Even with simple initial data, they rapidly yield configurations which seem random for practical purposes. Such behavior may well be the basis for the widespread validity of the Second Law of thermodynamics. One of its important consequences is that a probabilistic or statistical description should indeed be valid for many systems such as cellular automaton fluid models. Such a description would depend only on macroscopic average variables. This may explain why different microscopic models

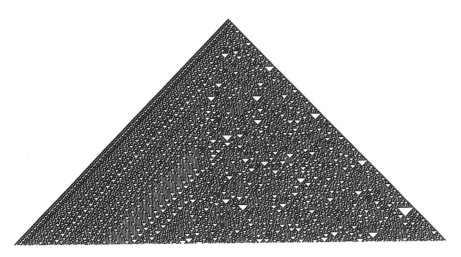

Figure 3. Pattern generated by a one-dimensional cellular automaton with two possible values at each site, and rule $a_i' = (a_{i-1} + a_i + a_{i+1} + a_i a_{i+1})$, starting from a single nonzero site. Despite the simplicity of its specification, many aspects of the pattern seem random. For example, the center column of site values passes all standard statistical tests of randomness. This cellular automaton illustrates the rather general phenomenon that simple processes can lead to complexity that is so great that many aspects of it seem random.

often yield the same macroscopic behavior. It is the basic reason that simple discrete dynamics can give essentially the same overall behavior as the full dynamics of physical molecules.

Statistical descriptions of cellular automaton fluid models are close in form to explicit finite difference approximations to partial differential equations. In both cases, each site on a grid carries a continuous variable which describes the average density and velocity of the fluid at that point. In practical computations with the finite difference method, this variable is typically represented directly as a floating-point number. In the cellular automaton method, the variable can be viewed as represented in a probabilistic or statistical fashion.

Following the usual development of statistical mechanics, a statistical description of a cellular automaton fluid can be obtained as an average over an ensemble of possible microscopic particle configurations. But an actual cellular automaton fluid simulation involves the evolution of just a single, specific, microscopic configuration. Nevertheless, following a fundamental assumption of statistical mechanics, one expects that suitable space or time averages of this specific configuration should yield results which are close to those obtained from averages over the whole ensemble.

This interpretation allows a comparison between cellular automata and discrete approximations to partial differential equations. In the latter case, ensemble average properties are considered, and their evolution is followed precisely. In the former case, just a single instance of the ensemble is considered, and macroscopic quantities are obtained as explicit averages over microscopic variables. If the fundamental assumptions of statistical mechanics are indeed valid, one expects that the cellular automaton method cannot fail to be more efficient than the finite difference method, because much of the information manipulated in the finite difference case is undoubtedly irrelevant to the macroscopic behavior of interest.

Some evidence for this conclusion comes from the fact that most fluid computations yield results which are accurate to, at most, the percent level. Yet in the finite difference approach, fluid velocities at individual grid points are typically stored to 16-decimal-digit accuracy. Presumably it is only the most significant few digits, and certain overall features of less significant digits, which affect the final results. In the cellular automaton method, all bits of information about microscopic particle configurations are equally important. The cellular automaton representation may thus be a more efficient encoding of the state of the fluid.

The cellular automaton approach to fluid dynamics is but one example of an expanding set of computational models which are based on the collective properties of large numbers of simple discrete components. Standard cellular automata with deterministic rules have been used as models for reaction-diffusion systems, dendritic growth processes, dynamic spin systems, aggregation processes, and many other phenomena (e.g., [3]). Intrinsically probabilistic rules can also be used, and their consequences deduced by Monte Carlo sampling. The resulting models have been used extensively in studying quantum fields and many other systems.

In practice the probabilistic elements of such models must be implemented on digital computers using pseudorandom number generation algorithms. The resulting complete computational model, including the pseudorandom number generator, must thus be entirely deterministic. And since even very simple deterministic cellular automata can yield a high degree of randomness, one expects that formally probabilistic models can be replaced by deterministic ones, often involving a smaller total number of steps. One example of this occurs for the Ising spin system model, which is conventionally studied by updating spins probabilistically, but for which a more efficient algorithm based on a simple deterministic cellular automaton is known [13].

In general, there may be many different cellular automaton models for any particular system. Although the microscopic rules are different, their large-scale or continuum behavior may be equivalent. In seeking the most efficient simulation algorithm for a particular system, one must find the "simplest" cellular automaton rules which yield the required large-scale behavior.

Most computational models are created by explicit construction. Like most computer programs, each step or feature of their construction is specifically designed to have particular known consequences, but in most cases, this methodology will not yield truly optimal programs. Instead, one may imagine defining particular goals or constraints, and then searching the space of possible programs for the optimal ones which achieve these goals (e.g., [14,15]). This approach is particularly promising for problems such as finding optimal cellular automation rules, in which the space of possible programs has a comparatively simple structure. Thus, for example, one may consider searching for the simplest cellular automaton rule which has a particular form of large-scale behavior. Typically the space of possible rules can be reduced by imposing certain constraints, such as microscopic conservation laws, but the suitability of any particular rule can usually be determined essentially only by explicit simulation. The randomness-generating rule of figure 3 was found by such a search-based method.

The problem of finding optimal automaton rules is analogous in many ways to problems such as the optimization of Boolean logic circuits, or the layout of large-scale integrated circuits. The overall goal is defined by the function to be implemented, but the most efficient circuits or rules can usually not be obtained by explicit construction. Instead one searches a large number of candidates, typically using a computer, and finds which of them is best.

Rather than performing an exhaustive search of possible circuits or rules, it is often better to use an iterative or adaptive procedure. One begins with a particular circuit or rule which has been constructed to satisfy the constraints that have been imposed. Then one makes a sequence of "moves" in the space of possible rules or circuits, with each move arranged so that the constraints are still satisfied. In the simplest cases, each move is chosen to yield a circuit or rule which is more optimal, or may be considered to have a lower "cost." But such a "gradient descent" method can find optima only when the "landscape" associated with the problem (whose height gives the cost for a circuit represented by a particular point) is essentially a smooth bowl. For many actual problems, the landscape seems closer to a "mountainous" or fractal one on which the gradient descent method will get stuck in local optima. Simulated annealing seems to be a more promising general technique for optimization in such cases [16]. With this method, randomness is introduced into the choice of moves. Initially, a high level of randomness is used, so that the moves are sensitive only to the gross features of the landscape. The randomness is progressively decreased, so that optimization is carried out with respect to smaller and smaller scale features of the landscape.

As one example of such "adaptive programming," I have recently been searching for the simplest one-dimensional cellular automaton rule that reproduces the diffusion equation in the large scale limit. (For another example, see ref. [17].) The rule must conserve a scalar additive quantity (analogous to particle number), but must generate randomness on a microscopic scale. In addition, the rules were chosen to be microscopically reversible, so that, analogous with real physical systems, the evolution of the system can be uniquely reversed. Figure 4 shows the behavior of a rule found by a search over a particular class of simple rules. Starting from a simple initial state, the rule generates progressively more random microscopic configurations. Although the simple initial conditions can in principle be recovered at any time by reversing the evolution, it becomes progressively more difficult to do so. As discussed above, this phenomenon may well illustrate the fundamental

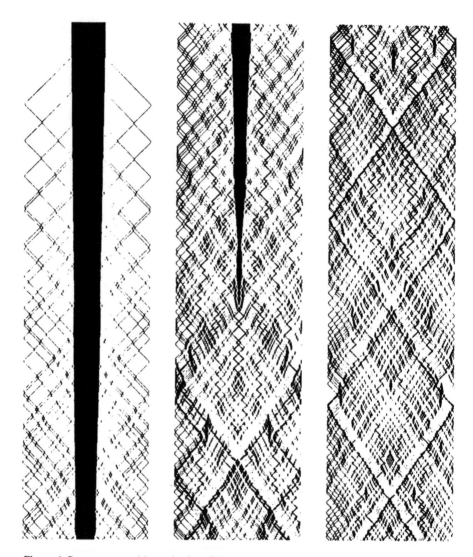

Figure 4. Pattern generated by a simple cellular automaton rule intended to mimic one-dimensional diffusion. Starting from a simple initial state, the reversible cellular automaton rule yields states that seem progressively more random. Such behavior corresponds to that expected from the Second Law of thermodynamics, and can form the basis for simple discrete cellular automata to show macroscopic average behavior which mimics continuum phenomena.

basis for the Second Law of thermodynamics. With the rule of figure 4, macroscopic average densities should follow the diffusion equation. As a result, slow spatial variations in density are, for example, damped on average according to the diffusion equation.

508

In this article I have discussed some new directions for computational modeling. The fundamental principle is that the models considered should be as suitable as possible for implementation on digital computers. It is then a matter of scientific analysis to determine whether such models can reproduce the behavior seen in physical and other systems. Such analysis has now been carried out in several cases, and the results are very encouraging.

References

1. S. Wolfram, "Computer software in science and mathematics," *Scientific American* (September 1984).

2. A. Sommerfeld, *Partial differential equations of mathematical physics* (Academic Press, 1955).

3. S. Wolfram (ed.), *Theory and applications of cellular automata* (World Scientific, Singapore, 1986).

4. S. Wolfram, "Cellular automata as models of complexity," *Nature* 311 (1984) 419.

5. W. D. Hillis, *The Connection Machine* (MIT Press, 1985).

6. U. Frisch, B. Hasslacher and Y. Pomeau, "Lattice gas automata for the Navier-Stokes equation," *Physical Review Letters* 56 (1986) 1505 [reprinted in ref. 3]; J. Salem and S. Wolfram, "Thermodynamics and hydrodynamics with cellular automata," in ref. 3; S. Wolfram, "Cellular automaton fluids 1: Basic theory," *Journal of Statistical Physics* 45 (1986) 471–526; B. Nemnich and S. Wolfram, "Cellular automaton fluids 2: Basic phenomenology," Center for Complex Systems Research/Thinking Machines Corporation Report, to appear.

7. D. J. Tritton, *Physical fluid dynamics* (Van Nostrand, 1977).

8. A. Sommerfeld, *Thermodynamics and statistical mechanics* (Academic Press, 1955).

9. S. Wolfram, "Random sequence generation by cellular automata," *Advances in Applied Mathematics* 7 (1986) 123.

10. S. Wolfram, "Origins of randomness in physical systems," *Physical Review Letters* 55 (1985) 449.

11. S. Wolfram, "Cryptography with cellular automata," in *Proceedings of CRYPTO '85* (Santa Barbara, August 1985).

12. J. Hopcroft and J. Ullman, *Introduction to automata theory, languages, and computation* (Addison-Wesley, 1979).

13. M. Creutz, "Deterministic Ising dynamics," *Annals of Physics* 167 (1986) 62; Y. Pomeau, "Invariant in cellular automata," *Journal of Physics* A17 (1984) L415.

14. D. Farmer, A. Lapedes, N. Packard, and B. Wendroff (eds.), *Evolution, games and learning*, *Physica D*, 1986.

15. S. Wolfram, "Approaches to complexity engineering," in ref. 2 and ref. 14.

16. S. Kirkpatrick et al., "Global optimization by simulated annealing," *Science* 220 (1986) 671.

17. N. Packard, "Adaptation in the space of cellular automaton rules," Center for Complex Systems Research Report, in preparation.

509

PART THREE

Appendices

Tables of Cellular Automaton Properties

1986

Introduction

This appendix gives tables of properties of one-dimensional cellular automata with two possible values at each site ($k = 2$), and with rules depending on nearest neighbours ($r = 1$). These cellular automata are some of the simplest that can be constructed. Yet they are already capable of a great diversity of highly complex behaviour. The tables in this appendix attempt to capture some of this behaviour, both pictorially and numerically.

There are 256 possible rules for $k = 2$, $r = 1$ cellular automata. Table 1 gives forms for these rules, together with simple equivalences among them.

Tables 2 and 3 show patterns produced by evolution according to all possible inequivalent rules, starting from "typical" disordered or random initial conditions. Several general classes of qualitative behaviour are seen (see pages 115–157 in this book):

1. A fixed, homogeneous, state is eventually reached (e.g. rules 0, 8, 136).

2. A pattern consisting of separated periodic regions is produced (e.g. rules 4, 37, 56, 73).

3. A chaotic, aperiodic, pattern is produced (e.g. rules 18, 45, 146).

4. Complex, localized structures are generated (e.g. rule 110). (This behaviour is clearly visible in the pictures of table 15.)

Much of the data in this appendix can be understood in terms of this classification.

The patterns produced with a particular rule by evolution from different disordered initial states are qualitatively similar. Nevertheless, changes in initial conditions can lead to detailed changes in the configurations produced. Table 4 shows the pattern of

Originally published in *Theory and Applications of Cellular Automata*, World Scientific Publishing Co. Ltd., pages 485–557 (1986).

differences produced by single-site changes in initial conditions. For class 1 rules, the changes always die out. For class 2 rules, they may persist, but remain localized. Class 3 rules, however, show "instability": small changes in initial conditions can lead to an ever-expanding region of differences. "Information" on the initial state thus propagates, typically at a fixed speed, through the cellular automaton. In class 4 cellular automata, such information transmission occurs irregularly, through motion of specific localized structures.

Table 6 gives the values of some statistical quantities which characterize some of the behaviour seen in tables 2, 3 and 4. The definitions of entropies and Lyapunov exponents for cellular automata (see pages 115–157 in this book) are closely analogous to those for conventional continuous dynamical systems.

Tables 2, 3, 4 and 6 concern the generic behaviour of cellular automata with "typical" disordered initial conditions. The generation of complexity in cellular automata is however perhaps more clearly illustrated by evolution from particular, simple, initial conditions, as in table 5. With such initial conditions, some cellular automaton rules yield simple or regular patterns. But other rules yield highly complex patterns, which seem in many respects random.

Tables 2 through 6 suggest that many different $k = 2$, $r = 1$ cellular automata exhibit similar behaviour. Table 1 gives some simple equivalences between rules. Table 7 gives equivalences arising from more complex transformations. Often different regions in a cellular automaton will form "domains" which show different equivalences.

Table 8 gives further relations between rules, in the form of factorizations which express one rule as compositions of others.

An important feature of cellular automata is their capability for "self organization". Even starting from arbitrary disordered or random initial conditions, their time evolution can pick out particular "ordered" states. Tables 9 through 11 give mathematical characterizations of the sets of configurations that can occur in the evolution of $k = 2$, $r = 1$ cellular automata. Table 9 concerns blocks of site values which are filtered out by the cellular automaton evolution.

The complete set of configurations produced after any finite number of time steps can be described in terms of regular formal languages (see pages 159–202 in this book). Tables 10 and 11 give the values of quantities which characterize the certain aspects of the "complexity" of these languages.

The behaviour of class 3 and 4 cellular automata often seems to be so complex that its outcome cannot be determined except by essentially performing a direct simulation. Tables 10 and 11 may provide some quantitative basis for this supposition. Table 12 gives a more direct measure of the difficulty of computing the outcome of cellular automaton evolution in the context of a simple computational model involving Boolean functions.

The results for most of the tables here are for cellular automata on lattices with an infinite number of sites. Tables 13 and 14 give some of the more complete results that can be obtained for cellular automata on finite lattices (or with spatially periodic configurations). Table 13 shows fragments of the state transition diagrams which describe the global evolution of finite cellular automata. Table 14 plots some of their overall properties.

Many of the $k = 2, r = 1$ cellular automata show highly complex behaviour. Such behaviour is probably most evident in rule 110. Table 15 gives some properties of the particle-like structures which are found in this rule. One suspects that with appropriate combinations of these structures, it should be possible to perform universal computation.

The final table shows patterns produced by reversible generalizations of the standard $k = 2, r = 1$ cellular automata. Qualitatively similar behaviour is again seen.

It is remarkable that with such simple construction, the $k = 2, r = 1$ cellular automata can show such complex behaviour. The tables in this appendix give some first attempts at characterizing and quantifying this behaviour. Much, however, still remains to be done.

Table 1: Rule Forms and Equivalences

rule number			boolean expression	dep	equivalent rules			min
dec	binary	hex			conj	refl	c.r.	
0	00000000	00	0	———	255	0	255	0
1	00000001	01	$(\bar{a}_{-1}\bar{a}_0\bar{a}_1)$	●●●	127	1	127	1
2	00000010	02	$(\bar{a}_{-1}\bar{a}_0 a_1)$	●●●	191	16	247	2
3	00000011	03	$(\bar{a}_{-1}\bar{a}_0)$	●●—	63	17	119	3
4	00000100	04	$(\bar{a}_{-1}a_0\bar{a}_1)$	●●●	223	4	223	4
5	00000101	05	$(\bar{a}_{-1}\bar{a}_1)$	●—●	95	5	95	5
6	00000110	06	$(\bar{a}_{-1}a_0\bar{a}_1) + (\bar{a}_{-1}\bar{a}_0 a_1)$	●●●	159	20	215	6
7	00000111	07	$(\bar{a}_{-1}\bar{a}_1) + (\bar{a}_{-1}\bar{a}_0)$	●●●	31	21	87	7
8	00001000	08	$(\bar{a}_{-1}a_0 a_1)$	●●●	239	64	253	8
9	00001001	09	$(\bar{a}_{-1}\bar{a}_0\bar{a}_1) + (\bar{a}_{-1}a_0 a_1)$	●●●	111	65	125	9
10	00001010	0a	$(\bar{a}_{-1}a_1)$	●—●	175	80	245	10
11	00001011	0b	$(\bar{a}_{-1}\bar{a}_0) + (\bar{a}_{-1}a_1)$	●●●	47	81	117	11
12	00001100	0c	$(\bar{a}_{-1}a_0)$	●●—	207	68	221	12
13	00001101	0d	$(\bar{a}_{-1}\bar{a}_1) + (\bar{a}_{-1}a_0)$	●●●	79	69	93	13
14	00001110	0e	$(\bar{a}_{-1}a_0) + (\bar{a}_{-1}a_1)$	●●●	143	84	213	14
15	00001111	0f	$(\bar{a}_{-1})$	○——	15	85	85	15
16	00010000	10	$(a_{-1}\bar{a}_0\bar{a}_1)$	●●●	247	2	191	2
17	00010001	11	$(\bar{a}_0\bar{a}_1)$	—●●	119	3	63	3
18	00010010	12	$(a_{-1}\bar{a}_0\bar{a}_1) + (\bar{a}_{-1}\bar{a}_0 a_1)$	●●●	183	18	183	18
19	00010011	13	$(\bar{a}_0\bar{a}_1) + (\bar{a}_{-1}\bar{a}_0)$	●●●	55	19	55	19
20	00010100	14	$(a_{-1}\bar{a}_0\bar{a}_1) + (\bar{a}_{-1}a_0\bar{a}_1)$	●●●	215	6	159	6
21	00010101	15	$(\bar{a}_0\bar{a}_1) + (\bar{a}_{-1}\bar{a}_1)$	●●●	87	7	31	7
22	00010110	16	$(a_{-1}\bar{a}_0\bar{a}_1) + (\bar{a}_{-1}a_0\bar{a}_1) + (\bar{a}_{-1}\bar{a}_0 a_1)$	●●●	151	22	151	22
23	00010111	17	$(\bar{a}_0\bar{a}_1) + (\bar{a}_{-1}\bar{a}_1) + (\bar{a}_{-1}\bar{a}_0)$	●●●	23	23	23	23
24	00011000	18	$(a_{-1}\bar{a}_0\bar{a}_1) + (\bar{a}_{-1}a_0 a_1)$	●●●	231	66	189	24
25	00011001	19	$(\bar{a}_{-1}a_0 a_1) + (\bar{a}_0\bar{a}_1)$	●●●	103	67	61	25
26	00011010	1a	$(a_{-1}\bar{a}_0\bar{a}_1) + (\bar{a}_{-1}a_1)$	●●●	167	82	181	26
27	00011011	1b	$(\bar{a}_0\bar{a}_1) + (\bar{a}_{-1}a_1)$	●●●	39	83	53	27
28	00011100	1c	$(a_{-1}\bar{a}_0\bar{a}_1) + (\bar{a}_{-1}a_0)$	●●●	199	70	157	28
29	00011101	1d	$(\bar{a}_0\bar{a}_1) + (\bar{a}_{-1}a_0)$	●●●	71	71	29	29
30	00011110	1e	$(a_{-1}\bar{a}_0\bar{a}_1) + (\bar{a}_{-1}a_0) + (\bar{a}_{-1}a_1)$	○●●	135	86	149	30
31	00011111	1f	$(\bar{a}_0\bar{a}_1) + (\bar{a}_{-1})$	●●●	7	87	21	7
32	00100000	20	$(a_{-1}\bar{a}_0 a_1)$	●●●	251	32	251	32
33	00100001	21	$(\bar{a}_{-1}\bar{a}_0\bar{a}_1) + (a_{-1}\bar{a}_0 a_1)$	●●●	123	33	123	33
34	00100010	22	$(\bar{a}_0 a_1)$	—●●	187	48	243	34
35	00100011	23	$(\bar{a}_{-1}\bar{a}_0) + (\bar{a}_0 a_1)$	●●●	59	49	115	35
36	00100100	24	$(\bar{a}_{-1}a_0\bar{a}_1) + (a_{-1}\bar{a}_0 a_1)$	●●●	219	36	219	36
37	00100101	25	$(a_{-1}\bar{a}_0 a_1) + (\bar{a}_{-1}\bar{a}_1)$	●●●	91	37	91	37
38	00100110	26	$(\bar{a}_{-1}a_0\bar{a}_1) + (\bar{a}_0 a_1)$	●●●	155	52	211	38
39	00100111	27	$(\bar{a}_{-1}\bar{a}_1) + (\bar{a}_0 a_1)$	●●●	27	53	83	27
40	00101000	28	$(a_{-1}\bar{a}_0 a_1) + (\bar{a}_{-1}a_0 a_1)$	●●●	235	96	249	40
41	00101001	29	$(\bar{a}_{-1}\bar{a}_0\bar{a}_1) + (a_{-1}\bar{a}_0 a_1) + (\bar{a}_{-1}a_0 a_1)$	●●●	107	97	121	41
42	00101010	2a	$(\bar{a}_0 a_1) + (\bar{a}_{-1}a_1)$	●●●	171	112	241	42
43	00101011	2b	$(\bar{a}_{-1}\bar{a}_0) + (\bar{a}_0 a_1) + (\bar{a}_{-1}a_1)$	●●●	43	113	113	43
44	00101100	2c	$(a_{-1}\bar{a}_0 a_1) + (\bar{a}_{-1}a_0)$	●●●	203	100	217	44
45	00101101	2d	$(a_{-1}\bar{a}_0 a_1) + (\bar{a}_{-1}\bar{a}_1) + (\bar{a}_{-1}a_0)$	○●●	75	101	89	45
46	00101110	2e	$(\bar{a}_{-1}a_0) + (\bar{a}_0 a_1)$	●●●	139	116	209	46

rule number			boolean expression	dep	equivalent rules			min
dec	binary	hex			conj	refl	c.r.	
47	00101111	2f	$(\bar{a}_0 a_1) + (\bar{a}_{-1})$	●●●	11	117	81	11
48	00110000	30	$(a_{-1}\bar{a}_0)$	●●−	243	34	187	34
49	00110001	31	$(\bar{a}_0\bar{a}_1) + (a_{-1}\bar{a}_0)$	●●●	115	35	59	35
50	00110010	32	$(a_{-1}\bar{a}_0) + (\bar{a}_0 a_1)$	●●●	179	50	179	50
51	00110011	33	$(\bar{a}_0)$	−○−	51	51	51	51
52	00110100	34	$(\bar{a}_{-1}a_0\bar{a}_1) + (a_{-1}\bar{a}_0)$	●●●	211	38	155	38
53	00110101	35	$(\bar{a}_{-1}\bar{a}_1) + (a_{-1}\bar{a}_0)$	●●●	83	39	27	27
54	00110110	36	$(\bar{a}_{-1}a_0\bar{a}_1) + (a_{-1}\bar{a}_0) + (\bar{a}_0 a_1)$	●○●	147	54	147	54
55	00110111	37	$(\bar{a}_{-1}\bar{a}_1) + (\bar{a}_0)$	●●●	19	55	19	19
56	00111000	38	$(\bar{a}_{-1}a_0 a_1) + (a_{-1}\bar{a}_0)$	●●●	227	98	185	56
57	00111001	39	$(\bar{a}_{-1}a_0 a_1) + (\bar{a}_0\bar{a}_1) + (a_{-1}\bar{a}_0)$	●○●	99	99	57	57
58	00111010	3a	$(a_{-1}\bar{a}_0) + (\bar{a}_{-1}a_1)$	●●●	163	114	177	58
59	00111011	3b	$(\bar{a}_{-1}a_1) + (\bar{a}_0)$	●●●	35	115	49	35
60	00111100	3c	$(a_{-1}\bar{a}_0) + (\bar{a}_{-1}a_0)$	○○−	195	102	153	60
61	00111101	3d	$(\bar{a}_{-1}\bar{a}_1) + (a_{-1}\bar{a}_0) + (\bar{a}_{-1}a_0)$	●●●	67	103	25	25
62	00111110	3e	$(\bar{a}_{-1}a_1) + (a_{-1}\bar{a}_0) + (\bar{a}_{-1}a_0)$	●●●	131	118	145	62
63	00111111	3f	$(\bar{a}_0) + (\bar{a}_{-1})$	●●−	3	119	17	3
64	01000000	40	$(a_{-1}a_0\bar{a}_1)$	●●●	253	8	239	8
65	01000001	41	$(\bar{a}_{-1}\bar{a}_0\bar{a}_1) + (a_{-1}a_0\bar{a}_1)$	●●●	125	9	111	9
66	01000010	42	$(a_{-1}a_0\bar{a}_1) + (\bar{a}_{-1}\bar{a}_0 a_1)$	●●●	189	24	231	24
67	01000011	43	$(a_{-1}a_0\bar{a}_1) + (\bar{a}_{-1}\bar{a}_0)$	●●●	61	25	103	25
68	01000100	44	$(a_0\bar{a}_1)$	−●●	221	12	207	12
69	01000101	45	$(\bar{a}_{-1}\bar{a}_1) + (a_0\bar{a}_1)$	●●●	93	13	79	13
70	01000110	46	$(\bar{a}_{-1}\bar{a}_0 a_1) + (a_0\bar{a}_1)$	●●●	157	28	199	28
71	01000111	47	$(a_0\bar{a}_1) + (\bar{a}_{-1}\bar{a}_0)$	●●●	29	29	71	29
72	01001000	48	$(a_{-1}a_0\bar{a}_1) + (\bar{a}_{-1}a_0 a_1)$	●●●	237	72	237	72
73	01001001	49	$(\bar{a}_{-1}\bar{a}_0\bar{a}_1) + (a_{-1}a_0\bar{a}_1) + (\bar{a}_{-1}a_0 a_1)$	●●●	109	73	109	73
74	01001010	4a	$(a_{-1}a_0\bar{a}_1) + (\bar{a}_{-1}a_1)$	●●●	173	88	229	74
75	01001011	4b	$(a_{-1}a_0\bar{a}_1) + (\bar{a}_{-1}\bar{a}_0) + (\bar{a}_{-1}a_1)$	○●●	45	89	101	45
76	01001100	4c	$(a_0\bar{a}_1) + (\bar{a}_{-1}a_0)$	●●●	205	76	205	76
77	01001101	4d	$(\bar{a}_{-1}\bar{a}_1) + (a_0\bar{a}_1) + (\bar{a}_{-1}a_0)$	●●●	77	77	77	77
78	01001110	4e	$(a_0\bar{a}_1) + (\bar{a}_{-1}a_1)$	●●●	141	92	197	78
79	01001111	4f	$(a_0\bar{a}_1) + (\bar{a}_{-1})$	●●●	13	93	69	13
80	01010000	50	$(a_{-1}\bar{a}_1)$	●−●	245	10	175	10
81	01010001	51	$(\bar{a}_0\bar{a}_1) + (a_{-1}\bar{a}_1)$	●●●	117	11	47	11
82	01010010	52	$(\bar{a}_{-1}\bar{a}_0 a_1) + (a_{-1}\bar{a}_1)$	●●●	181	26	167	26
83	01010011	53	$(a_{-1}\bar{a}_1) + (\bar{a}_{-1}\bar{a}_0)$	●●●	53	27	39	27
84	01010100	54	$(a_{-1}\bar{a}_1) + (a_0\bar{a}_1)$	●●●	213	14	143	14
85	01010101	55	$(\bar{a}_1)$	−−○	85	15	15	15
86	01010110	56	$(\bar{a}_{-1}\bar{a}_0 a_1) + (a_{-1}\bar{a}_1) + (a_0\bar{a}_1)$	●●○	149	30	135	30
87	01010111	57	$(\bar{a}_{-1}\bar{a}_0) + (\bar{a}_1)$	●●●	21	31	7	7
88	01011000	58	$(\bar{a}_{-1}a_0 a_1) + (a_{-1}\bar{a}_1)$	●●●	229	74	173	74
89	01011001	59	$(\bar{a}_{-1}a_0 a_1) + (\bar{a}_0\bar{a}_1) + (a_{-1}\bar{a}_1)$	●●○	101	75	45	45
90	01011010	5a	$(a_{-1}\bar{a}_1) + (\bar{a}_{-1}a_1)$	○−○	165	90	165	90
91	01011011	5b	$(\bar{a}_{-1}\bar{a}_0) + (a_{-1}\bar{a}_1) + (\bar{a}_{-1}a_1)$	●●●	37	91	37	37
92	01011100	5c	$(a_{-1}\bar{a}_1) + (\bar{a}_{-1}a_0)$	●●●	197	78	141	78
93	01011101	5d	$(\bar{a}_{-1}a_0) + (\bar{a}_1)$	●●●	69	79	13	13
94	01011110	5e	$(\bar{a}_{-1}a_0) + (a_{-1}\bar{a}_1) + (\bar{a}_{-1}a_1)$	●●●	133	94	133	94
95	01011111	5f	$(\bar{a}_1) + (\bar{a}_{-1})$	●−●	5	95	5	5

rule number			boolean expression	dep	equivalent rules			min
dec	binary	hex			conj	refl	c.r.	
96	01100000	60	$(a_{-1}a_0\bar{a}_1) + (a_{-1}\bar{a}_0a_1)$	•••	249	40	235	40
97	01100001	61	$(\bar{a}_{-1}\bar{a}_0\bar{a}_1) + (a_{-1}a_0\bar{a}_1) + (a_{-1}\bar{a}_0a_1)$	•••	121	41	107	41
98	01100010	62	$(a_{-1}a_0\bar{a}_1) + (\bar{a}_0a_1)$	•••	185	56	227	56
99	01100011	63	$(a_{-1}a_0\bar{a}_1) + (\bar{a}_{-1}\bar{a}_0) + (\bar{a}_0a_1)$	•○•	57	57	99	57
100	01100100	64	$(a_{-1}\bar{a}_0a_1) + (a_0\bar{a}_1)$	•••	217	44	203	44
101	01100101	65	$(a_{-1}\bar{a}_0a_1) + (\bar{a}_{-1}\bar{a}_1) + (a_0\bar{a}_1)$	••○	89	45	75	45
102	01100110	66	$(a_0\bar{a}_1) + (\bar{a}_0a_1)$	−○○	153	60	195	60
103	01100111	67	$(\bar{a}_{-1}\bar{a}_0) + (a_0\bar{a}_1) + (\bar{a}_0a_1)$	•••	25	61	67	25
104	01101000	68	$(a_{-1}a_0\bar{a}_1) + (a_{-1}\bar{a}_0a_1) + (\bar{a}_{-1}a_0a_1)$	•••	233	104	233	104
105	01101001	69	$(\bar{a}_{-1}\bar{a}_0\bar{a}_1) + (a_{-1}a_0\bar{a}_1) + (a_{-1}\bar{a}_0a_1) + (\bar{a}_{-1}a_0a_1)$	○○○	105	105	105	105
106	01101010	6a	$(a_{-1}a_0\bar{a}_1) + (\bar{a}_0a_1) + (\bar{a}_{-1}a_1)$	••○	169	120	225	106
107	01101011	6b	$(a_{-1}a_0\bar{a}_1) + (\bar{a}_{-1}\bar{a}_0) + (\bar{a}_0a_1) + (\bar{a}_{-1}a_1)$	•••	41	121	97	41
108	01101100	6c	$(a_{-1}\bar{a}_0a_1) + (a_0\bar{a}_1) + (\bar{a}_{-1}a_0)$	•○•	201	108	201	108
109	01101101	6d	$(a_{-1}\bar{a}_0a_1) + (\bar{a}_{-1}\bar{a}_1) + (a_0\bar{a}_1) + (\bar{a}_{-1}a_0)$	•••	73	109	73	73
110	01101110	6e	$(\bar{a}_{-1}a_0) + (a_0\bar{a}_1) + (\bar{a}_0a_1)$	•••	137	124	193	110
111	01101111	6f	$(a_0\bar{a}_1) + (\bar{a}_0a_1) + (\bar{a}_{-1})$	•••	9	125	65	9
112	01110000	70	$(a_{-1}\bar{a}_1) + (a_{-1}\bar{a}_0)$	•••	241	42	171	42
113	01110001	71	$(\bar{a}_0\bar{a}_1) + (a_{-1}\bar{a}_1) + (a_{-1}\bar{a}_0)$	•••	113	43	43	43
114	01110010	72	$(a_{-1}\bar{a}_1) + (\bar{a}_0a_1)$	•••	177	58	163	58
115	01110011	73	$(a_{-1}\bar{a}_1) + (\bar{a}_0)$	•••	49	59	35	35
116	01110100	74	$(a_0\bar{a}_1) + (a_{-1}\bar{a}_0)$	•••	209	46	139	46
117	01110101	75	$(a_{-1}\bar{a}_0) + (\bar{a}_1)$	•••	81	47	11	11
118	01110110	76	$(a_{-1}\bar{a}_0) + (a_0\bar{a}_1) + (\bar{a}_0a_1)$	•••	145	62	131	62
119	01110111	77	$(\bar{a}_1) + (\bar{a}_0)$	−••	17	63	3	3
120	01111000	78	$(\bar{a}_{-1}a_0a_1) + (a_{-1}\bar{a}_1) + (a_{-1}\bar{a}_0)$	○••	225	106	169	106
121	01111001	79	$(\bar{a}_{-1}a_0a_1) + (\bar{a}_0\bar{a}_1) + (a_{-1}\bar{a}_1) + (a_{-1}\bar{a}_0)$	•••	97	107	41	41
122	01111010	7a	$(a_{-1}\bar{a}_0) + (a_0\bar{a}_1) + (\bar{a}_{-1}a_1)$	•••	161	122	161	122
123	01111011	7b	$(a_{-1}\bar{a}_1) + (\bar{a}_{-1}a_1) + (\bar{a}_0)$	•••	33	123	33	33
124	01111100	7c	$(a_{-1}\bar{a}_1) + (a_{-1}\bar{a}_0) + (\bar{a}_{-1}a_0)$	•••	193	110	137	110
125	01111101	7d	$(a_{-1}\bar{a}_0) + (\bar{a}_{-1}a_0) + (\bar{a}_1)$	•••	65	111	9	9
126	01111110	7e	$(a_{-1}\bar{a}_1) + (\bar{a}_0a_1) + (\bar{a}_{-1}a_0)$	•••	129	126	129	126
127	01111111	7f	$(\bar{a}_1) + (\bar{a}_0) + (\bar{a}_{-1})$	•••	1	127	1	1
128	10000000	80	$(a_{-1}a_0a_1)$	•••	254	128	254	128
129	10000001	81	$(\bar{a}_{-1}\bar{a}_0\bar{a}_1) + (a_{-1}a_0a_1)$	•••	126	129	126	126
130	10000010	82	$(\bar{a}_{-1}\bar{a}_0a_1) + (a_{-1}a_0a_1)$	•••	190	144	246	130
131	10000011	83	$(a_{-1}a_0a_1) + (\bar{a}_{-1}\bar{a}_0)$	•••	62	145	118	62
132	10000100	84	$(\bar{a}_{-1}a_0\bar{a}_1) + (a_{-1}a_0a_1)$	•••	222	132	222	132
133	10000101	85	$(a_{-1}a_0a_1) + (\bar{a}_{-1}\bar{a}_1)$	•••	94	133	94	94
134	10000110	86	$(\bar{a}_{-1}a_0\bar{a}_1) + (\bar{a}_{-1}\bar{a}_0a_1) + (a_{-1}a_0a_1)$	•••	158	148	214	134
135	10000111	87	$(a_{-1}a_0a_1) + (\bar{a}_{-1}\bar{a}_1) + (\bar{a}_{-1}\bar{a}_0)$	○••	30	149	86	30
136	10001000	88	(a_0a_1)	−••	238	192	252	136
137	10001001	89	$(\bar{a}_{-1}\bar{a}_0\bar{a}_1) + (a_0a_1)$	•••	110	193	124	110
138	10001010	8a	$(\bar{a}_{-1}a_1) + (a_0a_1)$	•••	174	208	244	138
139	10001011	8b	$(\bar{a}_{-1}\bar{a}_0) + (a_0a_1)$	•••	46	209	116	46
140	10001100	8c	$(\bar{a}_{-1}a_0) + (a_0a_1)$	•••	206	196	220	140
141	10001101	8d	$(\bar{a}_{-1}\bar{a}_1) + (a_0a_1)$	•••	78	197	92	78
142	10001110	8e	$(\bar{a}_{-1}a_0) + (\bar{a}_{-1}a_1) + (a_0a_1)$	•••	142	212	212	142
143	10001111	8f	$(a_0a_1) + (\bar{a}_{-1})$	•••	14	213	84	14
144	10010000	90	$(a_{-1}\bar{a}_0\bar{a}_1) + (a_{-1}a_0a_1)$	•••	246	130	190	130
145	10010001	91	$(a_{-1}a_0a_1) + (\bar{a}_0\bar{a}_1)$	•••	118	131	62	62

| rule number | | | boolean expression | dep | equivalent rules | | | min |
dec	binary	hex			conj	refl	c.r.	
146	10010010	92	$(a_{-1}\bar{a}_0\bar{a}_1) + (\bar{a}_{-1}\bar{a}_0 a_1) + (a_{-1}a_0 a_1)$	●●●	182	146	182	146
147	10010011	93	$(a_{-1}a_0 a_1) + (\bar{a}_0\bar{a}_1) + (\bar{a}_{-1}\bar{a}_0)$	●○●	54	147	54	54
148	10010100	94	$(a_{-1}\bar{a}_0\bar{a}_1) + (\bar{a}_{-1}a_0\bar{a}_1) + (a_{-1}a_0 a_1)$	●●●	214	134	158	134
149	10010101	95	$(a_{-1}a_0 a_1) + (\bar{a}_0\bar{a}_1) + (\bar{a}_{-1}\bar{a}_1)$	●●○	86	135	30	30
150	10010110	96	$(a_{-1}\bar{a}_0\bar{a}_1) + (\bar{a}_{-1}a_0\bar{a}_1) + (\bar{a}_{-1}\bar{a}_0 a_1) + (a_{-1}a_0 a_1)$	○○○	150	150	150	150
151	10010111	97	$(a_{-1}a_0 a_1) + (\bar{a}_0\bar{a}_1) + (\bar{a}_{-1}\bar{a}_1) + (\bar{a}_{-1}\bar{a}_0)$	●●●	22	151	22	22
152	10011000	98	$(a_{-1}\bar{a}_0\bar{a}_1) + (a_0 a_1)$	●●●	230	194	188	152
153	10011001	99	$(\bar{a}_0\bar{a}_1) + (a_0 a_1)$	–○○	102	195	60	60
154	10011010	9a	$(a_{-1}\bar{a}_0\bar{a}_1) + (\bar{a}_{-1}a_1) + (a_0 a_1)$	●●○	166	210	180	154
155	10011011	9b	$(\bar{a}_{-1}\bar{a}_0) + (\bar{a}_0\bar{a}_1) + (a_0 a_1)$	●●●	38	211	52	38
156	10011100	9c	$(a_{-1}\bar{a}_0\bar{a}_1) + (\bar{a}_{-1}a_0) + (a_0 a_1)$	●○●	198	198	156	156
157	10011101	9d	$(\bar{a}_{-1}a_0) + (\bar{a}_0\bar{a}_1) + (a_0 a_1)$	●●●	70	199	28	28
158	10011110	9e	$(a_{-1}\bar{a}_0\bar{a}_1) + (\bar{a}_{-1}a_0) + (\bar{a}_{-1}a_1) + (a_0 a_1)$	●●●	134	214	148	134
159	10011111	9f	$(\bar{a}_0\bar{a}_1) + (a_0 a_1) + (\bar{a}_{-1})$	●●●	6	215	20	6
160	10100000	a0	$(a_{-1}a_1)$	●–●	250	160	250	160
161	10100001	a1	$(\bar{a}_{-1}\bar{a}_0\bar{a}_1) + (a_{-1}a_1)$	●●●	122	161	122	122
162	10100010	a2	$(\bar{a}_0 a_1) + (a_{-1}a_1)$	●●●	186	176	242	162
163	10100011	a3	$(\bar{a}_{-1}\bar{a}_0) + (a_{-1}a_1)$	●●●	58	177	114	58
164	10100100	a4	$(\bar{a}_{-1}a_0\bar{a}_1) + (a_{-1}a_1)$	●●●	218	164	218	164
165	10100101	a5	$(\bar{a}_{-1}\bar{a}_1) + (a_{-1}a_1)$	○–○	90	165	90	90
166	10100110	a6	$(\bar{a}_{-1}a_0\bar{a}_1) + (\bar{a}_0 a_1) + (a_{-1}a_1)$	●●○	154	180	210	154
167	10100111	a7	$(\bar{a}_{-1}\bar{a}_0) + (\bar{a}_{-1}\bar{a}_1) + (a_{-1}a_1)$	●●●	26	181	82	26
168	10101000	a8	$(a_{-1}a_1) + (a_0 a_1)$	●●●	234	224	248	168
169	10101001	a9	$(\bar{a}_{-1}\bar{a}_0\bar{a}_1) + (a_{-1}a_1) + (a_0 a_1)$	●●○	106	225	120	106
170	10101010	aa	(a_1)	––○	170	240	240	170
171	10101011	ab	$(\bar{a}_{-1}\bar{a}_0) + (a_1)$	●●●	42	241	112	42
172	10101100	ac	$(\bar{a}_{-1}a_0) + (a_{-1}a_1)$	●●●	202	228	216	172
173	10101101	ad	$(\bar{a}_{-1}a_0) + (\bar{a}_{-1}\bar{a}_1) + (a_{-1}a_1)$	●●●	74	229	88	74
174	10101110	ae	$(\bar{a}_{-1}a_0) + (a_1)$	●●●	138	244	208	138
175	10101111	af	$(\bar{a}_{-1}) + (a_1)$	●–●	10	245	80	10
176	10110000	b0	$(a_{-1}\bar{a}_0) + (a_{-1}a_1)$	●●●	242	162	186	162
177	10110001	b1	$(\bar{a}_0\bar{a}_1) + (a_{-1}a_1)$	●●●	114	163	58	58
178	10110010	b2	$(a_{-1}\bar{a}_0) + (\bar{a}_0 a_1) + (a_{-1}a_1)$	●●●	178	178	178	178
179	10110011	b3	$(a_{-1}a_1) + (\bar{a}_0)$	●●●	50	179	50	50
180	10110100	b4	$(\bar{a}_{-1}a_0\bar{a}_1) + (a_{-1}\bar{a}_0) + (a_{-1}a_1)$	○●●	210	166	154	154
181	10110101	b5	$(a_{-1}\bar{a}_0) + (\bar{a}_{-1}\bar{a}_1) + (a_{-1}a_1)$	●●●	82	167	26	26
182	10110110	b6	$(\bar{a}_{-1}a_0\bar{a}_1) + (a_{-1}\bar{a}_0) + (\bar{a}_0 a_1) + (a_{-1}a_1)$	●●●	146	182	146	146
183	10110111	b7	$(\bar{a}_{-1}\bar{a}_1) + (a_{-1}a_1) + (\bar{a}_0)$	●●●	18	183	18	18
184	10111000	b8	$(a_{-1}\bar{a}_0) + (a_0 a_1)$	●●●	226	226	184	184
185	10111001	b9	$(a_{-1}\bar{a}_0) + (\bar{a}_0 a_1) + (a_0 a_1)$	●●●	98	227	56	56
186	10111010	ba	$(a_{-1}\bar{a}_0) + (a_1)$	●●●	162	242	176	162
187	10111011	bb	$(\bar{a}_0) + (a_1)$	–●●	34	243	48	34
188	10111100	bc	$(a_{-1}a_1) + (a_{-1}\bar{a}_0) + (\bar{a}_{-1}a_0)$	●●●	194	230	152	152
189	10111101	bd	$(\bar{a}_0 a_1) + (a_{-1}a_1) + (\bar{a}_{-1}a_0)$	●●●	66	231	24	24
190	10111110	be	$(a_{-1}\bar{a}_0) + (\bar{a}_{-1}a_0) + (a_1)$	●●●	130	246	144	130
191	10111111	bf	$(\bar{a}_0) + (\bar{a}_{-1}) + (a_1)$	●●●	2	247	16	2
192	11000000	c0	$(a_{-1}a_0)$	●●–	252	136	238	136
193	11000001	c1	$(\bar{a}_{-1}\bar{a}_0\bar{a}_1) + (a_{-1}a_0)$	●●●	124	137	110	110
194	11000010	c2	$(\bar{a}_{-1}\bar{a}_0 a_1) + (a_{-1}a_0)$	●●●	188	152	230	152
195	11000011	c3	$(\bar{a}_{-1}\bar{a}_0) + (a_{-1}a_0)$	○○–	60	153	102	60

519

| rule number | | | boolean expression | dep | equivalent rules | | | min |
dec	binary	hex			conj	refl	c.r.	
196	11000100	c4	$(a_0\bar{a}_1)+(a_{-1}a_0)$	•••	220	140	206	140
197	11000101	c5	$(\bar{a}_{-1}\bar{a}_1)+(a_{-1}a_0)$	•••	92	141	78	78
198	11000110	c6	$(\bar{a}_{-1}a_0a_1)+(a_0\bar{a}_1)+(a_{-1}a_0)$	•○•	156	156	198	156
199	11000111	c7	$(\bar{a}_{-1}\bar{a}_1)+(\bar{a}_{-1}\bar{a}_0)+(a_{-1}a_0)$	•••	28	157	70	28
200	11001000	c8	$(a_{-1}a_0)+(a_0a_1)$	•••	236	200	236	200
201	11001001	c9	$(\bar{a}_{-1}\bar{a}_0\bar{a}_1)+(a_{-1}a_0)+(a_0a_1)$	•○•	108	201	108	108
202	11001010	ca	$(a_{-1}a_0)+(\bar{a}_{-1}\bar{a}_1)$	•••	172	216	228	172
203	11001011	cb	$(\bar{a}_{-1}a_1)+(\bar{a}_{-1}\bar{a}_0)+(a_{-1}a_0)$	•••	44	217	100	44
204	11001100	cc	(a_0)	—○—	204	204	204	204
205	11001101	cd	$(\bar{a}_{-1}\bar{a}_1)+(a_0)$	•••	76	205	76	76
206	11001110	ce	$(\bar{a}_{-1}a_1)+(a_0)$	•••	140	220	196	140
207	11001111	cf	$(\bar{a}_{-1})+(a_0)$	••—	12	221	68	12
208	11010000	d0	$(a_{-1}\bar{a}_1)+(a_{-1}a_0)$	•••	244	138	174	138
209	11010001	d1	$(\bar{a}_0\bar{a}_1)+(a_{-1}a_0)$	•••	116	139	46	46
210	11010010	d2	$(\bar{a}_{-1}\bar{a}_0a_1)+(a_{-1}\bar{a}_1)+(a_{-1}a_0)$	○••	180	154	166	154
211	11010011	d3	$(a_{-1}\bar{a}_1)+(\bar{a}_{-1}\bar{a}_0)+(a_{-1}a_0)$	•••	52	155	38	38
212	11010100	d4	$(a_{-1}\bar{a}_1)+(a_0\bar{a}_1)+(a_{-1}a_0)$	•••	212	142	142	142
213	11010101	d5	$(a_{-1}a_0)+(\bar{a}_1)$	•••	84	143	14	14
214	11010110	d6	$(\bar{a}_{-1}\bar{a}_0a_1)+(a_{-1}\bar{a}_1)+(a_0\bar{a}_1)+(a_{-1}a_0)$	•••	148	158	134	134
215	11010111	d7	$(\bar{a}_{-1}\bar{a}_0)+(a_{-1}a_0)+(\bar{a}_1)$	•••	20	159	6	6
216	11011000	d8	$(a_{-1}\bar{a}_1)+(a_0a_1)$	•••	228	202	172	172
217	11011001	d9	$(a_{-1}a_0)+(\bar{a}_0\bar{a}_1)+(a_0a_1)$	•••	100	203	44	44
218	11011010	da	$(a_{-1}a_0)+(a_{-1}\bar{a}_1)+(\bar{a}_{-1}a_1)$	•••	164	218	164	164
219	11011011	db	$(\bar{a}_0\bar{a}_1)+(\bar{a}_{-1}a_1)+(a_{-1}a_0)$	•••	36	219	36	36
220	11011100	dc	$(a_{-1}\bar{a}_1)+(a_0)$	•••	196	206	140	140
221	11011101	dd	$(\bar{a}_1)+(a_0)$	—••	68	207	12	12
222	11011110	de	$(a_{-1}\bar{a}_1)+(\bar{a}_{-1}a_1)+(a_0)$	•••	132	222	132	132
223	11011111	df	$(\bar{a}_1)+(\bar{a}_{-1})+(a_0)$	•••	4	223	4	4
224	11100000	e0	$(a_{-1}a_0)+(a_{-1}a_1)$	•••	248	168	234	168
225	11100001	e1	$(\bar{a}_{-1}\bar{a}_0\bar{a}_1)+(a_{-1}a_0)+(a_{-1}a_1)$	○••	120	169	106	106
226	11100010	e2	$(a_{-1}a_0)+(\bar{a}_0a_1)$	•••	184	184	226	184
227	11100011	e3	$(a_{-1}a_1)+(\bar{a}_{-1}\bar{a}_0)+(a_{-1}a_0)$	•••	56	185	98	56
228	11100100	e4	$(a_0\bar{a}_1)+(a_{-1}a_1)$	•••	216	172	202	172
229	11100101	e5	$(a_{-1}a_0)+(\bar{a}_{-1}\bar{a}_1)+(a_{-1}a_1)$	•••	88	173	74	74
230	11100110	e6	$(a_{-1}a_0)+(a_0\bar{a}_1)+(\bar{a}_0a_1)$	•••	152	188	194	152
231	11100111	e7	$(\bar{a}_{-1}\bar{a}_1)+(\bar{a}_0a_1)+(a_{-1}a_0)$	•••	24	189	66	24
232	11101000	e8	$(a_{-1}a_0)+(a_{-1}a_1)+(a_0a_1)$	•••	232	232	232	232
233	11101001	e9	$(\bar{a}_{-1}\bar{a}_0\bar{a}_1)+(a_{-1}a_0)+(a_{-1}a_1)+(a_0a_1)$	•••	104	233	104	104
234	11101010	ea	$(a_{-1}a_0)+(a_1)$	•••	168	248	224	168
235	11101011	eb	$(\bar{a}_{-1}\bar{a}_0)+(a_{-1}a_0)+(a_1)$	•••	40	249	96	40
236	11101100	ec	$(a_{-1}a_1)+(a_0)$	•••	200	236	200	200
237	11101101	ed	$(\bar{a}_{-1}\bar{a}_1)+(a_{-1}a_1)+(a_0)$	•••	72	237	72	72
238	11101110	ee	$(a_0)+(a_1)$	—••	136	252	192	136
239	11101111	ef	$(\bar{a}_{-1})+(a_0)+(a_1)$	•••	8	253	64	8
240	11110000	f0	(a_{-1})	○——	240	170	170	170
241	11110001	f1	$(\bar{a}_0\bar{a}_1)+(a_{-1})$	•••	112	171	42	42
242	11110010	f2	$(\bar{a}_0a_1)+(a_{-1})$	•••	176	186	162	162
243	11110011	f3	$(\bar{a}_0)+(a_{-1})$	••—	48	187	34	34
244	11110100	f4	$(a_0\bar{a}_1)+(a_{-1})$	•••	208	174	138	138

rule number			boolean expression	dep	equivalent rules			min
dec	binary	hex			conj	refl	c.r.	
245	11110101	f5	$(\bar{a}_1) + (a_{-1})$	•−•	80	175	10	10
246	11110110	f6	$(a_0\bar{a}_1) + (\bar{a}_0 a_1) + (a_{-1})$	•••	144	190	130	130
247	11110111	f7	$(\bar{a}_1) + (\bar{a}_0) + (a_{-1})$	•••	16	191	2	2
248	11111000	f8	$(a_0 a_1) + (a_{-1})$	•••	224	234	168	168
249	11111001	f9	$(\bar{a}_0\bar{a}_1) + (a_0 a_1) + (a_{-1})$	•••	96	235	40	40
250	11111010	fa	$(a_{-1}) + (a_1)$	•−•	160	250	160	160
251	11111011	fb	$(\bar{a}_0) + (a_{-1}) + (a_1)$	•••	32	251	32	32
252	11111100	fc	$(a_{-1}) + (a_0)$	••−	192	238	136	136
253	11111101	fd	$(\bar{a}_1) + (a_{-1}) + (a_0)$	•••	64	239	8	8
254	11111110	fe	$(a_{-1}) + (a_0) + (a_1)$	•••	128	254	128	128
255	11111111	ff	1	−−−−	0	255	0	0

Forms of rules and equivalences between rules.

The table lists all 256 possible rules for $k = 2$, $r = 1$ one-dimensional cellular automata. Such cellular automata consist of a line of sites, each with value 0 or 1. At each time step, the value a_i of a site at position i is updated according to the rule

$$a_i' = \phi(a_{i-1}, a_i, a_{i+1}).$$

This table lists the $2^{2^3} = 256$ possible choices of ϕ.

Each digit in the binary representation of the rule number gives the value of ϕ for a particular set of (a_{i-1}, a_i, a_{i+1}). The digit corresponding to the coefficient of 2^n in the rule number gives the value of $\phi(n_2, n_1, n_0)$, where $n = 4n_2 + 2n_1 + n_0$. Thus the leftmost digit in the binary representation of the rule number gives $\phi(1, 1, 1)$, the next gives $\phi(1, 1, 0)$, and so on, down to $\phi(0, 0, 0)$.

The table also gives the decimal and hexadecimal representations of the rule numbers.

Each ϕ can be considered a Boolean function of three variables, say a_{-1}, a_0 and a_1. The table gives the minimal disjunctive normal form representations for these Boolean functions. Boolean multiplication and addition are used (corresponding to AND and OR operations). Bar denotes complementation. In each case, the expression with the minimal number of components, using only these operations, is given.

The column labelled "dep" gives the dependence of $\phi(a_{-1}, a_0, a_1)$ on each of the a_{-1}, a_0 and a_1. The symbol − indicates no change in ϕ when the corresponding a_j is changed. The symbol ∘ denotes linear dependence of ϕ on the corresponding a_j: whenever a_j changes, ϕ also changes. The symbol • denotes arbitrary dependence of ϕ. Rules such as 90 in which only ∘ and − dependence occurs, are called additive, and can be represented as linear functions modulo two.

For each rule, the table gives rules equivalent under simple transformations. "conj" denotes conjugation: interchange of the roles of 0 and 1. "refl" denotes reflection. Rules invariant under reflection are symmetric. "c.r." denotes the combined operation of conjugation and reflection.

Many of the properties considered in this Appendix are unaffected by these transformations. The rules form equivalence classes under these transformations, and it is usually convenient to consider only the minimal (lowest-numbered) representatives of each class, as given by the last column in the table.

In some cases, further equivalences between rules can be used. Table 7 gives one important set of such further equivalences.

Some special rules are:

51	complement
170	left shift
204	identity
240	right shift

Table by Lyman P. Hurd (*Mathematics Department, Princeton University*). (Boolean expressions by S. Wolfram.)

Table 2: Patterns from Disordered States

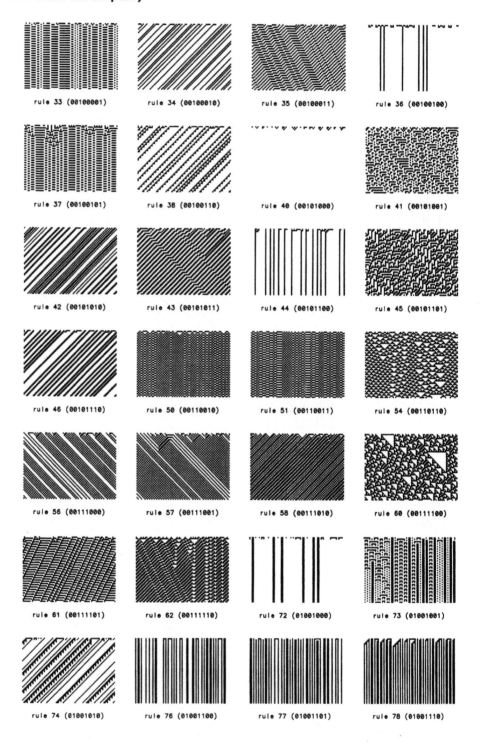

rule 33 (00100001)   rule 34 (00100010)   rule 35 (00100011)   rule 36 (00100100)

rule 37 (00100101)   rule 38 (00100110)   rule 40 (00101000)   rule 41 (00101001)

rule 42 (00101010)   rule 43 (00101011)   rule 44 (00101100)   rule 45 (00101101)

rule 46 (00101110)   rule 50 (00110010)   rule 51 (00110011)   rule 54 (00110110)

rule 56 (00111000)   rule 57 (00111001)   rule 58 (00111010)   rule 60 (00111100)

rule 61 (00111101)   rule 62 (00111110)   rule 72 (01001000)   rule 73 (01001001)

rule 74 (01001010)   rule 76 (01001100)   rule 77 (01001101)   rule 78 (01001110)

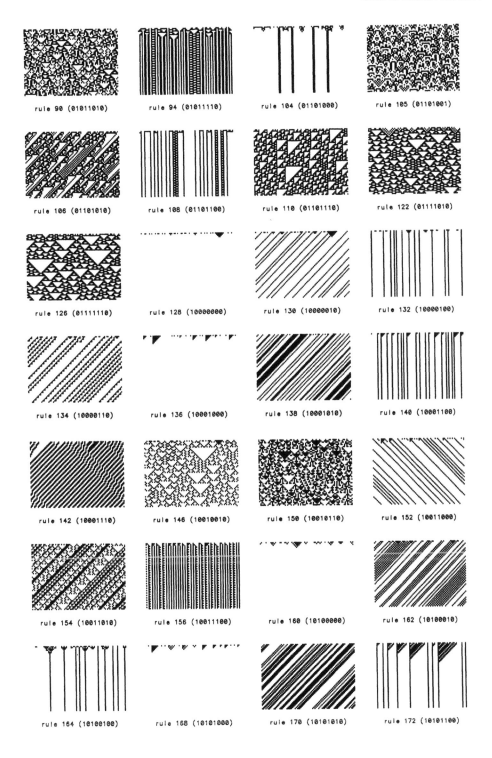

rule 178 (10110010)   rule 184 (10111000)   rule 188 (10111100)   rule 200 (11001000)

rule 204 (11001100)   rule 232 (11101000)

Patterns generated by evolution from disordered initial states.

Each picture is for a different rule. All the "minimal representative" rules of table 1 are included. (Other rules have patterns equivalent to those of their minimal representatives.)

Sites with values 1 and 0 are represented respectively by black and white squares. The initial configuration is at the top of each picture. The values of sites in it are chosen randomly to be 0 or 1 with probability 1/2. Successive lines are obtained by applications of the cellular automaton rule.

These pictures show the evolution of cellular automata with 80 sites for 60 time steps. Periodic boundary conditions were imposed on the edges.

Different specific initial configurations for a particular rule almost always yield qualitatively similar patterns. Different rules are however seen to give a wide variety of different kinds of patterns.

Table 3: Blocked Patterns from Disordered States

527

rule 33 (00100001)   rule 34 (00100010)   rule 35 (00100011)   rule 36 (00100100)

rule 37 (00100101)   rule 38 (00100110)   rule 40 (00101000)   rule 41 (00101001)

rule 42 (00101010)   rule 43 (00101011)   rule 44 (00101100)   rule 45 (00101101)

rule 46 (00101110)   rule 50 (00110010)   rule 51 (00110011)   rule 54 (00110110)

rule 56 (00111000)   rule 57 (00111001)   rule 58 (00111010)   rule 60 (00111100)

rule 61 (00111101)   rule 62 (00111110)   rule 72 (01001000)   rule 73 (01001001)

rule 74 (01001010)   rule 76 (01001100)   rule 77 (01001101)   rule 78 (01001110)

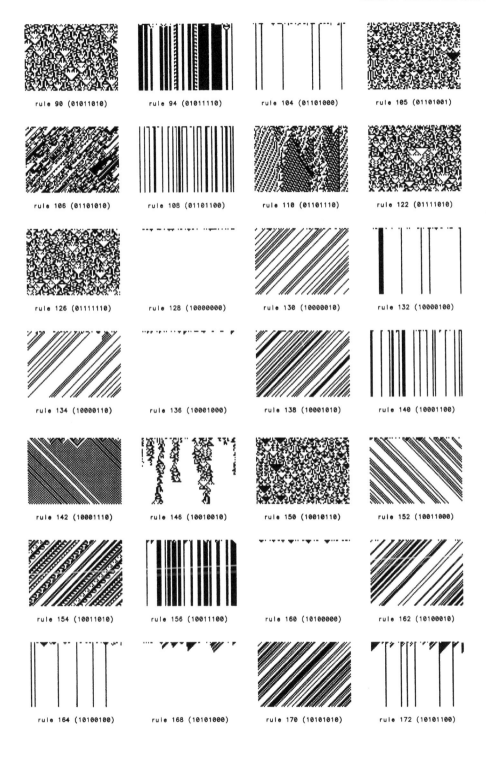

rule 90 (01011010)   rule 94 (01011110)   rule 104 (01101000)   rule 105 (01101001)

rule 106 (01101010)   rule 108 (01101100)   rule 110 (01101110)   rule 122 (01111010)

rule 126 (01111110)   rule 128 (10000000)   rule 130 (10000010)   rule 132 (10000100)

rule 134 (10000110)   rule 136 (10001000)   rule 138 (10001010)   rule 140 (10001100)

rule 142 (10001110)   rule 146 (10010010)   rule 150 (10010110)   rule 152 (10011000)

rule 154 (10011010)   rule 156 (10011100)   rule 160 (10100000)   rule 162 (10100010)

rule 164 (10100100)   rule 168 (10101000)   rule 170 (10101010)   rule 172 (10101100)

rule 178 (10110010)   rule 184 (10111000)   rule 188 (10111100)   rule 200 (11001000)

rule 204 (11001100)   rule 232 (11101000)

Blocks in patterns generated by evolution from disordered initial states.

The pictures in this table are analogous to those in table 2, but show only every other site in both space and time. Certain features become clearer in this "blocked" representation.

It is common for cellular automata to exhibit several "phases". The blocked representation often makes differences between these phases visible.

Table 4: Difference Patterns

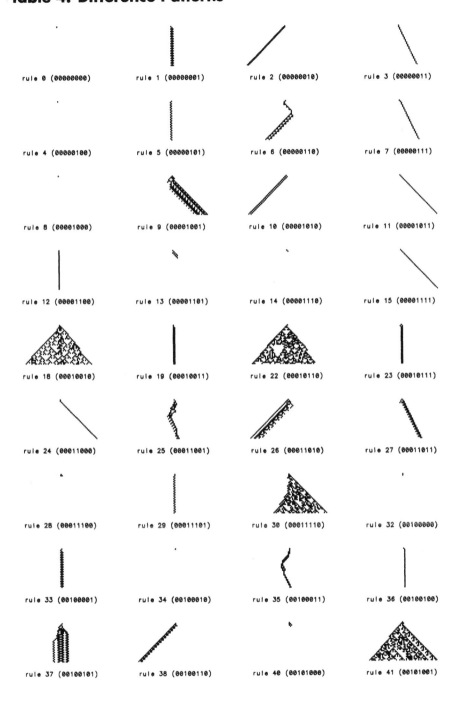

rule 0 (00000000)   rule 1 (00000001)   rule 2 (00000010)   rule 3 (00000011)

rule 4 (00000100)   rule 5 (00000101)   rule 6 (00000110)   rule 7 (00000111)

rule 8 (00001000)   rule 9 (00001001)   rule 10 (00001010)   rule 11 (00001011)

rule 12 (00001100)   rule 13 (00001101)   rule 14 (00001110)   rule 15 (00001111)

rule 18 (00010010)   rule 19 (00010011)   rule 22 (00010110)   rule 23 (00010111)

rule 24 (00011000)   rule 25 (00011001)   rule 26 (00011010)   rule 27 (00011011)

rule 28 (00011100)   rule 29 (00011101)   rule 30 (00011110)   rule 32 (00100000)

rule 33 (00100001)   rule 34 (00100010)   rule 35 (00100011)   rule 36 (00100100)

rule 37 (00100101)   rule 38 (00100110)   rule 40 (00101000)   rule 41 (00101001)

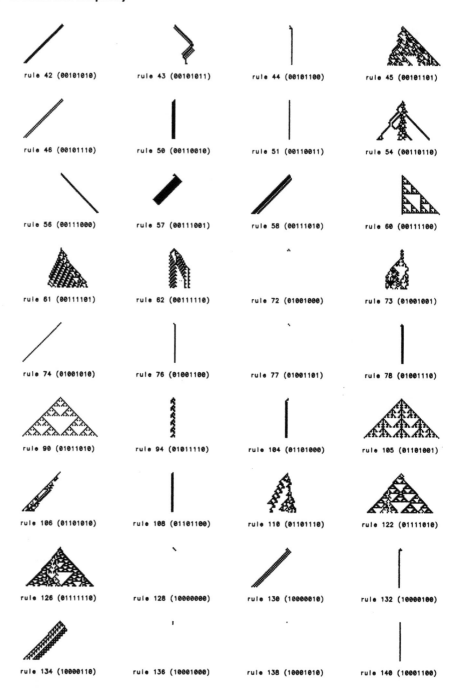

rule 42 (00101010)   rule 43 (00101011)   rule 44 (00101100)   rule 45 (00101101)

rule 46 (00101110)   rule 50 (00110010)   rule 51 (00110011)   rule 54 (00110110)

rule 56 (00111000)   rule 57 (00111001)   rule 58 (00111010)   rule 60 (00111100)

rule 61 (00111101)   rule 62 (00111110)   rule 72 (01001000)   rule 73 (01001001)

rule 74 (01001010)   rule 76 (01001100)   rule 77 (01001101)   rule 78 (01001110)

rule 90 (01011010)   rule 94 (01011110)   rule 104 (01101000)   rule 105 (01101001)

rule 106 (01101010)   rule 108 (01101100)   rule 110 (01101110)   rule 122 (01111010)

rule 126 (01111110)   rule 128 (10000000)   rule 130 (10000010)   rule 132 (10000100)

rule 134 (10000110)   rule 136 (10001000)   rule 138 (10001010)   rule 140 (10001100)

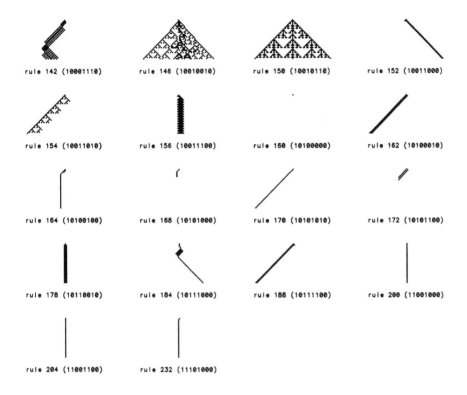

rule 142 (10001110)   rule 146 (10010010)   rule 150 (10010110)   rule 152 (10011000)

rule 154 (10011010)   rule 156 (10011100)   rule 160 (10100000)   rule 162 (10100010)

rule 164 (10100100)   rule 168 (10101000)   rule 170 (10101010)   rule 172 (10101100)

rule 178 (10110010)   rule 184 (10111000)   rule 188 (10111100)   rule 200 (11001000)

rule 204 (11001100)   rule 232 (11101000)

Differences in patterns produced by evolution from disordered states resulting from changes in single initial site values.

The evolution of small perturbations made in the initial configurations for all the "minimal representative" rules of table 1 are given. In each case, an initial configuration was chosen in which sites had value 0 or 1 with probability $1/2$, and the pattern obtained by evolution according to the cellular automaton rule was found. Then the value of the centre site in the initial configuration was complemented, and the resulting pattern obtained by cellular automaton evolution was found. The pictures show as black squares the site values that differed between the patterns found with these initial configurations. Evolution for 40 time steps is shown.

In some cases, the differences die out, or remain localized, with time. In other cases, the differences grow. The left and right growth speeds correspond to the left and right Lyapunov exponents λ_L and λ_R, given in table 6.

For some rules (such as 18), initial perturbations on some configurations may grow, but on others may die out. The pictures show results from a particular trial.

Table 5: Patterns from Single Site Seeds

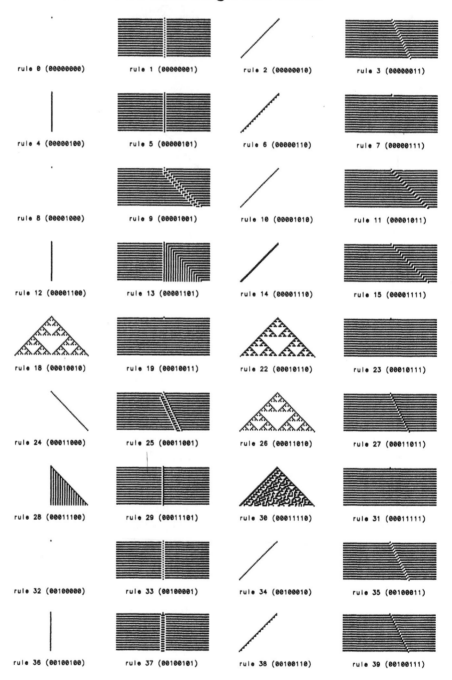

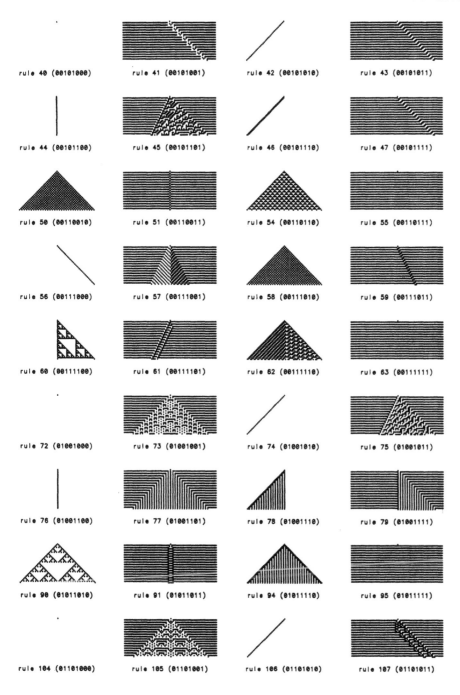

rule 40 (00101000) rule 41 (00101001) rule 42 (00101010) rule 43 (00101011)

rule 44 (00101100) rule 45 (00101101) rule 46 (00101110) rule 47 (00101111)

rule 50 (00110010) rule 51 (00110011) rule 54 (00110110) rule 55 (00110111)

rule 56 (00111000) rule 57 (00111001) rule 58 (00111010) rule 59 (00111011)

rule 60 (00111100) rule 61 (00111101) rule 62 (00111110) rule 63 (00111111)

rule 72 (01001000) rule 73 (01001001) rule 74 (01001010) rule 75 (01001011)

rule 76 (01001100) rule 77 (01001101) rule 78 (01001110) rule 79 (01001111)

rule 90 (01011010) rule 91 (01011011) rule 94 (01011110) rule 95 (01011111)

rule 104 (01101000) rule 105 (01101001) rule 106 (01101010) rule 107 (01101011)

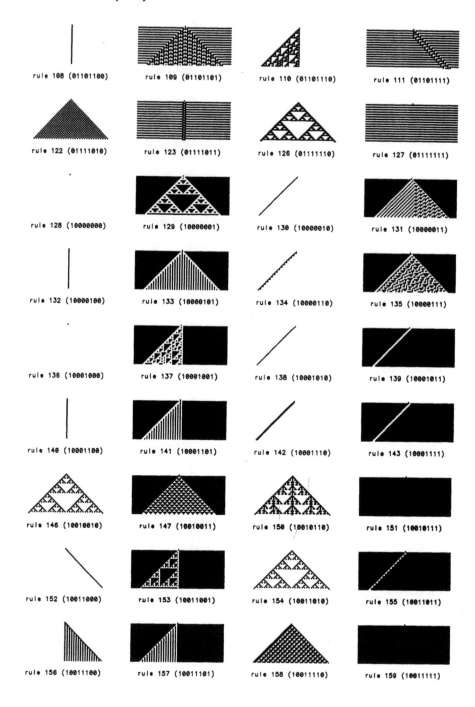

rule 108 (01101100)   rule 109 (01101101)   rule 110 (01101110)   rule 111 (01101111)

rule 122 (01111010)   rule 123 (01111011)   rule 126 (01111110)   rule 127 (01111111)

rule 128 (10000000)   rule 129 (10000001)   rule 130 (10000010)   rule 131 (10000011)

rule 132 (10000100)   rule 133 (10000101)   rule 134 (10000110)   rule 135 (10000111)

rule 136 (10001000)   rule 137 (10001001)   rule 138 (10001010)   rule 139 (10001011)

rule 140 (10001100)   rule 141 (10001101)   rule 142 (10001110)   rule 143 (10001111)

rule 146 (10010010)   rule 147 (10010011)   rule 150 (10010110)   rule 151 (10010111)

rule 152 (10011000)   rule 153 (10011001)   rule 154 (10011010)   rule 155 (10011011)

rule 156 (10011100)   rule 157 (10011101)   rule 158 (10011110)   rule 159 (10011111)

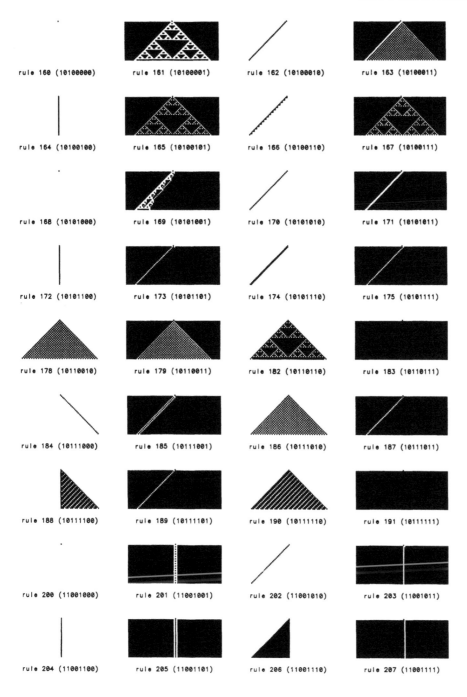

rule 160 (10100000)   rule 161 (10100001)   rule 162 (10100010)   rule 163 (10100011)

rule 164 (10100100)   rule 165 (10100101)   rule 166 (10100110)   rule 167 (10100111)

rule 168 (10101000)   rule 169 (10101001)   rule 170 (10101010)   rule 171 (10101011)

rule 172 (10101100)   rule 173 (10101101)   rule 174 (10101110)   rule 175 (10101111)

rule 178 (10110010)   rule 179 (10110011)   rule 182 (10110110)   rule 183 (10110111)

rule 184 (10111000)   rule 185 (10111001)   rule 186 (10111010)   rule 187 (10111011)

rule 188 (10111100)   rule 189 (10111101)   rule 190 (10111110)   rule 191 (10111111)

rule 200 (11001000)   rule 201 (11001001)   rule 202 (11001010)   rule 203 (11001011)

rule 204 (11001100)   rule 205 (11001101)   rule 206 (11001110)   rule 207 (11001111)

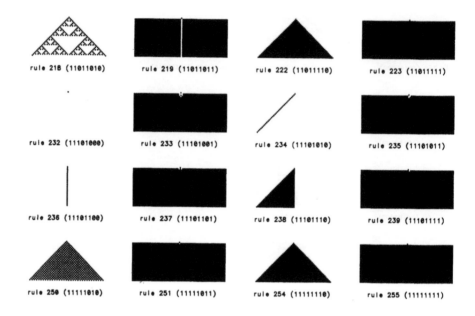

rule 218 (11011010)   rule 219 (11011011)   rule 222 (11011110)   rule 223 (11011111)

rule 232 (11101000)   rule 233 (11101001)   rule 234 (11101010)   rule 235 (11101011)

rule 236 (11101100)   rule 237 (11101101)   rule 238 (11101110)   rule 239 (11101111)

rule 250 (11111010)   rule 251 (11111011)   rule 254 (11111110)   rule 255 (11111111)

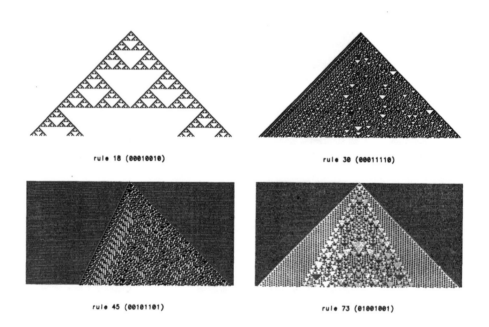

rule 18 (00010010)

rule 30 (00011110)

rule 45 (00101101)

rule 73 (01001001)

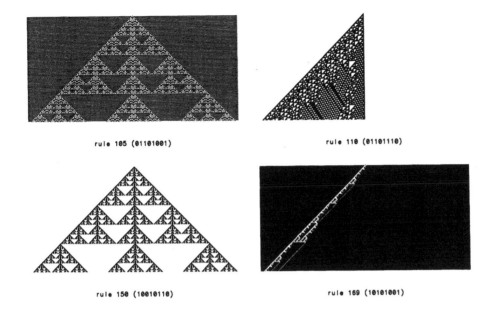

rule 105 (01101001)

rule 110 (01101110)

rule 150 (10010110)

rule 169 (10101001)

Patterns generated by evolution from configurations containing a single nonzero site.

The first part of the table shows pictures for all distinct rules. Since the initial configuration is not invariant under complementation, rules which differ by complementation can produce different patterns, and are shown separately. Only the minimal representative is shown for rules related by reflection. In all cases, the patterns correspond to evolution for 38 time steps.

Many rules are seen to yield equivalent patterns. The results of table 7 can often be used to deduce these equivalences.

Some rules (such as 122) yield asymptotically homogeneous patterns. Others (such as 90 and 150) yield asymptotically self similar or fractal patterns. (The fractal dimensions of the patterns obtained from rules 90 and 150 are respectively $\log_2 3 \simeq 1.59$ and $\log_2(1 + \sqrt{5}) \simeq 1.69$.) But some rules (such as 30 and 73) yield irregular patterns which show no periodic or almost periodic behaviour. The second part of the table gives some of the distinct patterns obtained by evolution for 360 time steps. Note that the structure on the right of the pattern generated by rule 110 eventually dies out, leaving an essentially periodic structure.

Table 6: Statistical Properties

	density	$h_\mu^{(x)}$	λ_L	λ_R	$h_\mu^{(t)}$	h_μ	$h_\mu^{[min]}$
0	0	0	—	—	0	0	0
1	1/8	.43536	0	0	0	0	0
2	1/8	.48752	1	-1	$h_\mu^{(x)}$	$h_\mu^{(x)}$	0
3	1/4	.70121	-1/2	1/2	$h_\mu^{(x)}/2$	$h_\mu^{(x)}/2$	0
4	1/8	.51771	0	0	0	0	0
5	7/16	.702 ± .001	0	0	0	0	0
6	.241 ± .001	<.573 ± .001	1	-1	$h_\mu^{(x)}$	$h_\mu^{(x)}$	0
7	.469 ± .001	<.502 ± .001	-1/2	1/2	$h_\mu^{(x)}/2$	$h_\mu^{(x)}/2$	0
8	0	0	—	—	0	0	0
9	.410 ± .001	<.264 ± .002	-1	1	$h_\mu^{(x)}$	$h_\mu^{(x)}$	0
10	1/4	.68872	1	-1	$h_\mu^{(x)}$	0	$h_\mu^{(x)}$
11	1/2	<.567 ± .001	-1	1	$h_\mu^{(x)}$	$h_\mu^{(x)}$	0
12	1/4	.68872	0	0	0	0	0
13	.437 ± .001	.378 ± .001	0	0	0	0	0
14	1/2	0	(-1, 1)	(1, -1)	0	0	0
15	1/2	1	-1	1	1.0	1.0	0
18	1/4	1/2	1	1	0.5	1.0	1.0
19	1/2	.62351	0	0	0	0	0
22	.35095 ± .00002	<.795 ± .001	.7660 ± .0002	.7660 ± .0002	.744 ± .003	<.9146 ± .0007	<.9146 ± .0007
23	1/2	.599 ± .001	0	0	0	0	0
24	3/16	.55081	-1	1	$h_\mu^{(x)}$	$h_\mu^{(x)}$	0
25	.447 ± .001	<.180 ± .001	-1/2	1/2	$h_\mu^{(x)}/2$	$h_\mu^{(x)}/2$	0
26	.386 ± .001	<.790 ± .001	1	-1	$h_\mu^{(x)}$	$h_\mu^{(x)}$	0
27	.531 ± .001	<.800 ± .001	-1/2	1/2	$h_\mu^{(x)}/2$	$h_\mu^{(x)}/2$	0
28	1/2	.500 ± .001	0	0	0	0	0
29	1/2	.86742	0	0	0	0	0
30	1/2	1	.2428 ± .0002	1	1	<1.15436	<.763141
32	0	0	—	—	0	0	0
33	.396 ± .001	<.637 ± .001	0	0	0	0	0
34	1/4	.68872	1	-1	$h_\mu^{(x)}$	$h_\mu^{(x)}$	0
35	.375 ± .001	<.645 ± .001	-1/2	1/2	$h_\mu^{(x)}/2$	$h_\mu^{(x)}/2$	0
36	1/16	.32483	0	0	0	0	0
37	.384 ± .001	.506 ± .001	0	0	0	0	0
38	9/32	.73733	1	-1	$h_\mu^{(x)}$	$h_\mu^{(x)}$	0
40	0	0	—	—	0	0	0
41	.372 ± .001	<.360 ± .001	-1	1	$h_\mu^{(x)}$	$h_\mu^{(x)}$	0
42	3/8	.85684	1	-1	$h_\mu^{(x)}$	$h_\mu^{(x)}$	0
43	1/2	0	(-1, 1)	(1, -1)	0	0	0
44	.167 ± .001	.528 ± .001	0	0	0	0	0
45	1/2	1	.1724 ± .0003	1	1	<1.13036	<.673893
46	3/8	.55081	1	-1	$h_\mu^{(x)}$	$h_\mu^{(x)}$	0
50	1/2	.601 ± .001	0	0	0	0	0
51	1/2	1	0	0	0	0	0

	density	$h_\mu^{(x)}$	λ_L	λ_R	$h_\mu^{(t)}$	$\mathbf{h}_\mu$	$\mathbf{h}_\mu^{[min]}$
54	$.49 \pm .01$	$<.2720 \pm .0005$	$.553 \pm .002$	$.553 \pm .002$	$<.250 \pm .002$	$<.250 \pm .002$	$<.250 \pm .002$
56	$.376 \pm .001$	$<.589 \pm .001$	-1	1	$h_\mu^{(x)}$	$h_\mu^{(x)}$	0
57	$1/2$	0	$(-1, 1)$	$(1, -1)$	0	0	0
58	$.625 \pm .001$	$<.332 \pm .001$	1	-1	$h_\mu^{(x)}$	$h_\mu^{(x)}$	0
60	$1/2$	1	0	1	1	2	2
62	$.644 \pm .002$	$<.262 \pm .001$	0	0	0	0	0
72	$1/8$	$.32483$	0	0	0	0	0
73	$.463 \pm .001$	$<.714 \pm .001$	0	0	0	0	0
74	$.318 \pm .001$	$<.629 \pm .001$	1	-1	$h_\mu^{(x)}$	$h_\mu^{(x)}$	0
76	$3/8$	$.85060$	0	0	0	0	0
77	$1/2$	$.599 \pm .001$	0	0	0	0	0
78	$.562 \pm .001$	$.377 \pm .001$	0	0	0	0	0
90	$1/2$	1	1	1	1	2	2
94	$.584 \pm .001$	$<.562 \pm .001$	0	0	0	0	0
104	$.068 \pm .001$	$.208 \pm .001$	0	0	0	0	0
105	$1/2$	1	1	1	1	2	2
106	$1/2$	1	1	$-.1335 \pm .0006$	1	<1.06985	$<.461366$
108	$5/16$	$.78025$	0	0	0	0	0
110	$4/7$	0	$(.26 - 5)$	$(-.27 - 0.)$	0	0	0
122	$1/2$	$1/2$	1	1	0.5	1.0	1.0
126	$1/2$	$1/2$	1	1	0.5	1.0	1.0
128	0	0	$-$	$-$	0	0	0
130	$.167 \pm .001$	$.525 \pm .001$	1	-1	$h_\mu^{(x)}$	$h_\mu^{(x)}$	0
132	$1/8$	$.599 \pm .001$	0	0	0	0	0
134	$.292 \pm .001$	$<.533 \pm .001$	1	-1	$h_\mu^{(x)}$	$h_\mu^{(x)}$	0
136	0	0	$-$	$-$	0	0	0
138	$3/8$	$.806 \pm .001$	1	-1	$h_\mu^{(x)}$	$h_\mu^{(x)}$	0
140	$1/4$	$<.678 \pm .001$	0	0	0	0	0
142	$1/2$	0	$(-1, 1)$	$(1, -1)$	0	0	0
146	$1/4$	$1/2$	1	1	0.5	1.0	1.0
150	$1/2$	1	1	1	1	2	2
152	$.185 \pm .001$	$.515 \pm .001$	-1	1	$h_\mu^{(x)}$	$h_\mu^{(x)}$	0
154	$1/2$	1	1	-1	1	1	0
156	$1/2$	$.502 \pm .001$	0	0	0	0	0
160	0	0	$-$	$-$	0	0	0
162	$.333 \pm .001$	$.667 \pm .001$	1	-1	$h_\mu^{(x)}$	$h_\mu^{(x)}$	0
164	$.083 \pm .001$	$.389 \pm .001$	0	0	0	0	0
168	0	0	$-$	$-$	0	0	0
170	$1/2$	-1	1	1	1.0	1.0	0
172	$1/8$	$.485 \pm .001$	0	0	0	0	0
178	$1/2$	$.599 \pm .001$	0	0	0	0	0
184	$1/2$	0	$(-1, 1)$	$(1, -1)$	0	0	0
200	$3/8$	$.70121$	0	0	0	0	0
204	$1/2$	1	0	0	0	0	0
232	$1/2$	$.599 \pm .001$	0	0	0	0	0

Statistical properties of evolution from disordered states.

Results are given for all the "minimal representative" rules of table 1. In all cases, initial configurations were used in which each site has value 0 or 1 with probability 1/2. Some properties of some rules remain unchanged with different kinds of initial configurations.

Rational numbers, or numbers without errors, are quoted whenever analytical arguments yield exact results. In a few cases, the rigour of these arguments may be subject to question.

The column labelled "density" gives the asymptotic density of nonzero sites. For some, but not all, rules this depends on the initial density, here taken to be 1/2. For most rules, the relaxation to the final density appears to be approximately exponential. For some rules (such as 18), in which particle-like excitations undergo random annihilation, the relaxation may be like $t^{-1/2}$, or slower. Rule 110 shows particularly slow relaxation.

The column labelled $h_\mu^{(x)}$ gives estimates for the asymptotic spatial measure entropy, as defined in pages 115–157 in this book. This quantity gives a measure of the "information content" of cellular automaton configurations. It is computed by breaking the configuration into blocks of sites, say of length X, then evaluating the quantity $-\frac{1}{X}\sum p_i \log_2 p_i$, where the sum runs over all 2^X possible blocks, which are taken to occur with probabilities p_i. $h_\mu^{(x)}$ is the limit of this quantity as $X \to \infty$. The values decrease monotonically with X, allowing upper bounds on the $X \to \infty$ limit to be derived from finite X results. Where errors are quoted, the values or bounds on $h_\mu^{(x)}$ given in the table were obtained after 400 time steps, with blocks up to length $X = 11$ considered. (More accurate results were obtained for rules 22 and 54.) Fits to values obtained as a function of X suggest that the exact $h_\mu^{(x)}$ for rules 22 and 54 may in fact be zero.

The definition of $h_\mu^{(x)}$ implies that it achieves its maximal value of 1 only when all possible sequences of site values occur with equal probability, so that each site has value 0 or 1 with independent probability 1/2. $h_\mu^{(x)} = 0$ if only a finite number of complete cellular automaton configurations can occur.

Results for $h_\mu^{(x)}$ given without errors in the table were obtained by explicit construction of probabilistic regular languages which represent the sets of configurations produced by cellular automaton evolution, as in table 11.

The quantities λ_L and λ_R are left and right Lyapunov exponents, which measure the rate of information transmission. They give the slopes of the left and right boundaries of the difference patterns illustrated in table 4. Thus they measure the rate at which perturbations in cellular automaton configurations spread to the left and right.

The notation — indicates that almost all changes in initial configurations die out, so that the $\lambda_{L,R}$ are not defined.

The notation $(-1, 1)$ indicates that the information propagation direction can alternate, typically as progressively more distant particle-like structures from the initial configuration are encountered. There is probably no definite infinite size limit for the $\lambda_{L,R}$ in such cases.

Rule 110 shows highly complex information transmission properties, associated with the particle-like structures of table 15. The values of $\lambda_{L,R}$ given in the table for this case are possible bounds associated with the fastest and slowest-moving particle-like structures.

The quantity $h_\mu^{(t)}$ is the temporal measure entropy, which measures the information content of time sequences of values of individual sites. It is evaluated by applying the same procedure as for $h_\mu^{(x)}$ but to sequences of values of a single site attained on many successive time steps. It can be shown (see pages 115–157 in this book) that $h_\mu^{(t)} \leq (\lambda_L + \lambda_R) h_\mu^{(x)}$.

The quantities $h_\mu^{(x)}$ and $h_\mu^{(t)}$ measure respectively the information content of spatial and temporal sequences that are one site wide. The quantity $\mathbf{h}_\mu$ gives the entropy associated with spacetime patches of sites of arbitrary width. (Nevertheless, for many rules, the exact value of $\mathbf{h}_\mu$ is in fact obtained from patches of width 1 or 2.) In general, $\mathbf{h}_\mu \leq 2h_\mu^{(t)}$, and $h_\mu^{(t)} \leq \mathbf{h}_\mu \leq (\lambda_L + \lambda_R) h_\mu^{(x)}$.

The quantity $\mathbf{h}_\mu$ is evaluated by considering spacetime patches of sites that extend in the time direction. The last column of the table uses a generalization in which the patches can extend in any spacetime direction. It gives the minimum value $\mathbf{h}_\mu$ obtained as a function of direction. (The actual bounds given in the table were obtained from vertical or diagonal patches; other directions may yield stricter bounds.)

Table by Peter Grassberger (*Physics Department, University of Wuppertal*).

Table 7: Blocking Transformation Equivalences

0	0:	00 10 (1)
1	0:	11 10 (2)
	200:	00 11 (2)
	204:	000 111 (2)
2	34:	00 10 (2)
	170:	000 100 (3)
	0:	1000 1100 (4)
3	0:	11 10 (2)
	240:	00 11 (4)
4	204:	00 10 (1)
	0:	00 11 (1)
5	200:	00 10 (2)
	204:	000 100 (2)
	0:	111 110 (2)
	51:	00010 11010 (1)
6	184:	00 10 (2)
	34:	00 11 (2)
	170:	0000 1000 (4)
	128:	0100 1100 (4)
	240:	1000 1010 (4)
	85:	10000 11000 (5)
	0:	11000 11100 (10)
7	192:	00 10 (2)
	0:	000 100 (2)
	240:	000 111 (6)
8	0:	00 10 (1)
9	0:	0010 1110 (2)
	170:	1000 0010 (6)
	34:	1000 0011 (6)
	204:	0100000 1100000 (5)
	240:	00000000 11010000 (8)
10	34:	00 10 (2)
	170:	000 100 (3)
	0:	010 110 (3)
11	240:	00 11 (2)
	0:	010 110 (3)
	15:	000 111 (3)
	128:	1100 1001 (4)
	170:	1100100 1001100 (7)
12	204:	00 10 (1)
	0:	100 110 (1)
13	192:	00 10 (2)
	0:	100 110 (1)
	204:	10100 10010 (1)
14	240:	10 01 (2)
	34:	00 11 (2)
	15:	010 101 (3)
	0:	1100 1000 (4)
	170:	0000 1100 (4)
	128:	1100 0110 (4)

15	240:	00 10 (2)
	15:	110 001 (3)
18	90:	00 10 (2)
	204:	11000 10100 (2)
	0:	10100 11100 (2)
19	51:	00 11 (1)
	0:	11 10 (2)
	204:	00 11 (2)
22	146:	00 10 (2)
	90:	0000 1000 (4)
	0:	11011000 11111000 (4)
23	51:	00 11 (1)
	128:	00 10 (2)
	204:	00 11 (2)
	0:	000 100 (2)
24	48:	00 10 (2)
	240:	000 100 (3)
	0:	100 011 (3)
25	0:	1101000 1111000 (7)
	240:	0000000 1101000 (14)
26	90:	00 10 (2)
	85:	010 110 (3)
	170:	100100 101100 (6)
	0:	10110100 10111100 (8)
	240:	11100100 10011100 (16)
27	48:	11 10 (4)
	85:	010 110 (3)
	240:	000 100 (6)
	0:	0100 1100 (8)
	170:	100100 101100 (6)
28	192:	00 10 (2)
	200:	10 01 (2)
	51:	100 110 (1)
	204:	100 110 (2)
	0:	1010 1100 (1)
29	204:	00 10 (2)
	200:	10 01 (2)
	51:	100 110 (1)
	0 :	1010 1100 (1)
30		
32	0:	00 11 (1)
	128:	00 10 (2)
33	132:	00 10 (2)
	200:	00 11 (2)
	0:	111 100 (2)
	204:	000 111 (2)
	128:	0000 1010 (4)
34	170:	00 10 (2)
	0:	100 110 (3)

35	240:	00 11 (4)
	0:	100 110 (3)
	170:	10100 10010 (5)
36	0:	00 11 (1)
	4:	00 10 (2)
	204:	000 100 (1)
37	200:	00 11 (2)
	0:	1111 1100 (2)
	204:	0000 1111 (2)
	170:	100000 110000 (6)
	240:	010000 110000 (6)
	128:	010000 111000 (6)
38	34:	00 10 (2)
	85:	100 110 (3)
	170:	0000 1000 (4)
	0:	1100 1110 (4)
40	128:	00 10 (2)
	0:	00 11 (2)
	170:	11010 10110 (5)
41	148:	00 10 (2)
	184:	0000 1000 (4)
	176:	0000 1010 (4)
	170:	11010 10110 (5)
	240:	00000000 10000000 (8)
	128:	10000000 10100000 (8)
	0:	11111000 11001000 (8)
	136:	01010000 10101101 (8)
42	170:	00 10 (2)
	34:	00 11 (2)
	0:	1100 1110 (4)
43	170:	10 01 (2)
	240:	00 11 (2)
	15:	000 111 (3)
	0:	1001 1000 (4)
	128:	1100 0110 (4)
44	12:	00 10 (2)
	204:	000 100 (1)
	0:	1000 1100 (1)
45		
46	34:	00 11 (2)
	0:	110 100 (3)
	170:	000 110 (3)
50	51:	10 01 (1)
	128:	00 10 (2)
	204:	10 01 (2)
	0:	1010 1000 (2)
51	51:	10 01 (1)
	204:	00 10 (2)

54	50:	00 10 (2)
	51:	1000 0010 (2)
	128:	0000 1000 (4)
	204:	1000 0010 (4)
	170:	1000 1110 (4)
	240:	0010 1110 (4)
	0:	000010 111010 (4)
56	240:	00 10 (2)
	128:	10 01 (2)
	184:	010 101 (3)
	0:	1010 0110 (2)
	34:	1010 1101 (4)
	170:	11010 10110 (5)
57	128:	10 01 (2)
	184:	010 101 (3)
	0:	1010 0110 (2)
	48:	1010 0100 (4)
	34:	1010 1101 (4)
	240:	10100 10010 (5)
	170:	11010 10110 (5)
58	128:	00 10 (2)
	0:	101 100 (3)
	240:	1100 1110 (8)
	170:	11010 10110 (5)
60	60:	00 10 (2)
62	240:	1100 1110 (8)
	204:	11000 11010 (3)
	0:	111110 100000 (3)
72	0:	00 10 (1)
	4:	00 11 (2)
	204:	000 110 (1)
73	204:	1100 0110 (2)
	51:	11000 11010 (1)
	0:	10110 10000 (2)
74	34:	00 10 (2)
	170:	000 100 (3)
	0:	10000 10100 (5)
	85:	1110000 1101000 (7)
76	204:	00 10 (1)
	0:	1010 1110 (1)
77	204:	10 01 (1)
	128:	00 10 (2)
	0:	1010 1000 (1)
78	0:	101 100 (1)
	204:	11010 10110 (1)
90	90:	00 10 (2)
94	90:	00 11 (2)
	0:	1010 1110 (1)
	204:	1010 0101 (2)
	51:	10010 11110 (1)
	136:	111100 110110 (6)
	192:	110110 011110 (6)

Rule		
104	128:	00 10 (2)
	4:	00 11 (2)
	0:	000 100 (1)
	204:	0000 1100 (1)
105	150:	00 10 (2)
106	170:	00 10 (2)
108	76:	00 10 (2)
	204:	000 100 (1)
	51:	10100 11100 (1)
	0:	10010 11110 (2)
110	0:	110100 101100 (9)
	240:	11000 100110 (9)
	170:	10011000 11111000 (16)
122	128:	00 10 (2)
	90:	00 11 (2)
	0:	1010 1000 (2)
	204:	11100 11110 (2)
126	90:	00 11 (2)
	204:	11100 11110 (2)
	0:	01110 10001 (2)
128	0:	00 10 (1)
	128:	00 11 (2)
130	34:	00 10 (2)
	170:	000 100 (3)
	0:	1000 1100 (4)
	128:	1000 1111 (4)
132	204:	00 10 (1)
	128:	00 11 (2)
	0:	000 110 (1)
134	184:	00 10 (2)
	162:	00 11 (2)
	170:	0000 1000 (4)
	128:	0100 1100 (4)
	240:	1000 1010 (4)
	85:	10000 11000 (5)
	0:	100000 101100 (6)
136	0:	00 10 (1)
	136:	00 11 (2)
138	34:	00 10 (2)
	170:	00 11 (2)
	0:	010 110 (3)
140	204:	00 10 (1)
	0:	100 110 (1)
142	240:	10 01 (2)
	170:	00 11 (2)
	15:	010 101 (3)
	0:	1100 1000 (4)
	128:	1100 0110 (4)
146	90:	00 10 (2)
	204:	11000 10100 (2)
	0:	10010 11110 (2)
150	150:	00 10 (2)
152	48:	00 10 (2)
	240:	000 100 (3)
	136:	1000 1111 (4)
	0:	01000 11000 (5)
154	90:	00 10 (2)
	85:	010 110 (3)
	170:	1100 0110 (4)
156	192:	00 10 (2)
	200:	10 01 (2)
	136:	10 11 (2)
	51:	100 110 (1)
	204:	100 110 (2)
	0:	1010 1100 (1)
160	128:	00 10 (2)
	0:	000 100 (1)
162	170:	00 10 (2)
	128:	10 11 (2)
	0:	100 110 (3)
164	90:	11 10 (2)
	128:	00 11 (2)
	204:	000 100 (1)
	0:	0000 1100 (1)
168	128:	00 10 (2)
	136:	00 11 (2)
	170:	10 11 (2)
	0:	000 100 (1)
170	170:	00 10 (2)
172	34:	11 10 (2)
	204:	000 100 (1)
	170:	110 111 (3)
	0:	1000 1100 (1)
178	51:	10 01 (1)
	128:	00 10 (2)
	204:	10 01 (2)
	0:	1010 1000 (2)
184	240:	00 10 (2)
	128:	10 01 (2)
	170:	10 11 (2)
	184:	010 101 (3)
	0:	1010 0110 (2)
200	0:	00 10 (1)
	204:	00 11 (1)
204	204:	00 10 (1)
232	204:	00 11 (1)
	128:	00 10 (2)
	0:	000 100 (1)

Equivalences between rules under blocking transformations.

When only particular blocks of site values occur, the evolution of one cellular automaton rule (say R) may be equivalent to that of another (say R'). Thus for example, the evolution under rule 1 of configurations consisting of the blocks 000 and 111 is equivalent to evolution under rule 204 in which 000 is replaced by 0, and 111 is replaced by 1. (Two time steps in evolution according to rule 1 are necessary to reproduce one time step of evolution according to rule 204.) Since rule 204 is the identity, this implies that configurations consisting only of the blocks 000 and 111 must be periodic under rule 1 (with period 2).

In general, one may consider replacing site values 0 and 1 in evolution according to rule R by blocks B_0 and B_1. In some cases, the resulting evolution may correspond to T time steps of another rule R'. Evolution according to rule R' can thus be "simulated" by evolution according to rule R, under the blocking transformation $0 \rightarrow B_0$, $1 \rightarrow B_1$. Such blocking transformations can be considered analogous to block spin transformations in the renormalization group approach.

The table gives possible simulations for all the "minimal representative" rules of table 1. The notation R': B_0 B_1 (T) indicates simulation of rule R' by replacing 0 with the block B_0, and 1 with B_1; T steps of rule R are needed to reproduce one step of rule R' evolution.

The table includes all simulations for block lengths up to 8. The blocks B_0 and B_1 are always assumed distinct. Only one representative set of blocks is given for each simulation. (Thus for example, only the blocks 00 and 10 are given for the simulation of rule 90 by rule 18; the blocks 00 and 01 would also suffice.) Simulations with block length 1 are not included; these correspond to transformations given in table 1. No simulations are found for rules 30 and 45 up to block length 8.

Many rules are seen to be equivalent under blocking transformations to simple rules, such as 204 (the identity), 170 (left shift), 240 (right shift), 51 (complementation) and 0. Equivalence is also often found to the additive rules 90 and 150. An important property of all these simple rules is that they simulate themselves under blocking transformations. This has the consequence that patterns generated by these rules are self similar. Fractal patterns are thus produced by evolution according to rules 90 and 150 from single site seeds, as shown in table 5.

The simulations given in the table occur when only particular blocks occur in the configuration of a cellular automaton. In disordered configurations, all possible blocks can occur. But since a cellular automaton under most rules is irreversible, only a subset of blocks may occur after a sufficiently long time. Often the subset of blocks that occur is, at least approximately, the blocks which correspond to a particular simulation. In this case, the behaviour of one cellular automaton may be considered "attracted" to that of another.

It is common to find "domains" in which only particular blocks occur. Within each such domain, the evolution may correspond to that of a simpler rule. The domains are separated by walls or "defects", whose behaviour is not reproduced by the simpler rule. In some cases, the defects remain stationary; in others, they execute random walks, and, for example, annihilate in pairs. In the latter cases, the sizes of domains grow slowly with time.

While a large subset of possible initial configurations for a cellular automaton may be attracted to a particular form of behaviour, there are usually some special initial states (typically occurring among disordered states with probability zero), for which very different behaviour occurs. Such special initial states may for example consist of blocks which yield a simulation to which the rule is not generically attracted.

The blocking transformations considered in the table represent one form of transformation between rules. Many others can also be considered. A general class, which includes the blocking transformations of the table, are those transformations which can be carried out by arbitrary finite state machines.

The blocking transformations used in the table have the property that they reduce the total number of sites. This is a consequence of the fact that the blocks used are always taken not to overlap. An alternative approach is to perform replacements for overlapping blocks, thus obtaining configurations with the same number of sites. An example of such a replacement is $00 \rightarrow 0$, $01 \rightarrow 1$, $10 \rightarrow 1$, $11 \rightarrow 0$. For some rules, the resulting transformed configurations show evolution according to other $k = 2$, $r = 1$ cellular automaton rules. Rules related in this way must have the same global properties, and must yield for example the same entropies. The minimal representative rules from table 1 equivalent under such transformations are:

15, 240	240
23, 232	132
43, 212	184
51, 204	204
77, 178	222
85, 170	170
105, 150	150
113, 142	226

Main table by John Milnor (*Institute for Advanced Study*). (Original program by S. Wolfram.) Second table by Peter Grassberger.

Table 8: Factorizations into Compositions of Simpler Rules

ϕ	ϕ_1	ϕ_2
0	0	0
	0	12
	0	48
	0	60
	0	192
	0	204
	0	240
	0	252
	17	0
	34	0
	34	192
	51	0
	68	0
	68	192
	85	0
	102	0
	119	0
	136	0
	153	0
	170	0
	187	0
	187	3
	204	0
	221	0
	221	3
	238	0
	255	0
	255	3
	255	12
	255	15
	255	48
	255	51
	255	60
	255	63

ϕ	ϕ_1	ϕ_2
1	17	192
	238	3
2	17	48
	238	12
3	17	240
	51	192
	204	3
	238	15
8	119	48
	136	12
12	34	48
	34	240
	51	48
	204	12
	221	12
	221	15
15	51	240
	204	15
18	17	60
	238	60
19	17	252
	238	63
24	102	48
	153	12
34	34	12
	34	204
	85	48
	170	12
	221	48
	221	51
36	102	192
	153	3

ϕ	ϕ_1	ϕ_2
46	34	60
	34	252
	221	60
	221	63
51	51	204
	85	240
	170	15
	204	51
60	51	60
	102	240
	153	15
	204	60
72	119	60
	136	60
90	102	60
	153	60
126	102	252
	153	63
128	119	3
	136	192
136	85	3
	119	51
	136	204
	170	192
170	85	51
	170	204
200	119	63
	136	252
204	51	51
	85	15
	170	240
	204	204

Factorizations into compositions of simpler rules.

The 256 rules in table 1 are stated as functions of three site values $\phi(a_{-1}, a_0, a_1)$. Of these, 48 depend only on two of the site values. Some other rules can be formed from compositions of these simpler rules. This table lists rules which can be formed by compositions according to

$$\phi(a_{-1}, a_0, a_1) = \phi_2(\phi_1(-, a_{-1}, a_0), \phi_1(-, a_0, a_1), -),$$

where – indicates that the value is irrelevant. Only minimal representative rules from table 1 are included. In each case, all possible compositions are listed. Note that most of the compositions do not commute.

Table by Erica Jen (*Los Alamos National Laboratory*).

Table 9: Lengths of Distinct Blocks of Sites Newly Excluded at Time *t*

rule	t = 1	2	3	4
0	1	—	—	—
1	3	—	—	—
2	2	—	—	—
3	3	—	—	—
4	2	—	—	—
5	5	—	—	—
6	3	6	7	7
7	4	5	5	6
8	2	1	—	—
9	4	7	9	9
10	3	—	—	—
11	3	5	7	9
12	2	—	—	—
13	4	4	6	6
14	3	5	7	9
15	—	—	—	—
18	3	11	12	13
19	3	3	—	—
22	8	7	11	9
23	5	6	7	8
24	2	3	—	—
25	5	6	8	8
26	4	10	8	11
27	4	6	6	9
28	3	6	6	8
29	4	—	—	—
30	—	—	—	—
32	2	4	6	8
33	4	7	6	6
34	2	—	—	—
35	4	6	7	9
36	3	2	—	—
37	9	8	9	8
38	4	3	—	—
40	3	4	5	7
41	5	9	8	9
42	3	—	—	—
43	5	7	9	11
44	4	4	6	6
45	—	—	—	—
46	3	3	—	—
50	3	5	9	11
51	—	—	—	—
54	5	9	9	7
56	3	4	6	8

rule	t = 1	2	3	4
57	6	5	5	7
58	4	5	5	6
60	—	—	—	—
62	5	7	8	7
72	3	3	—	—
73	6	6	7	14
74	4	6	6	7
76	3	—	—	—
77	5	6	7	8
78	4	4	6	5
90	—	—	—	—
94	5	7	11	11
104	8	8	8	7
105	—	—	—	—
106	—	—	—	—
108	5	4	—	—
110	5	10	11	11
122	5	7	8	10
126	3	12	13	14
128	3	5	7	9
130	4	6	7	10
132	4	5	6	7
134	5	6	6	8
136	3	4	5	6
138	3	—	—	—
140	4	5	6	7
142	5	7	9	11
146	6	6	8	8
150	—	—	—	—
152	5	5	6	6
154	—	—	—	—
156	6	7	7	9
160	5	7	9	11
162	4	6	8	10
164	9	9	8	9
168	4	5	6	7
170	—	—	—	—
172	4	5	6	7
178	5	6	7	8
184	4	6	8	10
200	3	—	—	—
204	—	—	—	—
232	5	6	7	8

Lengths of distinct blocks of sites newly excluded at time t.

Most cellular automaton rules are irreversible, so that even starting from all possible initial configurations, only a subset of configurations can occur after t time steps. In this subset of configurations, only certain blocks of site values can occur. The subset can be specified by giving the blocks which are excluded. In some cases (such as rule 128), the number of distinct excluded blocks is finite; in other cases, it is countably infinite. Irreversibility leads to an increase in the size of the set of excluded blocks with time.

The table gives the lengths of the shortest blocks which are newly excluded after exactly t time steps. Such blocks can occur in configurations up to time $t - 1$, but cannot occur at time t or after. The lengths $L(t)$ of the shortest blocks newly excluded at time t obey the inequality (see pages 159–202 in this book) $L(t) \geq L(t - 1) - 2$.

The notation — in the table indicates that no blocks are newly excluded at a particular time step. This implies that the rule has reached a stable set of configurations, which can occur after any number of steps. It should be noted, however, that this table takes no account of the probabilities with which different configurations may occur.

Table by Lyman P. Hurd (*Mathematics Department, Princeton University*). (Original program by S. Wolfram.)

Table 10: Regular Language Complexities

	$t=1$	$t=2$	$t=3$	$t=4$	$t=5$	$t>5$	∞
0	1 [1]					1 [1]	1 [1]
1	4 [6]					4 [6]	4 [6]
2	3 [4]					3 [4]	3 [4]
3	3 [5]					3 [5]	3 [5]
4	2 [3]					2 [3]	2 [3]
5	9 [15]					9 [15]	9 [15]
6	9 [16]	13 [22]	22 [37]	26 [44]	31 [52]		
7	4 [7]	7 [12]	12 [21]	14 [24]	16 [27]		
8	3 [4]	1 [1]				1 [1]	1 [1]
9	9 [16]	22 [40]	44 [80]	106 [198]	266 [500]		
10	4 [6]					4 [6]	4 [6]
11	3 [5]	7 [12]	10 [17]	12 [20]	14 [23]		
12	2 [3]					2 [3]	2 [3]
13	6 [11]	10 [17]	12 [19]	14 [21]	16 [23]		
14	3 [5]	7 [12]	10 [17]	12 [20]	14 [23]		
15	1 [2]					1 [2]	1 [2]
16	3 [4]					3 [4]	3 [4]
17	3 [5]					3 [5]	3 [5]
18	5 [9]	47 [91]	143 [270]				
19	3 [5]	5 [8]				5 [8]	5 [8]
20	10 [17]	21 [37]	32 [57]	37 [65]	50 [89]		
21	4 [7]	9 [16]	12 [21]	14 [24]	16 [27]		
22	15 [29]	280 [551]	4506 [8963]				
23	11 [20]	15 [26]	19 [32]	23 [38]	27 [44]		
24	2 [3]	3 [4]				3 [4]	3 [4]
25	6 [11]	26 [50]	55 [106]	114 [220]	333 [649]		
26	13 [25]	92 [179]	2238 [4454]				
27	10 [18]	14 [25]	18 [32]	21 [37]	24 [42]		
28	3 [5]	8 [14]	10 [17]	11 [18]	12 [19]		
29	4 [7]					4 [7]	4 [7]
30	1 [2]					1 [2]	1 [2]
32	2 [3]	5 [7]	7 [9]	9 [11]	11 [13]	$2t+1[2t+3]$	4 [6]
33	5 [9]	11 [20]	26 [47]	40 [68]	41 [68]		
34	2 [3]					2 [3]	2 [3]
35	4 [7]	7 [13]	9 [16]	10 [18]	12 [21]		
36	3 [5]	3 [4]				3 [4]	3 [4]
37	15 [29]	194 [376]	870 [1698]	3735 [7290]			
38	5 [9]	5 [8]				5 [8]	5 [8]
40	10 [17]	12 [19]	15 [22]	18 [25]	21 [28]		
41	14 [27]	128 [250]	1049 [2069]				
42	3 [5]					3 [5]	3 [5]
43	9 [16]	13 [22]	17 [28]	21 [34]	25 [40]		
44	4 [7]	11 [20]	18 [32]	23 [40]	27 [46]		
45	1 [2]					1 [2]	1 [2]
46	3 [5]	5 [8]				5 [8]	5 [8]
48	2 [3]					2 [3]	2 [3]
49	4 [7]	6 [10]	7 [11]	9 [14]	10 [15]		

	$t=1$	$t=2$	$t=3$	$t=4$	$t=5$	$t>5$	∞
50	3 [5]	8 [14]	10 [17]	12 [20]	14 [23]		
51	1 [2]	.	.	.	.	1 [2]	1 [2]
52	4 [7]	5 [9]	.	.	.	5 [9]	5 [9]
53	10 [18]	15 [25]	17 [28]	21 [33]	23 [36]		
54	9 [16]	17 [32]	94 [179]	675 [1316]			
56	3 [5]	5 [9]	7 [12]	9 [15]	11 [18]		
57	11 [20]	15 [27]	15 [26]	24 [42]	32 [55]		
58	10 [18]	20 [35]	33 [55]	55 [88]	76 [122]		
60	1 [2]	.	.	.	.	1 [2]	1 [2]
61	5 [9]	16 [30]	40 [76]	94 [177]	185 [350]		
62	5 [9]	21 [39]	61 [114]	81 [150]	129 [240]		
64	3 [4]	1 [1]	.	.	.	1 [1]	1 [1]
65	9 [15]	20 [35]	42 [75]	88 [157]	220 [401]		
66	2 [3]	3 [4]	.	.	.	3 [4]	3 [4]
68	2 [3]	.	.	.	.	2 [3]	2 [3]
69	5 [8]	10 [17]	12 [19]	14 [23]	16 [25]		
70	3 [5]	8 [14]	9 [15]	11 [19]	11 [19]		
72	5 [9]	5 [8]	.	.	.	5 [8]	5 [8]
73	15 [29]	82 [155]	390 [757]	1443 [2796]			
74	13 [25]	45 [85]	66 [123]	69 [125]	75 [135]		
76	3 [5]	.	.	.	.	3 [5]	3 [5]
77	11 [20]	15 [26]	19 [32]	23 [38]	27 [44]		
78	10 [18]	15 [27]	18 [30]	20 [34]	22 [36]		
80	4 [6]	.	.	.	.	4 [6]	4 [6]
81	3 [5]	7 [11]	9 [14]	11 [16]	13 [19]		
82	13 [25]	167 [331]	3134 [6257]				
84	3 [5]	7 [12]	9 [14]	11 [17]	13 [19]		
85	1 [2]	.	.	.	.	1 [2]	1 [2]
86	1 [2]	.	.	.	.	1 [2]	1 [2]
88	13 [25]	63 [117]	114 [210]	117 [213]	1288 [2106]		
89	1 [2]	.	.	.	.	1 [2]	1 [2]
90	1 [2]	.	.	.	.	1 [2]	1 [2]
92	10 [18]	14 [23]	18 [29]	18 [27]	22 [33]		
94	15 [29]	230 [455]	3904 [7760]				
96	9 [16]	11 [17]	14 [20]	17 [23]	20 [26]		
97	14 [27]	99 [195]	626 [1237]				
98	3 [5]	4 [6]	6 [9]	8 [12]	10 [15]		
100	5 [9]	11 [19]	17 [29]	18 [29]	22 [34]		
102	1 [2]	.	.	.	.	1 [2]	1 [2]
104	15 [29]	265 [525]	2340 [4647]	1394 [2675]	1542 [2913]		
105	1 [2]	.	.	.	.	1 [2]	1 [2]
106	1 [2]	.	.	.	.	1 [2]	1 [2]
108	9 [16]	11 [19]	.	.	.	11 [19]	11 [19]
110	5 [9]	20 [38]	160 [312]	1035 [2037]			
112	3 [5]	.	.	.	.	3 [5]	3 [5]
113	9 [16]	13 [22]	17 [28]	21 [34]	25 [40]		
114	10 [18]	20 [35]	33 [56]	50 [82]	72 [115]		
116	3 [5]	5 [8]	.	.	.	5 [8]	5 [8]
118	5 [9]	16 [29]	49 [92]	74 [139]	95 [175]		

	$t=1$	$t=2$	$t=3$	$t=4$	$t=5$	$t>5$	∞
120	1 [2]					1 [2]	1 [2]
122	15 [29]	179 [347]	5088 [9933]				
124	5 [9]	20 [38]	208 [407]	1356 [2672]			
126	3 [5]	13 [23]	107 [198]	2867 [5476]			
128	4 [6]	6 [8]	8 [10]	10 [12]	12 [14]	$2t+2[2t+4]$	3 [5]
130	9 [15]	14 [21]	18 [25]	22 [29]	26 [33]		
132	5 [9]	7 [12]	9 [15]	11 [18]	13 [21]		
134	14 [27]	44 [82]	99 [182]	125 [224]			
136	3 [5]	4 [6]	5 [7]	6 [8]	7 [9]	$t+2[t+4]$	3 [5]
138	3 [5]					3 [5]	3 [5]
140	4 [7]	5 [9]	6 [11]	7 [13]	8 [15]		
142	9 [16]	13 [22]	17 [28]	21 [34]	25 [40]		
144	9 [16]	16 [28]	20 [34]	24 [40]	28 [46]		
146	15 [29]	92 [177]	1587 [3126]				
148	14 [27]	68 [127]	113 [209]	188 [347]			
150	1 [2]					1 [2]	1 [2]
152	6 [11]	20 [37]	30 [55]	32 [59]	36 [65]		
154	1 [2]					1 [2]	1 [2]
156	11 [20]	20 [35]	24 [42]	28 [47]	34 [58]		
160	9 [15]	16 [24]	25 [35]	36 [48]	49 [63]	$(t+2)^2[(t+2)(t+4)]$	9 [15]
162	5 [8]	7 [10]	9 [12]	11 [14]	13 [16]		
164	15 [29]	116 [227]	667 [1310]	1214 [2363]			
168	4 [7]	5 [8]	6 [9]	7 [10]	8 [11]	$t+3[t+6]$	3 [5]
170	1 [2]					1 [2]	1 [2]
172	10 [18]	11 [20]	12 [22]	13 [24]	14 [26]		
176	6 [11]	8 [14]	10 [17]	12 [20]	14 [23]		
178	11 [20]	15 [26]	19 [32]	23 [38]	27 [44]		
180	1 [2]					1 [2]	1 [2]
184	4 [7]	6 [10]	8 [13]	10 [16]	12 [19]		
188	5 [9]	14 [25]	21 [36]	25 [43]	33 [56]		
192	3 [5]	4 [6]	5 [7]	6 [8]	7 [9]		
196	4 [7]	5 [8]	6 [9]	7 [10]	8 [11]		
200	3 [5]					3 [5]	3 [5]
204	1 [2]					1 [2]	1 [2]
208	3 [5]					3 [5]	3 [5]
212	9 [16]	13 [22]	17 [28]	21 [34]	25 [40]		
216	10 [18]	11 [19]	12 [20]	13 [21]	14 [22]		
224	4 [7]	5 [8]	6 [9]	7 [10]	8 [11]	$t+3[t+6]$	3 [5]
232	11 [20]	15 [26]	19 [32]	23 [38]	27 [44]		
240	1 [2]					1 [2]	1 [2]

Regular language complexities.

The set of configurations that can appear after t steps in the evolution of a one-dimensional cellular automaton can be shown to form a regular formal language (see pages 159–202 in this book). Possible configurations thus correspond to possible paths through a finite graph which represents the grammar for the regular language.

The table gives the minimum number of nodes in the graphs for such grammars; the number of arcs is given in brackets in each case. The notation . indicates that the regular language is the same as at the preceding time step.

Entries in the last column of the table give sizes of graphs for regular languages representing limiting sets of states that can be reached after any number of steps.

The size of a regular grammar gives a measure of the "complexity" of the set of configurations it describes. Notice that the grammar specifies merely which configurations can possibly occur; it does not account for the probabilities of different configurations.

The graphs used for the table represent possible sequences of site values that occur in configurations read from left to right. Rules related by reflection may in general yield different regular languages. The table thus includes minimal representatives for all rules from table 1 not related by complementation.

Entries in the table for $t \leq 5$ that have been left blank were not found. They are probably $\gtrsim 20000$. The growth of regular language complexities is bounded by $2^{2^{4t}} - 1$.

For some rules, it has been possible to find explicit forms for the regular languages produced after any number of time steps. Formulae for complexities in these cases are listed in the table. In many cases, it is however suspected that the limiting set does not form a regular language, and may in fact be non-recursive.

Table by Lyman P. Hurd (*Mathematics Department, Princeton University*). (Original program by S. Wolfram.)

Table 11: Measure Theoretical Complexities

rule	$t = 1$	$t = 2$	$t = 3$
0	0	0	0
1	0.8223	0.8223	0.8223
2	0.7356	0.7356	0.7356
3	0.9003	0.9003	0.9003
4	0.3768	0.3768	0.3768
5	1.8005	1.8005	1.8005
6	1.783	1.964	1.968
7	1.2707	1.670	1.933
8	0.7356	0	0
9	1.9135	2.598	3.303
10	1.1247	1.1247	1.1247
11	1.0434	1.3442	1.973
12	0.5623	0.5623	0.5623
13	1.4756	1.7879	1.713
14	0.9026	1.3607	1.927
15	0	0	0
16	0.7356	0.7356	0.7356
17	0.9003	0.9003	0.9003
18	0.9026	2.129	3.933
19	0.9026	1.1539	1.1539
20	1.7756	2.759	3.059
21	1.2707	1.8266	1.931
22	2.591	4.601	6.213
23	1.9862	2.153	2.330
24	0.5623	0.9216	0.9216
25	1.643	2.665	3.231
26	2.244	2.659	2.945
27	1.666	2.128	2.392
28	0.8305	1.6009	1.6645
29	1.2652	1.2652	1.2652
30	0	0	0
32	0.3768	0.4957	0.2346
33	1.2930	1.941	2.529
34	0.5623	0.5623	0.5623
35	1.1034	1.748	2.006
36	0.9003	0.4634	0.4634
37	2.518	4.435	5.410
38	1.298	1.256	1.256
40	1.775	1.547	1.127
41	2.332	4.134	5.471
42	0.9003	0.9003	0.9003
43	1.9584	2.269	2.483
44	0.9003	1.574	1.748
45	0	0	0
46	0.5623	0.9216	0.9216
48	0.5623	0.5623	0.5623
49	1.2512	1.451	1.455
50	0.8305	1.589	1.775
51	0	0	0

rule	$t = 1$	$t = 2$	$t = 3$
52	0.9003	0.9216	0.9216
53	1.906	2.057	2.016
54	1.7707	2.609	3.921
56	0.8305	1.3503	1.732
57	2.086	2.190	1.946
58	1.9132	2.132	2.106
60	0	0	0
61	1.430	2.065	3.076
62	1.341	2.406	3.720
64	0.7356	0	0
65	1.5310	2.268	2.970
66	0.5623	0.9216	0.9216
68	0.5623	0.5623	0.5623
69	1.0562	1.790	1.842
70	0.8305	1.7802	1.815
72	0.9003	0.4634	0.4634
73	2.604	3.685	4.473
74	2.461	2.713	2.755
76	0.8305	0.8305	0.8305
77	1.9862	2.153	2.330
78	1.7553	2.111	2.029
80	1.1247	1.1247	1.1247
81	1.0434	1.5837	1.716
82	2.460	3.823	5.375
84	0.8992	1.740	1.984
85	0	0	0
86	0	0	0
88	2.2441	3.081	3.605
89	0	0	0
90	0	0	0
92	1.9132	1.768	1.735
94	2.599	3.682	5.311
96	1.782	1.3689	0.995
97	2.491	3.470	4.846
98	0.8305	0.8932	0.979
100	1.298	1.667	1.632
102	0	0	0
104	2.591	4.379	4.969
105	0	0	0
106	0	0	0
108	1.7707	1.5093	1.5093
110	1.344	2.435	3.407
112	0.9003	0.9003	0.9003
113	1.957	2.266	2.482
114	1.754	2.344	2.825
116	0.5623	0.9215	0.9215
118	1.342	1.945	2.827
120	0	0	0
122	2.600	4.307	5.981

rule	$t = 1$	$t = 2$	$t = 3$
124	1.343	2.321	3.933
126	0.9003	2.049	3.914
128	0.8223	0.457	0.1986
130	1.533	1.263	1.025
132	1.292	1.459	1.637
134	2.496	3.050	3.010
136	0.9003	0.8223	0.641
138	1.0434	1.0434	1.0434
140	1.2512	1.5808	1.7551
142	1.957	2.266	2.482
144	1.913	1.988	2.056
146	2.604	3.742	5.350
148	2.328	3.589	3.815
150	0	0	0
152	1.644	2.626	3.031
154	0	0	0
156	2.083	2.506	2.563
160	1.805	1.633	1.281
162	1.0562	0.8654	0.7355

rule	$t = 1$	$t = 2$	$t = 3$
164	2.520	3.343	3.353
168	1.2707	1.3676	1.369
170	0	0	0
172	1.909	2.080	2.242
176	1.4757	1.723	1.863
180	0	0	0
184	1.2652	1.575	1.788
188	1.433	1.962	1.733
192	0.9003	0.8113	0.6415
196	1.1034	1.0302	0.9170
200	0.9003	0.9003	0.9003
204	0	0	0
208	1.0434	1.0434	1.0434
212	1.9584	2.269	2.483
216	1.667	2.033	2.065
224	1.2707	1.368	1.369
232	1.9862	2.153	2.330
240	0	0	0

Measures of the information content of regular grammars for sets of configurations generated by evolution from disordered initial states.

This table gives values of a probabilistic analogue of the regular language complexity of table 10, in which the nodes of regular language graphs are weighted with the probabilities that they are visited.

Starting from a disordered state in which all possible configurations occur with equal probability, irreversible cellular automaton evolution can lead to ensembles in which different configurations occur with different probabilities. These ensembles can be described by probabilistic analogues of regular languages.

All the configurations that can occur after t steps correspond to possible paths through the standard regular language graphs of table 10. To account for the different probabilities of different configurations, one may weight the nodes of the graph according to the probabilities P_i that they are visited. In terms of these probabilities, one may then compute a measure theoretical complexity $-\sum P_i \log_2 P_i$, where the sum runs over all nodes in the regular language graph. The table gives estimated values for this quantity. The last digit in each estimate is subject to statistical errors.

Table and concept by Peter Grassberger (*Physics Department, University of Wuppertal*).

Table 12: Iterated Rule Expression Sizes

rule	$t=1$	$t=2$	$t=3$	$t=4$	$t=5$
0	0 (0)	0 (0)	0 (0)	0 (0)	0 (0)
1	1 (1)	6 (4)	1 (1)	6 (4)	1 (1)
2	1 (1)	1 (1)	1 (1)	1 (1)	1 (1)
3	1 (1)	3 (2)	1 (1)	3 (2)	1 (1)
4	1 (1)	1 (1)	1 (1)	1 (1)	1 (1)
5	1 (1)	2 (2)	1 (1)	2 (2)	1 (1)
6	2 (2)	4 (4)	13 (10)	25 (21)	110 (50)
7	2 (2)	7 (4)	6 (6)	18 (8)	10 (10)
8	1 (1)	0 (0)	0 (0)	0 (0)	0 (0)
9	2 (2)	5 (5)	18 (11)	43 (31)	138 (53)
10	1 (1)	1 (1)	1 (1)	1 (1)	1 (1)
11	2 (2)	5 (4)	10 (6)	26 (12)	50 (16)
12	1 (1)	1 (1)	1 (1)	1 (1)	1 (1)
13	2 (2)	6 (4)	9 (5)	13 (7)	17 (8)
14	2 (2)	5 (5)	17 (10)	51 (24)	144 (48)
15	1 (1)	1 (1)	1 (1)	1 (1)	1 (1)
18	2 (2)	4 (4)	18 (18)	35 (26)	140 (108)
19	2 (2)	8 (5)	3 (3)	8 (5)	3 (3)
22	3 (3)	7 (7)	27 (26)	80 (62)	308 (206)
23	3 (3)	8 (5)	7 (7)	33 (9)	11 (11)
24	2 (2)	4 (4)	4 (4)	4 (4)	4 (4)
25	2 (2)	4 (4)	8 (8)	20 (16)	42 (27)
26	2 (2)	6 (6)	21 (17)	56 (43)	192 (100)
27	3 (2)	4 (4)	7 (5)	7 (6)	15 (7)
28	2 (2)	4 (4)	11 (7)	15 (11)	30 (12)
29	3 (2)	4 (4)	3 (2)	4 (4)	3 (2)
30	3 (3)	9 (7)	23 (17)	76 (41)	185 (105)
31	2 (2)	6 (4)	7 (6)	12 (10)	11 (9)
32	1 (1)	1 (1)	1 (1)	1 (1)	1 (1)
33	2 (2)	7 (7)	12 (12)	44 (23)	38 (24)
34	1 (1)	1 (1)	1 (1)	1 (1)	1 (1)
35	2 (2)	4 (3)	8 (4)	13 (8)	30 (11)
36	2 (2)	2 (2)	2 (2)	2 (2)	2 (2)
37	2 (2)	8 (7)	25 (17)	75 (47)	238 (109)
38	2 (2)	4 (3)	4 (4)	4 (3)	4 (4)
39	3 (2)	3 (3)	6 (5)	5 (5)	15 (7)
40	2 (2)	3 (3)	5 (5)	8 (8)	13 (13)
41	3 (3)	8 (8)	26 (24)	92 (69)	283 (218)
42	2 (2)	2 (2)	2 (2)	2 (2)	2 (2)
43	3 (3)	7 (6)	24 (12)	62 (27)	176 (55)
44	2 (2)	3 (3)	4 (4)	8 (5)	10 (6)
45	3 (3)	9 (8)	24 (20)	72 (53)	219 (118)
46	3 (2)	6 (4)	6 (4)	6 (4)	6 (4)
47	2 (2)	5 (4)	13 (6)	28 (8)	64 (16)
50	2 (2)	6 (4)	15 (6)	31 (8)	64 (10)
51	1 (1)	1 (1)	1 (1)	1 (1)	1 (1)
54	3 (3)	7 (6)	18 (15)	59 (38)	165 (85)
55	2 (2)	5 (3)	5 (5)	5 (3)	5 (5)

rule	$t=1$	$t=2$	$t=3$	$t=4$	$t=5$
56	2 (2)	6 (4)	14 (9)	38 (20)	103 (45)
57	3 (3)	7 (6)	17 (12)	41 (23)	130 (50)
58	2 (2)	6 (5)	15 (10)	34 (18)	80 (32)
59	2 (2)	4 (4)	9 (8)	14 (10)	34 (17)
60	2 (2)	2 (2)	8 (8)	2 (2)	8 (8)
61	3 (3)	4 (4)	14 (10)	21 (17)	60 (30)
62	3 (3)	6 (6)	20 (12)	56 (27)	137 (48)
63	2 (2)	2 (2)	2 (2)	2 (2)	2 (2)
72	2 (2)	2 (2)	2 (2)	2 (2)	2 (2)
73	3 (3)	8 (7)	36 (20)	90 (46)	276 (118)
74	2 (2)	5 (5)	13 (11)	30 (22)	77 (45)
75	3 (3)	8 (8)	24 (20)	81 (52)	241 (118)
76	2 (2)	2 (2)	2 (2)	2 (2)	2 (2)
77	3 (3)	7 (5)	14 (7)	32 (9)	57 (11)
78	2 (2)	5 (4)	8 (7)	20 (10)	21 (12)
79	2 (2)	7 (4)	9 (5)	13 (6)	20 (7)
90	2 (2)	2 (2)	8 (8)	2 (2)	8 (8)
91	3 (3)	9 (6)	26 (19)	82 (47)	255 (107)
94	3 (3)	8 (8)	26 (19)	106 (46)	276 (106)
95	2 (2)	2 (2)	2 (2)	2 (2)	2 (2)
104	3 (3)	6 (6)	15 (14)	27 (26)	49 (45)
105	4 (4)	4 (4)	16 (16)	4 (4)	256 (256)
106	3 (3)	5 (5)	25 (21)	46 (37)	192 (126)
107	4 (4)	10 (9)	37 (28)	108 (70)	390 (210)
108	3 (3)	5 (5)	9 (8)	5 (5)	9 (8)
109	4 (4)	10 (8)	31 (20)	91 (54)	268 (118)
110	3 (3)	7 (6)	15 (15)	40 (28)	95 (60)
111	3 (3)	7 (6)	21 (14)	57 (25)	139 (56)
122	3 (3)	9 (8)	27 (20)	88 (48)	264 (136)
123	3 (3)	8 (8)	22 (13)	51 (28)	81 (30)
126	3 (3)	8 (8)	22 (19)	103 (67)	221 (116)
127	3 (3)	3 (3)	3 (3)	3 (3)	3 (3)
128	1 (1)	1 (1)	1 (1)	1 (1)	1 (1)
129	2 (2)	9 (8)	26 (20)	93 (78)	250 (120)
130	2 (2)	2 (2)	5 (4)	5 (4)	8 (6)
131	2 (2)	6 (5)	13 (10)	36 (20)	88 (45)
132	2 (2)	3 (3)	4 (4)	5 (5)	6 (6)
133	2 (2)	8 (7)	23 (17)	74 (41)	216 (111)
134	3 (3)	6 (6)	20 (17)	46 (34)	174 (90)
135	3 (3)	9 (7)	22 (17)	66 (41)	202 (107)
136	1 (1)	1 (1)	1 (1)	1 (1)	1 (1)
137	2 (2)	8 (6)	14 (14)	39 (25)	111 (60)
138	2 (2)	2 (2)	2 (2)	2 (2)	2 (2)
139	2 (2)	5 (4)	6 (4)	6 (4)	6 (4)
140	2 (2)	2 (2)	2 (2)	2 (2)	2 (2)
141	2 (2)	7 (4)	10 (6)	22 (8)	28 (9)
142	3 (3)	7 (6)	18 (12)	52 (27)	151 (55)
143	2 (2)	5 (4)	12 (8)	32 (15)	86 (34)

rule	$t=1$	$t=2$	$t=3$	$t=4$	$t=5$
146	3 (3)	6 (6)	29 (29)	61 (48)	224 (193)
147	3 (3)	8 (7)	21 (18)	69 (39)	207 (79)
150	4 (4)	4 (4)	16 (16)	4 (4)	256 (256)
151	4 (4)	10 (7)	33 (29)	104 (63)	372 (201)
152	2 (2)	4 (4)	7 (7)	12 (11)	20 (19)
153	2 (2)	2 (2)	8 (8)	2 (2)	8 (8)
154	3 (3)	4 (4)	28 (15)	6 (6)	42 (19)
155	3 (3)	7 (4)	8 (5)	8 (4)	8 (5)
156	3 (3)	3 (3)	14 (8)	12 (8)	43 (13)
157	3 (3)	7 (4)	11 (7)	24 (8)	23 (10)
158	4 (4)	12 (8)	34 (20)	106 (37)	330 (92)
159	3 (3)	9 (7)	18 (14)	63 (24)	139 (55)
160	1 (1)	1 (1)	1 (1)	1 (1)	1 (1)
161	2 (2)	6 (6)	20 (18)	65 (43)	236 (140)
162	2 (2)	3 (3)	4 (4)	5 (5)	6 (6)
163	2 (2)	5 (5)	15 (9)	32 (15)	73 (25)
164	2 (2)	5 (4)	10 (10)	16 (13)	27 (21)
165	2 (2)	2 (2)	8 (8)	2 (2)	8 (8)
166	3 (3)	4 (4)	24 (15)	6 (6)	32 (19)
167	3 (3)	8 (7)	26 (19)	82 (43)	218 (104)
168	2 (2)	5 (4)	10 (8)	23 (16)	49 (32)
169	3 (3)	6 (5)	28 (21)	76 (37)	244 (124)
170	1 (1)	1 (1)	1 (1)	1 (1)	1 (1)
171	3 (2)	3 (2)	3 (2)	3 (2)	3 (2)
172	2 (2)	4 (3)	6 (4)	7 (5)	9 (6)
173	3 (3)	10 (7)	32 (14)	70 (27)	206 (46)
174	2 (2)	2 (2)	2 (2)	2 (2)	2 (2)
175	2 (2)	2 (2)	2 (2)	2 (2)	2 (2)
178	3 (3)	7 (5)	16 (7)	32 (9)	65 (11)
179	2 (2)	4 (4)	8 (6)	15 (8)	31 (10)
182	4 (4)	11 (9)	47 (33)	103 (55)	466 (162)
183	3 (3)	7 (7)	24 (22)	55 (22)	203 (69)

rule	$t=1$	$t=2$	$t=3$	$t=4$	$t=5$
184	2 (2)	5 (5)	15 (13)	48 (37)	161 (111)
185	3 (3)	5 (5)	24 (10)	51 (22)	149 (45)
186	3 (2)	8 (3)	20 (4)	43 (5)	88 (6)
187	2 (2)	2 (2)	2 (2)	2 (2)	2 (2)
188	3 (3)	6 (5)	24 (10)	39 (15)	125 (23)
189	3 (3)	10 (5)	8 (5)	8 (5)	8 (5)
190	4 (3)	5 (4)	23 (6)	21 (7)	91 (9)
191	3 (3)	3 (3)	3 (3)	3 (3)	3 (3)
200	2 (2)	2 (2)	2 (2)	2 (2)	2 (2)
201	3 (3)	10 (5)	23 (10)	10 (5)	23 (10)
202	2 (2)	4 (4)	7 (6)	11 (8)	16 (10)
203	3 (3)	7 (5)	17 (9)	41 (13)	66 (19)
204	1 (1)	1 (1)	1 (1)	1 (1)	1 (1)
205	3 (2)	3 (2)	3 (2)	3 (2)	3 (2)
206	3 (2)	5 (3)	8 (4)	11 (5)	15 (6)
207	2 (2)	2 (2)	2 (2)	2 (2)	2 (2)
218	3 (3)	9 (6)	40 (16)	92 (24)	158 (38)
219	3 (3)	10 (5)	10 (5)	10 (5)	10 (5)
222	4 (3)	10 (5)	19 (7)	31 (9)	46 (11)
223	3 (3)	3 (3)	3 (3)	3 (3)	3 (3)
232	3 (3)	8 (5)	19 (7)	33 (9)	58 (11)
233	4 (4)	10 (7)	39 (20)	112 (34)	307 (63)
234	2 (2)	6 (4)	17 (7)	44 (12)	106 (21)
235	4 (3)	11 (6)	28 (10)	62 (14)	134 (19)
236	3 (2)	3 (2)	3 (2)	3 (2)	3 (2)
237	4 (3)	7 (5)	7 (5)	7 (5)	7 (5)
238	2 (2)	4 (3)	6 (4)	9 (5)	12 (6)
239	3 (3)	1 (1)	1 (1)	1 (1)	1 (1)
250	2 (2)	3 (3)	4 (4)	5 (5)	6 (6)
251	3 (3)	6 (5)	10 (7)	19 (9)	28 (11)
254	4 (3)	11 (5)	24 (7)	45 (9)	76 (11)
255	1 (1)	1 (1)	1 (1)	1 (1)	1 (1)

Sizes of Boolean expressions representing functions corresponding to iterations of cellular automaton rules.

Cellular automaton rules with $k = 2$ and $r = 1$ can be expressed as Boolean functions of three variables, as in table 1. Iterations of these rules for t steps correspond to functions of $2t + 1$ variables, which may be expressed as Boolean expressions.

The minimal Boolean expressions obtained after one step were given in table 1. This table gives the numbers of terms in Boolean expressions obtained after t time steps. An increase in these numbers potentially reflects increasing difficulty of computing the outcome of more steps of cellular automaton evolution.

The first number in each case gives the number of prime implicants in the corresponding Boolean expression. The possible values of a set of n Boolean variables correspond to the vertices of a Boolean n-cube. The cases in which a Boolean function has value 1 then correspond to a region of the Boolean n-cube. The number of prime implicants is essentially the number of hyperplanes of various dimensions which must be combined to form this region.

Boolean expressions can conveniently be stated in a disjunctive normal form (DNF), in which they are written as a disjunction (OR) of conjunctions (ANDs). The number of prime implicants gives an upper bound on the number of terms needed in such a form.

Notice that complementation of a function has no simple effect on its DNF expression. As a result, the table includes minimal representatives for all rules from table 1 not related by reflection.

The general problem of finding an absolutely minimal DNF representation for a function appears to be computationally intractable. The table gives in parentheses the numbers of terms in minimal DNF expressions found by the *espresso* computer program (R. Rudell, Computer Science Department, University of California, Berkeley, 1985) which incorporates known algebraic and heuristic techniques. In most cases, the results given are probably absolutely minimal.

Table 13: Finite Lattice State Transition Diagrams

rule 0:
N=9:1x1
N=10:1x1
N=11:1x1

rule 12:
N=9:76x1
N=10:123x1
N=11:199x1

rule 1:
N=9:37x2
N=10:61x2
N=11:100x2

rule 13:
N=9:1x2,
12x1
N=10:1x2,
17x1
N=11:1x2,
22x1

rule 2:
N=9:3x9,
1x3,1x1
N=10:4x10,
1x5,1x1
N=11:6x11,
1x1

rule 14:
N=9:4x18,
1x7,1x6,
1x1
N=10:13x10,
5x5,3x1
N=11:9x22,
9x11,1x1

rule 3:
N=9:8x18,
1x9,1x3,
1x1
N=10:13x20,
1x2,1x5,
N=11:21x22,
2x11,1x2

rule 15:
N=9:28x18,
1x6,1x2
N=10:99x10,
2x1,1x2,
N=11:93x22

rule 4:
N=9:76x1
N=10:123x1
N=11:199x1

rule 18:
N=9:2x6,
9x2,1x1
N=10:5x6,
1x4,15x2,
1x1
N=11:2x11,
11x4,11x2,
1x1

rule 5:
N=9:73x2,
12x1
N=10:136x2,
17x1
N=11:232x2,
22x1

rule 19:
N=9:37x2
N=10:61x2
N=11:100x2

rule 6:
N=9:3x18,
1x1
N=10:11x10,
4x5,1x1
N=11:5x22,
1x1

rule 22:
N=9:9x4,
1x1
N=10:10x6,
15x4,3x1
N=11:2x11,
11x5,11x4,
1x1

rule 7:
N=9:2x18,
1x9,1x2
N=10:3x20,
2x7,1x2,
N=11:4x22,
1x11,1x2

rule 23:
N=9:37x2
N=10:61x2,
1x1
N=11:100x2

rule 8:
N=9:1x1
N=10:1x1
N=11:1x1

rule 24:
N=9:3x9,
1x3,1x1
N=10:4x10,
1x5,1x1
N=11:6x11,
1x1

rule 9:
N=9:1x18,
3x3,1x2
N=10:2x15,
1x2,2x5,
1x2
N=11:2x22,
1x2

rule 25:
N=9:1x18,
1x3,
1x2
N=10:2x20,
2x10,1x15,
1x2
N=11:1x22,
1x2,1x11,
1x2

rule 10:
N=9:8x9,
1x3,1x1
N=10:11x10,
2x5,1x1
N=11:18x11,
1x1

rule 26:
N=9:2x72,
1x18,1x6,
N=10:12x20,
2x6,1x5,
1x1
N=11:4x88,
1x22,2x11,
1x2,1x1

rule 11:
N=9:4x18,
1x3,1x2
N=10:12x10,
2x5,1x2
N=11:9x22,
1x11,1x2

rule 27:
N=9:9x18,
1x2,1x6,
N=10:13x20,
1x15,
1x2
N=11:21x22,
1x2,1x11

rule 28:
N=9:21x2, 1x1
N=10:30x2, 1x1
N=11:55x2, 1x1

rule 29:
N=9:121x2
N=10:221x2, 1x1
N=11:408x2

rule 30:
N=9:1x171, 1x2,1x1
N=10:2x15, 1x5,3x1
N=11:1x154, 11x17,1x1

rule 32:
N=9:1x1
N=10:1x2, 1x1
N=11:1x1

rule 33:
N=9:37x2
N=10:62x2
N=11:100x2

rule 34:
N=9:8x9, 1x3,1x1
N=10:11x10, 1x2,1x2
N=11:18x11, 1x1

rule 35:
N=9:8x18, 1x3,1x3, 1x1
N=10:13x20, 1x2,2x5, 2x2
N=11:21x22, 1x11,1x2

rule 36:
N=9:31x1
N=10:46x1
N=11:67x1

rule 37:
N=9:19x2, 3x1
N=10:26x2
N=11:2x33, 4x1

rule 38:
N=9:6x18, 1x2,1x1
N=10:19x10, 1x5,1x1
N=11:15x22, 1x1

rule 40:
N=9:1x1, 1x1
N=10:1x10, 1x5,1x2, 1x1
N=11:2x11, 1x1

rule 41:
N=9:2x36, 1x3,4x3,1x1
N=10:2x40, 1x18,1x5, 2x2
N=11:1x44, 3x11,1x2

rule 42:
N=9:26x9, 1x3,1x1
N=10:42x10, 1x1,1x2
N=11:74x11, 1x1

rule 43:
N=9:1x18, 1x3,2x1, 1x2
N=10:22x10, 4x5,2x2
N=11:9x22, 16x11,1x2

rule 44:
N=9:1x3, 1x1
N=10:46x1
N=11:67x1

rule 45:
N=9:1x504, 1x2, 1x1
N=10:1x430, 1x19,1x18, 1x2
N=11:1x979, 1x935,11x6, 1x2,11x5, 1x2

rule 46:
N=9:3x9, 1x3,1x1
N=10:4x10, 1x5,1x1
N=11:6x11, 1x1

rule 50:
N=9:39x2, 1x1
N=10:61x2, 1x1
N=11:99x2, 1x1

rule 51:
N=9:256x2
N=10:512x2
N=11:1024x2

rule 54:
N=9:2x27, 9x4,1x1
N=10:2x10, 15x4,1x1
N=11:2x99, 2x11,11x4, 1x1

rule 56:
N=9:9x9, 1x3,1x1
N=10:13x10, 1x5,1x2, 1x1
N=11:20x11, 1x1

rule 57:
N=9:2x9, 1x1
N=10:2x10, 1x5,2x2
N=11:4x11, 1x1

rule 58:
N=9:1x18, 1x1
N=10:1x20, 1x2,1x5, 1x1
N=11:2x22, 1x1

rule 60:
N=9:1x63, 1x1
N=10:9x10, 1x1
N=11:3x341, 1x1

rule 62:
N=9;1x18,
22x3;1x1,
N=10;1x30,
1x1;35x3,
N=11;2x23,
1x1;77x3,

rule 72:
N=9:31x1
N=10:46x1
N=11:67x1

rule 73:
N=9;18x2,
19x2;3x1,
N=10;10x2,
16x1;
N=11;11x12,
1x5;34x2,
11x1,

rule 74:
N=9;2x18,
4x1,
N=10;1x30,
1x5;1x1,
N=11;1x22,
6x11;
1x1,

rule 76:
N=9:241x1
N=10:443x1
N=11:815x1

rule 77:
N=9;1x2,
76x1,
N=10;1x2,
122x1,
N=11;1x2,
198x1,

rule 78:
N=9:13x1
N=10:18x1
N=11:23x1

rule 90:
N=9;36x7,
4x1,
N=10;40x6,
5x3;1x1,
N=11:33x31,
1x1,

rule 94:
N=9;9x3,
1x3;13x1
N=10;5x6,
10x2;40x2,
N=11;11x2,
4x1;11x1,

rule 104:
N=9:19x1
N=10:1x2,
26x1,
N=11:34x1

rule 105:
N=9:9x14,
1x3,
N=10:170x6,
5x3,
N=11:33x62,
1x1,

rule 106:
N=9;3x54,
1x1,
N=10;2x205,
1x1;1x1,
N=11;1x176,
18x11;1x1,

rule 108:
N=9;54x2,
76x1,
N=10;100x2,
121x1,
N=11;187x2,
199x1,

rule 110:
N=9;9x7,
3x3;1x1,
N=10;2x25,
2x19;16x5,
1x1,
N=11;1x10,
11x7;1x1,

rule 122:
N=9;9x6,
9x2;1x1,
N=10;5x6,
5x4;16x2,
1x1,
N=11;2x11,
1x4;21x1,
1x1,

rule 126:
N=9;9x6,
9x2;1x1,
N=10;5x6,
5x4;15x2,
1x1,
N=11;2x11,
1x4;21x2,
1x1,

rule 128:
N=9:2x1
N=10:2x1
N=11:2x1

rule 130:
N=9;3x9,
1x3;2x1,
N=10;4x10,
1x5;2x1,
N=11:6x11,
2x1,

rule 132:
N=9:77x1
N=10:124x1
N=11:200x1

rule 134:
N=9:3x18,
2x1,
N=10;11x10,
4x5;4x1,
N=11:5x22,
2x1,

rule 136:
N=9:2x1
N=10:2x1
N=11:2x1

rule 138:
N=9;17x9,
1x3;2x1,
N=10;26x10,
5x5;2x1,
N=11;44x11,
2x1,

rule 140:
N=9:77x1
N=10:124x1
N=11:200x1

rule 142:
N=9;4x18,
2x1;1x6,
N=10;22x10,
4x5;4x1,
N=11;9x22,
18x11;2x1,

State transition diagrams for cellular automata on finite size lattices.

A $k = 2$ cellular automaton on a finite lattice with N sites has a total of 2^N possible states. The complete evolution of such a cellular automaton can be represented by a finite diagram which shows the possible transitions between these states. Each node in the diagram corresponds to a complete configuration or state of the finite cellular automaton. A directed arc leads from each such node to its successor under one time step of cellular automaton evolution. The possible time sequences of configurations in the complete evolution of the cellular automaton then correspond to possible paths through the directed graph thus formed.

After a time of at most 2^N steps, a finite cellular automaton must always enter a cycle, periodically visiting a fixed set of states. In general, the complete state transition diagram contains a number of distinct cycles.

564

The table shows the fragment of the state transition diagram associated with the longest cycle, for all inequivalent $k = 2$, $r = 1$ rules. Results are given for lattices of sizes $N = 9$, $N = 10$ and $N = 11$. In all cases, the lattices are taken to have periodic boundary conditions, as if their sites were arrranged in a circle.

The table also gives the lengths and multiplicities of all the cycles for each rule. (The notation used is $g \times L$, representing g cycles of length L.) Notice that the state transition diagram fragments associated with different cycles of the same length may not be identical. When there are several cycles of maximal length, the fragment shown is the one involving the largest total number of states.

State transition diagram fragments have the general form of cycles fed by trees. The cellular automaton always reaches the cycle after a sufficiently long time. The trees represent transients, and contain states which can occur only after a limited number of time steps. Such transient phenomena are a manifestation of irreversibility in the cellular automaton evolution.

Some finite cellular automata, such as rule 13, are reversible, so that their state transition diagrams contain no transients, and all states are on cycles.

In some other cases, such as rule 90, highly regular state transition diagrams are obtained, containing for example only balanced trees (see pages 159–202 in this book). Many rules, however, yield complicated state transition diagrams.

In the pictures given, individual nodes are not indicated. Nevertheless, the arcs joining nodes are all of equal length in a particular diagram. The overall scale of each diagram can be deduced from the total cycle length given.

The constraint of equal length in some cases forces arcs to intersect in the diagram. In some cases, there are dense areas containing large numbers of arcs. For highly irreversible rules, such as rule 0, large numbers of arcs converge on a single node, and appear essentially as a filled black circle.

Notice that the results given here and in table 14 for cellular automata on finite lattices with periodic boundary conditions also apply to infinite cellular automata in which only spatially periodic configurations are considered.

Table by Holly Peck (*Los Alamos National Laboratory*).

Table 14: Global Properties for Finite Lattices

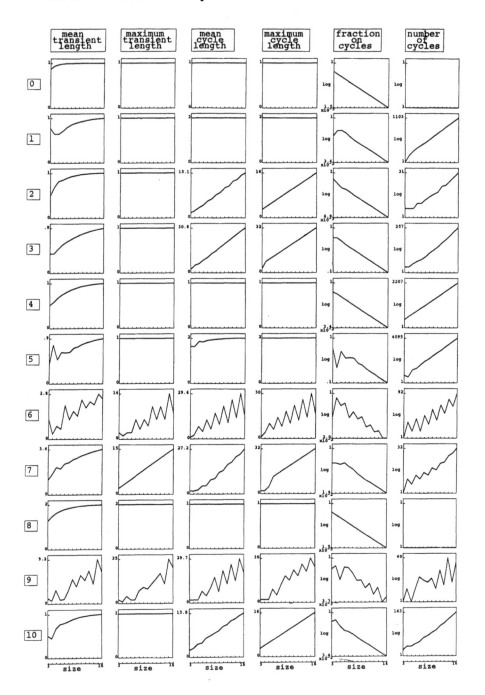

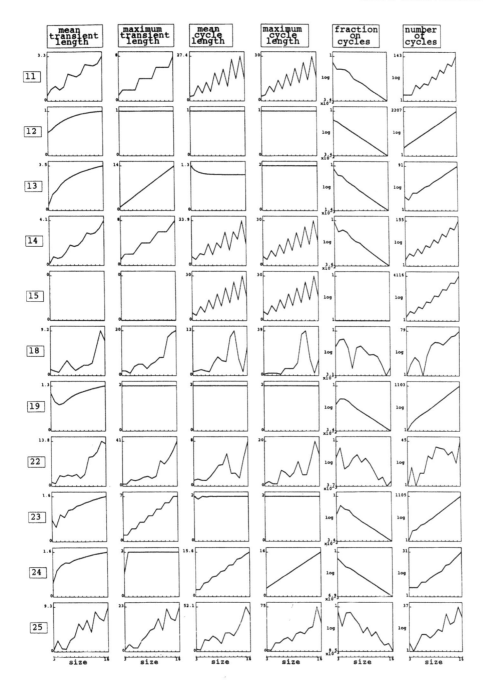

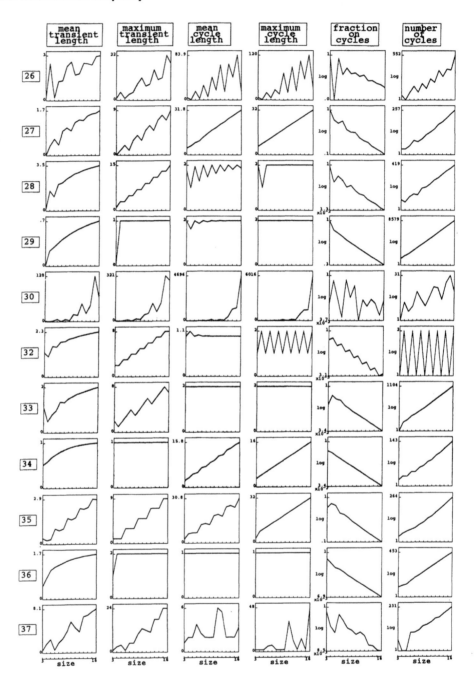

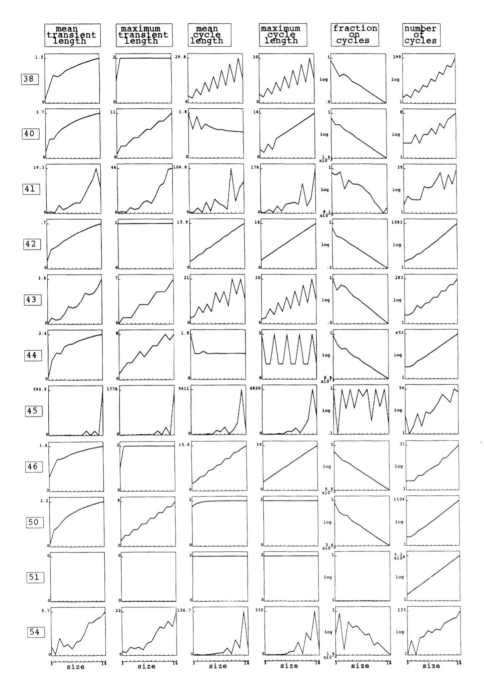

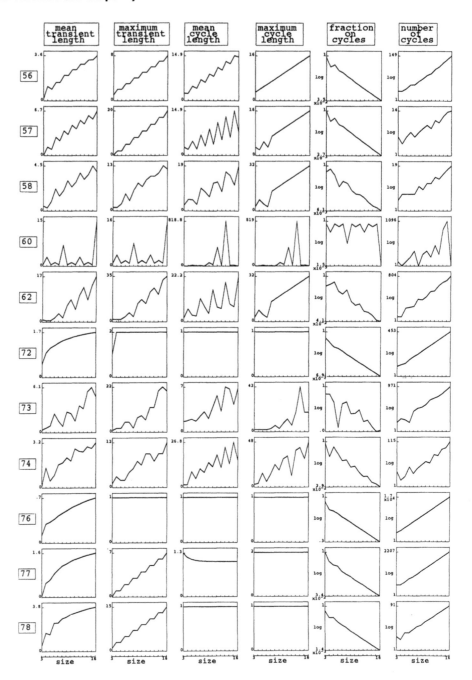

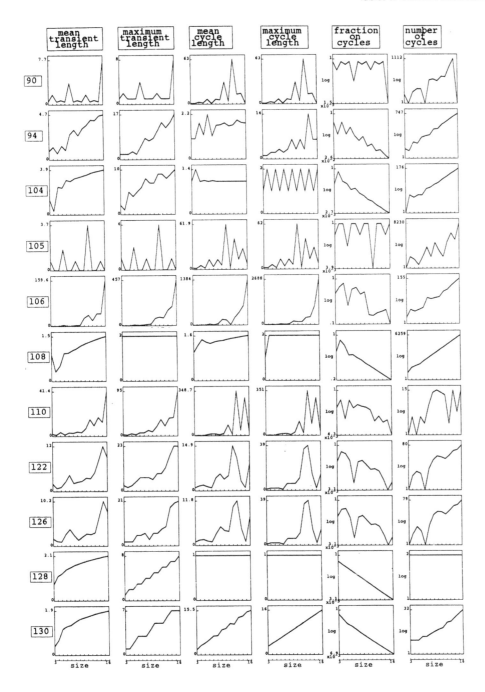

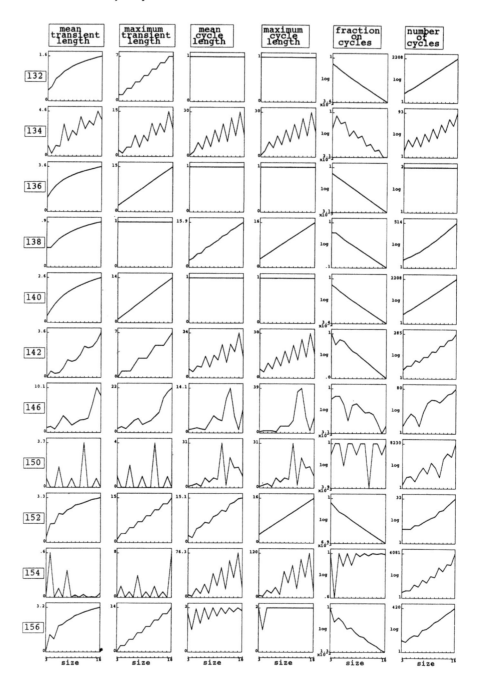

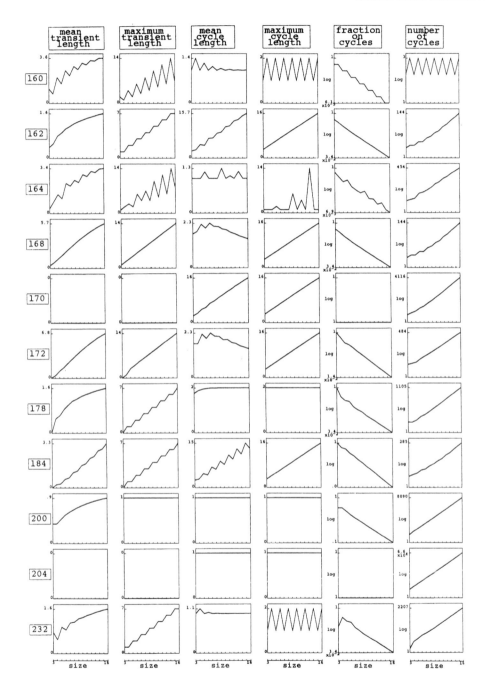

Global properties of cellular automata on finite lattices.

This table gives some properties of state transition diagrams for finite cellular automata. Each picture shows values plotted as a function of lattice size N, varying from 3 to 16. (Periodic boundary conditions are assumed for the cellular automaton evolution.)

The quantities shown are as follows. (In all cases, values for integer sizes N are shown joined by lines.)

"Mean transient length" represents the average number of steps necessary for any particular state to evolve to a cycle. "Maximum transient length" gives the maximum number of steps needed.

"Mean cycle length" gives the average length of the cycle on to which any particular state evolves. (Each cycle is thus weighted in the average with the number of states which evolve to it.) "Maximum cycle length" gives the longest cycle for each value of N. Some such cycles are shown in table 13.

"Fraction on cycles" (given in logarithmic form) represents the fraction of all 2^N possible states which appear on cycles, and thus can occur after a long time. This quantity is related to the set (topological) entropy for invariant set of the evolution.

The last picture for each rule gives the total number of distinct cycles (in logarithmic form). This can be considered as the number of possible distinct attractors for the evolution.

Table by Holly Peck (*Los Alamos National Laboratory*). (Original program by S. Wolfram.)

Table 15: Structures in Rule 110

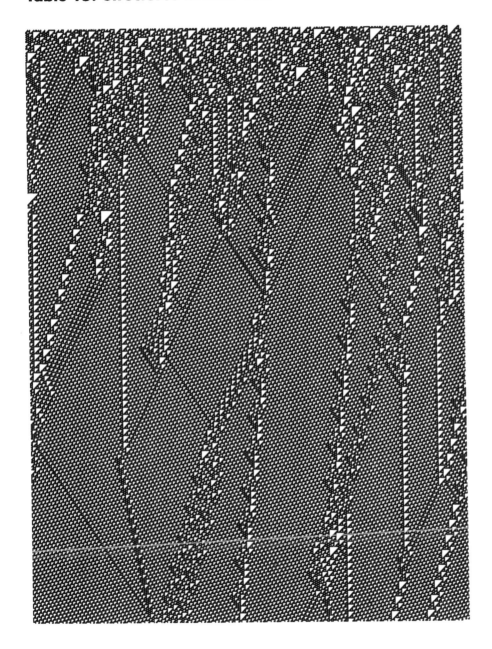

Structures in rule 110.

The previous two pages show patterns produced by evolution according to rule 110, starting from a disordered initial configuration. The first picture shows all sites on a size 400 lattice. The second picture shows every other site in space and time on a size 800 lattice.

The configurations produced after many steps can be represented in terms of particle-like structures superimposed on a periodic background. The background is found to have spatial period 14 and temporal period 7, and corresponds to repetitions of the block $\mathbf{B} = 10011011111000$. The configurations are then of the form $\cdots \mathbf{BBBBPBBB} \cdots$, where the particles $\mathbf{P}$ that have been found so far are:

velocity	P
−6/12	100011001110111111111000
−2/4	11111000
−14/42	111000011101111111111000
−8/30	100110011000111111000
−4/15	00000
−4/36	111011111111000
−8/20	1111000011000
0/7	11111111000
0/7	100011000
0/7	10011011111111000
2/10	11101011000
2/10	11101000110111111000
2/3	111000

The "velocity" is written as (spatial period)/(temporal period).

One may speculate that the behaviour of rule 110 is sophisticated enough to support universal computation.

Table of particles by Doug Lind (*Mathematics Department, University of Washington, Seattle*).

Table 16: Patterns Generated by Second-Order Rules

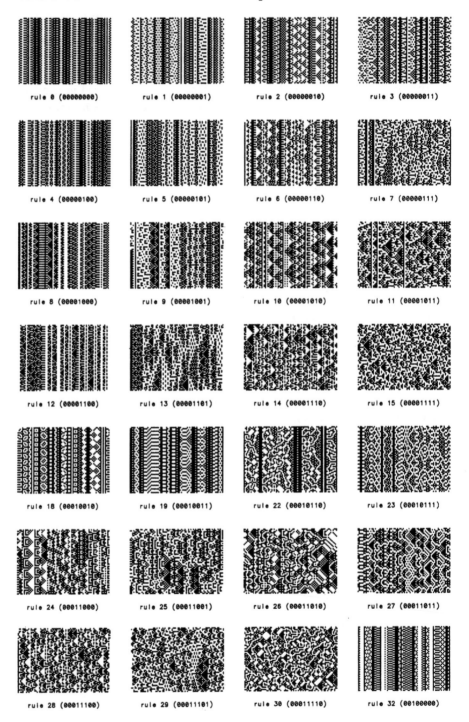

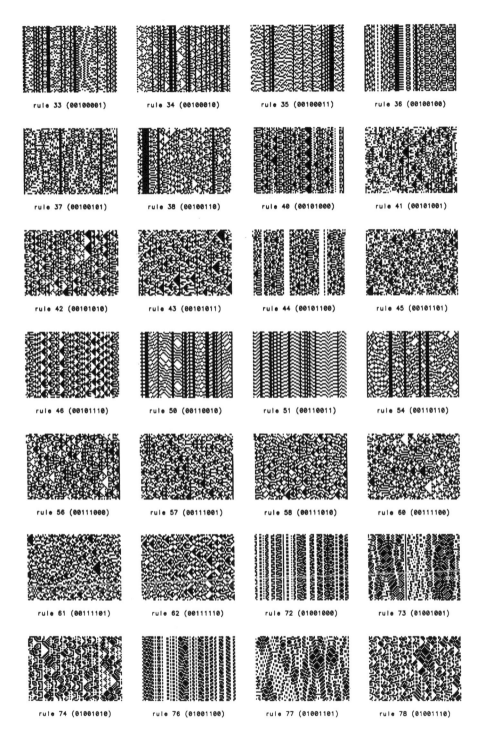

rule 33 (00100001)   rule 34 (00100010)   rule 35 (00100011)   rule 36 (00100100)

rule 37 (00100101)   rule 38 (00100110)   rule 40 (00101000)   rule 41 (00101001)

rule 42 (00101010)   rule 43 (00101011)   rule 44 (00101100)   rule 45 (00101101)

rule 46 (00101110)   rule 50 (00110010)   rule 51 (00110011)   rule 54 (00110110)

rule 56 (00111000)   rule 57 (00111001)   rule 58 (00111010)   rule 60 (00111100)

rule 61 (00111101)   rule 62 (00111110)   rule 72 (01001000)   rule 73 (01001001)

rule 74 (01001010)   rule 76 (01001100)   rule 77 (01001101)   rule 78 (01001110)

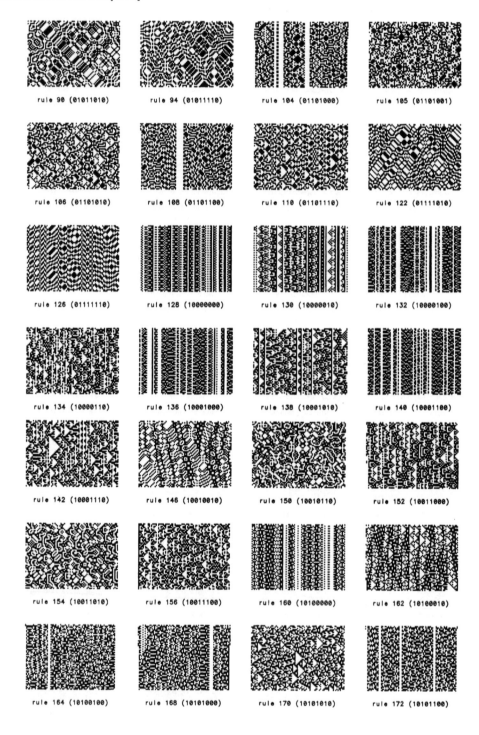

rule 90 (01011010)   rule 94 (01011110)   rule 104 (01101000)   rule 105 (01101001)

rule 106 (01101010)   rule 108 (01101100)   rule 110 (01101110)   rule 122 (01111010)

rule 126 (01111110)   rule 128 (10000000)   rule 130 (10000010)   rule 132 (10000100)

rule 134 (10000110)   rule 136 (10001000)   rule 138 (10001010)   rule 140 (10001100)

rule 142 (10001110)   rule 146 (10010010)   rule 150 (10010110)   rule 152 (10011000)

rule 154 (10011010)   rule 156 (10011100)   rule 160 (10100000)   rule 162 (10100010)

rule 164 (10100100)   rule 168 (10101000)   rule 170 (10101010)   rule 172 (10101100)

rule 178 (10110010)   rule 184 (10111000)   rule 188 (10111100)   rule 200 (11001000)

rule 204 (11001100)   rule 232 (11101000)

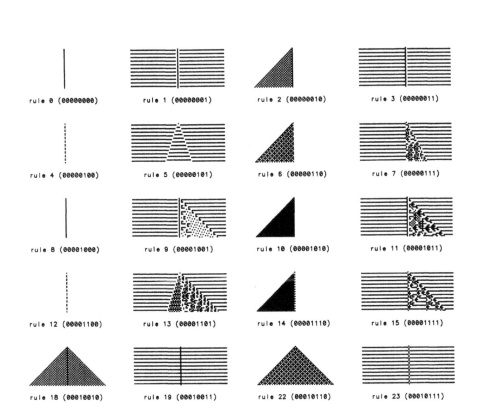

rule 0 (00000000)   rule 1 (00000001)   rule 2 (00000010)   rule 3 (00000011)

rule 4 (00000100)   rule 5 (00000101)   rule 6 (00000110)   rule 7 (00000111)

rule 8 (00001000)   rule 9 (00001001)   rule 10 (00001010)   rule 11 (00001011)

rule 12 (00001100)   rule 13 (00001101)   rule 14 (00001110)   rule 15 (00001111)

rule 18 (00010010)   rule 19 (00010011)   rule 22 (00010110)   rule 23 (00010111)

581

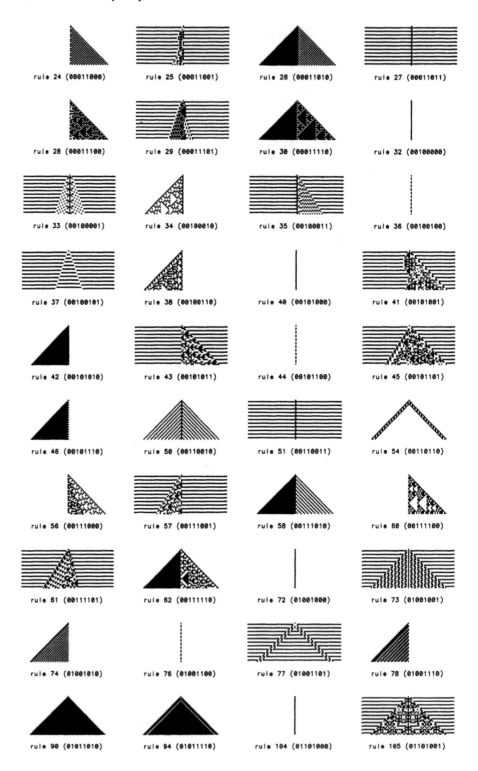

rule 24 (00011000)   rule 25 (00011001)   rule 26 (00011010)   rule 27 (00011011)

rule 28 (00011100)   rule 29 (00011101)   rule 30 (00011110)   rule 32 (00100000)

rule 33 (00100001)   rule 34 (00100010)   rule 35 (00100011)   rule 36 (00100100)

rule 37 (00100101)   rule 38 (00100110)   rule 40 (00101000)   rule 41 (00101001)

rule 42 (00101010)   rule 43 (00101011)   rule 44 (00101100)   rule 45 (00101101)

rule 46 (00101110)   rule 50 (00110010)   rule 51 (00110011)   rule 54 (00110110)

rule 56 (00111000)   rule 57 (00111001)   rule 58 (00111010)   rule 60 (00111100)

rule 61 (00111101)   rule 62 (00111110)   rule 72 (01001000)   rule 73 (01001001)

rule 74 (01001010)   rule 76 (01001100)   rule 77 (01001101)   rule 78 (01001110)

rule 90 (01011010)   rule 94 (01011110)   rule 104 (01101000)   rule 105 (01101001)

582

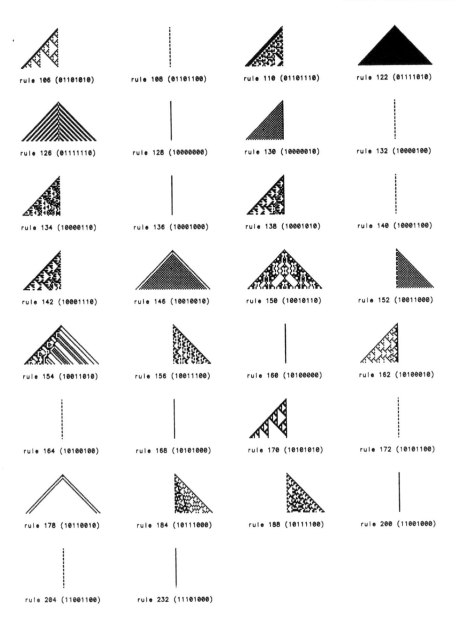

rule 106 (01101010)   rule 108 (01101100)   rule 110 (01101110)   rule 122 (01111010)

rule 126 (01111110)   rule 128 (10000000)   rule 130 (10000010)   rule 132 (10000100)

rule 134 (10000110)   rule 136 (10001000)   rule 138 (10001010)   rule 140 (10001100)

rule 142 (10001110)   rule 146 (10010010)   rule 150 (10010110)   rule 152 (10011000)

rule 154 (10011010)   rule 156 (10011100)   rule 160 (10100000)   rule 162 (10100010)

rule 164 (10100100)   rule 168 (10101000)   rule 170 (10101010)   rule 172 (10101100)

rule 178 (10110010)   rule 184 (10111000)   rule 188 (10111100)   rule 200 (11001000)

rule 204 (11001100)   rule 232 (11101000)

Patterns generated by second-order reversible rules.

This table shows patterns produced by second-order generalizations of the $k = 2$, $r = 1$ cellular automata considered above. The rules are of the form

$$a_i^{(t+1)} = \phi(a_{i-1}^{(t)}, a_i^{(t)}, a_{i+1}^{(t)}) + a_i^{(t-1)} \bmod 2,$$

where ϕ is a standard $k = 2$, $r = 1$ function, as listed in table 1. Such rules determine the configuration at time $t + 1$ in terms of the configurations both at time t and time $t - 1$. The rules have the special feature that they are reversible: given configurations at times t and $t + 1$, the configuration at time $t - 1$ can be deduced uniquely according to the rule

$$a_i^{(t-1)} = \phi(a_{i-1}^{(t)}, a_i^{(t)}, a_{i+1}^{(t)}) + a_i^{(t+1)} \bmod 2.$$

The first set of patterns were generated with disordered initial configurations at times 0 and −1. In the second set of patterns, the configurations at time 0 and −1 were both taken to be 1.

The forms of behaviour produced by these reversible rules are qualitatively similar to those from standard $k = 2$, $r = 1$ cellular automata, shown in tables 2 and 5. Evolution to a homogeneous, fixed, pattern is however impossible for reversible systems.

Scientific Bibliography
of Stephen Wolfram

1975

"Hadronic Electrons?" *Australian Journal of Physics* 28 (1975) 479–487.

A new form of high energy electron-hadron coupling is examined with reference to the experimental data.

1976

"Neutral Weak Interactions and Particle Decays", *Nuclear Physics* B117 (1976) 109–133.

Decays in which neutral weak interaction effects may be observed are discussed, concentrating on decays of pseudoscalar mesons.

1977

"Lepton Energy Spectra in e^+e^- Annihilation and Other Processes" (with M. Gronau, C.H. Llewellyn Smith, T.F. Walsh and T.C. Yang), *Nuclear Physics* B123 (1977) 47–60.

An analysis is presented of the single inclusive lepton spectra expected from the decay of charmed hadrons or heavier particles with new flavors produced in e^+e^- annihilation.

"The Decoupling of Axial Mesons from Currents", *Lettere al Nuovo Cimento* 20 (1977) 10–12.

It is demonstrated that conventional quark models require axial vector mesons to decouple from local currents.

1978

"Transverse-Momentum and Angular Distributions of Hadroproduced Muon Pairs" (with E.L. Berger and J.T. Donohue), *Physical Review* D17 (1978) 858–861.

We study the angular distribution of muons in the dimuon rest frame from $pp \rightarrow \mu\bar{\mu}X$ at high energy and large dimuon mass.

"Quantum-Chromodynamic Estimates for Heavy-Particle Production" (with J. Babcock and D. Sivers), *Physical Review* D18 (1978) 162–181.

The associated production of hadrons containing heavy quarks is studied in the framework of a model based on QCD.

"Positivity Constraints on Quark and Gluon Distributions in QCD" (with C.H. Llewellyn Smith), *Nuclear Physics* B138 (1978) 333–344.

Arguments are given that quark and gluon momentum distributions generated by asymptotic freedom formulae from distributions which are positive at $Q^2 = Q_0^2$ inevitably become negative for

$Q^2 < Q_0^2$. Momentum distributions should be chosen so that this occurs only for very small Q^2, thus constraining the partition of momentum between the quarks and gluons and the shapes of their momentum distributions.

"The Effective Coupling in QCD", Caltech report CALT-68-690 (1978) 13 pages.

The use of an effective coupling in QCD is investigated in the context of a simple class of processes which depend on only one kinematic invariant.

"Observables for the Analysis of Event Shapes in e^+e^- Annihilation and Other Processes" (with G.C. Fox), *Physical Review Letters* 41 (1978) 1581–1585.

We present a set of rotationally invariant observables which characterizes the "shapes" of events, and is calculable in quantum-chromodynamics perturbation theory for final states consisting of quarks and gluons.

1979

"Abundances of New Stable Particles Produced in the Early Universe", *Physics Letters* 82B (1979) 65–68.

The standard model of the early universe is used to estimate the present abundances of possible absolutely-stable hadrons or charged leptons more massive than the proton. It is found that experimental limits on their present abundances indicate that no such particles exist with masses below about 16 GeV/c^2. Forthcoming experiments could increase this limit to masses up to around 300 GeV/c^2.

"Heavy-Particle Production by Cosmic Rays" (with N. Isgur), *Physical Review* D19 (1979) 234–238.

We calculate the production of heavy charged or strongly interacting particles by cosmic rays and find that any sufficiently light stable ones should be detectable in static terrestrial searches.

"Electromagnetic Radiative Corrections to Deep-Inelastic Neutrino Interactions" (with R. Barlow), *Physical Review* D20 (1979) 2198–2206.

The electromagnetic-radiative-correction factors to deep-inelastic neutrino-nucleon scattering are estimated using the parton model.

"Event Shapes in e^+e^- Annihilation" (with G.C. Fox), *Nuclear Physics* B149 (1979) 413–496.

We present a complete set of rotationally invariant observables (H_l) which characterizes the 'shapes' of final states in e^+e^- annihilation.

"Tests for Planar Events in e^+e^- Annihilation" (with G.C. Fox), *Physics Letters* 82B (1979) 134–138.

We present a new class of observables which distinguish events containing two or three hadron jets from those containing a larger number.

"Weak Effects in Σ^0 Decay", *Physical Review* D20 (1979) 345.

It is pointed out that weak interactions should admix a small parity-violating $E1$ component into the $M1$ electromagnetic decay $\Sigma^0 \to \Lambda\gamma$.

"Bounds on Particle Masses in the Weinberg-Salam Model" (with H.D. Politzer), *Physics Letters* 82B (1979) 242–246.

Various conditions necessary for the self-consistency of the Weinberg-Salam model are used to place constraints on fermion and Higgs Boson masses. We find that spontaneous symmetry breakdown cannot generate fermion masses in excess of about 100 GeV.

1980

"Event Shapes in Deep Inelastic Lepton-Hadron Scattering" (with G.C. Fox and T.-Y. Tse), *Nuclear Physics* B165 (1980) 80–92.

The structure of hadronic final states in deep inelastic scattering expected from QCD is analyzed in terms of the shape parameters H_l and C_l.

"The Development of Baryon Asymmetry in the Early Universe" (with E.W. Kolb), *Physics Letters* 91B (1980) 217–221.

> The development of an excess of baryons over antibaryons due to *CP* and baryon number violating reactions during the very early stages of the big bang is calculated in simple models using the Boltzmann equation.

"A Model for Parton Showers in QCD" (with G.C. Fox), *Nuclear Physics* B168 (1980) 285–295.

> A Monte Carlo model for the development of parton jets in QCD is described. Explicit low-order calculations are supplemented by leading logarithmic approximations for higher orders.

"Spontaneous Symmetry Breaking and the Expansion Rate of the Early Universe" (with E.W. Kolb), *Astrophysical Journal* 239 (1980) 428–432.

> Gauge theories for weak interactions which employ the Higgs mechanism for spontaneous symmetry breakdown imply that there should exist a large vacuum energy associated with the Higgs scalar field condensate. A cosmological term in Einstein's field equations can be arranged to remove the unobserved gravitational effect of this vacuum energy in the present universe. However, in the early universe, the spontaneously broken symmetry should have been restored, leaving the cosmological term uncanceled. In this paper we investigate the conditions necessary for the uncanceled cosmological term to be dynamically important in the early universe. We find that if certain mass relations are satisfied, then for a brief period, the expansion rate of the universe will be determined by the uncanceled cosmological term prior to symmetry breaking, and the universe would have undergone a period of nonadiabatic expansion.

"Baryon Number Generation in the Early Universe" (with E.W. Kolb), *Nuclear Physics* B172 (1980) 224–284.

> The generation of an excess of baryons over antibaryons in the very early universe due to *CP*- and *B*-violating interactions is described. The Boltzmann equation is used to perform detailed calculations of the time development of such an excess in several simple illustrative models.

"Two- and Three-Point Energy Correlations in Hadronic e^+e^- Annihilation" (with G.C. Fox), *Zeitschrift für Physik* C4 (1980) 237–256.

> Correlations between the energies incident on two or three detectors around e^+e^- annihilation events are considered as a probe of the QCD structure of the events.

"Parton and Hadron Production in e^+e^- Annihilation" *Proceedings of the 15th Recontre de Moriond* (Les Arcs, France, March 9–21, 1980) ed. J. Tran Thanh Van, volume 1, 549–587.

> The production of showers of partons in e^+e^- annihilation final states is described according to QCD, and the formation of hadrons is discussed.

1981

"Cosmological Constraints on Heavy Weakly Interacting Fermions" (with J.A. Harvey, E.W. Kolb and D.B. Reiss), *Nuclear Physics* B177 (1981) 456–460.

> The masses and lifetimes of very heavy weakly interacting fermions which appear in many grand unified gauge models are constrained by the requirement that their decays in the hot big bang early universe should not generate excessive entropy which would dilute n_B/n_γ below its observed value.

"Weak Decays", *Nukleonika* 26 (1981) 273–309.

> Weak decays of strange, charmed and heavier mesons are discussed in the context of QCD.

"Cosmological Baryon-Number Generation in Grand Unified Models" (with J.A. Harvey, E.W. Kolb and D.B. Reiss), *Physical Review Letters* 47 (1981) 391–394.

> Methods for complete calculation of cosmological baryon-number generation in the hot big-bang early universe are outlined and are applied to several SU(5) models. Effects of several baryon-number-nonconserving bosons and the presence of nonthermalizing modes are treated.

"SMP—A Symbolic Manipulation Program" [*SMP Summary, SMP Primer, SMP Reference Manual*] (approx. 600 pp.), First Edition (July 1981).

"SMP: A Symbolic Manipulation Program" (with C.A. Cole), in *SYMSAC '81: Proceedings of the 1981 ACM Symposium on Symbolic and Algebraic Computation* (Snowbird, Utah, August 5–7, 1981) ed. P. Wang, 20–22.

The structure and implications of the new computer program SMP for algebraic manipulation of symbolic expressions are discussed.

1982

"QCD Expectations for High-Energy Hadronic Collisions" (with S. Pokorski), *Zeitschrift für Physik* C15 (1982) 111–114.

It is argued that perturbative QCD effects could be important in low-momentum transfer hadronic collisions at very high energies and should then give rise to several distinctive phenomena.

"Calculation of Cosmological Baryon Asymmetry in Grand Unified Gauge Models" (with J.A. Harvey, E.W. Kolb and D.B. Reiss), *Nuclear Physics* B201 (1982) 16–100.

Features of grand unified gauge models relevant to cosmology are discussed. Several SU(5) and SO(10) models are considered in detail. Comparison with observation places constraints on possible grand unified models.

1983

"Properties of the Vacuum. 1. Mechanical and Thermodynamic" (with J. Ambjørn), *Annals of Physics* 147 (1983) 1–32.

Casimir energies are calculated for quantized fields in cavities with a variety of forms. Consequences for models of the vacuum state are considered. The possibility of negative mass systems is discussed. Results on energy and entropy of finite quantum systems at non-zero temperature are given.

"Properties of the Vacuum. 2. Electrodynamic" (with J. Ambjørn), *Annals of Physics* 147 (1983) 33–56.

The behaviour of a charged scalar field in an external electric field is discussed. Instabilities encountered in the external field approximation are absent when back reaction effects are included through a self-consistent semiclassical procedure.

"A QCD Model for e^+e^- Annihilation" (with R.D. Field), *Nuclear Physics* B213 (1983) 65–84.

A QCD model for e^+e^- annihilation is presented, and its consequences are compared with experimental data. The model involves production of a shower of partons described by a simple approximation to QCD perturbation theory, and decay of colour singlet clusters of produced partons into hadrons through a simple phase space process. The model reproduces most known theoretical features of QCD, and, with certain choices of parameters, appears to correspond well with experimental results.

"Cosmology with Very Large Gauge Models" (with J.A. Harvey and E.W. Kolb), *Physical Review* D27 (1983) 315–321.

Several theoretical principles suggest the existence of large numbers of very massive particles. Such particles have negligible effect in the present universe, but may have been important in the very early universe. It is shown that under some circumstances their presence could completely change the equation of state and expansion rate of the very early universe, and could have important effects on baryon-number generation. Possible cosmological constraints on the complexity of grand unified gauge models are discussed.

"Statistical Mechanics of Cellular Automata", *Review of Modern Physics* 55 (1983) 601–644.

See pages 3–69 of this book.

"Cellular Automata", *Los Alamos Science* (Fall 1983) 2–21.

See pages 411–437 of this book.

1984

"Computing: A New Tool for Fundamental Physics", in *Supermicrocomputers: Proceedings of the Workshop on the Applications of Supermicrocomputer Workstations in Physics and Astronomy* (Chapel Hill, January 20–22, 1984) ed. S.M. Christensen.

"Geometry of Binomial Coefficients", *American Mathematical Monthly* 91 (1984) 566–571.

See pages 451–456 of this book.

"Algebraic Properties of Cellular Automata" (with O. Martin and A. Odlyzko), *Communications in Mathematical Physics* 93 (1984) 219–258.

See pages 71–113 of this book.

"Universality and Complexity in Cellular Automata", *Physica* 10D (1984) 1–35.

See pages 115–157 of this book.

"Preface to Cellular Automata: Proceedings of an Interdisciplinary Workshop", *Physica* 10D (1984) vii–xii.

"Computation Theory of Cellular Automata", *Communications in Mathematical Physics* 96 (1984) 15–57.

See pages 159–202 of this book.

"Cellular Automata as Models for Complexity", *Nature* 311 (1984) 419–424.

Natural systems from snowflakes to mollusc shells show a great diversity of complex patterns. The origins of such complexity can be investigated through mathematical models termed 'cellular automata'. Cellular automata consist of many identical components, each simple, but together capable of complex behaviour. They are analysed both as discrete dynamical systems, and as information-processing systems. Here some of their universal features are discussed, and some general principles are suggested.

"Computer Software in Science and Mathematics", *Scientific American* 251 (September 1984) 188–203.

See pages 439–449 of this book.

1985

"Symbolic Mathematical Computation", *Communications of the ACM* 28 (1985) 390–394.

Standard programming languages are inadequate for the kind of symbolic mathematical computations that theoretical physicists need to perform. Higher mathematics systems like SMP address this problem.

"Twenty Problems in the Theory of Cellular Automata", *Physica Scripta* T9 (1985) 170–183.

See pages 457–485 of this book.

"Two-Dimensional Cellular Automata" (with N. Packard), *Journal of Statistical Physics* 38 (1985) 901–946.

See pages 211–249 of this book.

"Undecidability and Intractability in Theoretical Physics", *Physical Review Letters* 54 (1985) 735–738.

See pages 203–209 of this book.

"Origins of Randomness in Physical Systems", *Physical Review Letters* 55 (1985) 449–452.

See pages 251–258 of this book.

"Cellular Automata and Condensed Matter Physics", in *Scaling Phenomena in Disordered Systems: Proceedings of a NATO Advanced Study Institute* (Geilo, Norway, April 8–19, 1985) ed. R. Pynn and A. Skjeltorp, 249–277.

Cellular automata are mathematical models for systems containing many identical components with local interactions. These notes describe some of their properties, and discuss applications to condensed matter physics.

"Cryptography with Cellular Automata", in *CRYPTO '85 Proceedings: Advances in Cryptology* (Santa Barbara, California, 1985); *Lecture Notes in Computer Science* 218, 429–432.

See pages 487–490 of this book.

"Thermodynamics and Hydrodynamics with Cellular Automata" (with J. Salem), Thinking Machines Corporation technical report (1985).

See pages 259–265 of this book.

1986

"Analytical and Empirical Mathematics with Computers", in *Supercomputers: Algorithms, Architectures, and Scientific Computation*, ed. F.A. Matsen and T. Tajima (University of Texas Press, 1986) 456–464.

> Some of the practical, methodological and theoretical implications of computation for the mathematical sciences are discussed.

"Random Sequence Generation by Cellular Automata", *Advances in Applied Mathematics* 7 (1986) 123–169.

> *See pages 267–307 of this book.*

"Approaches to Complexity Engineering", *Physica* 22D (1986) 385–399.

> *See pages 309–328 of this book.*

"Cellular Automaton Fluids 1: Basic Theory", *Journal of Statistical Physics* 45 (1986) 471–526.

> *See pages 359–408 of this book.*

Theory and Applications of Cellular Automata (including selected papers 1983–1986), (World Scientific Publishing Co. Pte. Ltd., 1986).

"Cellular Automaton Hydrodynamics" (with B. Nemnich), Thinking Machines Corporation (1986).

> The theory and phenomenology of cellular automaton fluids have been studied. Simulations of two-dimensional cellular automaton fluids have been carried out on a 65536 processor Connection Machine computer. Flows at Reynolds numbers of a few hundred have been obtained. Detailed studies are underway of flows in channels, and flows past simple geometrical objects, such as circular cylinders. Preliminary results indicate good agreement with experiments and existing calculations. At the highest Reynolds numbers investigated, flows past cylinders are observed to make a transition to aperiodicity, corresponding to weak turbulence.

1987

Complex Systems (ISSN 0891-2513), ed. S. Wolfram, volume 1– (1987–).

> A journal devoted to the rapid publication of research on the science, mathematics, and engineering of systems with simple components but complex overall behavior.

1988

"Complex Systems Theory", in *Emerging Syntheses in Science: Proceedings of the Founding Workshops of the Santa Fe Institute*, volume 1, (Addison-Wesley, 1988) 183–189.

> *See pages 491–497 of this book.*

"Cellular Automaton Supercomputing", in *High-Speed Computing: Scientific Applications and Algorithm Design*, ed. R.B. Wilhelmson, (University of Illinois Press, 1988) 40–48.

> *See pages 499–509 of this book.*

Mathematica: A System for Doing Mathematics by Computer (Addison-Wesley, 1988).

1991

Mathematica: A System for Doing Mathematics by Computer, Second Edition (Addison-Wesley, 1991).

Index

\# (bit count function), 19, 49, 286

Adaptive programming, 331, 507
Adaptive systems, 314
Additive rules, 9, 73, 147, 451
    two-dimensional, 96, 216
Adjacency matrices, 175
Algebraic analysis of cellular automata, 71
Algebraic computation, 443
Algebraic numbers, 178
Algorithmic information content, 253
Algorithmic probability, 155
American Mathematical Monthly, 451
Analog computing, 209
Asynchronous cellular automata, 318, 471
Attractors, 36, 73, 310, 473
Autocorrelation, 24
Automata, finite, 165
Automorphism group, of graph, 179, 473
Automorphisms, 303
Autoplectic behavior, 252
Average density
    of nonzero sites, 18, 542
    of particles, 261, 363

Basins of attraction, 290, 312, 574
BBGKY hierarchy, 365
Bessel functions, 395
Bijective rules, 303, 584
Binary representation, 19, 453, 521
Binomial coefficients, 12, 20, 49, 62, 77, 451
Biological systems, modelled by cellular automata, 6, 416, 447
Bit count function, 19, 49, 286
Block cellular automata, 332
Block entropies, 132
Blocking transformations, 48, 190, 435, 466, 547

Boltzmann equation, 4, 23
    in cellular automaton fluids, 364
Boltzmann H theorem, 392
Boolean functions, 312, 320
    for cellular automaton rules, 285, 521, 559
Boundary conditions, 17
Bravais lattices, 380

Cantor set, 38, 120, 133, 425
Capacity, 133
Catastrophe theory, 4, 314
Cellular automata, basic definition of, 5, 412
Chaos theory, 4, 145, 251
Chapman-Enskog expansion, 368
Characteristic polynomials, 75, 176
Characters, in groups, 378
Chemical reactions, 207
Chi squared, 298
Chomsky classification, 163
Chromatic polynomials, 179
Church-Turing thesis, 208, 431
Circulant matrices, 387
Class 1 rules, 140
Class 2 rules, 141, 232
Class 3 rules, 143, 239, 271
Class 4 rules, 150, 205, 240, 430
Classification of cellular automata, 116, 161, 423, 460
    in two dimensions, 232
Coarse-grained entropy, 353, 464
Code 20, persistent structures in, 152
Code numbers, for totalistic rules, 118
Codes, 267, 345, 487
Coin tossing, 269
Collision operator, 365
Communications in Mathematical Physics, 71, 159

Complex systems theory, 491
Complexity, regular language, 173, 472, 554
Composition of rules, 117, 549
Compression of information, 252
Computation theory, 160, 203, 474
Computational complexity theory, 204, 481
Computational irreducibility, 203, 313, 448, 481
Computers, cellular automata as, 47, 150, 309
Confidence intervals, 298
Congruential pseudorandom generators, 13, 258, 268
Connection Machine computer, 259, 305, 500
Conservation laws, 332, 366, 464
Context-free languages, 163, 194, 473
Context-sensitive languages, 166
Continuity equation, 366
Continuum systems, 262, 329, 362, 469
Convergence, 312
Correlation functions
    in cellular automata, 24, 463
    in fluids, 364
Cryptanalysis, 253
Cryptography, 267, 487
Crystal growth, 215
Crystallographic symmetries, 377
Cubic lattice, 97, 380
Curvature, of interfaces, 239
Cycles, in cellular automata, 41, 71, 291
Cyclotomic polynomials, 88, 108

Data analysis, 199, 207
de Bruijn graph, 169, 174
Defects, 102, 239
Dendritic crystals, 6, 215, 493
Density, 18, 542
Dependence, of values in rule 30, 287
Determinism, 253
DFAs, 165
Difference patterns, 35, 127, 233, 278, 533
Differential equations, 330, 359, 441, 469
Diffusion coefficient, 401
Diffusion equation, 331, 401
Digit count function, 19, 49, 286
Digit sequences, 254, 299
Dimension
    of ensemble, 38, 126, 542
    of ensemble in two dimensions, 241
Diophantine equations, 154, 204
Dipolynomials, 75
Disordered systems, 3
Dissipation
    as source of attractors, 310
    origins of, 345

Distribution functions, in cellular automaton fluids, 364
Divergences, in viscosity, 371
DNA sequences, 324
DNF, 285, 320, 560
Domains, in cellular automata, 22, 102, 190, 233, 467
Dominoes, 244
Dynamical systems theory, 126, 160, 241, 474

Einstein equations, 207
Elementary rules, 8
    grammars for, 175
    properties of, 513
Encoding, of one cellular automaton in another, 48, 190, 435, 466, 547
Encryption, 267, 345, 487
Engineering, complexity, 309
Ensembles, 33, 363
Entropy
    estimation of, 135
    in two dimensions, 241
    of configuration ensemble, 38, 83, 126, 464, 542
    of physical systems, 3, 260, 345
    of rule 30, 276
Equilibrium
    approach to, 388
    thermodynamic, 260, 329
Equilibrium density, 24
Ergodicity, 85, 252, 317
Error-correcting codes, 311
Euler phi function, 109, 292
Excluded blocks, 182, 551
Expansive maps, 189
Experimental mathematics, 440, 460, 494

Fabric design, 7
Factorization of rules, 549
Feature extraction, 324
Feedback shift registers, 15, 44, 73, 268, 299
Fermat's theorem, 109
Fermi-Dirac distribution, 394
Ferrofluids, 239
Fibonacci sequence, 26
Filtering, by class 2 rules, 141
Finite automata, 165
Finite cellular automata, 33, 71, 290
Finite difference methods, 261, 319, 500
Finite fields, 107
Fitness functions, 321
Fixed points, 140, 190, 315
Fluctuations, 24, 358

Fluctuations (continued)
in entropy estimates, 135
Fluids, cellular automaton models for, 259, 359, 502
Fokker-Planck equation, 400
Formal languages, 163, 468
Four manifolds, 207
Fourier transforms, 463
Fractals, 13, 26, 144, 217, 451
Free energy, 467
Fugacity relation, 392

Galilean invariance, 370
Galois fields, 107
Garden of Eden configurations, 38, 78
Gases, cellular automaton models for, 259, 359
Gaussian fluctuations, 24, 239
Generating functions, 75, 103, 178, 195
Genetic algorithms, 322
Geometrical constructions, 13
Geometry
of cellular automaton configuration space, 462
of two-dimensional cellular automata, 215
Gliders, 61, 152, 240, 577
Global properties of cellular automata, 33, 160
in two dimensions, 241
Goal-oriented systems, 309
Godel's theorem, 167
Golden ratio, 13, 26, 104, 149, 377
Grammars, formal, 163
Graphs, of regular languages, 182, 473, 554
Gravity, 207
Green's functions, 139, 245
Group theory, 378
Growth
circular, 229
conditions for, 12, 119
in two dimensions, 222
Growth inhibition, 6, 119, 228

H theorem, 392
Halting probability, 153
Hamiltonian systems, 464
Hamming distance, 34, 311
Hard sphere gases, 207, 264
Hausdorff dimension, 134
Heat, 261
Heat bath, 401
Hexagonal lattice, in cellular automaton fluids, 259, 361
Homoplectic behavior, 251
Huygens' principle, 240
Hydrodynamics, 259, 359

Icosahedral symmetry, 265, 377
Identity rule (rule 204), 19
Image processing, 141, 311
Immiscible fluids, 402
Infinite size limit, 479
Information content, 252
Information propagation, 138
in two dimensions, 246
Inhomogeneous cellular automata, 315
Initial conditions, 18
Interfaces, 229, 239
Invariant configurations, 188, 243, 284
Invariants, 332, 366, 464
Invertible rules, 303, 584
Irreducibility, computational, 203, 313, 448, 481
Irreversibility, 3, 38, 77, 147
in two dimensions, 241
Ising model, 6, 207, 319
Isotropy, in cellular automaton fluids, 374
Iterated maps, 5, 255
corresponding to cellular automata, 34
in complex plane, 312

Julia sets, 312

Kinetic theory, 261, 364, 504
Kinks, 22, 102, 239
Kolmogorov entropy, 133

Labyrinthine patterns, 239
Landscapes, 321
Languages, formal, 163, 468
Lattice dynamics, 317
Lattice gases, 259, 359
Lattice spin systems, 6
Lattices, 212, 379
Law of large numbers, 24
Learning, 326
Lie groups, 381
Life, Game of, 7, 61, 150, 213, 240
Light cones, 245
Limits sets, structure of, 192, 473
Linear rules, 9, 49, 71
Liouville's theorem, 38
Logic circuits, 320, 507
Logical depth, 155
Lognormal distribution, 134
Long-range connections, 318
Long-range order, 30
Los Alamos Science, 411
Lyapunov exponent, 139, 462, 542
in two dimensions, 245
of rule 30, 278

Mach number, 261
Macroscopic dynamics, 362
Mappings
    corresponding to cellular automata, 34
    random, 45
Markov chains, 166
Markovian approximation, 24
Martin, Olivier, 71
Master equation, 23, 360
Maxwell-Boltzmann distribution, 394
Mean field theory, 18, 23, 469
Mean free path, 401
Measure entropy, 131
Membranes, 49, 126
Metastable states, 319
Metric entropy, 131
Minimization
    of Boolean expressions, 286, 321
    of functions, 314
Modelling, general issues of, 207
Modularity, 318
Modulo 2 rule (rule 90), 13, 20, 71
Modulo k rules, 49, 90
Molecular dynamics, 259, 359
Mollusc, 416
Momentum conservation, 367
Monte Carlo method, 267
Multiple scale analysis, 364
Multiple scale cellular automata, 309
Multiplicative order function, 109

NAND gates, 63
Natural selection, 324
Navier-Stokes equations, 261, 370, 502
n-body problem, 207
NC complexity class, 256
NDFAs, 169
Necklaces, 292
Networks, 73, 291, 471, 564
    Boolean, 45
Newtonian fluids, 265
Noise, 32, 46, 470
Noise diodes, 269
NP-completeness, 168, 206, 289, 312, 482
    approximate solutions with, 323
    of tiling, 244
Number theory, 107, 476
Numerical analysis, 469, 500

Odlyzko, Andrew, 71
ord function, 110

Packard, Norman, 211
Parallel computation, 259, 483

Parent configurations, 78, 148, 180
Parsers, 163
Partial differential equations, 330, 359
Particle dynamics, 259, 329
Pascal's triangle, 12, 451
Pattern recognition, 311
Penrose tilings, 382
Pentagonal lattices, 382
Percolation, 471
Periodic configurations, 153
Periodic points, 182, 189
    in rule 30, 284
Permutation rules, 332
Perron numbers, 179
Persistent structures
    in class 2 rules, 141
    in class 4 rules, 150, 479, 577
Phase transitions, 33
Phases, 190, 234, 467, 530
Phenotypes, 314
Physica D, 115, 309
Physical laws, 411, 439
Physical Review Letters, 203, 251
Physical systems
    computation in, 203
    modelled by cellular automata, 6
Pi, digits of, 254, 299
PLAs, 320
Platonic solids, 377
Poincare recurrence time, 41
Point groups, 378
Polymers, conformation of, 207
Polynomial time computations, 168
Polynomials
    for cellular automaton configurations, 75
    over finite fields, 108
Polytopes, 227, 377
Porous media, 402
Power spectra, 24, 463
Predecessors
    in one dimension, 78, 148, 180
    in rule 30, 291
    in two dimensions, 242
Prediction, general problem of, 203
Pressure, 467
Prime numbers, 7, 85, 91, 453
Probabilistic behavior, 251
Probabilistic cellular automata, 33
Probability measures, 36, 132
Production rules, 163
Propagator, 139
Pseudorandom generators, 254, 267
PSPACE-completeness, 197, 204, 293, 481
Psychology, 325

Quantum field theory, 208
Quantum mechanics, 257
Quasicrystals, 383
Quiescence condition, 8

Random cellular automata, 32
Random mappings, 45, 106, 290
Random number generators, 13
Random walks, 22, 279, 331
Randomness
        in computations, 504
        in rule 30, 267
        origins of, 251, 345
Reaction-diffusion equations, 211, 469
Real numbers, 208
Recursive sets, 473
Reflection symmetry, 225
Regular languages, 142, 163, 313, 554
Renormalization group, 48, 190, 435, 466, 547
Repeatability, of experiments, 254
Reversible cellular automata, 41, 98, 260, 303,
        332, 464, 584
Reviews of Modern Physics, 3
Reynolds number, 261, 503
Rice's theorem, 479
Rotational symmetry, 226, 229, 377
Roulette, 269, 312
Round-off errors, 469
Rule 4, density in, 20
Rule 18
        correlations in, 25
        density in, 22
        domains in, 22
        global evolution of, 37, 101
        grammar for, 172, 186
        temporal sequences in, 149
Rule 22
        density in, 23
        grammar for, 187
Rule 30
        encryption with, 487
        randomness in, 252, 267, 487, 505
Rule 36, density in, 20
Rule 45, randomness in, 274
Rule 50, density in, 9, 19
Rule 72, grammar for, 188
Rule 76, grammar for, 169
Rule 90
        cycles in, 43, 71, 85
        density in, 20
        form of, 9
        geometrical construction in, 13, 20
        global evolution of, 37, 73
        sensitive dependence in, 35

Rule 94, grammar for, 187
Rule 110, structures in, 577
Rule 126
        cycles in, 42
        global evolution of, 37, 46
        grammar for, 185
        sensitive dependence in, 35
        triangle density in, 28
Rule 128, grammar for, 184
Rule 150, 13, 21
        cycles in, 86
        density in, 21
        geometrical construction in, 13, 21
Rule 182, density in, 18, 22
Rule 204, density in, 19
Rule 254, density in, 19
Rules
        additive, 9, 73, 147, 451
        asynchronous, 471
        bijective, 303, 584
        Boolean form of, 8, 273, 521
        deduction from data, 471
        elementary, 8, 513
        factorization of, 549
        finding minimal, 330
        illegal, 14
        inhomogeneous, 315
        injective, 148, 180
        legal, 8, 118, 521
        linear, 9, 90
        multidimensional, 96
        multiscale, 315
        numbering of, 8, 117, 212, 521
        outer totalistic, 213
        particle, 259, 359
        reversible, 41, 98, 260, 303, 332, 464, 584
        second-order, 41, 98, 584
        solidification, 221
        space of, 466
        surjective, 180
        symmetry of, 8, 521
        totalistic, 117, 212, 419
        two-dimensional, 59, 96, 211, 259

Salem, James, 259
Sampling, in entropy computations, 134
Santa Fe Institute, 491
Satisfiability, 168, 312, 482
Scale invariance, 13, 143, 466
Scattering processes, 386
Scientific American, 439
Search processes, 322, 506
Second law of thermodynamics, 3, 329, 508
Self-organization, 4, 38

Self-reproduction, 61, 207, 433, 479

Self-similarity, 13, 26, 133, 144, 220, 258, 451

Sensitive dependence, 5, 35, 139, 233

Shapes, from two-dimensional cellular automata, 221

Shell, 416

Shift map, 251

Shift registers, 15, 44, 73, 487

Shift rules, 522

Shock waves, 371

Short cuts, computational, 199, 203

Simulated annealing, 322

Simulation, of one cellular automaton by another, 48, 190, 435, 465, 547

Slow growth, 231

SMP, 404, 443

Snowflakes, 493

Sofic system, 142

Solidification, 221, 493

Solitons, 469

sord function, 110

Sound, in cellular automaton fluids, 260, 371

Spatial chaos, 140

Spatial entropy, 126, 145, 542

    computation of, 178

    in two dimensions, 224

Speed of information propagation, 140

Sphere packing, 380

Spin glasses, 207, 314

Square roots, digits of, 254, 299

Stack, 166

State transition graphs, 73, 291, 564

Statistical mechanics, 330

    of cellular automata, 3

Statistical tests of randomness, 255, 296

Stochastic systems, 469

Strange attractors, 15, 148

String manipulation, 167, 204

Subadditivity, 131

Subset construction, 170

Subshifts of finite type, 180, 190

Supercomputing, 499

Superposition principles, 9, 49

Supersonic flow, 371

Surface tension, 239

Surjectivity, 180

    in two dimensions, 243

    of rule 30, 276

Symbolic dynamics, 33, 160

Symbolic representations, 285, 325, 483

Symmetry

    in cellular automaton fluids, 369

Symmetry (continued)

    of rules, 8, 213

Synchronous updating, 471

Systolic arrays, 318

Tapestry design, 7

Temporal chaos, 140

Temporal entropy, 136, 149, 277, 542

Tensors, 373

Textile design, 7

Thermodynamic limit, 480

Thermodynamics, 3, 259, 329, 363, 464, 504

Tiling problem, 243

Time series, 136, 149

Topological defects, 247

Topological entropy, 131

Topological invariants, 179

Totalistic rules, 117

Totient function, 109, 292

Transitions, in space of possible rules, 466

Transport coefficient, 390

Transport equations, 365

Tree searching, 322

Trees, 81, 291

Triangles, density of, 25, 143

Turbulent fluids, 253, 262

Turing machines, 47, 63, 204, 243

Two-dimensional cellular automata, 59, 96, 211, 259

Two-point functions, 25

Undecidability, 167, 197, 203, 478

    in two dimensions, 242

Unit cells, 379

Universal cellular automata, 63, 152, 205, 477

Universal computation, 63, 152, 166, 255

Universality, in cellular automata, 115

Unix system, random generator in, 13, 299

Unpredictability, 251, 267

Unreachable configurations, 38, 78

Viscosity, 261, 370

Visual system, 310

Voronoi polyhedra, 380

Vortex street, 503

Walsh transforms, 463

Wigner-Seitz cell, 380

Zeta functions, 182, 463